The NEW BOOK of
·APPLES·

PLATE 1: BLOSSOM (clockwise from top left): SANDRINGHAM, GREEN BALSAM, WINTER QUEENING, GRAVENSTEIN

The NEW BOOK of
·APPLES·

JOAN MORGAN
AND ALISON RICHARDS
WITH PAINTINGS BY
ELISABETH DOWLE

The watercolours, which were commissioned to illustrate this book, were painted by the distinguished botanical artist Elisabeth Dowle. They took four years to complete as each fruit and blossom was painted from life. The varieties are shown both as they grow on the tree, and as they can appear when fully ripe.

Ebury Press · London

First published in 1993

This revised and updated edition published in 2002

5 7 9 10 8 6 4

Editors: Irene Slade and Margot Richardson
Designer: David Fordham
Line drawing of apple by Philip Hood

First published in the United Kingdom in 1993 by
Ebury Press
Random House, 20 Vauxhall Bridge Road, London SWIV 2SA

Random House Australia (Pty) Limited
20 Alfred Street, Milsons Point, Sydney, New South Wales 2061, Australia

Random House New Zealand Limited
18 Poland Road, Glenfield, Auckland 10, New Zealand

Random House South Africa (Pty) Limited
Endulini, 5A Jubilee Road, Parktown 2193, South Africa

The Random House Group Limited Reg. No. 954009
www.randomhouse.co.uk

A CIP catalogue record for this book is available from the British Library

ISBN 9780091883980

Typeset in Electra by MATS, Southend-on-Sea, Essex
Printed in Singapore by Tien Wah Press

Papers used by Ebury Press are natural recyclable products made from wood grown
in sustainable forests.

CONTENTS

PLATE 2: BEAUTY OF BATH

FOREWORD

BY THE EARL OF SELBORNE, KBE, FRS
APPLE GROWER AND COLLECTOR OF OLD APPLE VARIETIES

THE PUBLICATION OF *The New Book of Apples* is testimony to the invaluable place that the original *Book of Apples* occupies in the library of every apple enthusiast, student of our fruit heritage and, indeed, the whole conservation movement both in Britain and further afield. We are all deeply indebted to Joan Morgan and Alison Richards for this magnificent record of our heritage of apples. This book has become the Apple Bible, giving new insight into the numerous and unsuspected ways that the growing and eating of apples has shaped our lives. The culture, countryside and cuisine of lands far beyond our own all bear traces of the importance of apples in making their different histories.

The Book of Apples also brought alive the little-known treasures of Britain's National Apple Collection. It highlighted the Collection's importance as a genetic resource of great diversity – containing over 2,000 living varieties and including long-forgotten garden favourites, local specialities, as well as many varieties discovered and raised overseas. Prized apple varieties from English villages sit proudly beside those that formed the basis of the first orchards in America and Australia. Apples whose ancestry is obscured by centuries can be studied alongside novelties that have been developed for the demands and tastes of the new millennium.

Since *The Book of Apples* was first published in 1993, I am delighted to say, nurserymens' lists have expanded to reflect the burgeoning interest in growing and enjoying more varieties of apples. Regional fruit groups are flourishing and actively seeking out local specialities and encouraging their planting. Conservation orchards are being established not only in Britain, but all over Europe and in America and Australia. Many more people are now looking to buy and appreciate good home-grown produce through local markets, while government and private sponsorship is available for the restoration of lost orchard landscapes and old fruit gardens.

The Book of Apples has become both an indispensable guide and a constant source of inspiration to all those interested in discovering and celebrating the world's rich apple heritage. *The New Book of Apples* reflects the new discoveries and developments of the last decade and comes at a time when interest in apples is growing rapidly. The revised Directory has added many fresh details to existing entries and I am pleased to have been able to provide some new information on the old English apple, Ashmead's Kernel. The Directory also includes some 100 new varieties accessed over the last decade; a number of these may appeal to both commercial and amateur growers, as they carry some immunity to disease – the key to successful organic cultivation. The directory also reflects the on-going verification work – the important task of ensuring that all the varieties are true to name. In this, Joan has collaborated with the Scientific Curators, Imperial College at Wye, to resolve many queries. This painstaking and fascinating work has greatly enhanced the value of the Collection and ensured that *The New Book of Apples* maintains the position of definitive guide to the world's apple varieties, as well as being a fascinating history of our best loved fruit.

Selborne .

Note from the Authors

We would like to express our thanks to Department for Environment, Food and Rural Affairs for allowing us to continue to use and study the National Apple Collection, also to the Scientific Curator, Imperial College at Wye, and to the Brogdale Horticultural Trust, which provides the home for the Collections.

The National Apple and other Fruit Collections are now owned and funded by the Department for Environment, Food and Rural Affairs, which in 2001 replaced the Ministry and Agriculture, Fisheries and Food. The Brogdale Horticultural Trust continues to provide the public access to the Collections.

We would also like to thank the many pomologists, both professional and amateur, in this country and abroad, who have helped in the revisions and verification work. Their generous assistance, specialised knowledge and not least their libraries have brought clarification to difficult areas and enabled many tricky questions to be resolved.

JOAN MORGAN AND ALISON RICHARDS

There is no fruit in temperate climates so universally esteemed, and so extensively cultivated, nor is there any which is so closely identified with the social habits of the human species as the apple. Apart from the many domestic purposes to which it is applicable, the facility of its cultivation, and its adaptation to almost every latitude, have rendered it, in all ages, an object of special attention and regard.

DR ROBERT HOGG, *The Apple*, 1851

Chapter 1
The
FRUIT of PARADISE

Once long, long ago there was an old man named Arstanbap. He had a box and in that box there grew a tree of pure gold with a nightingale singing on it. If the nightingale whistled when Arstanbap opened the box, a pistachio tree would appear. If it trilled, an almond tree or an apricot tree would rise from the ground. And if it warbled, nut trees would grow. And the nightingale sang on and on . . .
V. VITKOVICH: *Kirghizia Today*, 1960

IRGHIZIA LIES AMONG THE SLOPES of the Tien Shan or Heavenly Mountains, which form the boundary between western China and the former Soviet Union. It was in this remote area that the shepherds used to tell the story of the forests of wild fruit trees which, legend claimed, were the remains of a great orchard that stretched in antiquity from China to the shores of the Caspian. It is here that the origins of the domestic apple are to be found. In the 1920s, when the Russian plant geneticist, Nicolai Vavilov first surveyed the forests of wild fruit trees, he found scenes evocative of 'the Garden of Paradise' with fruits so good that 'they might be merely removed to an orchard'. The travel writer Vitkovich, in 1960, after visiting Kirghizia, described camping on the edge of apple groves, pursuing mountain turkeys through clusters of nut trees, and watching porcupines disappear into plum thickets. 'In a word', he wrote, 'this was life in a marvellous garden of wonders such as are described in fairy tales, a marvellous garden where apples and pears look down on you from the trees and beg to be eaten, where a magic wind brings you showers of nuts, where birds are radiantly feathered and animals trustful and the imprints of bears' paws to be seen on the paths'. Even today large areas of wild fruit trees can still be found in the foothills of the Caucasus, the Kopet-Dag mountains in Turkmenistan, in the Pamirs and especially in the Tien Shans where there are areas in which wild apple trees dominate the landscape.

The homeland of our domestic apple lies in the fruit forests of Kazakhstan and neighbouring regions. In 1929, when Vavilov visited the capital Alma Ata, which appropriately means 'Father of Apples', he cast his geneticist's eye 'around the town and a great distance along the mountain slopes,' where he saw 'the overgrowths of wild apples creating massive forests'. He concluded: 'It is possible to see with our own eyes that we are in the centre of origin of the cultivated apple'. A decade later, Vavilov fell victim to Stalin's rejection and purge of the science of genetics. He was imprisoned in 1940 and died in 1943. His work on wild apple trees had, however, inspired

Aimak Dzhangaliev, a young Kazakh botanist. For the next fifty years Dzhangaliev and his colleagues at the Kazakhstan Academy of Sciences in Alma Ata have been gathering data that does, indeed, show that the apples of Kazakhstan closely resemble our cultivated varieties with a similarly extensive range of colours, forms, sizes and eating qualities.

The wild fruit trees grow on the mountain slopes at altitudes between 800 and 2000 metres. Extensive forests are found in the Zailiiskii mountains, in whose northern foothills lies Alma Ata, and also some 400 kms to the north in the Djungarskii mountains, which form the most northerly extent of the Tien Shans. The northern limit of the apple forests is some further 400 kms to the north in the Tarbagatai range of the western Altay mountains, again right on the border with China. The most southerly extent of the apple forests and of Dzhangaliev's studies is in the more arid regions of the Talasskii and Karatu mountains which extend into Kyrgyzstan, Uzbekistan and Tajikistan.

The best trees and the best-quality fruits are found in the semi-moist conditions of the mountain slopes and river valleys of the Zailiiskii and Djungarskii ranges. Here the apple trees bloom in late April and May as they do in England and produce a range of fruits familiar to apple growers everywhere: apples ready to eat from the tree in August, others that do not ripen until the autumn, and late keeping apples that can be stored even to March. The climate is continental with a long, warm growing season and winters sufficiently cold to break the dormancy of the fruit buds, but not so harsh that the trees are damaged. Trees a hundred years old are not uncommon and still fruitful.

The main and in many places only apple species in the Kazakhstan forests is *Malus sieversii*. As Dzhangaliev and his colleagues have found, this species shows an immense diversity in the apples that it produces. These can range from large to small, red to yellow and green, and sweet to sharp; the whole spectrum of our domestic apple is seen in its fruits. *Malus sieversii* extends over other mountain ranges of the Tien Shans, the Pamirs and over into the Kopet Daghs and probably into areas of the Caucasus as well. Modern genetic studies also suggest that *Malus sieversii* is the principal ancestor of the domestic apple we know today. Other species, in particular *Malus orientalis*, which is the main species in the Caucasus, and *Malus sylvestris*, which is native to an area extending from Britain to the Balkans and northern Turkey, do not seem to have contributed significantly to today's apples. *Malus orientalis* produces very late-keeping, but rather bitter fruits and *Malus sylvestris* bears small, astringent greenish yellow fruits. Although earlier researchers, including Vavilov, thought these species would have hybridised with *Malus sieversii*, the evidence, so far, indicates that this is unlikely.[1]

The wild apple trees in the forest spread out as the fruits were eaten by birds, animals and humans whose stomachs carried them further afield. Both animals and humans chose the largest, juiciest fruits and their gastric juices help break down the dormancy of the seeds which encourages germination. In addition to the indirect transport of pips, local people bartered apples for other goods brought by traders and travellers, who then took the apples yet further afield. The wild forests lay close to ancient trade, nomadic and migration routes, which meant apples were carried way beyond their homeland. The Djungarian gate at the eastern end of the Djungarskii mountain range is the wide depression through which, from prehistoric times, waves of nomadic peoples have passed through to the Kazakh lowlands and on to the Caspian Ural gateway to Europe. From remote antiquity the foothills of the Zailiiskii mountains and

Ferghana Valley in Kyrgyzstan were also part of the network of caravan routes between Europe, China, India and Persia.

The apple, like many fruits, does not breed true from seed so every pip is potentially a new kind of apple. Some would have been an improvement on their parents, while others would have been inferior, but with human intervention the best would have been retained. Good fruit trees would have been marked or taken directly from the forest and planted near settlements. *Malus sieversii* readily throws up suckers and its branches will root. In the wild, whole stands of apple trees may be formed of suckers connected to the mother tree by an underground network of roots and the first orchardists were able to take advantage of these properties to propagate their chosen trees. Some cultivated varieties, such as the Burr Knot apple, have retained the ability to root from cuttings and bear 'burrs' at the base of branches that will take root. In this respect, it resembles the Chiloé apple, which Charles Darwin found growing along the coast of Chile in 1835. The inhabitants, Darwin wrote in his diary, 'possess a marvellously short method of making an orchard. At the lowest part of almost every branch, small, conical brown wrinkled points project; these are always ready to change into roots, as may sometimes be seen, where any mud had been accidentally splashed against the tree'.

Most apple varieties, however, do not root easily from cuttings and need to be grafted. Natural grafts can occur when two branches rub together and eventually fuse, but it is impossible to say how soon farmers were inspired to follow the examples they found in the wild. Binding a cutting from a favoured tree into the cleft of a sapling is regarded as a very ancient art in China, and it is on the discovery of grafting that much of the subsequent history of the domestic apple depends. It gave growers the ability not only to reproduce useful trees and establish them elsewhere, but also to transform worthless trees. Expertise in grafting has ensured that the best varieties have been conserved for centuries, and enables us to grow apples enjoyed by our ancestors.

FARMING BEGAN IN THE Middle East about 8000 years ago with the domestication of cereals in the 'Fertile Crescent' – the arc of rich land flanked by mountain slopes where wild wheat and barley grew. As farming spread, and areas of fruit forest were cleared for cereal cultivation the best trees were spared and left standing in the middle of fields or acting as boundary markers. Orchards would also have been carved out of fruit forests, young trees transplanted and seedlings raised. Apple remains have been found at Jericho in the Jordan Valley, and at Catal Hüyük in Anatolia, where they were dated to about 6500 BC. But whether this fruit was gathered from the wild species growing to the north, or a superior sweeter apple from the east, or taken from husbanded trees growing near the settlements is not known.

Unlike their ancestors who moved in search of food, these farmers had to stay in the same place from seed time to harvest. Their communities became larger and more settled. They began to build in brick and stone, and to think beyond annual crops to the possibilities offered by vineyards and orchards which take several years to mature. The new irrigation techniques, which from 5000 BC onwards enabled water to be channelled from rivers to semi-desert areas,

11

brought farming communities down from the hills into the plains, and in the basins of the Nile, China's Yellow River, the Indus, and the Tigris–Euphrates the first great civilisations began to take shape. Food production and long-distance trade developed accordingly, and between 3500 and 1500BC a network of trading centres and city states sprang up in a great arc stretching from the Eastern Mediterranean to the Indus. Knowledge and supplies of fresh and dried fruit and the techniques for producing it, began to spread rapidly. Along with copper, grain, lapis lazuli, carnelian, obsidian, pearls, tin, timber, ivory and textiles, merchants and prospectors almost certainly dealt in plants and fruit as well. Eleven charred apple rings, which appear to have been dried, and, perhaps, threaded on a string were found on saucers in the grave of Queen Pu-abi at Ur near present-day Basra in southern Iraq.[2] The tomb dates from 2500 BC and scholars believe that the fruit had been brought some several hundred miles, from the north, where the cooler climate was better suited to apple growing. An Ur text of this period does, however, refer to vineyards in this region interplanted with figs and 'hashur', which is usually translated as 'apples'. It seems unlikely, nevertheless, that apples would have fruited in Sumer. The trees would have received no winter chilling, although it is possible that Sumer gardeners may have discovered that by picking off all the leaves, as they do today in tropical Indonesia, they could achieve the same effect.

Apples were certainly being cultivated in Georgia, Armenia, Anatolia, northern Mesopotamia and Persia from about this time and the value often placed upon the trees suggests that the apples may have been of the sweeter type. In sheltered valleys of Anatolia, the Hittites, who rose to dominance in the second millennium, appear to have invested in fruit to the extent that on one estate alone, 40 apple trees were recorded.[3] Penalties for the destruction of orchards and vineyards were set out in their Law Codes:

> *If anyone sets [brushwood(?)] on fire and [leaves] it there and the fire seize a vineyard, if vines, apple trees, pomegranates, and pear trees(?) burn up, for one tree he shall give [six] shekels of silver and re-plant the Plantation. If he be a slave, he shall give three shekels of silver.*

In northern Mesopotamia, a tablet found at the Assyrian city of Nuzi records that in about 1500 BC Tupkitilla of Nuzi sold an orchard he had inherited, for a considerable sum, namely three sheep.[4] It seems likely that apples were cultivated as far west as the Aegean coast of Turkey, which was part first of Minoan and then Mycenaean Greece. They may also have been grown – though with difficulty – in Egypt. Rameses II (1298–1235 BC) was reported to have planted apple trees in his Nile delta garden, but recent research has cast doubt on this and the old Egyptian saying, 'what does a fella know of apples?' may refer to the problems of fruiting apple trees in Egypt.[5]

BY THE FIRST MILLENNIUM BC, the cultivation and enjoyment of orchard fruits, including apples, had become an essential part of civilised life. In the *Odyssey*, written in the ninth or eight century BC, the Greek poet Homer retells how the Mycenaean hero Odysseus sought refuge at the court of King Alcinous.

Outside the courtyard but stretching close up to the gates, and with a hedge running down on either side, lies a large orchard of four acres, where trees hang their greenery on high, the pear and the pomegranate, the apple with its glossy burden, the sweet fig and the luxuriant olive . . . In the same enclosure there is a fruitful vineyard, in one part of which is a warm patch of level ground, where some of the grapes are drying in the sun, while others are gathered or being trodden, and on the foremost rows hang unripe bunches that have just cast their blossom or show the first faint tinge of purple.

A 19th-century impression of the closing stages of a Persian feast. Fruit and wine are laid out on the richly decorated cloth; scented water is being carried in.

In ninth century BC Mesopotamia, the Assyrian king's new royal palace at Nineveh boasted orchards stocked with 42 different fruit and gum producing trees – many of which may have been brought back from military campaigns – and the inaugural banquet in 879 BC featured dish upon dish of fruit and nuts.[6] Further north, the rival kingdom of Urartu, which occupied present Armenia and eastern Anatolia, had a long-established reputation for its orchards and vineyards. When Rusa II (685–645 BC) extended and improved his summer residence on the banks of the river Razdan, near today's capital, Yerevan, he proudly recorded that 'I planted this vineyard, fields with their crops, orchards did I create there and with them I surround the towns.' Archaeologists have excavated a citadel whose storerooms still contained the remains of great quantities of fruit including grapes, plums, pomegranates and also cut, and possibly dried, apples.[7]

It was, however, with the coming together of the Persian Empire that the enjoyment and celebration of fruit reached its ancient climax. At its height under Darius, from about 512 BC, the Empire stretched from the Aegean coast of Turkey, across Iran and Afghanistan to the Indus. It extended north to the edge of the Caucasus and up into Central Asia. It reached southwards to include the Middle East, and around the Mediterranean coast to Egypt. Rich, fertile and politically stable, with an efficient system of communications, the Empire and its rulers encouraged goods, crops, people, and ideas to cross old boundaries. Sesame was introduced to Egypt, rice to Mesopotamia and peaches and oranges were brought in from China. Vineyards and orchards were planted everywhere that climate and the new underground irrigation systems allowed, and in 495 BC Darius was able to compliment one of his officials in Ionia, on the Aegean coast, on 'improving my country by the transplantation of fruit trees from the other side of the Euphrates, in the furthest parts of Asia'.[8]

As well as its orchards, every royal and provincial palace had an enclosed garden of fruit trees watered by canals and designed as much for beauty as gastronomic pleasure. The trees provided blossom, fragrance and shade as well as an autumn display of jewel-like fruits. In 401 BC such a garden inspired the Greek essayist and historian Xenophon to lay out a similar pleasure ground on his estate in Greece and to introduce a new word into the Greek language. This transliteration of the Persian *pairidaeza*, or walled garden, became the Roman *paradisus*, and the English word – and image – of *paradise*.

The Persian dining tables were also devoted to the most exquisite pleasures the Empire could offer. 'For the benefit of the King', Xenophon reported, 'they go about the entire country

13

in search of something that he may like to drink, and countless persons devise dishes that he may like to eat.' Many of these dishes would have included fresh or dried fruit and the feast would end with a magnificent array of the finest fresh fruit.

Ancient, pre-Islamic Persia followed a dualist religion which aimed to achieve equilibrium or harmony through the balance of opposites. Apples would have been prized, then as now, for their refreshing combination of sugar and acid and, while the finest would have been reserved for the grand finale, less perfect specimens would have featured as ingredients in the preceding savoury dishes. Traditional Persian cookery opposes sweet and sour, vinegar and sugar, strong and mild. Apples, fresh or dried, together with other fruits, nuts, vegetables, honey, spices and pomegranate juice, are commonly combined with meat and pulses and, like quinces, are sufficiently large and firm to be stuffed and braised. It is probable that the final course of fruit would have been served in much the same way as it is in the Middle East today.[9] After the tables have been cleared and guests have rinsed their hands in perfumed water, plates piled high with fruits in season are carried in. The fruit is usually served whole, and each person dips his choice in an individual bowl of ice-cold water. A simple compôte of stewed fruit, or a mixture of grated fruit, crushed ice and scented water may also be on offer. When fresh fruit is not available, dried fruit, nuts, jams and compôtes are served instead.

This kind of display inevitably made the choice and quality of the fruit served a matter of fashion and prestige. Discerning diners began to distinguish between different sources and varieties. It was later asserted, for example, that the finest apples served at the Persian court were those which came from Georgia. Home-grown fruit soon became an indicator of the host's taste, and the resources he was able to invest in the selection and management of his orchards and vineyards. These associations of luxury and exclusivity, connoisseurship and social display, have clung to the fresh-fruit dessert ever since, and it is partly due to the apple's long history of cultivation and appreciation as a dessert fruit that we owe its extraordinary diversity. Ever since those first feasts in Darius's kingdom, gardeners and *bon viveurs* have striven to produce apples of ever greater novelty and distinction. In Britain alone, our Victorian forebears had access to hundreds of different apple varieties and the connoisseurs among them would compare varieties, regions and vintages with the same subtlety and intensity with which they discussed the claret.

IN 313 BC ALEXANDER THE GREAT completed his conquest of Persia, and Persian culture – including its pleasure in fruit and fruit trees – diffused through the Greek world. Greek banquets closed with a finale known as 'the second tables' which was said to be 'the stimulant of the sated appetite' and consisted of 'delicious flat cakes' and fruit such as 'pears, luscious apples, pomegranates and grapes'. Particular varieties of apple were distinguished and sought after. Writing in about 300 BC the Greek philosopher Theophrastus – who is credited with founding the science of botany – clearly distinguishes between astringent wild crab apples and cultivated apples, and mentions a number of varieties including one he calls the Epirotic apple,

PLATE 3: DISCOVERY

which may have come from Epirus in Albania. A particularly rich area for apples according to Theophrastus was 'about Panticapæum . . . [where] pears and apples are abundant in a great variety of forms and are excellent'.[10] Panticapæum is today's Kerch', on the northern shores of the Black Sea in the Crimea, which is still renowned for its fruit. By the first century AD, the Greek historian Plutarch makes a point of drawing attention to the apple's special place in Greek affections. 'No other fruit,' he recalled a guest observing at a banquet, 'unites the fine qualities of all fruits as does the apple. For one thing, its skin is so clean when you touch it that instead of staining the hands it perfumes them. Its taste is sweet and it is extremely delightful both to smell and to look at. Thus by charming all our senses at once, it deserves the praise that it receives.' [11]

With the rise of the Roman Empire, the fruits, customs and combined horticultural skills of the Greeks and the Persians moved westwards. At its height, the Roman Empire extended from Armenia in the east to Britain in the west and encircled the entire Mediterranean. The network of trade routes collectively known as the Silk Road linked Rome with China and, via its various side branches, with India, the Baltic, and beyond. Goods and cultures moved around the Empire, and were brought in from outside. Silks and spices from the east reached all provinces, feeding and encouraging the taste for luxury, novelty and display. The caravans inevitably passed through or close to the wild fruit belt, and through regions where both the cultivation and enjoyment of fruit was at its most refined. The northern loop of the Silk Route passed through the foothills of the Zailiiski mountains with a caravanserai at Almatu near Alma Ata. The main route crossed through from Kashgar to the Hindu Kush and into Afghanistan, with another route through Kyrgyzstan and the Ferghana Valley to Samarkand and across Persia and Syria, or traversed Armenia and Georgia, to the Mediterranean.

Gathering sour apples: an illustration from the Tacuinum Sanitatis in Medicina, *or* Table of Health in accordance with Medical Science, *compiled in northern Italy in the 14th century.*

Novel or superior fruits, and new horticultural skills were eagerly brought back to Roman gardens and orchards. Among them were sweet cherries from Cerasus on the Black Sea; peaches, which were known as Persian apples; apricots, which they called the Armenian plum; and the Median apple, which was actually the orange. Superior quinces came from Crete, new grapes from Armenia, figs from Egypt and apple varieties from Greece and further east. According to the poet Horace, writing in the first century BC, Italy became one vast orchard, and fruit trees even acquired their own deity, the goddess Pomona.

> *N*o other Latin wood nymph could tend a garden more skilfully than she, none was more devoted to the cultivation of the fruit trees from which she derived her name. She did not care for woods or rivers, but loved the countryside, and branches loaded with luscious apples. Instead of weighing down her hands with heavy javelins, she used to carry a curved knife with which at one time she would cut away growth that was too luxurious, and prune back branches that were spreading in different directions . . . Her garden was her passion and her love . . . But she was afraid of being violently attacked by some rustic wooer, and so she fenced herself inside her orchards, to prevent the men she shunned from reaching her.
>
> OVID: *Metamorphoses, BOOK XIV*

Rome probably had more varieties of apple in cultivation than it did of any other fruit. Over 20 different varieties are mentioned by Pliny in his *Natural History* of the first century AD. These included the Greek 'Epirotic', the 'Syrian Red', and many which had arisen in Italy. First of the season came those which needed to be eaten straight from the tree and soon spoiled, such as the 'Honey apple' and, as its name suggests, the 'Flour apple' which 'is the earliest to come and hastens to be picked', if it was not to become mealy. Then came the 'little Greek' and the late keeping 'Armeria', named after a town in Umbria. Other late keepers, such as the 'ruddy Appian' which 'smelt of quinces' and the 'Scaudian' were sufficiently well thought of to have been named after distinguished families, and the recently introduced 'small Petisian', Pliny also found, had 'a most agreeable flavour'.[12]

Apples would sometimes feature in prepared dishes such as the combination of apple and pork laced with coriander, liquamen and honey that is recorded in Apicius's cookery book, but were more likely to appear at the end of the meal. According to Horace, the perfect Roman meal began with eggs and ended with fruit – *ova ad malum*. While the diners reclined on deep cushions specially imported from Carthage, the low cedar wood tables from the Lebanon would be cleared and in would come lavish bowls of fresh figs, grapes and plump, glossy apples, just as they appear on a villa wall at Pompeii.

One reason – or at least rationalisation – for the custom of serving fresh fruit, and particularly apples, at the end of the meal was the digestive qualities attributed to them. Galen, the Greek physician who lived in second-century Rome, and his medical and philosophical predecessors such as Hippocrates, subscribed to the theory that the human body was composed of four elements or humours, namely blood, phlegm and black and yellow bile. These elements were hot, cold, dry and moist respectively, and each was associated with different human temperaments and disorders. Foods were similarly classified. This gave a system whereby physicians could work out when, in what combinations, and by whom, various foods should be eaten in order to maintain or correct the balance of the bodily humours. As apples were cool and moist they made the perfect counterbalance to the heat of foods such as red meat which made up the main part of the meal.

Sour apples, however, were distinctly unwholesome, although they had specific medical applications such as the treatment of syncopes (fainting) and constipation. When the Roman satirist Juvenal invites guests to dinner, he is at pains to reassure them about the quality of his home-grown apples:

> D*on't worry, winter's ripened them, the cold's dried out
> Their green autumnal juices, made them quite safe for dessert.*
>
> JUVENAL: *Satire XI*

These ancient dogmas concerning the physiological properties of apples have had a profound influence on the way they have been regarded ever since. Handed on to the modern world in the 12th century by the teachings of the medical school at Salerno near Naples, the therapeutic applications of apples in relation to disturbances of the bowels, lungs and nervous system – to mention just a few – initiated a long and venerable tradition of apple cures which was still going strong in the 19th century. Apples became a celebrated medieval laxative, a common ingredient of 17th-century cough cures, and in the centuries before anti-depressants, a cup of apple juice or cider would be prescribed. The Elizabethan herbalist, Gerard, also recommends a poultice of apples for hot swellings, and records that 'the pulp of roasted apples mixed to a froth in water and drunk by the quart, has benefited those with gonorrhoea'. Chapped skin was treated to a mixture of lard and chopped apples and, like their ancestors in the Middle Ages, Victorian farm workers relied on rags soaked in crab-apple vinegar – verjuice – to relieve their aches and pains. The fact that we still combine cooked apples with warming spices such as cloves and cinnamon similarly has its origins in the dietary prescriptions of the ancient world.

Roman advice on growing apples also holds good today. Well versed in the arts of grafting – which they claimed had been handed down by the ancients – they propagated esteemed varieties in the spring, by inserting a scion into a cleft of a stock tree or slipping it between the bark and the hard wood. Alternatively, a new tree might be raised in the summer by budding. This involved taking a sliver of bark containing a bud from the favoured tree and fitting it like a patch on to a young tree - rootstock - from which a corresponding piece had been removed. Such skills attracted considerable kudos, suggesting as they did, yet another dimension to Roman imperialism: control over nature herself.

The classical image of Hercules, beneath the apple tree in the garden of the Hesperides, may be a source of the tradition that the fateful tree in Eden was an apple.

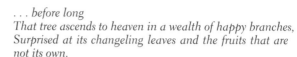

. . . before long
That tree ascends to heaven in a wealth of happy branches,
Surprised at its changeling leaves and the fruits that are
not its own.

VIRGIL: *Georgics*, BOOK 2

One of the highest compliments a Roman host could pay his guests was to present them with fruits he had grown himself. World weary Augustans increasingly took refuge from the squalid streets and frenetic social round of the metropolis to cultivate the orchards, vineyards and olive groves of their country estates. Here, with armies of slaves to call upon, they could concentrate on producing fruits and vintages finer than anything available in the markets and shops of the city. The fact that the practical orchardist and discriminating host were often one and the same created the optimal circumstances in which the skills of growing fine fruit could develop, and these epicurean estate owners produced a voluminous literature of remarkably sound advice and instruction on everything from grafting and pest control, to picking and storage. Juvenal rightly points out, for example, that late maturing varieties of apple can be left to mellow and sweeten in 'a cool dry place, laid on straw', where they will keep quite sound until the following spring. The orator and agricultural writer, Columella, recommends that the choicest specimens should receive special attention. He advocates storing them in 'chests of beech or even lime wood such as those used for storing official robes . . . in a very cold dry loft, to which neither smoke nor foul odour can penetrate'. An even better strategy was to store the apples in a purpose-built, shaded, but airy fruit house with north facing windows. These should be open but shuttered 'to keep the fruit from loosing its juice when the wind blows steadily'. To make it cooler, ceilings, walls and floors could be coated with marble cement. Varro, another horticultural authority, noted that

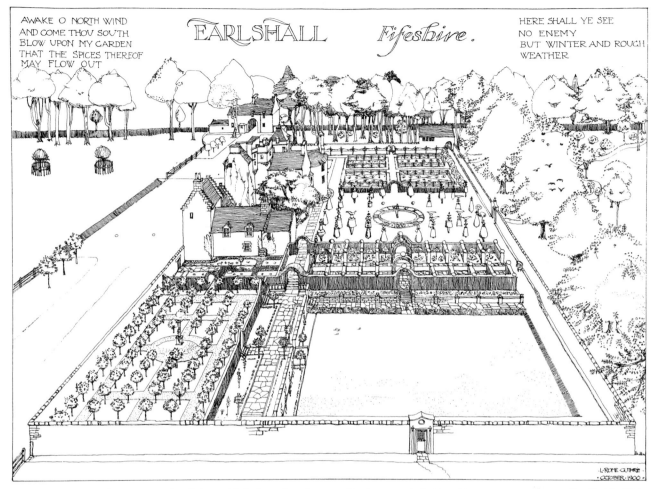

Some people even spread a dining table in it to dine there; and in fact, if luxury allows people to do this in a picture gallery, where the scene is set by art, why should they not enjoy a scene set by nature, in a charming arrangement of fruit? Provided always that you do not follow the example set by some of buying fruit in Rome and carrying it to the country to pile it up in the fruit-gallery for a dinner party.[13]

A fashionable attempt to recreate Earlshall's 17th-century fruit gardens was made in the late 19th century. As depicted here by H. Inigo Triggs in 1902, the elaborate, formal enclosures with their trained trees, walks and statuary may also owe something to an enthusiastic imagination.

Like the Persians, the Greeks and Romans also saw fruit trees as ornament for the pleasure ground. Apart from their obvious decorative properties, this was closely linked to the desire of earthly gardeners to create their own patch of what we have come to call paradise. As in earlier and later cultures, the basic human longing for a time or place where men and women are not locked in a constant battle with nature for food and shelter was symbolised by the image of a magical island or garden where, in Homer's words, the earth 'brings forth her fruit in abundance and without toil' and 'pear after pear, apple after apple, cluster on cluster of grapes, and fig upon fig are always coming to perfection'. In Greek mythology, Gaia, or Mother Earth, presented a tree of golden apples to the god Zeus and his bride Hera on their wedding day. Guarded by a serpent, Ladon, who never slept, it grew in the garden of the Hesperides, who were the daughters of the Evening Star. One of the 12 labours of Hercules was to steal these magic apples from the

19

garden – and no Greek or Roman garden was complete without its tubs of apple trees and groves of fruit. It is an association which, like the apple's role as a dessert fruit, has shaped its history ever since. Medieval gardens were graced by the golden fruits of the small growing Paradise apple and whole orchards of such miniature trees became popular with Renaissance princes. Rows of apple trees were planted along walks in the formal gardens of Stuart England, and were set in pots around the parterres, while the Edwardians delighted in arbours, arches and tunnels of apple trees.

Like the quest for new dimensions of flavour, this continued search for novel decorative effects has contributed to the diversity and characteristics of the fruit trees we grow today. Not only do apple varieties exhibit an enormous range of different blossom and fruit colour, for example, but it was interest in the ornamental possibilities of miniature Paradise trees that first pointed growers in the direction of the dwarfing rootstocks on which modern orchards are based, and improvements in the quality of the fruit produced by decorative 'fan' trees, which first suggested the benefits of training and pruning to the fruit grower.

Classical fruit gardening also encouraged – and was in turn encouraged by – the custom of outdoor dining. Eating out among the fruit trees was the nearest thing to the paradisial picnic envisaged by the 17th-century English poet, Andrew Marvell:

Outdoor dining in 16th-century Germany: turfed seats, a wooden trellis, vines and fruit trees create an illusion of pastoral freedom in which the constraints of everyday life cease to exist. Apples, along with other fruits, were invested with powerful aphrodisiac properties, and this was possibly one reason for their consumption at the end of a meal, as they ushered in the more relaxed pleasures to come.

What wond'rous Life is this I lead!
Ripe Apples drop about my head;
The Luscious Clusters of the Vine
Upon my Mouth do crush their Wine;

The Nectaren, and curious Peach,
Into my hands themselves do reach;
Stumbling on Melons, as I pass,
Insnar'd with Flow'rs, I fall on Grass.

CHAPTER 1

APPLES IN THE
CLASSICAL
WORLD

The Greeks enclosed the Persian garden and turned it into a courtyard; the Romans went one step further to create dining rooms under the skies. Banquets were served amidst living apple and other fruit trees in tubs, which could be wheeled around, and the walls were often painted with trees and birds. There might be a pool or fountain in the centre, and a view of further gardens beyond. Centuries later, the Tudor and Stuart courts were to defy the English weather by adjourning to separate banqueting houses in the garden to partake of fresh fruit, sweetmeats and other delights, while in France Louis XIV indulged a taste for extravagant outdoor festivals complete with real and artificial fruit trees, fountains, magnificent displays of fresh and preserved fruits, and servants dressed as gardeners and shepherdesses.

The sensuous freedom of such occasions was underlined by the sexual connotations of fresh fruit, which constitute another long-standing influence on the way we regard and use apples. Along with pears, pomegranates, figs, prunes, quinces, almonds and oranges, Greek and Roman apples were, like Persian ones, closely associated with the arts of love and invested with powerful aphrodisiac properties. The Greek geographer, Strabo, informs us that a Persian girl on her wedding night would be allowed to eat nothing but apples and camel's marrow, and, in one aphrodisiac concoction, nine apple pips are to be pounded up with

> *a little shaving from the head of a man who has died a violent death . . . seven grains of barley that had been buried in a grave . . . the blood of a worm, of a black dog, and of the second finger of your left hand . . . Mixed with semen and added to a cup of wine . . . you make the woman drink it.*

As the fruit sacred to the Greek love goddess Aphrodite and her Roman counterpart Venus, there are numerous references in classical literature and mythology to the apple's role in the rituals of courtship and marriage. In ancient Greece the unattached would hurl apples at each other as a sign of romantic interest, and spitting apple pips at the ceiling was a way of foretelling success or failure in love. There was even a kind of 'lovers' catch', in which an apple rather than a ball was used – and was kissed by the two players as they tossed it back and forth.

The myth of Atalanta and Hippomenes tells how the love-lorn Hippomenes uses magic apples to win the hand of the beautiful and fleet-footed Atalanta. Dedicated to a life of perpetual chastity, Atalanta has challenged all potential suitors to a running race, with the promise that she will marry the first one who beats her. The price of losing, however, is death. In despair, Hippomenes prays to the goddess Aphrodite for help, and she provides him with three golden apples from the Garden of the Hesperides. The race begins and every time Atalanta overtakes him Hippomenes casts one of his golden apples before her. So tempting is the fruit, that she pauses each time to pick it up, loses the race, and becomes Hippomenes' bride.

The Fall of Troy was also precipitated by an apple. Miffed at being excluded from the marriage celebrations of Peleus and Thetis – who became the parents of Achilles – Eris, goddess of Strife, hit on the idea of enlivening the nuptials by hurling down a golden apple inscribed with the provocative words 'To the Fairest'. Paris, son of Priam, King of Troy, was nominated to make the choice between the three goddesses present, and after much bribery, chose Aphrodite

PLATE 4: YELLOW INGESTRIE

who promised him the hand of the most beautiful woman in the world, namely Helen, wife of the Spartan king. Paris sailed for Sparta, abducted Helen, and carried her back to Troy, the act which provoked the Trojan war.

The sexual and romantic connotations of the apple was almost certainly another reason for its customary position at the end of a meal. The ancient Greeks acknowledged that, quite apart from its digestive benefits and the pleasure to be derived from the contrast between the fresh, clean, sweetness of fruit and the richer dishes which precede it, the 'second tables' had mainly been 'invented to prolong the party'. Fresh fruit, like the nuts and sweetmeats which often accompanied it, allowed guests to sustain themselves while free from the attentions of hovering servants, and the aphrodisiac properties of apples were still adding piquancy to the after-dinner proceedings in 17th-century England. The Victorians would have shuddered to realise that the magnificent dessert of fresh fruit they presented at the end of the meal once marked the transition from the formal entertainments to an altogether more relaxed kind of pleasure.

It has to be said, however, that most Greek and Roman references to apples are not specific. The Greek *mēlon* and Latin *malum* refer not only to an apple but to any round fleshy fruit, and in most cases it is impossible to know whether it is apples, rather than pomegranates or quinces, to which the author refers. As apples and apple-like fruits are connected with love and fertility in all European, Asian and North American folklore, it seems likely that the association may have a common source far back in time, with the fruit in question varying as the stories spread across geographical and climatic boundaries. As the botanic apple is the only candidate with a species native to Greece, it is tempting to speculate that classical accounts may reflect older stories genuinely associated with 'real' apple trees, but it is impossible to know.

The more important point, as far as the subsequent history of the apple in the Western world is concerned, is that European and British cultures interpreted the classical references in terms of the fruit they knew best. In this way the apple acquired all the symbolic baggage of the mythical fruit – its associations of sexuality and temptation; its place on a tree guarded by a serpent in a paradise garden; and its tendency to invite chaos and destruction. These contributed directly to the belief from at least the 13th century that the unspecified fruit which tempted Eve with such catastrophic results was an apple – a conclusion which was to have profound consequences for the apple's future.

Iᴛ ɪs ᴀʟᴍᴏsᴛ ᴄᴇʀᴛᴀɪɴʟʏ through the Romans that the domestic apple came to Europe and Britain. Until then the inhabitants seem to have made good use of their native crab apples. There is evidence for the gathering and storing of fruit in Switzerland during the Iron Age, and a possibility that the Celts or their predecessors in central Europe were making cider. Although fruit of the wild crab apple is mouth-puckeringly sharp, fermenting, drying or cooking the fruit would have rendered it more palatable as the large quantities devoured in Viking York centuries later indicate. The importance of the apple in northern mythology and custom also suggests that regardless of how or where the traditions originated, crab apples must have played an important role long before the arrival of the cultivated fruit. As in classical mythology, Celtic stories identify the apple with love and fertility.[14] The Irish hero Blamain is conceived when his mother

A 3rd-century mosaic from St Romain-en-Gal, Vienne, France, showing fruit picking. Roman skills in grafting and pruning produced superior fruit and it is to their pleasure in good apples that we probably owe the establishment of the first orchards in France and Britain.

eats the single apple growing on the tree in her father's garden. During the Celtic festival of Samhain or summer's end, when it was believed that the vegetation deities retreated to the underworld and the barriers between that world and this were temporarily opened, apples played a significant part in the rituals of thanksgiving and divination. The old custom of wassailing apple trees and Hallowe'en games such as 'bobbing' for apples, and using strips of apple peel, apple pips or an apple under the pillow to peer into the romantic future, are almost certainly fossilised relics of Celtic ceremonies. Decorating Christmas trees with baubles similarly recalls the ancient practice of hanging apples on to evergreen boughs as symbols of hope and renewal in the depths of mid-winter, and in parts of Britain it was usual to make a 'Kissing Bough' of evergreens, mistletoe and apples at Christmas. In the 1950s it was still a custom in the Forest of Dean, Gloucestershire, to offer 'Apple Gifts' made of apples, nuts and yew to welcome in the new year, and apple boughs and garlands are again becoming fashionable Christmas decorations.

Celtic stories also link apples with paradise and immortality. Avalon, where the dying King Arthur is laid to rest, is a heavenly 'Isle of Apples'. The Irish hero Bran is beckoned on his journey by a branch of apple blossom from Emain Ablach, an island where the apple trees bloom and fruit at the same time, which lies in a marvellous archipelago beyond the sea. Apple trees are frequently sources of magical sustenance, and of everlasting youth. On the Island of Trees, the legendary seafarer Mael Dúin cuts himself a rod which grows into a tree bearing three apples, each of which sustains him for 40 nights.

In Scandinavian legend, the gods owe their continuous youth and immortality to eating the magic apples carried around by the goddess Idun and, like those of the Hesperides, these, too, are stolen and returned. In Sweden and Denmark traditional stories tell of a hillman or dwarf king who fills the lap of his earthly mistress with golden apples, and so compels her to return with him to dwarfland.

The native crab apples were not, however, to the taste of the incoming Romans. They liked to take their home comforts with them wherever they went, and it is to the conquerors' yearning for the pleasures of fresh fruit that we owe the establishment of orchards in north-west Spain, France, and Britain itself. Roman mosaics at St Romain-en-Gal, Vienne, France, which date from the third century, show the grafting of trees and the harvesting of apples.[15] Tools such as pruning knives have been found at a number of Romano-British sites and what have been described as pruning hooks have been unearthed at East Grimstead in Wiltshire. At Watts Wells, a Romano-British villa near Ditchley in Oxfordshire, there are ordered rows of tree pits, which suggest the existence of an orchard or vineyard.[16] Certainly east Kent, with its continental climate, deep, fertile soils, and good road and water links with London and the major Roman city of Colchester, in Essex, would have been the perfect site for the first British orchards and is still Britain's main apple growing area today. Classical and northern apple beliefs inevitably became intertwined as Roman festivals adapted native customs. The feasts of Samhain and

Pomona coincided, and both were in effect taken over by the Christian Church as the feast of All Souls and the Eve of all Saints – which is, of course, Hallowe'en.

As long as the Roman Empire was secure, cultivated fruits, including apples, were grown and enjoyed across much of northern and western Europe. When the Roman Empire began to collapse, however, much (but not all) of the fruit growing went down with it. Towards the end of the fourth century huge armies of Asiatic nomads from east and north of the Gobi desert moved westwards into Europe, unseating and pushing before them waves of Germanic, and then Slavonic peoples who poured in through the borders of Rome and overwhelmed the western empire within three hundred years. In Britain, Roman orchards probably did not survive the withdrawal of the legions and the subsequent influx of Jutes, Angles and Saxons from northern Germany. In Gaul, the Franks who came in from the Rhineland, and gave France its name, seem to have been concerned for the vineyards and orchards but probably not for the finer points of horticulture, including skills such as grafting. As a result many varieties and the techniques of growing them might have been lost. What saved them from obscurity was the orcharding traditions of the Christian church on the one hand, and the rise of Islam on the other.

THE CHRISTIAN MONASTIC TRADITION had begun with the 'Desert Fathers' of Egypt (AD 250–500). With them originated the first notions of a communal, rather than solitary, religious life which was developed in the East by the Greek theologian, Basil the Great (c330–379 AD), who set down the first monastic rule, and in the West by Benedict of Nursia (c480–543 AD). Through the practical as well as ideological commitment to supporting their communities, monasteries became the effective guardians of a whole body of cultural and intellectual skills and artefacts, which were preserved and transmitted through the religious network. From the outset monastic orders had been involved in feeding themselves. Anthony of Egypt, one of the founding 'Desert Fathers', is one of the patron saints of gardening; St Phocas, from Sinope on the Black Sea, is another. Within the boundary walls of the monastery of another founding father, Abbot Isidore, near Thebes, there were 'well watered gardens, and all the fruits and trees of Paradise; it provided in abundance all that was necessary, so that the monks who lived there never needed to go outside for anything'.[17]

Strongly influenced by the eastern example, the Benedictines in Europe similarly placed a strong practical emphasis on gardening and the cultivation of fruit. When the Roman Empire was overrun, monasticism and its skills were able to survive as part of the larger world of eastern and Celtic Christianity. Legend has it that in the middle of the sixth century St Teilo travelled from Llandaff in South Wales to visit a former fellow student, Samson of Dol, in Brittany. His stay extended to several years, during which he planted an orchard of fruit trees three miles across – which was apparently still there two centuries later! When the Irish missionary St Columban visited Burgundy in the seventh century he was reputedly served cider of a quality to rival the finest white wine.

With the coming to power of the Frankish ruler, Charlemagne, in 771, western Europe entered a period of relative peace and prosperity, and the opportunity arose again for the considered cultivation and appreciation of fruit. Charlemagne's vast kingdom extended through Germany, and east to Bavaria; it went southwards through France, across the Alps into Italy, and south-west to the Pyrénées. The Franks had converted to Christianity in 497 and Charlemagne was crowned Holy Roman Emperor about AD 800. Monastic skills and learning were encouraged, both within the holy precincts and beyond. Charlemagne's *Capitulare de Villis* laid down what should be planted on the crown lands of every city including apples, cherries, plums, peaches and pears. Among the varieties of apples listed are the 'Gozmaringa', which may take its name from a village in Württemberg, southern Germany, an area still renowned for its apples. Mentioned also were the 'Geroldinga, Crevadella' and 'Spirania', or perfumed apples, as well as sweet apples and sour apples, and early and late keeping apples.

In England, the continued disruption caused by the Danish and Viking invasions meant that the horticultural outlook did not begin to improve until the tenth century, when Brithnoð, Abbot of Ely, is reported to have been 'skilled in planting gardens and orchards'. He 'planted choice fruit trees in regular and beautiful order' so that in time they 'appeared at a distance like a wood loaded with the most excellent fruits in great abundance and they added much to the commodiousness and beauty of the place'.[18] The Laws of Hywel Dda, a tenth-century Welsh prince, suggest that fruit cultivation was also being more generally encouraged. They set the value of a sweet apple at double that of a sour apple and placed an increasing value on an 'imp' – a graft or young sapling – as it grows to maturity and bears fruit. The oldest extant manuscript containing this law, however, is dated to 1250 and it is impossible to know how far back this regulation goes.[19] Nevertheless, an 11th-century Anglo-Saxon document gives the preparation of orchards as one of the bailiff's winter duties, and the grafting of new trees as a spring task.[20]

The Norman Conquest of 1066 brought closer contact with the Continent and, among other things, fired England with the Norman enthusiasm for cider. But the main impetus to the renewed cultivation of apples came in the 12th century, and affected Britain and mainland Europe alike: this was the spectacular expansion of the Cistercian order of monks under St Bernard of Clairvaux. Founded in 1098 by a breakaway group of Benedictines, it put increased emphasis on the value of manual labour and the cultivation of abbey lands. At Clairvaux, on the borders of Burgundy and Champagne, were laid the foundations of the region's reputation as a wine producer, and there was 'within the precinct . . . a wide level area containing an orchard of many different fruit trees, like a little wood'.[21]

Apples followed the abbeys. At St Bernard's death, there were over 300 Cistercian abbeys, and the order had spread to Scotland, Sweden, Portugal and the eastern Mediterranean. New fruits were introduced, and the monks took pains to propagate and distribute good varieties. In Normandy, near Rouen, on an island at Léry in the River Eure, a tributary of the Seine, land was leased by the abbey of St Oeun specifically for a garden of grafted pears and apples. From Burgundy, the Cistercians sent apple varieties to northern and eastern Germany and grafts were sent from Paris to Denmark. Tradition has it that the Arbroath and Melrose apples were raised in Scottish Cistercian monasteries.

The effect was to encourage monastic fruit-growing in general. In the 12th century William of Malmesbury described the Benedictine abbey of Thorney, near Peterborough in Cambridgeshire, as being set in a paradise 'filled with apple-bearing trees'.[22] In Kent, at

Plate 5: Devonshire Quarrenden

Canterbury, a map of 1165 shows there to have been an orchard or *pomarium* on the outer edge of the cultivated land of the Benedictine Christchurch Abbey, and there may also have been a second orchard as in 1170 Thomas à Becket is reported to have fled from the murderous knights by a different route from 'the usual passage through the orchard to the west end of the Church'.

APPLES AND
ISLAM

As the Western Roman Empire and its orchards struggled against the Barbarians, the Eastern Empire – Byzantium – continued to flourish. But during the seventh and eighth centuries AD, it too was overrun. The enemy this time was the newly emergent force of Islam. In the two hundred years following the prophet Mohammed's death in 632, his armies occupied Palestine, Syria, Egypt and North Africa as well as the new Persian Empire which had grown up east of the Euphrates. Unlike the northern invaders, however, these conquerors were strictly disciplined and under orders to preserve crops and orchards. Peaceful government was restored as soon as possible and the skills of Byzantium and Persia were assimilated into Islamic cultural life. In keeping with the exhortations of the Koran, and the richness of its acquired inheritance, the Muslim world encouraged all manner of scholarship, gardening, fruit growing, fine craftsmanship and the arts. Greek and Roman botanical works were translated, updated and expanded, and the opportunity was taken to test, acclimatise and introduce new fruits, varieties and crops wherever possible. In the shady groves and tranquil pools of Persian gardens, Islam also found the earthly image of the Koranic paradise, and set about recreating it throughout the Empire. In this way, the horticultural wisdom of the Christian and classical world was not only preserved but expanded, and through the Arab conquests of Spain and the Mediterranean islands was given a sanctuary and a route back into western Europe.

In the advanced horticultural areas of the Low Countries, fruit sellers with their large baskets of ripe, good-sized fruit were a feature of every town and village square.

Moorish Spain in particular became a centre of horticultural expertise. In the tenth century, while the rest of Europe was just beginning to revive its fruit-growing skills, sophisticated botanic gardens were being established at Toledo and Seville under the auspices of the sultans. At both were noted pleasure gardens, but the sultans also maintained collections of plants for study and experiment. The *Book of Agriculture*, written around 1080 by Ibn Bassal, a superintendent at Toledo, contains information on all kinds of fruits including the apple, with sections on pruning, grafting, planting, soils and manure. Despite the determination of the Christian rulers in northern Spain to rid the Peninsula of the 'Infidels'- which they finally achieved in 1492 – the intervening centuries gave many opportunities for the exchange of ideas, plants and scholars between east and west. Alphonso X of Castile (1252–1284) had Arabic horticultural works, including Ibn Bassal's, translated, and through his sister Eleanor, who seems to have shared his enthusiasm, passed some of this expertise through to England. Eleanor married Edward I, and immediately imported gardeners from Spain to help her create a pleasure ground at Kings Langley in Hertfordshire. The conquest of Saracen Sicily by the Normans in 1072 also made this Arabic heritage available to the political, cultural and religious networks of western Christendom.

The modern transport revolution enabled fruit from all over the British Empire to reach the home market. This image from an English newspaper advertisement of 1929 was designed to promote the qualities of Canadian apples.

FROM THE THIRTEENTH CENTURY onwards, apples were grown increasingly widely in Europe. The number of named and esteemed varieties soared, and pips from these were taken to almost every corner of the expanding world. During the 16th and 17th centuries, French fur traders and missionaries planted apple pips in what is now Canada, and an apple tree still growing in Vancouver dates from 1825 when a group of young gentlemen, just arrived from England in the employ of the Hudson's Bay Company, discovered that the pips from the apples eaten at their farewell dinner had been slipped into their waistcoat pockets by their sisters and sweethearts.

The Protestant settlers who left England, Holland, Germany, France, Scandinavia and other European countries in increasing numbers throughout the 16th and 17th centuries planted orchards

A young Kent orchard of the 1960s.

29

Plate 6: Reverend W. Wilks

all along the eastern seaboard of America, and from these the pioneers took pips and scions to found orchards in the Midwest and along the Pacific coast. The Spanish and the Portuguese took apples to South America, and apples proliferated wherever the climate allowed. When Darwin made landfall in Chile in 1835 he found apple trees growing not only on the island of Chiloé, but all along the coast, and the port of Valdiva was 'completely hidden in a wood of Apple trees'. The 'streets are merely paths in an orchard', he observed on disembarking. 'I never saw this fruit in such abundance.' Members of the Spanish Mission had carried the apple northwards to California, where the Mission's orchards were among the wonders of this new land which greeted the first pioneers from the east in the 1850s. By 1900 the United States was the largest apple producer and exporter in the world, with Canada a close second.

The apple was first taken to South Africa by the Dutchman Jan van Riebeeck, who founded the Netherlands East India Company's trading centre at Cape Town. He brought apple trees in 1654 from St Helena, which the Dutch had seized from the Portuguese, but these were swiftly followed by trees from the Netherlands. He made fruit growing a virtual requirement for the first settlers, not only to feed themselves, but to supply the trading boats on their journey eastwards. When the Cape vineyards were devastated by phylloxera at the end of the last century, Cecil Rhodes, founder of the new British State of South Africa, turned to apples as an alternative crop and, with the help of Californian fruit expertise, his new venture formed the basis of today's flourishing Cape industry.

Australia reputedly owes its first apple plantings to Captain Arthur Phillip, who in 1788 established the first English settlement at Port Jackson, later to become the city of Sydney in New South Wales. How many of these survived, however, is not known and the first recorded crop of apples grown in New South Wales relates to six fruits produced in 1791 on a tree which had been brought from the Cape of Good Hope. Apples were first taken to Tasmania, which later became known as the Apple Isle, by the notorious Captain Bligh who anchored his ship the *Bounty* off Bruny Island in 1788. The log records that Dr Nelson, the ship's botanist, planted three apple seedlings and several apple and pear pips. Four years later it was reported that one apple tree was making good growth but feared that the others had been disturbed by aborigines.

As the settlements of Australia took off, so did orchards, and in 1814 a party of English missionaries set off to deliver both the word of God and apples to New Zealand. Led by the Reverend Samuel Marsden, who had made his home in Sydney, they established a settlement in the North Island and introduced not only apples but also peaches and oranges. News of a French colonisation scheme led to the islands being swiftly annexed to Australia, and later the fruit-growing area around Hawkes Bay became known as the 'Apple Bowl'. Like South Africa, Australia and New Zealand were able to take advantage of the seasonal opposition of southern and northern hemispheres to supply apples during the summer months in Europe.

In the twentieth century, new apple varieties developed in North America were planted throughout the world. They have helped form the basis of the apple industries of South America, enabling Argentina, Brazil, Chile and Uruguay to join the southern hemisphere fruit exporters. Western apples have been taken to China, Korea, Japan, India and Pakistan. Up until about 1870, *Malus prunifolia* and its cultivated form *Malus asiatica* were apparently the only source of apples in these regions. The fruits were small and sharp, and were turned into preserves. With the introduction of large, sweet Western varieties apple production soared. Apple growing is now a major industry in Japan, which grows four times as many apples as England. India produces more than twice as much as Canada; and China has become the world's largest apple grower. Over the last decade China's crop has increased to over 22 million tonnes of apples, which is double that of the whole of Europe and nearly five times that of the United States.

WHILE THE APPLE'S GASTRONOMIC, symbolic, religious, decorative and patriotic appeal all contributed to this pomological colonisation, particularly at the beginning, it is ultimately the apple's remarkable adaptability which has given it dominion. Horticulturally speaking, apple trees can tolerate a wide range of summer and winter temperatures, many different kinds of soils, and are indifferent to day-length, which means they can be cultivated at any latitude where a suitable micro-climate exists. To extend its northerly limit, varieties have been grafted on to rootstocks of the Siberian crab apple, and new varieties raised by crossing with this exceptionally hardy species have the capability to withstand Arctic temperatures. The major factor which used to limit the apple's spread into warmer regions was its need for a winter chilling but many subtropical countries, such as India and Pakistan, have mountainous regions which provide a niche for the apple and, even where the winter is not cold enough to satisfy fully the chilling requirement, apples can still be grown satisfactorily by removing the trees' leaves either by hand or by using chemicals. The development of new tropical varieties – crosses between old Mediterranean and modern varieties – is also opening up the possibilities of growing apples right across the African continent, and in parts of the Middle East and Asia which were formerly out of bounds.

The stocks of western apples are also about to be increased with new seedlings from their original homeland in the Tien Shan mountains. In 1989 contacts were established between botanists at the Kazakhstan Academy of Sciences in Alma Ata and Cornell University, Geneva, New York. This has resulted in four expeditions to Central Asia by western groups. Seeds from the wild apple forests were brought back to Cornell's Apple Collection and also some scion wood from trees that had already been selected for the quality of their fruits by the centre in Alma Ata. This has allowed seeds to be distributed to fruit breeders in all the main apple growing countries and now thousands of seedlings are being screened and evaluated for a range of useful traits. One of the main objectives is to find disease resistance for breeding new varieties, and researchers also hope to find seedlings with drought tolerance, hardiness, low demands for winter chilling and possibly even new flavours. The other objective is to increase the genetic diversity of the west's collections and broaden the genetic base of our apples. By gaining more knowledge of the parent species – *Malus sieversii* – it may also be possible to help in the conservation of the wild apple forests of Kazakhstan, which are under threat from timber felling, mining and urban development.

The apple possesses a quite remarkable diversity of colours, forms, flavours and textures as a dessert fruit, and in culinary terms ranks among those prized ingredients such as eggs and olive

WORLD ANNUAL APPLE PRODUCTION 2000
Production in thousand metric tonnes

Afghanistan 17.5

Albania.............12.0

Algeria...............87.0

Argentina 1056.0

Armenia 50.6

Australia 334.35

Austria 409.6

Azerbaijan 115.0

Belarus 205.0

Belgium–
Luxembourg ...562.4

Bhutan 5.5

Bolivia..............9.75

Bosnia
Herzegovina......18.0

Brazil.............1136.3

Bulgaria...........109.9

Canada............500.0

Chile 919.2

China 22059.7

Croatia 66.7

Cyprus..............12.8

Czech
Republic..........290.0

Denmark...........67.0

Ecuador 12.4

Egypt...............410.0

Estonia 7.0

Finland 10.5

France 2500.0

Georgia 115.0

Germany.......1800.0

Greece 320.0

Grenada 0.5

Guatemala 27.5

Honduras 0.1

Hungary 444.5

India 1320.6

Iran...............2200.0

Iraq 75.0

Ireland................8.0

Israel...............102.9

Italy 2416.2

Japan 941.4

Jordan...............33.5

Kazakhstan........58.0

Kenya 0.8

Korea, DPRP ..650.0

Korea
Republic of 490.5

Kyrgyzstan.........86.0

Latvia 24.4

Lebanon..........120.0

Libya 46.0

Lithuania 33.0

Macedonia........61.7

Madagascar 6.6

Malta...................0.3

Mexico 477.1

Moldova,
Republic of 215.0

Morocco...........300.0

Netherlands 518.0

New Zealand ..504.0

Norway..............10.1

Pakistan 377.3

Paraguay..............0.6

Peru.................171.0

Poland 1280.0

Portugal...........235.0

Réunion 0.1

Romania...........365.0

Russian
Federation.....1200.0

Saint Vincent
Grenadines 0.6

Slovakia.............80.0

Slovenia 81.2

South Africa 650.1

Spain 734.8

Sweden.............65.0

Switzerland 380.0

Syria 320.0

Tajikistan 96.0

Tunisia 100.0

Turkey...........2500.0

Turkmenistan ...22.0

Ukraine 1325.0

United
Kingdom 248.0

United States
of America.....4843.0

Uruguay 61.0

Uzbekistan 407.0

Yemen 2.0

Yugoslavia,
Fed Rep of 185.0

Zimbabwe............5.9

World 60,126.3

F.A.O. 2001 *Food and Agriculture Organisation of the United Nations, Rome.*

oil which function as both stars and staples. Combining multiplicity of use with prodigality of supply, apples are easy to harvest and supremely portable. Quick to prepare, they can be eaten raw or cooked, peeled or unpeeled, hot or cold. Baked, roasted, poached, steamed or fried, they may be enjoyed whole, stuffed, sliced, chopped, grated or puréed; either on their own, or as the perfect foil for other ingredients. Sweet without being cloying and acid without being coarse, they relieve the fattiness of meat, the saltiness of fish, the blandness of cereals. Stuffings, salads, stews, pies, tarts, cakes, loaves, fools, jellies, creams, jams, pickles, sweetmeats, juices, liquors . . . there is hardly a culinary technique that cannot be applied to the apple. It is also a useful source of fibre and essential nutrients, including vitamin C, antioxidants and potassium. With health experts recommending increasing the amount of fruit and vegetables that should be consumed, apples have a clear role to play in keeping the doctor away.

In the chapters that follow we explore in detail how this familiar but extraordinary fruit came to occupy its unique position in the world's imagination and on its table, and why in Britain and North America particularly, the apple is regarded with special affection as a part of our national heritage.

Chapter 2
For PLEASURE
MEATE and MEDICINE

*D*igestive cheese and fruit, there sure will be . . .
and a pure cup of rich Canary wine.'
BEN JONSON: *Inviting a Friend to Supper*, 1616

*F*ruit trees created a 'perpetual tapestry covered in
spring with flowers, in summer and autumn with
fruits and foliages and beautiful even in winter with
bare branches laced together in cunning artifice'.
ANTOINE LE GENDRE: *The Manner of Ordering Fruit*
Trees (Trans. 1655)

OST OF THE APPLES grown in Europe during the Middle Ages were destined for the cider press. Some found their way into the kitchen and the equivalent of the medicine chest, while also acting as ornamental trees in the pleasure ground. Relatively few apples seem to have been eaten fresh for pleasure. There are several possible reasons for this. One was almost certainly the dearth of good eating varieties. Once the cultivated orchards of the Romans fell into neglect, sweet apples may have largely disappeared in parts of northern Europe. Although some new seedlings would have retained the sweet eating qualities of their parents, the absence of a network of gardeners to repropagate them meant they remained in obscurity, and most of the apples produced were astringent and too sour to eat.

As the medieval physician's bible was still the Salerno School's *Regimen Sanitatis*, or *Prescription for Health*, with its classical precepts and distrust of sour apples, contemporary medical wisdom helped reinforce a rather negative view of the raw fruit, which seems to have adhered even to more edible specimens. Apples were banned completely for children and wet nurses, and were among the first suspects in any case of a 'bad stomach', flux or fever. As eating sour fruit and being tempted to over-indulge in the occasional edible variety would both cause gastric havoc, a certain caution was, perhaps, understandable. As late as 1541 the Englishman Sir Thomas Elyot observed that 'All fruits are generally noyful to man, and do engender ill humours, and be oft times the cause of putrefied fevers, if they be much and continually eaten.'

Unease may also have been encouraged by the identification of the apple as the cause of Eve's downfall. The depiction of the 'Expulsion from Paradise' in the *Book of Hours* painted for the Duc de Berry in the early 15th century shows Eve holding what seems to be a golden Paradise apple in her hand, and the famous 'Fall of Man' (*c*1470), by the Flemish artist Hugo van der Goes depicts Adam and Eve beneath an unmistakable apple tree. At least one popular English lyric suggests that the association might have given the would-be-consumer pause for thought:

'The Fall of Man', c1470, by Hugo van der Goes. The apples and the tree are both depicted in realistic detail.

An apple I took of the tree,
God had it forbidden me,
Wherefore I should be damned.

The practice of cooking apples, however, seems to have calmed both medical and theological nerves. A 14th-century Italian manuscript, the *Table of Health in Accordance with Medical Science*, which is based upon the works of the Arab doctor, Ibn Botlan, points out that apart from the benefits of heat and spices in offsetting their cold negative properties, apples 'taste better if cooked under the embers, sprinkled with sugar, and served with candied anise, or followed by rose or cinnamon sugar'. Apples cooked in this way were eaten at all social levels, although only the better off would have been able to lace them with spices. In the 14th-century English poem *The Vision of Piers the Plowman* 'the poor folk came with peascods and brought . . . baked apples by the lapful' to appease the figure of Hunger. Baked apples mixed with 'a syrup of liquorice, starch and sugar' could also be 'administered successfully for chest pains twice a day before meals', while grand dinners closed with a digestive of spiced wine, wafers and roasted apples or pears 'for your stomak to ese'. It is still the custom at Trinity College, Cambridge, to serve baked apples and caraway at the end of the Audit Feast.

Apples were also a useful ingredient in the medieval kitchen. Sharp apples of all kinds were used to enliven cereal porridges, and to make a kind of purée or broth, which seems to have been a precursor of the later 'appulmos' – a more refined dish made from boiled sieved apples mixed with almonds, honey, breadcrumbs and spices. Apples were used to add interest, or bulk out 'brown' meat dishes, and baked or sieved apples seem to have been a regular accompaniment to fat autumn roasts. They also turned up in the combinations of white meat, wine, and spices which were put into side dishes such as tarts, custards, pancakes and fritters. In Britain, apples would also be turned into verjuice: sharp, slightly fermented apple juice that would be used in much the same way as cider or wine vinegar for flavouring sauces and preserving pickles. 'Be suer of vergis (a gallon at least), so good for the kitchen, so needful of beest', advised the agricultural writer and poet Thomas Tusser in about 1560.

Raw apples could, however, be safely eaten in the case of particular ailments or situations. Ten raw apples a day were prescribed to keep English monastic bowels open during Lent, and an apple eaten raw at the beginning of a meal was deemed to be a good way to 'open the stomach'. Large, sweet smelling and, above all, ripe apples could be eaten to 'stimulate the heart' even though they were 'harmful to the nerves'.

APPLES IN THE
PLEASURE
GROUND

APPLES COULD ALSO, as Albert, Count of Bollstädt, pointed out, be left on the tree for 'delight rather than fruit'. Albert was a well-connected and much travelled German churchman whose treatise *On Vegetables and Plants*, compiled in about 1260, gave detailed instructions for

Plate 7: Worcester Pearmain

the design and layout of pleasure gardens, based on his impressions and observations of gardens all over Europe. As the main aim of the pleasure ground should be 'the delight of the two senses viz: sight and smell', a fine sward of grass planted with 'sweet trees, with perfumed flowers and agreeable shade, like grapevines, pears, apples, pomegranates, sweet bay trees, cypresses and suchlike' was essential. As 'shade was more sought after than fruit' the Count warns gardeners that 'not much trouble should be taken to dig about and manure [the trees], for this might cause great damage to the turf'. His ideas were reproduced some 50 years later in an exhaustive Italian gardening manual by Pietro de Crescenzi which became the standard reference work on garden layout for the next two centuries, and undoubtedly helped promote the medieval fashion for ornamental gardening.[1]

This was influenced both indirectly and directly by what amounted to the rediscovery by western Europe of the ancient Persian fruit garden, the *pairidaeza*. The medieval pleasure garden took its immediate inspiration from the biblical descriptions of Eden before the Fall and of the enclosed garden of the 'Song of Solomon'. These in turn had been inspired by eastern gardens, and when trade and the military expeditions of the crusades in defence of the Holy Land brought the West into increasing contact with Arab culture, they found their images of biblical gardens made manifest.

All this provided a powerful stimulus to horticultural ambition. Eleanor of Castile, wife of Edward I, introduced the Blandurel apple to England in about 1280 when she procured French grafts of the variety for her Hertfordshire pleasure ground, and when Charles VI of France laid out his new pleasure ground at the Hôtel St Pol in Paris, in 1398, he planted 1000 cherry trees, 150 plum trees, 115 grafts of pears, 100 common apple trees and a dozen of the decorative, miniature Paradise apple.[2] These were easily propagated from suckers or cuttings and came into bearing more quickly than other apples, but cost nearly four times as much.

At a more artless level, medieval cider orchards seem to have been valued for their aesthetic contribution as well as their alcoholic benefits. Gerald of Wales, in his lyrical recollections of Manorbier Castle, near Pembroke, where he was born in 1147, remembered 'a fine fish pond under the walls, as conspicuous for its grand appearance as for the depth of water, and a beautiful orchard on the same side enclosed on one part by a vineyard and on the other by a wood . . . of hazel trees on a rocky eminence'.[3]

ALTHOUGH MOST OF THE APPLES that ended up in the cider press or the markets were the random products of seedling orchards, it was inevitable that the revival of fruit-growing skills brought with it a new interest in identifying and repropagating varieties of particular value. The Continent led the way in this respect. It has been estimated that 32 named varieties were grown in medieval France and a number of these found their way to market.[4] In the 13th century, the fruit sellers of Paris cried out the charms of the scarlet Rouviau and pale yellow Blanc-Duriau or Blandurel, which had so entranced Queen Eleanor on a visit to Ponthieu in Aquitaine that she ordered them to be sent to her in England from Paris, together with other luxuries such as pears and '7 dozen cheese of Brie'. Capendu, which was probably Court Pendu Gris, was also on sale in Rouen market and the golden Paradise apple was acclaimed as one of the most appreciated varieties in Normandy.

At first English fruiterers relied on France for their supplies of apples. Costards, for example, which became the most popular and widely grown apple in medieval England and gave their name to the costermongers, or fruit sellers of London, were being imported from France in 1292 when an order for 300lb sent to Edward I at Berwick Castle appears in the Royal accounts. They cost 12 pence a hundred compared with a mere 3 pence for unnamed varieties.

The demand for apples seems to be one of the reasons that English fruiterers were among the first entrepreneurs going between growers and consumers. Records in 1292 refer to the activities of 'Tree Fruiterers', and the Fruiterers are one of the oldest of the City of London Companies, being first mentioned as a corporate body in 1463. Fruiterers dealt not only in fresh fruit but also in dried fruits and nuts. The expansion of the population and the growth of towns meant that local supplies could not always meet demand and, instead, fruiterers turned to Normandy for apples. In London a system of renting gardens and orchards in which to grow market produce began on Tower Hill and, by the 14th century, had extended to Billingsgate, East Cheap, Lombard Street and Bow Lane. The principal market was at Saint Augustine's Gate at the west end of Watling Street, near St Paul's Cathedral. It soon became so crowded as to 'hinder persons passing on foot and horseback', and such was the 'scurrility, clamour and nuisance of the gardeners and their servants' that, in 1355, the protests of local residents and the priest who conducted mass at St Paul's persuaded the mayor and aldermen to move the market to a more suitable place; accordingly it was transferred to a site near Blackfriars, south of the cathedral.[5]

Apples being harvested, from a late 15th-century French translation of Pietro de Crescenzi's gardening manual. There is a carefully tended and protected nursery of young trees visible on the right. It may even be a trial ground for evaluating new varieties.

The growing demand for named varieties for important occasions also made it clear that establishing English orchards of the most popular varieties would not just be a way of acquiring prestige but also made sound commercial sense. When shrewd landowners such as the Earl of Lincoln began to purchase grafts of Costards, for example, the subsequent demand became so great that relatively limited supplies allowed one Oxford nurseryman in 1325 to charge 3 shillings for a mere 29 trees – compared with a shilling per 100 a few years later. Other French varieties established in England during the 13th and 14th centuries included the red flushed Ricardons and the Pomewater, both mentioned by the poet Lydgate.

Neither of the most often mentioned varieties – Costard and Blandurel – were, however, good to eat straight from the tree. Costards were sharp and usually cooked, and Blandurels were a late keeping apple which had to be stored for three months after picking to mellow and sweeten. The author of *Le Ménagier de Paris*, a French household instruction book written in about 1393, records ordering 'two hundred blaudrel apples' for a wedding feast in May.

The first named variety to have originated in England seems to have been the Pearmain, which was primarily a cider apple. It appears for the first time in an Exchequer account of 1290 which confirms that one Robert Evermere paid his annual rent of 200 Pearmains and four

PLATE 8: EMNETH EARLY (EARLY VICTORIA)

hogsheads of Pearmain cider for the petty serjeanty of Runham in Norfolk.[6] At about the same time rents paid in Pearmain cider start to be recorded in Normandy at Rouen and Caen, and as the French regarded the Pearmain as an English variety, it suggests that at least the trade in apple scions was not entirely a one-way affair. The Queening, or Quoining, which Lydgate also mentioned, took its name from its angular shape – *coin* or *quoin* signified a corner – and may also have been an English variety. 'Quarendouns' may have been introduced from Normandy, but they are mentioned in a Middle English alliterative poem on plant names, which was written *c*1450 and believed to be of west country provenance.[7]

THE CLOUD OF SUSPICION surrounding fresh fruit did not really begin to lift in England until the early 16th century but, in the meantime, developments elsewhere in Europe paved the way both for the revival of apples as a dessert fruit, and for the discovery of a new way of enjoying apples altogether: as preserved sweetmeats.

Sugar, which was probably first domesticated in New Guinea, had been known to the Mediterranean world since at least 325 BC when Alexander's admiral, Nearchos, returning from India, reported that he had seen 'reeds [which] produce honey, although there are no bees'. For centuries, however, sugar was regarded as a medicine rather than a foodstuff. The Romans, for example, imported it from the east along with other spices, and used it as a warming medicine, good for stomach, bowels, kidneys and bladder. It was the Persian Empire and then its Arab conquerors who pioneered the growing and refining of sugar on a larger scale, and began to exploit its properties in cookery as well as medicine. There were sugar refining centres in Persia by about AD 400 and from the seventh century onwards sugar followed the Koran to North Africa, Sicily and Spain.

The combination of fruit and sugar in medicine was logical and inevitable. Not only was sugar a 'warming spice' and therefore useful to counteract the 'cold phlegmatic' properties of fruit but it was a great improvement on honey both medically and as a preservative. Whereas honey was believed to 'heat' the blood and was therefore only really suited to 'those with cold, moist temperaments, and to the elderly', sugar was 'good for the blood and therefore suitable for every temperament, age, season and place'. On the practical side, refined sugar was a less variable substance than honey. It made a sterile syrup, whose strength could be accurately controlled, and provided a preserving medium which prevented spoilage. In the right proportion it reacted with fruit acid and pectin to produce a 'wet' paste, jam or jelly and it could also create 'dry' fruit pastilles or crystallised fruits.

At first all these techniques were used to preserve and deliver the therapeutic properties of fruit rather than to make sweetmeats for enjoyment. It is in Arab pharmacopoeias rather than cookery books that the first recipes for fruit pastes, candies, jams, cordials, syrups and sherbets are to be found. It was clear, however, that if it was available in sufficient quantity, sugar could also take over and expand the role played by honey in cookery. As well as all its other advantages, the colourless transparency of a sugar syrup made it a more attractive culinary preservative than honey; it also tastes of nothing but itself, whereas honey added its own distinctive flavour. In addition, the crystals of sugar caught the light bringing a magical glitter to the preserves, and the advantages of 'dry' sweetmeats which could easily be stored and eaten out of hand were obvious.

The Arab world seems to have begun to exploit these culinary possibilities in the eighth century AD, if not earlier, and by the 14th and 15th centuries they were being taken to Europe. Spice ships sailing into London from the east-west entrepôts of Venice and Genoa brought supplies of *chardequynce*, a thick quince paste, made first with honey and later with sugar, which as *marmelada* was also imported from Portugal and Spain, along with barrels of succade – a preserve of citrus peel in syrup. By the end of the 15th century, the customary English digestive of hippocras (spiced wine), wafers, roast apples and other spices, had expanded to include these fashionable and expensive novelties. The list of provisions for the feast when George Neville was enthroned as Archbishop of York in 1467 ends with 'spices, sugared delicates and wafers plenty', while John Russell's *Boke of Nurture*, written between 1440 and 1470, suggests closing on 'Fygges, reysons, almandes, dates, buttur, chese, nottus, apples & peres, Compostes & confites, chare de quynces, white & green gyngere.'[8] Comfits were spices such as caraway seeds coated with sugar, while a compost was a kind of sweet chutney made of root vegetables, apples, pears, raisins and currants preserved in a spiced syrup of sweet wine, honey and vinegar. In England, this final course was usually referred to as the 'voyde' from the French *voider*, meaning to clear either a table or a room, indicating that it would be brought in after the table had been cleared or, following the French fashion on grand occasions, might be taken in a separate room – often with considerable ceremony and a display of the host's silver plate.

This emergence of a separate and predominantly sweet final course was of crucial importance to the future of the English apple. Over the next two hundred years, as sugar became cheaper and more widely available, the 'voyde' expanded further to become a separate and increasingly magnificent fruit banquet, at which fresh and preserved fruit was presented in every conceivable guise. Out of this fruit banquet developed the 18th-century dessert which became the magnificent fresh-fruit dessert of Victorian times, preceded by a separate sweet pudding course. It was the unique importance which Britain, and later America, accorded to both these sections of the meal that gave the apple its supreme showcase as a dessert and culinary fruit. The spectacular transformation of the medieval wine and wafers into the full glory of the 17th-century fruit banquet and its descendants depended, however, upon several further developments both in fruit confectionery and in attitudes towards fresh fruit, and these took place not in England itself, but in Italy and France.

THE FIRST SIGNS OF A SHIFT in European attitude towards the eating of fresh fruit appeared in Renaissance Italy, with the rediscovery of the spirit and values of classical humanism. Inspired both by their Roman mentors, and their knowledge of the Arab world, princes, cardinals and merchant families such as the de' Medici began to spend the enormous wealth Italy had derived from trade on the joys of cultivated living. Palaces were decorated with the richest textiles, the most splendid metal work and all that was finest in art and furniture. The garden, as in Roman times, became an extension of the house, and the setting for lavish outdoor feasts.

Wherever they looked, whether backwards or eastwards, these new Italians found enticing images of fruit and fruit trees, and Renaissance paintings, gardens and banquets explode with the joy of rediscovery. Apples and other fruits are depicted by artists such as Botticelli (1444–1510) and Bellini (1400–1471), with a new, almost luminous realism. Frescos on the walls of the *Salone* or reception room of the de' Medici villa of Poggio a Caiano near Florence, show the goddess Pomona in her orchard, and Hercules with the apple tree in the Garden of the Hesperides. All around, apples, citrus fruits and grapes glow with colour and loving detail against a background of green foliage.

In the pleasure ground, fruit trees appear everywhere. In a letter of 1460 the humanist writer Bartolomeo Pagello of Vicenza observes that, although he does not wish his villa to be as grand as those of Pliny the Younger, he desires a library and portico running down to the garden which should contain 'many apples, pears and pomegranates, damascene plums and generous vines'.[9] Dwarf Paradise apple trees were particularly prized and were used to adorn balconies and loggias and to make miniature orchards in the pleasure ground. Formal woods of fruit trees were also planted, often in the 'quincunx' pattern, which had the effect of creating vistas in every direction, and it was recommended that some of the fruit should be left hanging on the tree to allow the colours and forms to be enjoyed for as long as possible.[10]

On the table, fine fruit became as in Roman times, a symbol of status, reflecting the resources and discrimination of the host. Gardeners were exhorted to recover classical skills and concentrate on improving both the quality and range of fruit available. Outdoor dining rooms where fruit and garden could be enjoyed together became the height of fashion. At the Villa d'Este at Tivoli outside Rome, Cardinal Ippolito II had a great loggia built at the end of the main terrace in 1569. It contained a dining room, and a route to the kitchens. Arches on the three open sides gave guests a spectacular view over the gardens and countryside, while an internal stairway led up to a balustraded terrace from which the vistas were even more magnificent.[11] Specially constructed arbours for dining in were also popular and the Duke of Milan even had an entire 'portable orchard in small carts, whereon the trees, laden with fruit, were brought to his table and his room.'

'Autumn' from Botticelli's 'The Four Seasons'.

A 16th-century Florentine artist's impression of a banquet in ancient Rome. According to Horace, the perfect Roman meal began with eggs and ended with fruit – ove ad malum – and fresh fruit, including apples, grapes and melons, is much in evidence here.

Reflecting its high value, fruit was presented not just as food but as spectacle. The theatrical modelling, gilding and artificial colouring which had given splendour to the medieval feast now gave way to dazzling displays designed to show off the magnificence of the fruit, sweetmeats and other prized delicacies. These were not the responsibility of the kitchen, but of an entirely separate department, and instead of being held back until the end of the feast, dominated the proceedings throughout from a side table called the *credenza*.

An August lunch menu composed by Bartolomeo Scappi, the most influential of the Renaissance cooks, serves the first course which included red and white melons, cheese, prunes, cherries, apricots, mirabelle plums and figs upon vine leaves from the *credenza*; and also the last course which consisted of apples, pears, biscuits, tartlets, cakes, artichokes, pears in wine and sugar, whipped sugared fresh cream, Parmesan cheese and fresh almonds.'[12] The glowing colours of the fruit would be set off to maximum advantage by tall stemmed dishes – tazzas – made from the new and quite exquisite Venetian glass, and mirrored in great silver platters. On the most elaborate occasions, a servant dressed as Pomona would help guests to their choice.

Apples were highly useful fruits for display purposes. Their shades of red, gold and green provided greater variety of colour than any other fruit, and outlasted them in season. The firm globes would gleam in the candlelight, and unlike smaller, softer fruits, they lent themselves to large and elaborate arrangements. The great authority on the *credenza* was Christoforo di Messisbugo, who belonged to the court of the d'Este family, the Dukes of Ferrara. Among the apples he records serving were the 'Paradise', and 'Deci', while Scappi mentions '*mele Appie*' or 'Appia' apple, and '*mele Rosa*'. Great value was attached to the idea that these were the same varieties that had been enjoyed in ancient Rome. The esteemed 'Appia' was said to be the 'Appian' apple mentioned by Pliny. Others considered the Paradise apple to be his 'Petisian' apple. The 'Deci' or '*melo d'Ezio*' was claimed to have been named after General Ezio, who had brought it north from Rome when he journeyed to battle with Attila the Hun near Padua in about AD 450.[13] The Court Pendu Gris or 'Capendu' of medieval Normandy was also known to the Italians, and, according to the French botanist Charles Estienne, was grown around Rome and Bologna. Everyone admired its good taste, perfume and late keeping qualities and it, too, was proclaimed to be Roman in origin.

The Renaissance also saw an explosion of interest in the possibilities of preserved fruit. Once sugar began coming in from Brazil and the Azores, where it had been planted by the Portuguese in the 1420s, and from Spanish colonies in the West Indies, it became an affordable luxury. The possibility of preserving the newly fashionable – but often all too perishable – fruits for eating over the winter was irresistible. So, too, was the prospect of the new dimension of colour, translucence, taste and exclusivity that sugar cookery offered. Scappi's menus usually led up to a triumphant removal of the cloth, a ceremonial washing of the hands, and the opportunity to eat 'as much as you like' of *Conditi* – candied fruit – served with gilded knives and forks, and *Confettioni* – a range of fruit conserves presented with spoons. These were accompanied by all sorts of sugar coated spices such as caraway and fennel seeds, nuts, bunches of flowers, rose water and silver toothpicks.

The prestigious and specialised art of sugar cookery soon spawned a whole new genre of books devoted to sweetmeats and confectionery, and the choice of fruit and technique became a question of ever finer judgement. *Mele Appie*, for example, were generally used to make jellies and conserves while *mele Rosa* were recommended for the fruit tarts made from apples, sugar and spices. Significantly, sugar and fruit still retained their medicinal roles. The syrup prepared from *mele Appie* was said to be especially good for colds and illnesses generally, and Court Pendu was later said to serve much better than oranges and lemons for sick people.

The Italian fashion for fruit gardens, and for fresh and sugared fruit spread almost immediately to France. In 1495, the French king, Charles VIII, invaded Italy, ushering in four decades of close military, diplomatic, religious and horticultural contact between the two countries. As he advanced towards Naples he was apparently so enraptured by what he saw that he wrote to Cardinal de Bourbon.

> Y*ou would not credit the fine gardens I have seen in this town; For by my faith, only Adam and Eve seem lacking to make them an Earthly Paradise, so beautiful are they and so full of good and remarkable things.*[14]

When he returned to France after his short-lived occupancy, he took fruit trees and a bevy of Italian gardeners and artists back with him. He had an orangery made at his château of Amboise on the Loire, and in 1517 an Italian visitor to the royal palace of his successor Louis XII, at Blois, reported that 'Almost all fruits that grow in cultivated ground can be seen there'. The Florentine Catherine de' Medici, who married the future Henri II in 1533 was also, not surprisingly, an enthusiastic planter of fruit trees. Five hundred went into the ground at her Palace of the Tuileries in 1566 and the number was doubled again in the spring of 1570. Plantings almost certainly included the Paradise apple as well as 100 pears, 50 almonds, 150 cherries and 150 plums. Diane de Poitiers, Henri's mistress, seems to have been particularly fond of miniature apple trees, and had 300 planted at the Château Chenonceaux in 1557.[15]

The new fruit confectionery was embraced with equal enthusiasm. Italian cookery books were rapidly translated into French and, as in Italy, sugar cookery acquired a department to itself – *l'office*. Before long, the *collation*, an extravagant refreshment featuring fresh fruit, sweetmeats and cold delicacies, began to equal, if not surpass, anything the Italians had staged. It provided just the right kind of pastoral, portable, showy food that grand *al fresco* feasts required, and first Catherine de' Medici and then Louis XIV took the outdoor festival to its extremes.

Catherine de' Medici, after the death of her husband and during

The activities of the airy, high-ceilinged office, from Nicholas de Bonnefond's Le Jardinier Français, *c1651. Raw fruit is being peeled, cooked or dried, while fruit paste is rolled out and a preserve skimmed, or tested for a set.*

her time as regent for her sons, used every means in her power to increase the strength and popularity of the Crown. At one entertainment mounted at Bayonne on 24 June, 1565, guests floated along a river in boats, past a series of allegorical happenings, to an island where an octagonal banqueting house was set among the trees. Here the guests were served a magnificent *collation*, which may have included the new fruit ices which Catherine had introduced from Italy. They were waited upon by troops of 'diverse shepherdesses' who performed dances in between courses. At a meal offered to the court in 1571, dinner was followed by dancing and a *collation* which featured fruit jams, jellies and syrups; dry preserves including candied fruit; a variety of sugared nuts, fruit pastes, marzipans, sweet biscuits and 'every kind of fruit in the world'.[16]

BY THE EARLY 16TH CENTURY, continental fashion had begun to erode the English distrust of raw fruit. Figs were reintroduced and the apricot and possibly the greengage all arrived about this time. In 1533 Henry VIII's fruiterer, Richard Harris, established what was probably England's first large fruit collection at Teynham, in east Kent. He 'fetched out of France a great store of graftes, especially pippins, before which time there were no pippins in England. He fetched also out of the Lowe Countries cherrie grafts and Pear grafts of diverse sorts'.[17] Pippins, or *Pépyns*, were grown in Normandy and were on sale in the markets. By the end of the century Harris's collection had become 'the chief mother of all other orchards' in England and in time the term 'pippin' came to be synonymous with fine-flavoured late keeping English varieties.

As in France and Italy, it was the status of fruit as a princely luxury which was a main attraction. After Henry's split with Rome in 1534 he used every means at his disposal to proclaim his power and exalt the monarchy. Orchards, and fine gardens replete with fruit trees, were high on the list of palatial additions he financed out of the immense wealth derived from the dissolution of the monasteries. At Hampton Court some of the varieties purchased for the Great Orchard cost 6 pence each, as compared with less than a penny for each forest sapling.[18]

Royal example helped to stimulate consumption in fashionable circles, which in turn led to the establishment of new orchards to supply the metropolis. Tudor agriculture was shaking off its feudal ties and the new spirit of commercial enterprise provided the opportunity for men like Harris to set up thriving businesses. His 105 acres at Teynham became the basis of Kent's infant fruit industry and as the London market expanded, an increasing number of fruiterers seized the chance to buy or lease their own orchards. Others began to travel down each year before harvest to purchase the prospective crops of cherries, pears, wardens (pears), medlars, pippins, and other apples, while they were still on the tree. Soon Kent could claim to be 'the garden of England'. When William Lambard, a London lawyer, who became a Kent Justice of the Peace, was travelling through the coastal area between Rochester and Canterbury in 1570, he found

> orchards of apples and gardens of cherries and those of the most delicious and exquisite kindes
> that can be, no part or the realm (that I know) hath them in either such quantity and number
> or with such art and industry set and planted.[19]

Wealthy landowners began to follow the Royal example and lay out gardens and orchards. This was even more the case when Elizabeth I came to the throne. In contrast to her father, she abstained from ambitious building schemes and kept her finances firmly in check, preferring to live at her subjects' expense when away from London. When the Queen announced her intention to visit Kenilworth, home of Robert Dudley, Earl of Leicester, in 1575, a new garden of over an acre in size was immediately constructed. Robert Laneham, a court official, recounts that the grass terrace surrounding the house ended in 'a bower, smelling of sweet flowers and trees'. In the garden below, which was crossed by grassy paths

flowering plants, procured at great expense, yield sweet scent and beauty . . . their colours and many kinds betraying a vast outlay; then fruit trees full of apples, pears, and ripe cherries – a garden, indeed, so laid out that either on or above the lovely terrace paths, one feels a refreshing breeze in the heat of summer, or the pleasant cool of the fountain.[20]

The reigns of James I and Charles I saw English fruit growing continuing to gather momentum. In 1611, John Tradescant, gardener to Robert Cecil, Earl of Salisbury, and later to Charles I, visits Delft in Holland, to buy apple, cherry, quince, medlars, pear trees and currants. A few years later he travels as far as Algeria and even Russia where he 'wonderfully laboured to obtain all the rarest fruits he can hear of in any place of Christendom, Turkey, yea the whole World'. By 1629, the Royal botanist and apothecary John Parkinson reports that 'in most men's houses of account, where, if they grow any rare or excellent fruit, it is then set forth to be seen and tasted.'

His observation reflects not only the recovery of fresh fruit from centuries of suspicion, but the attention now being given to the importance of novel varieties. Parkinson's *Paradisus Terrestris* lists over 60 varieties of apples he considers worth growing and of these only a few are not home raised. There were some five or six pippins; a Great and a Summer Pearmain and 'the very well relished Harvey Apple'. The Green and the Grey Costard and the Catshead were for baking, and could have been used in pies and 'stewed with Rose water and Sugar, and Cinnamon or Ginger Caste upon'. The Kentish Codling was 'the best to coddle' for the famed 'Codlins and Cream'. It could also be hung above a fire on a piece of string and roasted until the soft juicy pulp started to fall into the waiting cup of warm ale to make the festive 'Lammas Wool'.

Like the rediscovered pleasure of fresh fruit, this increased concern with varieties was also a recent continental import. It depended on the careful comparison and cataloguing of plants, which in turn was underpinned by the assembly of specialised plant collections. The first European botanic garden of this kind, apart from those of Moorish Spain, had been set up at Padua in Italy in 1545, and among those which followed were Leyden in the Netherlands in 1587, and Montpellier, France, in 1593. England's

A plate from Parkinson's Paradisus Terrestris, *1629. It includes Pomewater (3); Pearmain (5); Queen Apple (6); Pound Royall (8); Kentish Codling (9) from a list of over 60 varieties of apples.*

first botanic garden had been set up at Oxford in 1621 just a few years before the publication of Parkinson's list.

The philosophy behind these gardens was complex and itself contributed to the growing importance of fresh fruit, and apples in particular. In the Middle Ages it had been thought that the Garden of Eden had survived the Flood and was still to be found on some remote eastern hilltop. This had now foundered in the new age of geographic discovery, to be replaced by the idea that at the Fall the contents of Eden had been scattered over the four corners of the globe. Every newly discovered plant or animal, therefore, represented a piece of pre-lapserian jigsaw puzzle which one day might be reassembled. Botanic gardens and the menageries often associated with them were an attempt to do just this. The finding and piecing together of Eden's fragments was seen not only as a way to recover lost knowledge, but also as a way to regain Adam's power over nature.[21] It suggested the tantalising prospect that, like Adam's garden, such collections might transcend the limitations of climate and season, and dramatically raised the status of fresh fruit by recognising it as the first food given to man.

The immediate consequence of all this was that the energy now directed towards collecting, cataloguing and exchanging plants was immense. In Germany, Valerius Cordus, who died in 1542, recorded 31 different apples known in Hessen and Saxony; over 20 were mentioned by the agricultural writer Augusto del Riccio in Italy in the late 16th century, and the lists compiled by French enthusiasts make it clear that many new varieties were being recognised for the first time. Charles Estienne writing in 1540 mentions 11 kinds of apple of which five do not occur in medieval lists, while in 1598 Jean Bauhin, driven by protestant persecution from the Montpellier Medical School to Montbéliard on the Swiss-German border, where he established another botanic collection, describes 60 apples of which 18 are mentioned for the first time. Olivier de Serre, who was horticultural adviser to Henri IV of France and lived at Pradel in the Rhône Valley, catalogued 48 apples in 1608, of which 32 seem to be new varieties. Taken together, these catalogues suggest that there may have been as many as 120 recognised varieties of apple in Europe in cultivation at this time, and that they were being exchanged between collectors in France, Switzerland, southern Germany, northern Italy and probably elsewhere. Jean Bauhin, for example, said that he had received grafts of the Paradise apple from both Montpellier and Lyon.

Gardeners also began to explore the possibilities of extending the growing season of plants both by employing new techniques, such as the use of shelter, and of focusing on the possibility of developing earlier and later ripening varieties. Del Riccio, for example, undertook a careful comparison of the storage properties of different apples, and was subsequently able to order his fruit into the months in which they matured. He also evaluated different methods of storing apples. Those for long keeping were to be individually wrapped in leaves or placed on straw and left undisturbed in a cool place, such as a cellar. Summer apples, on the other hand, were encouraged to ripen earlier by spraying with warm water.

The establishment of specialist fruit collections rapidly became high garden fashion. When Charles I purchased Wimbledon Manor in 1639 for his French wife, Henrietta Maria, she set about creating one of the most lavish gardens England had yet seen. Henrietta Maria was the daughter of Henri IV and Marie de' Medici. Her mother had made her own piece of Italy at the Luxembourg Gardens in Paris, and Henrietta Maria drew on all the resources of both her native

PLATE 9: ELLISON'S ORANGE

and adopted country. Tradescant was employed to collect plants for the king and over 1000 fruit trees were planted. As well as a magnificent 'Orange Garden' complete with winter shelter for her new citrus collection, Henrietta Maria planted the main formal pleasure ground with '150 fruit trees of divers kinds of apples, pears, pleasant and profitable' as well as '119 cherry trees . . . of great growth'. On the walls were 53 fruit trees of 'most rare and choice fruits', including 'muskadine Vines . . . bearing very sweet grapes'.[22] There was also a ten-acre walled orchard which certainly contained apple trees, which were probably planted as Parkinson directed in the quincunx pattern so that . . .

> *your trees are here set in such an equal distance from one another every way, &as fittest for them, that when they are grown great, the greatest branches shall not gall or rubbe against another . . . and give you way sufficient to pass through them, to prune, loppe or dress them, as need shall require, and may also bee brought (if you please) to that graceful delight, that every alley or distance may be formed like an arch, the branches of either side meeting to be enterlaced together.[23]*

Along the paths there may also have been 'a low hedge' either of Paradise apples themselves, or of Paradise rootstocks on which had been grafted other varieties, as this was considered a 'pretty way to have Pippins, Pomewaters, or any other sort of Apples'.

With the new varieties and careful management of the fruit store, it was now possible to defy the seasons and have apples on the table from the end of July right through until the following spring or later. The first to be picked were the tiny Jenettings; which, in the philosopher Francis Bacon's ideal garden, would be ripe when the pinks flowered beneath. Also ripe at the end of July was the Margaret apple, which the garden writer John Rea described as 'a fair and beautiful fruit, yellow and thickly striped with red . . . of a delicate taste, sweet scent and best eaten off the tree'.[24] Later in the season would come various pippins – the 'Kirton', 'Bridgewater' and 'Carlisle' – which were all considered fine table fruits. The 'Royal Pearmain', another autumn apple, was 'much bigger and better tasting than all the other kinds of pearmain, and the best of all the reinettes was the 'Lincoln Rennet'. The late keepers, which would not be ready to eat until after Christmas, but stored well until the following May, or even until 'apples came again', including the John Apple and the newer Deux Ans. There were also some new kitchen apples such as the 'Gyant Apple' and the 'Good Housewife', which were excellent to 'coddle or bake in tarts'.

THE EXPANSION OF THE MEDIEVAL *VOYDE* into a separate sweet course, the new interest in fruit and sweetmeats coming in from Europe, and the continuing fall in sugar prices – as supplies started to come in from the New World, and then from England's colonies in the West Indies – all set the stage for the emergence of the Tudor, and the increasingly elaborate Stuart, fruit banquet. These were glittering arrays of fresh fruit and fruit sweetmeats as well as some other sweet dishes, which were displayed and eaten in a special apartment known as a banqueting house. This was a separate building out in the garden or up on the roof of the mansion to which guests would adjourn after the main feast. Here, protected from the English weather but surrounded by garden views, vistas of fruit trees, and the sound of bird song, they could sip sweet wine and toy with fruit and 'delicates' in an enchanted – if temporary – world of pastoral plenty and romance.

By the mid-17th century participants would be dazzled by a display of 'banquetting stuffe' which included preserved 'fruits, flowers and roots in their numerous different guises'; stiff fruit-flavoured jellies and sugar pastes; fruit tarts and creams, jams and butters. Their colours glowed and sparkled like jewels among the biscuit breads, jumballs, knots and other light cakey or biscuity confections, and were set off to perfection by the new Venetian glass, which reached London after 1575, the silver plate and the gilded marchpane sculptures. It was essentially an occasion for conspicuous consumption. Guests would eat their delicates with a special implement which comprised a spoon at one end and a fork at the other, from small silver dishes or tiny, wafer-thin trenchers of sycamore or beech, which were elaborately decorated with emblems or scenes and usually had a verse or 'posie' in the middle. The side cupboard of gold and silver dishes added to the show and there might also be music, games and dancing lasting far into the night.[25] A further piquancy would be added to the proceedings by the supposed aphrodisiac qualities of many of the items, including apples.

Banqueters ate from the plain backs of gilded and painted trenchers. The silver sucket fork on the left dates from c1660; that on the right from the 1680s.

It is, perhaps, not surprising that 'banquetting stuffe' with its sweetness and colour held such an attraction for the English. In a climate in which southern fruits such as apricots and figs rarely ripened and many apples were green and sharp, the ability of sugar cookery to transform the most prolific and variable of English fruits into a glittering range of exotic sweetmeats, was little short of magical. Continental cookery books were translated into English, and home grown volumes such as John Partridge's *The Treasures of Commodious Conceites and Hidden Secrets* (1584) and Sir Hugh Plat's *Delightes for Ladies* (1605) spelled out the mysteries of sugar confectionery to an eager audience, not just in noble households but, as sugar became more widely available, in the households of the gentry and prosperous yeomen.

In *A Queen's Delight*, published in 1655, and reputedly a collection of recipes compiled in Henrietta Maria's household, there are a number of recipes for apple sweetmeats as well as for a cough remedy based upon the pulp of roasted apples mixed with 'Elecampane' root and sugar. Candying or crystallising apples to make dry sweetmeats receives particular attention.[26] The simplest method was for the apples to be 'baked plump but not broken' and then laid in an oven to dry. Alternatively they could be simmered in syrup to increase the sugar content before being dried, or having been 'pricked full of holes with a bodkin' be steeped in a 'sweet wort for three or four days'. Once the prepared apples were in the oven, you had to 'strew on Sugar three or four times in the drying'. In a more elaborate recipe the fruit is sugared as it bakes in 'an oven as hot as for manchet [white bread]' and sprinkled with rosewater when the juices start to run. The apples would then be transferred to a cooling bread oven for the final drying. Candied like this 'you may keep them all year'.

For wet sweetmeats, apples could be preserved whole, or in small pieces in a lightly jellied syrup. Apple jelly was also used to preserve and make other rarer and more exotic fruits, such as apricots, peaches and oranges, go further. These wet 'suckets' – the term derives from their resemblance to the imported citrus 'succades' – could be poured directly into moulds to be turned out and decorated with candied fruit, or stored in pots and jars and served in little glass dishes. Apples could also be pulped and boiled with sugar to produce a stiff 'marmalade', which could be set in wooden or tin moulds, or cut into 'what forms you please'. These were stiff enough to be stored in boxes and were usually flavoured in the Arab way with rosewater and musk. They could also be dried to form 'Pippin Cakes'.

All these preserves could be produced in a range of colours: white through to deep amber; from the palest to the brightest green; or in shades from pink to claret. For a preserve of 'White Pippins' the apples would be peeled and briefly boiled in a sugar syrup, to give a lightly set jelly which could be flavoured with cinnamon, cloves, rosewater or slivers of Seville orange peel. This plain 'Pippin Jelly' enlivened with shreds of orange rind is probably the direct ancestor of the orange marmalade we are familiar with today. For a green preserve the fruit would be picked early, while it was small, green and sour. If it was first scalded, the thin outer skin readily slipped off, and the finished apples would be bright green. To make 'Red Pippins' a long gentle simmering would turn the fruit the deep pink of pomegranates. Apple marmalade could be produced in any shade of green, ruby or amber. Cut into neat pieces it would be arranged on plates in elaborate patterns which made the most of the contrasting colours. Strips could be twisted and tied into 'Pippin Knots' and A *Queen's Delight* suggests cutting green pippin paste into oak leaf shapes, and arranging them with round 'plums' coloured with barberries.

Apples were also used to make tart fillings. Either raw fruit and sugar would be baked in a pastry case or 'coffin', which would then have the crust removed and replaced with an elaborate cut-work lid made of puff-pastry, or the pastry case would be filled with apple purée, known as 'tart stuff'. These red, gold or yellow apple tarts with their lids cut in

the form of 'Beasts, Birds, Arms, Knots, Flowers and suchlike', would be complemented by tarts with black (prune), yellow (custard), or white (milk and almond) fillings. Sometimes the entire palette would be combined in a single magnificent tart. In *The English Hus-Wife*, Gervase Markham describes a tart depicting a heraldic beast with black eyes, white teeth and scarlet talons, a bird with variegated tail feathers, a coat of arms and a knot of green and yellow ribbons.

To produce a variety of butters, creams and fools, apples were mixed with cream, sugar and eggs. For 'Pippin Cream' the apples would be cooked with orange flower water and sugar, then sieved and mixed with eggs and cream, and poured into little glasses. A sliver of orange on top provided the finishing touch. Alternatively the sweetened apple pulp might be thickened only with egg yolks, flavoured with rosewater, and given a sprinkling of ginger and cinnamon for decoration. In a more elaborate recipe the sliced pippins are pre-cooked in a red wine syrup flavoured with lemon peel and ginger. They are then laid over a nutmeg flavoured cream in a pastry case. The simplest cream of all was syllabub. 'Fill your Syllabub pot with Cyder (for that is the best for a Syllabub) and good store of sugar, nutmeg . . . put in as much thick cream as hard as you can as though you milk it in . . . and let it stand for two hours.'

WITH FASHIONABLE ATTENTION focused on fruit, the 17th century saw concerted efforts to maximise quality. The collections of fruit that had been built up in botanic, royal and private gardens provided a practical resource as well as a personal paradise. Within their walls different varieties of fruit could be compared, and evaluated in terms of their yields, quality and suitability for a particular purpose or locality.

The lead for improving quality came from Europe. It was partly a natural consequence of the increased attention paid to fresh fruit on the dining table, but there were more complex social reasons too. In France, the desire to cultivate fruit trees and keep a good table was spreading downwards from court circles to the enlarging class of families who, having made their fortunes in commerce or industry, aspired towards the lifestyle of the nobility. The effect was to raise standards in general. The aristocracy were spurred on to keep the competitive edge and the new market gardeners and estate owners with surpluses for sale had to meet the demands of a growing, and ever more discerning, body of customers. The recipes of Nicholas de Bonnefonds, for example, who was *valet de chambre* to the young Louis XIV and also a Paris seedsman, are notable for their stress on particular varieties of fruit, high quality, and careful preparation.

The foundations for improving the quality of French apples and other fruits had been laid early in the century with the careful collection of good varieties made by Le Lectier, who was *Procureur Royale* at Orléans, and in his spare time a keen pomologist. To Olivier de Serre's catalogue of 48 apple varieties compiled in 1608, he added a further 35 in 1628. The list now included a number of France's most celebrated varieties including the Api, which was said to have arisen in the ancient Forest of Api in Brittany, the Fenouillet which tasted of fennel, Court Pendu Rouge and a number of different Calvilles, including the large, ribbed Calville Blanc d'Hiver, which was particularly favoured for making wet sweetmeats. The term 'Reinette' was applied then as now to a number of good flavoured, late keeping apples.

Then in 1652, Antoine le Gendre, Curé d'Hénonville, who had been almoner and superintendent of the Royal gardens under Louis XIII, published his detailed instructions for managing trees.[27] Drawing on his observations and experience of 'mural' trees trained against walls or pallisades for decoration, he had realised that the same methods could also be used for the benefit of the fruit itself. If, as he made his 'tapestry of greens' or 'set about nailing to the wall', the gardener spaced the branches 'like the fingers of a man's hand or the ribs of a fan' it allowed more sun to reach the ripening fruits, giving them more colour and a sweeter, better flavour. He advised gardeners to 'turn aside the leaves' for the same reasons. The usual practice of clipping fruit trees with shears in the manner of box or yew was condemned, and Le Gendre insisted that they be carefully pruned with the knife to preserve the fruit buds and create a productive tree. He also advised thinning the developing crop in order to increase the size of the individual fruits.

Fan training is not satisfactory for apple or pear trees because they do not usually bear fruit on the young growth which forms a large proportion of the fan's structure, and instead Le Gendre advocated the use of dwarfing rootstocks to produce smaller, more manageable trees. Apples were usually grafted on to seedling stocks, often raised from the residue – 'pomace' – of the cider press. This resulted in large, vigorous trees which bore heavy crops, but the fruit was often small and of poor quality. Grafting on to the Paradise apple, or the slightly stronger growing Doucin variety produced a smaller, and more easily managed tree. Careful pruning and even thinning became possible and, as with the trained forms, more sun and light could reach the buds and ripening fruit.

After 1661 the demand for handsome, high quality fruit became even more urgent. Louis XIV took control of national affairs and, in a series of the most spectacular court festivals yet seen in Europe, exploited every possibility of fresh and preserved fruit as part of his visual language of power. On the second evening of the festival held at the Palace of Versailles celebrating the Conquest of Franche Comté his stewards built an immense confectionery structure 'on which foods were served with extraordinary magnificence'. It consisted of 'all kinds of fruits ingeniously arranged in one hundred dozen little porcelain dishes which served as the solid body of the agreeable building'. There were columns and cornices, flowers, candles, and a 'large number of crystal vases used for the ices and liqueurs'.[28]

In addition, the king's doctor had advised him to eat as much fruit as possible for the sake of his health, and for the Sun King, only the very finest and rarest were good enough. When, in 1673, Jean de la Quintinye became *Jardinier en Chef* of the *Potager du Roi* at Versailles, Louis found a man with the skills and ambition to take fruit gardening into a new age. Between 1677 and 1683 he laid out 23 acres of new fruit and vegetable gardens divided up into numerous walled enclosures. Here he grew the entire range of temperate and Mediterranean fruits. Peaches, nectarines, cherries, plums and pears were trained along walls, and lines of fan-trained trees filled the open ground. Apples were grown around the gardens as small bush trees on the Paradise stock, and in this way he found they gave 'little encumbrance and . . . besides, is blest with the advantage of producing great increase'.

La Quintinye took great pride in his fruit and was determined that his choice, carefully nurtured specimens should reach the royal table in perfect condition and be given the honour that was due when they arrived. The new gardens included large, specially constructed fruit stores for the apples and pears, with walls two feet thick to keep out the frost and maintain a

stable, cool temperature. The buildings were well-ventilated and La Quintinye believed it was always a good idea to 'allow some secret entrance' for cats in order to keep out the mice. The fruit was arranged on shelves around the room, and carefully placed on a bed of sand, rather than being thrown down and left to 'lye pell mell any how'. Each section was labelled 'with respect to kind and maturity in relation to the sequel of months'.[29]

The walled potager at Versailles in the late 17th century showing the degree of labour and care La Quintinye devoted to the picking and storing of fruit. The well-ordered shelves of the fruit store are depicted on the left.

At the entrance to the stores was a room devoted to assembling the baskets and bowls of fruit for the dining room. From at least 1650 *service en pyramide* had been a fashionable way of presenting food in the high-ceilinged rooms of the period, and fruit lent itself particularly well to this method of display. Oranges and apples, with their shiny, fine-coloured, globe-shaped fruits, made especially tall, magnificent structures, and were built up in layers with moss, china or metal plates in between to secure the edifice. Such displays, together with flowers and candelabra, would sit on the table throughout the meal, and despite their inherent problems – Madame de Sévigné recalled a dinner party when a pyramid of fruit 20 layers high toppled over – remained in fashion through to the first quarter of the 18th century.

Even more important to La Quintinye than these mainly ornamental displays, however, was the fruit that was actually eaten. The pyramids were always accompanied by 'a pretty basket well filled with choice eating fruits all fair and goodly, and perfectly ripe'. He went on to reflect that the 'honour of the pyramid was to come off always whole and entire without suffering the least breach or ruffle neither in its constitution nor symmetry so I pretend that on the contrary the honour of the basket consists in returning always empty'.

The pioneering achievements of Le Gendre and La Quintinye's work began to have an impact in England by the late 17th century, by which time John Evelyn had translated both the great men's books. In the meantime, however, it was the newly formed country of the Netherlands that became France's major equal in the production of high-quality fruit. The new middle class of what was now the major trading nation of Europe was eager to cultivate its country estates and make a profit into the bargain. Their commercial wit, and the light, easily worked soils of the polder land, combined with the horticultural skills of the Flemish and Huguenot families who had fled to this safe haven of religious tolerance, put fruit growing close to the heart of the economy. Orchards in the vicinity of towns produced apples, pears, plums and nuts, and the intricate network of canals made marketing cheap and efficient. Melons, oranges and grapes were abundant and, by the last third of the century, relatively inexpensive. Every delicacy such as black cherries, strawberries, and even peaches, apricots and pineapples were grown locally.[30] Dried and fresh fruit featured in some form at every meal, and minced ox tongue and green apple sauce was a great favourite.

Fruit growing was also close to the centre of Dutch sensibility. Overflowing baskets of high-quality, exquisitely rendered fruit – including a wide variety of apples – are a recurrent image in contemporary Dutch painting. Like the opulent interiors of the same period, they are exuberant celebrations of the pleasures of prosperity. At the same time, however, fruit was both a time-honoured and topical symbol of the moral ambiguity of such riches. At its ripest and sweetest, fruit is on the edge of decay and, in the Netherlands, the very orchards which produced it could, in theory, be reclaimed at any time by the sea.

New varieties and horticultural skills undoubtedly formed a small but significant part of the cultural baggage which the Dutch William of Orange and his retinue brought with them to England in 1689, but by this time a whole new chapter had already opened in the history of the apple; it had been elevated to the status of Protestant fruit *par excellence*.

Chapter 3
For
GOD and COUNTRY

The Lord is good to me,
And so I thank the Lord:
For giving me the things I need
The sun and the rain, and the apple seed.
The Lord is good to me.

ANON: *A Grace*

A<small>T LEAST SOME OF THE REASONS</small> for the apple's special position in the economy, traditions and cookery of Britain, America, parts of northern and eastern Europe and Scandinavia lie in the religious history of 16th-century Europe. When Protestantism appeared in Germany early in the century it set off a series of religious struggles which eventually changed not only the moral and political geography of the Western world but also gave apples a new significance as the fruit favoured by 'God's Elect'. As a new pattern of national identities linked to religious belief began to emerge, countries which were predominantly protestant saw the apple as the fruit not only of God but of country, too.

T<small>HE FIRST PROTESTANTS</small> repudiated the corruption and many of the doctrines of the Roman Catholic Church and condemned the materialism and debauchery of the fallen world. They espoused an alternative lifestyle based on close study of the Bible, hard work, thrift and self-denial; and in anticipation of the second coming of Christ sought to repair as far as possible the scattering and corruption of nature which had occurred when Adam and Eve were expelled from the Garden of Eden. As it was widely held that, before the Fall, the blessed pair ate only fruit and seeds – while the beasts ate the herbs of the field – the planting and tending of orchards was high on the list of improving activities. Ralph Austen, an influential English Calvinist, included eight 'Divine Arguments' for the cultivation of orchards in his thoroughly practical *Treatise of Fruit Trees* (1653), and also appended a devotional tract, *The Spiritual Use of an Orchard*, in which he proposed that it was the separation of stock from their grafts at the Fall which had left man in a world of inferior crab apples. The grafting of good new varieties, therefore, made both

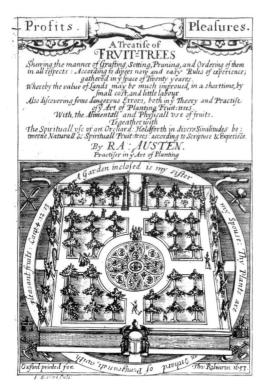

The title page of Ralph Austen's Treatise: *gardeners tend an immaculate fruit garden in anticipation of the Second Coming.*

horticultural and spiritual sense. There was also a tradition of Christian thought which held that the real damage in Eden had been done when God made Eve from Adam's rib, so dividing nature into two sexes. As fruit trees were deemed to reproduce asexually, orchards were also seen to communicate the ideal of undivided nature.[1]

Everywhere that Protestantism took root and climate allowed, orcharding followed and of all the tree fruits, it was the apple that appealed to Protestant sensibilities most. Apple trees were neither greedy nor temperamental. A few trees were well within the means of even the humblest cottager, and would reward his diligence a hundredfold by providing 'a good and whole food' for the 'sick and sound, for young and old, rich and poor all year long'. The flesh was firm and refreshing with none of the lush, courtly associations of Mediterranean fruits, and not a scrap of core or peel need be wasted. In addition to being eaten fresh and cooked, apples could be dried, made into jellies and preserves, and turned into cider and vinegar. Orchard prunings could go to the fuel store or the joiner's bench. Further up the social scale, the benefits were the same, but with the added advantage that the intensive nature of orchard work offered plenty of scope for employing the poor.

By happy coincidence, apples were also one of the fruits that grew best in the areas of northern and eastern Europe where Protestantism first took hold, and on the eastern seaboard of America where many of the Elect later found a new homeland. The orcharding traditions of northern Germany, the Rhine, the Palatinate and Bohemia, Switzerland, Poland, Scandinavia, northern France, Flanders, the Netherlands and England were all enthusiastically encouraged during the 16th and early 17th centuries, and when the Catholic backlash sent Protestants fleeing to the safe havens of Holland (the erstwhile Spanish Netherlands), England and the east coast of America, they took their skills and love of apple trees with them.

These original Protestant strongholds also included some of the most horticulturally advanced regions of the period. Protestantism was closely associated with the new spirit of critical enquiry, which had begun to replace the unquestioning acceptance of traditional teaching in everything from geography and physiology to agriculture and horticulture. Consequently, every aspect of the cultivation of fruit trees began to come in for scrutiny and experiment.

IN ENGLAND, apple growing benefited both directly and indirectly from the Protestant movement. Charles I's wife, Henrietta Maria, had returned to the Continent before her husband's execution, but the example she had set in her Wimbledon and other gardens remained an inspiration to Royalist grandees. Like Sir Thomas Hanmer – who married one of the queen's maids of honour – many of those who did not leave the country during the 11 years of Puritan government took refuge on their country estates. Here they channelled their energies into cultivating their gardens, an activity which provided a satisfying combination of solace and harmless subversion. Sir Thomas's family seat was at Bettisfield on the Welsh Marches, and he had soon established a notable collection of fruit trees and tulips. Against the walls of his 'Great Garden', and in the adjacent 'Small Garden', grew peaches, apricots, pears, vines and other fine fruits including Jenetting and Margaret apples. Given the shelter of a wall these gave him ripe, well-coloured fruit in July, after which the range of carefully selected varieties in the orchards beyond kept him supplied for the rest of the season.[2]

John Rea, a fellow fruit and tulip enthusiast who lived in Shropshire, put the fruit garden second only to the flower garden in importance, and emphasised its distinction from the kitchen garden whose lowly leaves and roots many considered fit only for foreigners and the poor. Two acres was deemed the right size for a nobleman's fruit garden, which should also contain tulips and auriculas, roses, evergreen and flowering shrubs. Even a mere gentleman needed a one-acre fruit garden, and nurseries were soon being established to meet the fashionable demand.

In the meantime, the Puritan government was also preoccupied with plans to plant fruit trees. Thanks in part to the arrival of refugees from Flanders, whose 'art of Gardening [had begun] to creep . . . into Sandwich [Kent], Surrey, Fulham and other places', the number and efficiency of market gardens, which often grew fruit trees as well as vegetables, had been increasing since the beginning of the century in southern England. With the expansion of London, market gardening had spread down the Thames into Middlesex, and a steady progression of barges carried their produce up-river to the City, returning with refuse from London's breweries, tanneries, streets and stables to enrich the land. Such thrift and diligence exemplified the Puritan ideal of spiritual and material health, and government advisers pressed for the advancement of horticulture and the planting of fruit trees as an essential component of the New England.

At the head of the fruit lobby was Samuel Hartlib who had come from Elbing in Poland to study at Cambridge University in 1625, and stayed on to become Cromwell's 'Agent for the Advancement of Universal Learning'. He gathered around him a distinguished group of fruit and cider men, including the Herefordshire parson and cider authority Dr John Beale, and the aforementioned Ralph Austen, who was a member of Oxford University's Experimental Philosophy Club and a nurseryman. The reformers believed that every landowner should be persuaded to plant fruit trees, and that every piece of spare ground should be filled with apples, pears, quinces and walnuts. In 1652 Hartlib published *A Designe for Plentie, By an Universall Planting of Fruit-Trees* 'for the relief of the poor, the benefit of the rich, and the delight of all'. The following year he called for a law to make this compulsory and enforceable in every district by two officers, to be known as 'Fruiterers' or 'Woodwards'.[3] The suggestion appeared as part of the advertisement for Austen's *Treatise of Fruit Trees*, which with its rational approach and emphasis on good husbandry was the most advanced fruit manual yet produced in England. Half the available copies of the book were sent to members of parliament and Austen followed them up with a petition to Cromwell outlining an extensive scheme for planting fruit and timber trees. A second edition appeared in 1657, when Beale produced a study of Hereford's orchards with the aim that they should be held up as an example for all of England. Their grander schemes were never fully realised, but the momentum created by the Puritan philosophy helped foster an enormous investment in orchards during the latter part of the 17th century.

PLATE 10: GRAVENSTEIN

B Y THE TIME CHARLES II RETURNED from exile in 1660, mainstream agriculture had begun to go into decline as production started to outstrip demand, causing prices and incomes to fall. Landowners, farmers and smallholders looked for alternative crops and turned amongst other things to fruit. Apples, in particular, appeared to offer farmers the best alternative source of revenue. As the agricultural commentator and cider expert, John Worlidge, pointed out, the fruit was not only extremely versatile 'being both meat and drink', but highly adaptable, there being 'scarce a country parish in England but in some part or other it will not thrive'.[4] In addition, Charles's exposure to the continental enthusiasm for fruit had fired his imagination and the new government put all its weight behind the fruit-growing option. As farmers began to plant orchards, the regional pattern of commercial apple growing that was to hold until the end of the 19th century began to take shape. In particular, these years saw the founding of the West Country cider industry, and of table-fruit growing in the Vale of Evesham in Worcestershire, which had an expanding and accessible market in the city of Birmingham. In Kent, improvements in water transport allowed orchards to expand beyond the well-established coastal East Kent belt into the Medway Valley and the centre of the county. The completion in 1664 of a series of locks and a tow path along the River Medway gave Maidstone access to the London markets by allowing apples to be carried down to Rochester on the coast by river boats or 'hoys', from whence they could be shipped to the City along with the East Kent fruit. Orchards were accordingly planted to the south of Maidstone, where the greensand ridge of deep, free-working soil was ideal for fruit trees, and by the 1670s Maidstone had its own market.[5]

A key figure in the advancement of horticulture and fruit cultivation at this time was the diarist, John Evelyn. The son of a wealthy landowner, he had spent his youth and much of the civil-war years travelling in France and Italy. On his return to England his interest in chemistry, anatomy, gardening and fruit growing brought him into contact with the 'Scientific Clubs', in which Hartlib and other Puritans of a horticultural bent played a prominent role. After the Restoration, these clubs formed the basis of Britain's most prestigious scientific institution, the Royal Society. Founded in 1662, it numbered Evelyn and several of the old Hartlib circle among its first Fellows, and discussions at the Society formed the core of Evelyn's weighty *Sylva: or, A Discourse of Forest-Trees, And the Propagation of Timber in His Majesty's Dominions*, published in 1664 and produced for the navy which needed wood for ships. Although mainly devoted to timber trees, it also contained a *Pomona*, or section on fruit trees, chiefly concerned with cider production, and a *Gardener's Almanack*, in which Evelyn detailed successions of varieties for each month of the year to guide both the market and private gardener. Evelyn also produced translations of the leading French works on fruit growing by Le Gendre, Bonnefonds and La Quintinye, and was the patron of Brompton Park Nursery, whose proprietors George London and Henry Wise were acknowledged fruit experts, and probably responsible for introducing a number of new varieties of apples.

Throughout the second half of the 17th century apple trees did indeed prove to be the farmer's saviour, producing gallons of excellent cider, while table fruit became an increasingly profitable crop for the market gardener. When the indefatigable Miss Celia Fiennes travelled through Herefordshire in the 1690s she found the county 'full of fruit trees, the apple and pear trees so thick even in corn fields and hedgerows', [6] while throughout the country there were many 'very good gardens' which 'boasted an abundance of fruit of all sorts'.

At the same time, apples continued to play an important part in the pleasure ground. When Charles had returned from Europe, he had also brought back with him gardening ambitions inspired by Louis XIV's great formal gardens at Versailles, and the splendours of his table. Royal gardeners were sent to Versailles and Charles even tried – unsuccessfully – to persuade La Quintinye to come and work for him. The accession of William of Orange to the English throne added Dutch ideas and expertise to the melting pot of horticultural fashion – particularly the practice of growing apples on the dwarfing Paradise rootstocks – and by the end of the century apple trees adorned gardens everywhere. When Miss Fiennes visited a house in the fashionable watering spa of Epsom, in Surrey, she found long grass walks flanked by apples grown as 'dwarf trees . . . and flowers and greens in all shapes, intermixed with beds of strawberyes for ornament and use'.[7] In another Epsom garden there was an apple hedge sheltering orange and lemon trees. Like the citrus fruits, apple trees themselves could also be grown in pots and were considered to be a 'very great embellishment of entrances, parterres, cabinets etc'. It was also fashionable to include a more irregular section – or wilderness – in the pleasure ground, and early in the next century the garden designer and seedsman Stephen Switzer proclaimed that in such a setting 'no flowering shrub excels, if equals, that of a peach or apple'. Batty Langley, another garden designer, considered that their 'various agreeable Mixtures of Leaves and Fruits, may not only delightfully entertain the Eyes of the Beholders, as they pass thr' the several Meanders therof, but their Taste also'; and an avenue of apple trees running out across a meadow was considered a fine addition to the demesne.[8]

The enthusiasm for dwarf apple trees by the end of the 17th century was such that they were even planted in geometric parterres in front of the house.

Enthusiasm for the walled fruit garden also continued undimmed, and it was here that the continental techniques for improving the quality of fruit could be explored. When Sir Thomas Coke, Vice Chamberlain to Queen Anne, succeeded to the estate of Melbourne Hall in Derbyshire in 1692, he ordered 'dwarf apples grafted on Paradise stocks from Brompton Nursery',[9] and, according to the horticulturist Richard Bradley, writing in the early 18th century, these dwarf trees had become widely available.

The espalier, as a means of training apples and pears in a similar way to the fan, was also coming into favour about this time. It consisted of a vertical main stem and branches trained at right angles, which gave a permanent framework to bear the fruit and was ideally suited to apples and pears which fruit on old wood. Gardeners had realised that the further a branch was inclined towards the horizontal the more likely it was to produce fruit buds, rather than leafy shoots. This had been exploited in fan training, and now apples and pears could also be closely trained and consequently benefit from the increased exposure to sunlight and more careful attention.

Espaliers delighted the eye, as well as bearing crops of excellent fruit. Such arrangements, claimed Bradley, 'give air in winter to flowers growing in borders where they stand, and shade in summer' and it was common practice to 'plant a great part of our gardens with hedges of fruit' which on account of their 'flowers and fruit were preferable to evergreens'. It was even suggested that espalier 'hedges' should consist of alternating apple and pear trees so that the apple blossom would 'make a regular and beautiful red Colour against the fine White of the Pear Blossom; which Contrast will yield a very delightful Prospect, especially when the Apples are justly placed and well managed'.[10]

So in love with apples had the English upper classes now become that they even found their way into the furnishings of the drawing room and bedchamber. Together with grapes and pineapples they adorn the plaster and carved wood festoons which decorate late 17th-century ceilings, overmantles and doorways in palaces such as Hampton Court, and grand houses such as Petworth in Sussex and Sudbury Hall in Derbyshire.

BELIEF IN THE SUPERIOR QUALITY of English apple varieties emerged early on in the century. Like the similarly patriotic enthusiasm for plain English cooking, the target for unfavourable comparison was often the French. Stephen Switzer was prepared to grant in 1724 that 'there are some good Sorts of Fruit in the Countries over-against us of Normandy and Brittany etc and perhaps some about Paris'. But he continued:

> what are their Cousinots, Oregans, Francatuses, Fennilets, Calvilles and Haute Bontes but an abundance more of hard Names compared with our Permains, Gilliflowers, Pippins, Reynets and the like not to mention the Nonpareil and the several sorts of Ruffetings.[11]

The French, on the other hand, had a strong preference for pears. La Quintinye had considered only seven varieties of apple worthy of the royal walled garden at Versailles, while admitting 67 sorts of pears; and in 1775 the famous fruit collection of the monks of Les Chartreux, Paris, still contained only 14 varieties of apples, but 75 pears.

Certainly, the apple grew as well in the England of the 18th century as it does today. It was also available in abundance due to the widespread investment in orchards as Richard Bradley observed in 1718:

> There is no kind of Fruit better known in England than the Apple, or more generally cultivated. It is that Use, that I hold it almost impossible for the English to live without it, whether it be employed for that excellent Drink we call Cider, or for the many Dainties, which are made of it in the Kitchen. In short, were all other Fruits wanting us, Apples would make amends.[12]

The status of apples was also enhanced by the new enthusiasm for all things deemed to represent the best of traditional country life. Political power had now been shifted away from the Crown and into the hands of the landed aristocracy, making the country estate rather than the metropolitan court the arbiter of fashionable taste. In marked contrast to the situation in France, country life and pursuits in England began to enjoy enormous prestige, and promoted a sense of national identity bound up with notions of simplicity, continuity, honourable toil and good plain English fare.

A suggested layout for the dessert, taken from The Complete Practical Cook, 1730, *by Charles Carter, lately cook to the Duke of Argyll, Earl of Pontefract, and Lord Cornwallis.*

A dessert of fresh and preserved fruit, as published for the guidance of the middle classes by Menon, chief cook at the French court in the mid-18th century.

By this time 'the dessert', as the finale to dinner was now called, had replaced the fruit banquet but in content it was broadly the same. On grand occasions it would still be taken in an adjoining room, but at other times the dishes of the main course would be cleared, the cloth removed and the finery of the dessert laid out on the polished dining table. Setting off the seasonal display of fruit and sweetmeats would be little dishes made out of the recently invented Ravenscroft or lead crystal glass, whose spectacular refractive brilliance was the perfect complement to the shimmering fruit jellies and syrups. The fashion for pyramids had now crossed the Channel, and the glasses would often be arranged on glass salvers, and stacked one on top of another to form glittering tiers of jewel-like sweetmeats.

Apples were usually an important feature of the dessert. They are the fruit mentioned most often in Lady Grisell Baillie's *Household Book*, which covers the period of 1692–1733, and includes records of meals eaten both at home in Scotland, and in the houses of the great, notably those of the Prince of Wales. On their tables, fresh dessert apples and wet and dry apple sweetmeats seem to have been accorded the same honours as the more exotic delicacies on offer. Indeed, the pomologist Dr Robert Hogg observed over a century later that the transparent green sweetmeat, which can be made from small and immature codlins, has an exquisite flavour 'resembling that of a green apricot'. At a dinner given for ten by Lord Mountjoy on 15 March, 1727, Lady Grisell notes that there were several elaborate pyramid arrangements down the centre of the table while guests could help themselves from the 'Aples in sawcers, Frensh figs and plumbs', pistachios, almonds, sliced oranges and 'aples in cyrop', which were arranged down each side.[13]

Apples are also frequently specified in the sample dessert menus given by Hannah Glasse in *The Compleat Confectioner*. On one menu, Golden Pippins and Nonpareils feature alongside 'Filberts [hazelnuts], Chestnuts, Plums, Large oranges slic'd and sugar strewn over, ice cream, Whip'd Syllabubs, Bloomage', and she gives recipes for drying Golden Pippins and preserving them in jelly; preparing 'Green Codlins'; and making 'Pippin Knots' from red and green coloured apple marmalade.

The national fondness for fine apples was not, however, confined to those with a country estate. As the century wore on, commercial suppliers responded to the increasing demand for fine table fruit, and market gardeners, especially those of London, acquired a reputation for selling the best quality. They grew fruit trees as the 'upper crop' with vegetables or fruit bushes in between as the 'under crop'. By 1800, market gardens formed a continuous belt of cultivation along the Thames from Bow in the east to Hampton in the west.[14] Regency ladies particularly liked to take a drive out from Kensington to Turnham Green past the orchards to see the fine fruit piled up waiting to be taken to the City. This would be sold not through costermongers' barrows – 'those moveable shops that run on wheels attended by ill-looking fellows', whose apples were cheap but of dubious quality – but would either be sold directly to those customers who liked to lay down a winter supply in their cellar, or find its way to Covent Garden which was now London's main fruit market, and to the specialist fruit shops which could be found adjacent to some of London's most elegant squares. One of the most famous of these was the fruiterers on St James's Street run by Mrs Elizabeth Neale. She was known to everyone as Betty and when she died in 1797, a flattering obituary appeared in the *Gentleman's Magazine*:

> She had first pre-eminence in her occupation, and must justly be called the Queen of Apple-women . . . She was a woman of pleasing manners and conversation and abounding with anecdote and entertainment.[15]

The number of good varieties of eating apples was now increasing steadily. At the beginning of the 18th century, Mrs Glasse's and Stephen Switzer's preference for Golden Pippins and Nonpareils would have been shared by most of their compatriots. The Golden Pippin was a small, golden, multi-purpose apple, with yellow flesh and an intense flavour. Available from the time of the Restoration, John Evelyn rated it amongst the very best, and such was its renown that by the 1770s English Golden Pippins were purchased by the Empress of Russia at a guinea a bushel. The Nonpareil, which appears to have been first recorded in the Brompton Park Nursery list of 1696, has an intense, sweet-sharp flavour when at its best in the new year. The ease with which it could be stored and its strong flavour in the early spring were obvious virtues and, if given enough autumn sunshine, it would develop a prettily coloured cheek. Switzer said it presented 'a beautiful aspect in the middle of the dessert', and the Nonpareil would almost certainly have featured in Hannah Glasse's centre-piece, 'A large dish of fruit of all sorts, piled up, and set out with green leaves'.

Also high on the list for dessert eating were Golden Reinettes which could be eaten fresh, or like the Golden Pippin, made 'excellent dainties'. Mrs Eales, Confectioner to Queen Anne, recommended them for 'Apple Jelly', on whose clarity, flavour and consistency so many other sweetmeats depended. 'Aromatic Russetting', considered by John Worlidge to possess 'a most delicate haute gust' was popular, as were La Quintinye's choices, particularly the tiny, bright red

Api. It played 'its part wonderfully well in all Winter Assemblies' he claimed, and advised eating it 'greedily, and at a chop; that is to say without ceremony and with its coat all on, for none have so fine and delicate a skin as this and . . . the perfume lies in the skin'.[16] The russetted Fenouillet or Anise Apple, whose juice was 'perfumed with a little smack of those plants from which it derives its name', and the Court Pendu Plat, which could be 'eaten with pleasure from December until February', also found favour in England.

Later on in the century, however, two new varieties, Ribston Pippin and Margil, began to challenge the superiority of many of these. Much sweeter than the Golden Pippin and with a more complex flavour than any of the other established varieties, they both have the intense aromatic flavour which characterises the finest apples. Margil may have been introduced from France by George London, following his visit to Versailles, while Ribston Pippin was reputedly raised from the pip of an apple enjoyed by Sir Henry Goodricke while visiting Rouen on his grand tour of Europe, and brought home to Ribston Hall in Yorkshire.

PERHAPS THE MOST DISTINCTIVE EXPRESSION of the 18th-century fondness for apples at dinner time was, however, in its enthusiasm for apple puddings. The original medieval pudding was a plain unsweetened suet dough wrapped in animal gut and boiled alongside the meat in a cauldron suspended over the fire. It could, however, only be made at slaughtering time. With the invention of the pudding cloth or bag in the early 17th century, puddings became a year-round possibility and recipes and ingredients began to proliferate. A Frenchman, Monsieur Misson, visiting England in the 1690s, wrote:

> *The Pudding is a Dish very difficult to be describ'd because of the several Sorts there are of it; Flower, Milk, Eggs, Butter, Sugar, Suet, Marrow, Raisins etc. etc. are the most common ingredients of a pudding. They bake them in an Oven, they boil them with Meat, they make them fifty several ways.*[17]

The development of fruit puddings in the 18th century was partly a consequence of the notorious English sweet tooth. Far more sugar was consumed in England than in France,[18] and the price of sugar continued to drop for most of the century as Britain's sugar islands in the West Indies became the world's major source. But it was also a reflection of fashion. Boiled puddings, together with robust pies, solid custards and thick batters, had been a mainstay of English cookery for centuries, and now enjoyed the high prestige of country things in general. Cookery writers like Mrs Glasse and, later, Elizabeth Raffald railed against French extravagance and complexity, and upheld as their models this native tradition of economic and homely fare. When these traditional techniques were fused with the newer methods of fruit and sugar cookery, the fruit pudding was the – apparently quite remarkable – result. Continued M. Misson:

> *Blessed be he that invented pudding, for it is a Manna that hits the Palates of all Sorts of People; a Manna better than that of the Wilderness, because the people are never weary of it . . . To come in Pudding Time is as much as to say, to come in the most lucky moment in the World.*

Among the earliest and simplest fruit puddings were boiled apple dumplings, which were enjoyed at all social levels. In 1754, William Ellis, the agricultural writer, recorded that apple dumplings and bacon or pickled pork were the most common foods of farming folk, while according to Dr Johnson apple dumplings were the staple upon which a certain clergyman of small income brought up a whole family. Wrapped in a hefty hot water paste and baked to a hard crust, dumplings became splendidly portable, and Bedfordshire 'Clangers' – with meat at one end and chopped apple at the other – made a complete meal for farm workers to take to the fields.

At the other end of the social scale, they were also reputedly among George III's favourite puddings. A variation on the dumpling theme was the 'Fruit Hat' in which a large quantity of chopped apples, rather than an individual fruit, was wrapped in dough and boiled in a basin. For a less rustic effect, both hats and dumplings were made using lighter, short-crust or puff pastry. Hannah Glasse uses puff and sends her elegant dumplings to table accompanied by 'good fresh butter melted in a Cup, and fine beaten Sugar in a Saucer'.[19] Other refined touches were to flavour the pudding with 'shreds of lemon peel boiled in rose or orange-flower water', or even to abandon pastry altogether and cover the apple with a light sponge mixture in which case it became 'Eve's Pudding'.

For many, though, the best apple pudding of all was not a true pudding but a pie. 'Good apple pies are a considerable part of our domestic happiness' wrote Jane Austen to her sister Constance in 1815, while George IV's foreign secretary, Lord Dudley, was said to have complained audibly throughout a meal if apple pie did not feature on the menu. Flavourings could include cloves, lemon peel, orange-flower water or a few slices of quince, which gave 'a charming relish to the Pie'. The final touch was cream, which would often be poured in under the lifted lid.

'Apple Florentine' was also much in vogue in the 18th century, particularly at Christmas time. It consisted of a very large dish filled with whole apples, sugar and lemon and covered with a pastry crust. When baked, the lid was removed 'by a skillful hand, divided into portions, to be again returned to the dish, ranged around by way of garnish, when to complete the mess, a full quart of well-spiced ale was poured in "hissing" hot'.

Eighteenth-century apple tarts also had lids, but were filled with fresh or preserved apple purée. For Miss Fiennes, the West Country version was one of the highlights of her trip:

> An apple pye with a Custard all on top, it is the most acceptable entertainment it Could be made me. They scald their Creame and milk in most parts of these countys and so it's a sort of Clouted creame as we call it, with a little sugar, and soe put on top of the apple Pye; I was much pleased with my Supper.[20]

Fresh or preserved apple purée or 'marmalade' could be cooked with egg, butter and breadcrumbs to become 'Apple Pupton'; topped with sponge it was another version of Eve's Pudding, and combined with beaten egg whites it made 'Apple Snow'. Chopped apples covered with an apple marmalade, studded with blanched almonds and then baked was the delightfully named 'Apple Hedgehog'. Poured into a round mould lined with slices of buttered bread, then baked and turned out, the 'marmalade' was the *raison d'être* of 'Apple Charlotte'. This famous dish may have originated in the kitchens of George III, and hence been named after his stout queen of maternity-hospital fame. Others connect the title 'Apple Charlotte' to Goethe's heroine in *The Sorrows of Young Werther*.

Various kinds of baked apple were still enormously popular. Apples toasted in front of the fire at the end of a long spike and served with caraway seeds had been a staple since the Middle Ages, as had apples cooked in bread ovens after the bread had been removed, and left to become soft and succulent. These were made at home and in bakers' shops, and hawked around by street traders. 'Hot Codlins Hot' appears in the earliest extant version of the 'Cries of London', which is dated to about 1600. The term 'codlin' originally described an apple which cooked easily, that is, 'coddled', but it probably also came to refer to any apple that was picked green and as a result quickly baked to a pulp. These early take-aways were carried off and eaten with special 'apple scoops' carved out of mutton bones. Certain varieties were baked as particularly local delicacies. In Norwich, the late keeping Norfolk Beefing was baked to make Norfolk Biffins, which were still being made in the 1950s. These were sent off, from the late 18th century, in boxes as presents and in Dickens' *A Christmas Carol*, Scrooge spies some in a London fruiterers 'entreating to be bought and carried home to be eaten with sugar and cream'.

A slightly more elaborate version of baked apples was a dish known as 'Black Caps', which consisted of apples halved and baked, cut side downwards, with a sprinkling of sugar and lemon juice or wine over the skins causing them to blacken during cooking. Apples were also baked whole, with the cores removed and the cavities stuffed with sugar, butter and dried fruit, a recipe which is also popular today. In all their guises, they were not confined to meal times, but could be eaten at any time of the day. Jane Austen wrote of her fictional namesake in *Emma*:

Dear Jane makes such a shocking breakfast . . . but about the middle of the day she gets hungry, and there is nothing she likes so well as the baked apples, and they are extremely wholesome.

Apples were also chopped or puréed and mixed with eggs or batter to make various pancakes and omelettes. Combined with bacon, they were a popular filling for the pancake-like 'fraze', and apple rings fried in butter then combined with eggs made an excellent 'tansy', or sweet omelette. Sliced, dipped in batter, and fried in hot fat, they were transformed into 'Apple Fritters'. These could be as plain or as rich as purse and inclination allowed, and were considered a particular delicacy. One 18th-century recipe uses a quart of cream, a pint of sack and three quarters of a pint of ale to make the batter, and a noted gourmet apparently declared that the meal he would choose above all others was whitebait, woodcock and apple fritters.

One of the earliest mechanical devices for peeling and coring apples. It did both simultaneously.

For all these puddings, cooks would have employed whichever apple varieties were available in the local market or their own gardens. In the south-east these might have been Lemon, Golden, Holland, French or Kentish Pippins, or later in the season Winter Queening or Quoining. In Lancashire Minchul Crab appeared in the markets, and Dr Harvey and Hubbard's Pearmain were well known in East Anglia, but all over the country numerous other unnamed table and cider apples were bulked together and sold as 'pot' fruit. It is almost certainly to the English abundance of apples, many of which were astringent, that we owe Britain's infinite range of apple puddings, and the unique English preference – which still persists – for tart cooking apples.

PLATE 11: MOTHER (AMERICAN MOTHER)

The 17th and 18th centuries also saw the apple become the first fruit of America. For the successive waves of protestants who left various parts of Europe to found a new Jerusalem across the Atlantic, establishing apple orchards was almost as important as putting a roof over their heads. Quite apart from the spiritual uplift of the enterprise, apples were a staple commodity. Fearing to drink the water, the colonists wanted to make cider. To preserve fruits and vegetables for the winter, they needed cider vinegar, and the cider spirit called applejack, which also came in useful as a medicine, antiseptic and anaesthetic. Apples themselves were easily turned into jams, sauces and jellies with the simplest kitchen equipment, and could be used to moisten, sharpen or extend any basic foodstuff. As in England, apples also had a wide range of medical applications from skin cream to cough syrup, which in communities poorly served by qualified medical men had an even greater importance than at home. And what the humans did not consume in one way or another, the pigs enjoyed.

The first orchard in Massachusetts was planted some time before 1625 by a dissident Church of England clergyman, William Blaxton (or Blackstone), on Boston's Beacon Hill, which was then called Shawmutt by the Indians. Blaxton was an eccentric, bookish recluse, who saddle-trained a bull, and from its back distributed apples and flowers to his friends. In 1630, however, Blaxton's peace and seclusion were shattered by the governor of the newly established Massachusetts Bay Colony, who claimed and renamed Shawmutt, granting Blaxton a mere 50 acres of land. Insulted by the offer, Blaxton sold his holding back to the governor and moved to Rhode Island in 1635. Here he raised its first orchard, and America's first named variety, Blaxton's Yellow Sweeting.[21]

It was not long before every colonial homestead had its own orchard. Edward Johnson's *Wonderworking Providence of Sion's Saviour in New England*, which was published in London in 1654, described Massachusetts in 1642 as transformed from the

> wigwams, huts and hovels the English dwelt in at their first coming, into orderly fair well built homes, well furnished with many of them, together with orchards filled with goodly fruit trees and gardens with a variety of flowers.

He estimated 'near a thousand acres of land planted for Orchards and Gardens, besides their fields are filled with garden fruit'.[22] Aided by English herbals they brought with them and the brisk trade in recipe collections from across the Atlantic, colonial housewives made the most of apples in what was the equivalent of a brew house, apothecary's shop and kitchen combined.

By the time the colonies made their bid for independence in 1775, orchards of apples had been planted all along the Atlantic seaboard as far south as northern Georgia, wherever the climate allowed. Apples were 'the prime article of culture' in New England. There were 'goodly orchards' along the Hudson and Mohawk Rivers in New York, and the milder conditions of New Jersey were 'so favourable to fruit [that] even the poor have plenty of apples, peaches, cherries, gourds and water melons'. Finer still was the fruit of Pennsylvania where Dutch, German, Swedish, British, and Swiss immigrants had cleared acre upon acre of forest to make way for apple orchards that did not merely reproduce, but surpassed in their crops, those they had left behind in Europe.[23]

With the coming of independence, the apple orchards began to spread westwards. They flourished in their new territories, continuing both the protestant tradition and, like maize, wheat and cotton, making efficient use of local resources to meet the demands of an expanding but still small population, with large surpluses left over for the hungry world markets beyond.

Every encouragement was given to the pioneers to plant apple trees. The 'requirements' for settling Ohio lands in 1787–88 included the stipulation that:

> within three years the settler must have set out at least fifty apple or pear trees and twenty peach trees; within five years he must have erected a dwelling house . . . At the end of five years he must have 15 acres for pasture and five acres ready for corn . . . With these conditions met the settler would receive a deed, at the end of five years, to his Donation lot.

In Virginia, a lease made by George Washington of Mount Vernon in 1774 requires that the prospective tenant must establish an orchard of 100 apple and 100 peach trees.[24]

As in the days of the first colonists, the claiming of the land by apple orchards symbolised the vision and the achievement of the pioneers. The legendary Johnny Appleseed, who for 46 years wandered through the wildernesses of Pennsylvania, Ohio and Illinois planting apple pips as he went, became an almost saintly folk hero.[25] Born Jonathan Chapman in Leominster, Massachusetts, in 1744 he devoted much of his life to preaching, planting apple pips, warning settlers of Indian raids, and befriending men, birds and animals. Exactly when he began his pilgrimages is not known, but in 1806 he loaded a boat with apple seeds from the cider mills of West Pennsylvania and floated off down the River Ohio to Wellsburg in West Virginia, where he founded a nursery with his brother. From there he paddled down to the Muskingum River and into the very centre of Ohio, handing out seeds as he went to any settler who promised to plant and care for them. Here and there he planted seeds himself, protecting them with brushwood and inviting farmers to help themselves from the young trees. Often he travelled on foot, wearing only a coarse sack. As he moved westwards into Indiana he left a trail of infant orchards in his wake, and died in 1847, having travelled more than a hundred thousand miles. Johnny Appleseed trees were still common in north central Ohio in 1900, and one tree is reputed to have remained on a farm in Ashland County until the 1960s.

Alongside the more eccentric aspects of his career, 'Johnny Appleseed' was also a professional nurseryman and owner of 1,200 acres of planted land.

Serious apple growing finally reached the other side of the new republic when Henderson Lewelling and William Meek established the west's first nursery in Oregon in about 1847.

A S IN ENGLAND, the creation and consumption of apple puddings was encouraged by an abundance of apples and the need for substantial, warming dishes in the long, cold winters. Plain-living protestants found their economy and simplicity appealing, while the elegant Virginians further down the coast saw them as an aspect of fashionable English life that they could reproduce in their new homeland. But like the English language, the colonial apple

Plate 12: Egremont Russet

pudding, having started out from the same point, proceeded to follow its own course, giving rise to a distinctive American tradition of apple cookery. The challenge of new ingredients such as cranberries, pumpkins, maple syrup and cheap molasses encouraged experiment, while settlers from other parts of Europe introduced their own recipes and variations. Even the challenge of cooking on the open hearth – which remained the main technique for longer in America than England – and of preparing food on the move during the great trek across America, had its effect on shaping America's culinary legacy.

The English can probably take the credit for contributing apple pie to the collection, although it is now considered a quintessentially American dish. The early pies were as robust as the first colonists. 'House pie in country places is made of apples neither peeled nor freed from their cores, and its crust is not broken if a wagon wheel goes over it', wrote a Swedish parson named Dr Acrelius in 1758. 'Apple-pie is used throughout the whole year', he continues, 'and when fresh apples are no longer to be had dried ones are used.[26] In the cold New England winters apple pies would be made in quantity and put outside to freeze. They were eaten at any time of day, and were a favourite breakfast dish. Fruits of the new continent, such as cranberries, were added to give extra flavour, and apples were often mixed with pumpkin.

The settlers also seem to have extended the old west-country custom of serving an apple pie with cheese. The first cheese they made resembled the Cheddar cheese of Somerset, and apple pies were made with Cheddar pastry or served with cheese sauce. The settlers made dumplings, fritters and pancakes of every description. They also extemporised around the simple theme of spiced apples, sweetened with molasses and topped with various kinds of scone, biscuit or pastry dough to make 'Apple Grunt', 'Cobbler', 'Pandowdy' and 'Slump', which have become part of the national heritage. Louisa May Alcott, the author of *Little Women*, was so fond of them that she named her house in Concord, Massachusetts, Apple Slump! Apple Pandowdy derives its name from the custom of 'dowdying' or cutting the crust into the apple filling, while 'Apple Betty' mixes the fruit with breadcrumbs and butter rather than dough. Fresh apples would be used when available, but throughout the winter housewives would also turn to the pounds of dried and preserved apples stored away in the cellars and outbuildings.

Large quantities of colonial apples tended to be conserved in this way for a number of reasons. The early American orchards were raised from pips, rather than by grafting trees of known varieties. Many of the resulting seedlings bore early-season apples which did not keep, and even the late keepers needed a good cellar dug into the ground to survive the hard New England winters. Drying, pickling, cider-making or boiling them down into 'Apple Butter', 'Apple Sauce', jam or jelly not only ensured a winter supply, but also concentrated a large volume into a small space.

Apple slices could be strung up in the kitchen to dry, put in ovens, or simply left out in the sunshine. In Orange County, New York, where John de Crèvecoeur, the French author of a famous series of letters on rural life had settled about 1760, every October saw neighbouring women gather together in each other's houses to hold a 'paring bee'.

A New England harvest. Every colonial homestead had its own apple orchard. The whole family would be involved, and as most of the fruit was to be turned into cider or preserves, the actual picking was an easy-going affair.

A basket of apples is given to each of them, which they peel, quarter and core. These peelings and cores are put in another basket and when the intended quantity is thus done, tea, a good supper, and the best things we have are served up . . . The quantity I have thus peeled is commonly twenty baskets, which gives me about three of dried ones.[27]

The next morning a stage would be erected on which to lay the apples out to dry.

Strong crotches are planted in the ground. Poles are horizontally fixed on these, and boards laid close together . . . When the scaffold is thus erected, the apples are thinly spread over it. They are soon covered with all the bees and wasps and sucking flies of the neighbourhood. This accelerates the operation of drying. Now and then they are turned. At night they are covered with blankets . . . By this means we are enabled to have apple-pies and apple-dumplings almost all year round.

Other fruit would also be dried, but it was apples which were staple fare.

My wife's and my supper half of the year consists of apple-pie and milk. The dried peaches and plums, as being more delicate are kept for holidays, frolics, and such other civil festivals as are common among us. With equal care we dry the skins and cores. They are of excellent use in brewing that species of beer with which every family is constantly supplied, not only for the sake of drinking it, but for that of the barm [yeast formed on fermenting liquors] without which our wives could not raise their bread.

Drying apples and preserve-making were a particular speciality of the Pennsylvania Dutch, who came mainly from Germany – their name is a corruption of 'deutsch' meaning German or German speaking – with a strong admixture of Dutch, Swiss and Moravians. They called dried apples *schnitz* from the German meaning 'cut', and there were both sweet and sour *schnitz*, depending upon the apples from which they were made. The dried apple slices were plumped up with water or cider and used in the same way as fresh apples, but also gave rise to dishes in their own right, such as *Schnitz-un-gnepp*, in which the slices of dried apple are cooked with a piece or pieces of ham or pork and served with dumplings. *Schnitz* was even used as a chewing gum and was so much part of Pennsylvania Dutch life that it is even on the map. The story goes that a farmer upset a wagon load of dried apples in a creek; the slices swelled up in the water causing it to flood the whole valley, which is known today as Schnitz Creek.

Apple Butter and its more refined relative Apple Sauce were the other most popular methods of preserving apples. Apple Butter was made from chopped whole apples, but the Sauce would be made by 'stewing pared and sliced apples in new cider until the whole is soft and pulpy'. After further cooking and reducing, this was stored away in earthenware jars or even barrels. For a superior product 'the best, the richest of our apples' were used and a 'due proportion of quinces and orange peels is added'. 'In our long winters', wrote de Crèvecoeur, it 'is a great delicacy and highly esteemed by some people. It saves sugar, and answers in the hands of an economical wife more purposes than I can well describe.'

Apples were also frequently combined with pork. The settlers kept pigs which foraged in the woods, and were ready to be killed when 'the first frost was in the ground', and apples were at their most plentiful. Not only did apples relieve the fattiness of the meat, but like cured pork, were one of the staples of the winter diet. When the pioneers began to trek westwards, combinations of cured pork and preserved apple were cooked in skillets over an open fire.

As in Britain, cooks would use whatever apples were available, but unlike 19th-century Britain, in America they did not select out a range of sharp apples specifically for use in the kitchen. As a result, American varieties have remained multi-purpose and in comparison with English cookers, the apples used in American puddings are much sweeter and cook more firmly. According to the Reverend Henry Ward Beecher, a leading figure in early Indiana horticulture, the variety Esopus Spitzenburg, which in England has always been regarded as an eating apple, exemplified the perfect American cooking apple. 'Anointed with sugar and butter and spices' in a pie, it formed a 'glorious unity' with a crust that let

A 'paring-bee'. Note the peeling and coring device on the table. The girl throws her unbroken apple peeling over her shoulder – superstition held it would fall into the shape of her future husband's initial.

> *the apple strike through and touch the papillae with a mere effluent flavor . . . the sugar suggesting jelly, yet not jellied, the morsels of apple neither dissolved not yet in original substance, but hanging as it were in a trance between the spirit and the flesh of applehood . . . then, O blessed man, favored by all the divinities! eat, give thanks, and go forth, 'in apple-pie order!'*[28]

THE NATIONAL IMPORTANCE OF apple growing in England and America helped to encourage – and was in turn encouraged by – the concerted attempt towards the end of the late 18th century to improve and extend the range and quality of apples and put the identification and evaluation of fruit on a more scientific footing. In England, as on the Continent, the impetus came both from estate owners eager to improve their tables, and from commercial growers and nurserymen who were keen to supply the expanding markets of the new industrial towns. In America, the new republic's need to develop its apple industry as efficiently as possible across enormous geographic and climatic diversity required hard information on the performance of different varieties without delay.

On the English country estate, a shift in economic and philosophical climate was directing more attention to the quality of fruit by the middle years of the 18th century. Prosperity had begun to return to mainstream agriculture and this, together with new developments in land use and stock rearing, persuaded many landowners to become more closely involved in improving and running their estates. They began to see themselves as new Augustans and to subscribe to the same ideals of fine, home-grown produce as their classical mentors. At Brambleton Hall, declares Matthew Bramble, in Tobias Smollett's *Humphrey Clinker* of 1771:

> *My salads, roots and pot herbs, my own garden yields in plenty and perfection . . . The same soil affords all the different fruits which England may call her own, so that my dessert is every day fresh-gathered from the tree.*

At the same time the craze for replacing formal gardens with 'Capability' Brown vistas of water, wood and meadow had begun to grip the upper classes. Flower beds and fruit trees disappeared from the pleasure ground and were banished to the walled fruit and vegetable garden. Here, teams of gardeners, relieved from the endless chores of clipping, watering and maintaining the formal garden, could concentrate their energies and skill on produce for the dining room. They developed the techniques of forcing fruit and vegetables, extended their range to include pineapples, and exploited the advantages of espalier training to give finer quality apples and pears for the dessert. They and their employers also became increasingly keen to identify, compare and evaluate new varieties in the search for ever greater novelty and refinement.

Their commercial counterparts, on the other hand, were faced with a kind of knock-on effect. Not only were their markets expanding in size, but a new middle class made wealthy by trade and manufacturing increasingly sought to emulate the taste of their landed superiors. The more keenly apples were appreciated in the mansion, the greater was the demand in the market-place, and commercial growers badly needed to identify attractive, good flavoured and heavy cropping varieties, that would also withstand the rigours of transport.

THE HEAVY INVESTMENT IN ORCHARDS in the late 17th century meant that many new seedlings arose. Seedlings sprang up from the discarded pomace of the cider press, from pips dropped by birds at the edge of a woodland, or from seeds that may have been been carefully planted in gardens. Some of these new varieties acquired local fame, and then through networks of gardeners and nurserymen were spread all over the country. Scions of varieties were not only eagerly exchanged within the British Isles but further afield. In 1759, for example, the American scientist and diplomat Benjamin Franklin, who was staying in London, received a box of

PLATE 13: LORD DERBY

Newtown Pippins from home. They raised so much interest that grafts were obtained and by 1768 the Brompton Park Nursery was offering its eager customers trees of 'the Newtown Pippin of New York', while America's foremost nursery, that of Robert Prince at Flushing Landing, Long Island, had begun to offer to send fruit trees to Europe. In the other direction, England's Golden Pippin and Nonpareil were sent from London in 1794 for George Washington to plant in his Mount Vernon garden in Virginia. The 1790s also saw the beginning of controlled breeding of apples, when Thomas Andrew Knight, a Herefordshire squire, with the help of his daughter Frances, produced the first varieties of known parentage by patiently cross fertilising the Golden Pippin with other varieties.

The main problem facing the pomologists, however, was that hundreds, if not thousands of different apple varieties existed by the end of the 18th century, and very little was known about any of them. In England, the early lists compiled by men such as John Parkinson, John Worlidge and Stephen Switzer gave little more than a name. Batty Langley had included line drawings of a few varieties in his *Pomona* of 1729, but in general, the information available was insufficient to positively distinguish between similar varieties. As a result, a particular apple was often known by a different name not only in another country or county, but in an adjacent village; old varieties were reintroduced under a different name and claimed as new improvements; and inferior, but similar looking apples might masquerade as an esteemed variety. Until the confusion of identities could be resolved the proper evaluation of varieties could not begin.

The first attempts at a detailed record of fruit characters for the purpose of identification had probably been those of Jean Bauhin of Switzerland, who in 1598 had accompanied his brief descriptions with line drawings. Nearly a century later the paintings commissioned by Cosimo III de' Medici brought a significant advance in fruit description. The paintings were designed to record his renowned fruit collection in sufficient detail both to enable traditional Tuscan produce and new introductions to be recovered, should a variety be lost, and make it possible for 'their correct names, which had previously often been confused, to be clarified by their labelled representation in painting'. The artist Bartolomeo Bimbi accordingly produced a series of huge canvases, each of which was devoted to a different fruit – apples, grapes, figs, cherries and citrus fruits among them – with each variety discreetly identified by means of a tiny number corresponding to a key beneath. In England, in 1730, the London nurseryman Robert Furber also sought to record as well as promote novel and established fruits in his lavishly illustrated catalogue. Each month's selection was carefully labelled and the guidance of 'a great many, both of Gentlemen and Gardeners' had been sought to ensure that no name had been inserted 'but by their unanimous Opinion and Consent'.

Paintings on their own could not, however, provide either a comprehensive or sufficiently analytical system of reference, and it was the greater precision of Linnaean botany combined with the studies of botanists at the French court which laid the foundations for a universal system for defining the constant features of a fruit. These characters were first defined by the French botanist Henri Louis Duhamel du Monceau in his *Traité des Arbres Fruitiers*, which appeared in 1768 and elevated the study of fruit into a branch of applied botany. French studies were halted during the Revolution, but detailed inventories of the national fruits of Holland, Austria, Switzerland, Germany and Italy were published before the century was out, and it was the new systematics which underpinned the studies of the London Horticultural Society which was founded in 1804.

Among the aims of the Society, which became the Royal Horticultural Society in 1861, was to improve the quality and range of all produce, including apples, both in the private garden and the market-place, and it provided an international focus for pomological research. At its gardens in Chiswick, the Society began to build up fruit collections in order to resolve the confusion of identities and to encourage its Fellows to seek out or raise new varieties. Old and new varieties came in from all over the British Isles, Europe and beyond to be examined at the Society's meetings; grafts of promising ones were then sought and trees planted up in the Chiswick gardens. In his first *Catalogue of Fruits* of 1826, the Society's Fruit Officer, Robert Thompson, was able to list the 'true names of each of the upwards of 1200 apples' and there were 400 more about which he was uncertain. Five years later the list had risen to 1396, and between 1828 and 1830, in collaboration with the Society's secretary John Lindley, Thompson produced the finely illustrated, three-volume *Pomological Magazine*, which carried descriptions as full as those of Duhamel and an evaluation of the main English and 'foreign' varieties. These had also been classified into dessert varieties – those of good fresh eating quality – and culinary apples for the kitchen.

A month from Robert Furber's 'Twelve Plates with Figures of Fruit'. Bound in with his 'Twelve Months of Flowers' and published 1730–2, these were botanically accurate and painted in the manner of Dutch still-lifes.

As a result of the Society's work the merits of Ribston Pippin, King of the Pippins, Golden Noble and Blenheim Orange were widely recognised, and together with Dumelow's Seedling – which was exhibited in 1815 and later renamed Wellington in honour of the hero of the Battle of Waterloo – they were subsequently planted all over Britain in both private and commercial orchards. The qualities of a number of new varieties were also quickly recognised. These included Kerry Pippin and Irish Peach which were sent over by nurseryman John Robertson of Kilkenny; Keswick Codlin which had been found on a rubbish heap at Gleaston Castle near Ulverston, and Pitmaston Nonpareil raised by John Williams of Pitmaston, Worcestershire. The Society was also able to recommend and supply varieties to its Fellows and sister societies abroad. Scions of the hardy Russian varieties – Duchess of Oldenburg, Emperor Alexander, Red Astrachan and Tetofsky – were sent over to America in 1832 for trial in the colder northern states, and Blenheim, Ribston and Gravenstein were introduced to the Annapolis Valley where they helped found the Nova Scotian and Canadian apple industry.

Another result of all this activity was the appearance of sumptuous fruit books or *Pomonas*, which contained some of the most exquisite paintings of apples and other fruits ever produced. Although designed as botanic record books, their use of colour illustration reflected the fashion for paintings of plants which had been growing steadily throughout the 18th century. Top botanic artists such as Georg Dionysius Ehret, Pierre Joseph Redouté and Francis Bauer, were patronised by royalty and fêted by the aristocracy; society ladies took lessons in botanic painting and artists began to produce botanical drawing books which contained a few coloured illustrations and a series of uncoloured plates for students to practise on. Images of fruit and flowers also spilled over on to dinner services, decorative porcelain, curtains and dress fabrics.

PLATE 14: McINTOSH

The public interest in fruit and flower painting was equalled only by the related passion for collecting and classifying the real thing and, as a result, the Horticultural Society and various other horticultural patrons were inspired to meet the demand for both in the publication of these beautifully illustrated volumes. The largest and showiest of the English fruit books was Brookshaw's *Pomona Britannica*, 'A Collection of the Most Esteemed Fruits at present cultivated in Great Britain; selected principally from the Royal Gardens at Hampton Court, and the Remainder from the most celebrated Gardens around London'. The 60 plates were engraved in aquatint and stipple, printed in colour and finished with water-colours, and illustrated 256 varieties of 15 kinds of fruits. The book was dedicated to the Prince Regent (later George IV) and occupied 'nearly ten years constant attention and labour'. In 1812 it cost 59 pounds and 18 shillings, and weighed a good 28 pounds.

The finest and most accurate paintings of apples, however, were those of William Hooker in his *Pomona Londinensis* of 1813–18, and Elisabeth Ronalds, who painted the fruits for her father's *Pyrus Malus Brentfordiensis* of 1831. Hooker was a pupil of Bauer, and was commissioned by the Horticultural Society to paint some of its finest specimens. He was said 'to know apples better than any of us' and among the varieties he recorded with glowing vitality were Ribston Pippin, King of the Pippins, Margil and Blenheim Orange, as well as some of the handsome new arrivals from abroad: Emperor Alexander and Borsdorfer, which was Queen Charlotte's favourite apple. His *Pomona* contained 'coloured Representations of the best Fruit cultivated in British Gardens'. It had 49 aquatints engraved and coloured by Hooker, and fruit descriptions compiled with the assistance of Thomas Knight and Society Fellows. Hugh Ronalds was a nurseryman of Brentford in the heart of the Thames Valley market-gardening area and his *Pomona* was a 'descriptive Catalogue of Apples of those varieties which I have thought most excellent'; each of the 179 varieties featured possessed 'some singularity of appearance rendering them pleasing to the eye'. The illustrations were reproduced by the recently invented technique of lithography, which was seen at its best in the *Pomona*'s rendering of the apple's delicate flushes and cinnamon russets.

Thomas Andrew Knight, 1759–1838.

On the other side of the Atlantic, the baton of systematic pomology was taken up by William Coxe. A descendant of the first governor of New Jersey, he planted all the varieties he could find in his own experimental orchards and set out his experiences and observations in his fruit book of 1817 entitled:

A View of the Cultivation of Fruit Trees, and the Management of Orchards and Cider, with Accurate Descriptions of the Most Estimable Varieties of Native and Foreign Apples, Pears, Peaches, Plums and Cherries, Cultivated in the Middle States of America; Illustrated by Cuts of two hundred kinds of Fruits of the natural size; Intended to Explain Some of the errors which exist relative to the origin, popular names, and character of many of our fruits; to identify them by accurate descriptions of their properties and correct delineations of the full size and natural formation of each variety; and to exhibit a system of practice adapted to our climate.

81

This brought to wider popularity many good native apples, and in recognition of his efforts the London Horticultural Society elected him a Fellow. Like Knight and Ronalds, he called upon the assistance of his family in the furtherance of his studies. Each of his three daughters was given fruit to paint and these little-known paintings, which in the end were never published, record 104 apples among the 154 fruits illustrated.

A year later the first American horticultural society was founded in New York with Dr David Hosack – who owned a renowned estate and garden on the Hudson River – as its President, and nurseryman William Prince as another founder member. Other regions soon followed this example. The Massachusetts Horticultural Society was founded in Boston in 1829, and like its counterpart in London, established experimental gardens, built up a library, held meetings, published reports, and staged exhibitions to present the best horticultural products to the public. Its first president was the indefatigable Robert Manning. In his 'Pomological Garden' he attempted to do for Massachusetts what Coxe had done for New Jersey, by bringing together all the fruits that would thrive in his region. Manning was a nurseryman, with correspondents in England, France and Belgium from whom he received grafts of the best fruits, and he built up an immense collection which he used to compile his *Book of Fruits* of 1838.

The subsequent success of the American apple industry can in a large part be attributed to the work of Coxe, his followers, and the societies across America which highlighted and promoted the best of the apples raised. By the middle of the century the industry had begun to take off in earnest, and the criteria by which new varieties were judged and developed were almost entirely those of the commercial grower. Back in Britain, however, the situation turned out rather differently and in Victorian England the demands of the private garden and country house dining table remained the major determining factor in selecting and promoting new varieties of apples.

Chapter 4
APPLES for the FEW

No fruit is more to our English taste than the Apple. Let the Frenchman have his Pear, the Italian his Fig, the Jamaican may retain his farinaceous banana and the Malay his Durian, but for us the Apple.

EDWARD BUNYARD: *The Anatomy of Dessert*, 1929

DISCERNING VICTORIANS OF THE 1890S DISCUSSED the flavour of their apples as passionately as they debated the finer points of wine. Savouring regional specialities and rejoicing in the quality of the good years, they reminisced about varieties and seasons with the same attention to detail as they devoted to the chateaux and vintages of their clarets. They had hundreds of varieties to choose from, most of which had been selected specifically for the dessert over the last 60 or 70 years, and in addition they could direct their attention to a whole new group of culinary apples, which had been bred especially for the kitchens.

It was a uniquely British phenomenon. Nowhere else has the status of the apple ever been elevated to such heights, and in no other country has such a wide spectrum of flavours and qualities been developed. It has had an enduring influence on pomological tastes and perceptions. The British still take a special pride in 19th-century favourites and benefit from the Victorian connoisseur's devotion to the fine flavour of Cox's Orange Pippin which, despite many commercial problems, became and remains a main market apple. Nor has love wavered for a good cooker, particularly Bramley's Seedling: Britain is the only country in the world where a culinary apple is a major market variety.

Commercial needs and the enormous technical developments in every aspect of horticulture were among the forces which gave impetus to the proliferation of apple varieties in 19th-century Britain, but another was the prominence and importance given to fresh fruit on the Victorian dining table. By the middle of the century, the Industrial Revolution had created a sizeable new class of wealthy merchants and manufacturers, who wished to gain access to the elite social and political circles of the landed aristocracy. The essential admission ticket was a country house and estate of at least 500 acres, which enabled one to advertise one's economic, social and aesthetic credentials and participate in all the rituals of upper-class life, including the seasonal round of house parties and sporting events at which the real business of running the country took place. In the face of this threat to its exclusivity, the small world of 18th-century Society had made the route to acceptance a social obstacle course. Every detail of dress, manners and taste counted. This put enormous pressure on new landowners to make every aspect of their houses, gardens, dining tables and entertainments the finest possible, not only to impress their upper-class neighbours, but also to outdo their social rivals. This in turn made the old established families pour money back into their estates in order to try to keep one step ahead. In this upward spiral of social barter, the dessert of fresh fruit became a potent expression of wealth and discernment.

Since the early 18th century, fruit enthusiasts had been expanding their growing repertoire to include exotic fruits such as pineapples. With the development of Victorian greenhouse technology, the tenderest and most outlandish fruits could be grown all year round in the artificial climates made possible by heating boilers and acres of glass. As this required a huge investment in buildings, coal and labour, the ability to offer a dessert of pineapples, grapes, peaches, passion fruit and so on was a dramatic statement of the host's resources. With the introduction of dining *à la russe* in fashionable circles in the 1860s, this became even more the case. As in today's restaurant service, dishes were now served in succession rather than all being laid out at once, which left plenty of space in the centre of the table for magnificent arrangements of fruit and flowers. These formed the dinner-table decorations, which were carefully set out in advance by the gardeners, and

A fashionable dinner table of the 1870s. A mirrored panel down the centre of the table reflects the épergnes *of fruit at either end, and the decorative surround of apples and flowers.*

so greeted the guests when they came into dinner, and remained in place until the grand finale of the dessert. Fresh fruit did not, therefore, only bring the meal to its glorious conclusion, it signalled the splendour and generosity of the occasion at the outset. Like the pheasant shoot it was a luxury that only ownership of a country estate could provide. Mere city dwellers who had to buy in their dessert fruits could not begin to match this paradisiacal plenty, which put even the finest sweetmeats in the shade.[1]

As might have been expected, hardy fruits such as apples were for a while overshadowed by their more exotic table-mates, but they nevertheless bathed in the reflected glory brought to all fresh fruit by the enhanced status of the dessert. Apples also benefited from the growing gastronomic awareness of the qualities of choice produce, which had first appeared in post-Revolution France and swiftly crossed the Channel. In 1824 the horticulturist John Claudius Loudon, in his *Encyclopaedia of Gardening*, had defined the perfect dessert apple as one 'characterised by a firm juicy pulp, elevated poignant flavour, regular form and beautiful colouring'. As one gentleman put it in 1876:

> Strawberries, delicious while they last, eventually pall on the palate . . . whilst grapes that princely fruit of almost perennial virtues cannot wholly satisfy the cravings for fruit, but something more substantial is desired – something crisply sweet and pleasantly brisk and these qualities are almost exclusively combined in apples.[2]

As nothing less than a very large fruit collection would ensure a continuity of choice apples throughout the season, the perfect quality of those brought to the table remained an important indicator of social refinement and substance.

PLATE 15: GOLDEN NOBLE

As the estate owners and their head gardeners were the nurseryman's best customers and occupied the most influential positions in the horticultural establishment, it was their criteria rather than those of the market grower which shaped the apple's development during this period. It was they who sat on the Royal Horticultural Society's Fruit and Vegetable Committee – the only official body appraising new varieties – and it was their fruit collections that provided the main trial grounds for new material. As a result, while more pragmatic virtues such as good crops and resistance to disease were not ignored, the emphasis was on qualities such as complexity of flavour, delicacy of colouring and diversity for as long a season as possible. Victorian apples therefore acquired a range and subtlety which did indeed bear comparison with wine, but did little to benefit commercial growers who needed a small number of reliable, marketable varieties, not a vast selection.

The apple season opened with tiny Jenettings, followed in mid-August by the new Irish Peach, which brought an unaccustomed richness to the taste of early apples. It 'possessed all the rich flavour of the winter varieties, with the abundant and refreshing juice of the summer fruits', claimed the eminent pomologist Dr Robert Hogg in his encyclopaedic *Fruit Manual*. The September selection now included Summer Golden Pippin and the tortoiseshell flushed Kerry Pippin – which were both intensely flavoured like the Golden Pippin – and the Pine Golden Pippin which hinted of pineapple. As the season progressed so the interest increased. For October and November there was Ribston Pippin, 'so well known as to require neither description nor ecomium' and the most splendid of the new varieties, Cox's Orange Pippin. Probably an offspring of the Ribston, it had arisen in the 1820s, and possessed the same richness and complexity of flavour as its presumed parent, but was sweeter, more delicate, and of better shape and size for the dessert. Before Christmas, Blenheim Orange provided a restrained yet 'nutty' flavour, while the newer Claygate Pearmain offered a firmer, more aromatic flesh, and a definite taste of walnuts. There was also Cornish Gilliflower, which by December had acquired the ethereal scented flavour that its name evokes. In the new year came a number of Nonpareil seedlings. These included Ashmead's Kernel, which possessed a 'more sugary juice' than its parent, but still had the same strong sweet-sharp flavour; Rosemary Russet, Duke of Devonshire, and D'Arcy Spice, which lived up to its name until May. Finally, there was Sturmer Pippin, a cross between the Nonpareil and the Ribston, raised in the village of Sturmer in Suffolk. This was still sound when all the others were past, and with its crisp refreshing taste provided a welcome contrast to the soft flesh of forced Easter grapes and strawberries.

The high value accorded to novelty on the dining table also gave the new American varieties a fascination for Victorians. They tended to be more highly coloured than English apples, particularly the scarlet Northern Spy, deep rose pink Esopus Spitzenburg, and dark maroon Baldwin. Most American varieties, however, did not thrive in the English climate. Even when given the benefits of a wall, the prized Newtown Pippins did not possess that 'peculiar rich aroma

The fruit garden at Liscard Hall, near Liverpool. Some of the trees are trained outwards and downwards to minimise the effects of the wind off the sea.

which characterises the imported fruit'. Not to be beaten, however, English gardeners grew them under glass in 'that charming luxury', the orchard house. Introduced by nurseryman Thomas Rivers in the 1850s, it was an unheated, airy greenhouse in which fruit trees in large pots enjoyed 'the climate of Toulouse or Nice, but without its biting winds'. In this drier, warmer atmosphere, Washington Strawberry acquired a fine perfume and a flesh 'that melts like a peach so that the juice runs down the knife', while the Melon apple of New York ripened to a primrose colour flushed with crimson. A number of continental varieties, which failed to fulfil their promise in the English climate, also lived up to their reputations if grown in an orchard house. These included the dark crimson Mela Carla of Italy, which had a 'perfume like that of roses'. Gardeners in the north of England and Scotland would also grow English varieties such as Sturmer Pippin under glass.

George Wilson, an owner of Wisley in Surrey, before it became the home of the Royal Horticultural Society, won many prizes for the apples he grew in orchard houses like this.

As many as a hundred different varieties of apple might be grown at any one establishment in order to ensure continuity and diversity, and to guard against failures. Most apples, of course, were grown outside, and the best dessert and culinary varieties given the privilege of a place in the walled kitchen garden. They were often grafted on to the now enormously popular dwarfing rootstocks and trained as bushes or pyramids. At Trentham Park, the Staffordshire estate of the Duke of Sutherland, several of the walks in the kitchen garden were bordered by apple and pear trees 'trained as umbrella or bell-shapes, and pyramidal or cylindrical shapes alternately'.[3] While the aim was to produce the best-quality fruit, gardeners still made the most of the apple's decorative qualities. In 1865, a visitor to the kitchen gardens at Archerfield, the East Lothian estate of the Nesbit-Hamilton family, found 'Dwarf apple trees at regular distances of 20 feet, alternating with standard Roses and the spaces in between them occupied with Mignonette'.[4] A ribbon border of bedding plants lay between the trees and the edging of clipped box, and this was backed by tall, sweetly-scented stocks. Such arrangements also tactfully protected visitors from the less seemly aspects of kitchen gardening, such as the sight and smell of organic mulch.

The espalier was another popular way of growing apple trees and could make a single-tiered edging to a path, or be extended to cover a wall or form a six-foot-high screen, and from the 1860s onwards cordon training also became popular. Cordons consisted of a single fruit-bearing stem, and as they could be planted close together, enabled a wide range of varieties to be grown in a small space. The invention of the eminent French pomologist, du Breuil, they reputedly took their name from the resemblance of the fruit, clustered along the stem, to a cord or chain, and were usually trained diagonally at an angle of 45 degrees to the horizontal. At Barham Court, outside Maidstone in Kent, in 1877 the head gardener had six walled gardens filled with trained apples and pears. The walls supported diagonal cordons, while low horizontal cordons bordered the walks. Rows of espaliers filled the quarters and the 'veritable lines of beauty' made a spectacular advertisement for the advantages of close training. The head gardener, Charles

Victorian gardeners exploited the decorative possibilities of trained apple trees in hedges and screens. This diamond design was the invention of the head gardener D. T. Fish, and was made up of twin cordons diagonally trained in opposite directions.

Section through the ideal Victorian or Edwardian fruit store.

Haycock, grew apples of a 'colour, size and general high quality' that made his 'competitors hold their breath and covet the possession of such skill, soil, and situation'.[5]

Careful management of the fruit on the tree had to be followed by equally careful picking and storing. The early varieties, picked in August and early September, were best eaten ripe straight from the tree and would go straight to the fruit room at the back of the kitchen garden. The mid-season and late varieties were stored somewhere 'similar to that which preserves milk – one with a cool atmosphere free from all vapours, bad smells and stagnant air, for apples like milk are easily tainted'. A shed might suffice but a cellar was better, and the best stores were properly designed and purpose-built. To maintain a low, constant temperature, they would be sited either against a north-facing wall, or set partly underground and well insulated. Double entrance doors also helped minimise temperature fluctuations, but on a frosty day the doors would be open for a while to 'let in the cold'. Good ventilation avoided a build up of gases, and the atmosphere was prevented from becoming too dry by an earth floor, which could if necessary be sprayed with water. There would often be an outer room, where labelled examples of the varieties in season were laid out on a table so that the host and his guests could see what was available for the dessert that evening.

Dessert apples would usually be stored on slatted shelves around the walls of the store, while the culinary varieties would be bulked together in bins underneath. Mid-season apples, picked in September, would keep for several weeks in store, but these were kept apart from the late maturing varieties so as not to accelerate their ripening. The late keepers, picked throughout October, completed their ripening in store, and the gardener depended on his careful records, built up over the years, to chose the moment to serve each fruit at its prime.

It was not just the apple's gastronomic virtues that made it a prized dessert fruit. The magnificent display of flowers and fruit that adorned the dining table was always carefully integrated into a scheme appropriate to the occasion and the time of year, and apples – and at Easter, garlands of apple blossom – played an important role. An autumn dinner party for 20 called for a prominent arrangement at either end of the table which would often take the form of an apple pyramid. In addition, small dishes of perfect fruit would be set along the table for guests to sample. The warm, bright colours – which ranged through all the shades of gold and pink to scarlet and russet – gave the table a richly festive air and their burnished cheeks gleamed in the candlelight, defying the approach of winter. The apples might be further embellished with rose hips, the red bracts of poinsettias, or sprays of cotoneaster or pyracantha. Crimson vine leaves and the colourful autumn foliage of some pear varieties were also favoured additions. In the flower vases, bronze and yellow chrysanthemums would complete the scheme, with a garland of Virginia creeper laid on the cloth itself drawing the whole design together. Sometimes the apples would even be given a patterned surface. The skin of some varieties is very sensitive to light, so gardeners would mask off designs with paper as the fruit ripened on the tree.

Gardeners ensured there was a succession of apples and flowers to meet these decorative needs. In August, pelargoniums and the crimson Devonshire Quarrenden were ideal partners and the bright red Fearn's Pippin and new, even more brilliantly coloured Worcester Pearmain, would light up the autumn table. The tiny red Api apple appeared at Christmas and was always grown in a warm spot in the garden to give the skin a good colour. It would be used in the table decorations and also wired on to the branches of yew and box that made up the wreaths, garlands and kissing boughs which festooned the house itself. Small red apples would also be roasted and floated in the Wassail bowl of hot spiced ale or cider, which was offered to guests and visitors. In the new year vermilion-cheeked Court Pendu Plat enlivened the displays.

For occasions demanding particularly spectacular displays of fruit, gardeners would also employ large, well-coloured culinary apples. Lady Henniker, for example, which developed a good red flush was always useful for buffets during the fox-hunting season. Bold colour could also be added with the Nottinghamshire variety, Bess Pool, which turned a pretty rose pink in store, and Norfolk Beefing, which became strikingly flushed in scarlet over gold.

ENGLISH
COOKING
APPLES

A SUCCESSION OF GOOD COOKING APPLES was almost as important as the dessert selection. The popularity of fruit puddings continued unabated into the 19th century, despite the adoption of French *haute cuisine*, and apple pies, charlottes, tarts, snows and so on appeared in the final course from the kitchens, which preceded the grand finale of the fresh fruit dessert.

The perfect culinary apple, like the perfect dessert apple, had been defined by Loudon at the beginning of the century. It should be distinguished, he wrote, by 'the property of falling as it is technically termed, or forming in general a pulpy mass of equal consistency when baked or boiled, and by a large size'. Until this time, most kitchen apples had been regarded as multipurpose. The Forge apple of Sussex, for example, was 'the cottager's friend', used for making cider, cooking, and by Christmas, when it was sweet and slightly perfumed, for serving as an eating apple. Now the old varieties, such as Lemon Pippin and Winter Queening, which 18th-century cooks had favoured for pies, and Black Caps, were overtaken by a new range of large, sharp culinary apples that would never be eaten raw. With their high acidity they softened readily and were ideally suited to many apple dishes. The filling inside a dainty pastry dumpling would be juicy and perfectly cooked by the time the pastry was baked, and apple tarts could be filled with fresh chopped apples rather than a cooked purée. The gentlemen, too, approved of the brisk, smooth sauce they made for goose and roast pork. The succession of new varieties also made it possible to enjoy these refreshing qualities right through the apple season, rather than only at the beginning.

The Victorian fruit men aimed to provide the cook with as long a culinary apple season as possible, and with flavours that needed neither the embellishment of spices, nor the addition of lemon juice to give them interest and zest. For each new variety, the acid test was quite literally the quality of the unadorned baked apple. The first apples of the season were the summer codlins, such as Keswick Codlin. These cooked to a soft, juicy fluff, which rose up like a soufflé and were ideal for folding into stiffly beaten egg whites to make Apple Snow. Then came juicy Hawthornden, followed by Golden Noble or the large Warner's King and, finally, Dumelow's Seedling, which retained its briskness and flavour until the first rhubarb or even the first gooseberries were ready. The Rev. A. Headley, who contributed to the *Journal of Horticulture* under the pen name of the Wiltshire Rector, recorded on 5 May, 1882, that he had 'a pudding of Dumelow's Seedling, and in spite of a warm November and a not sufficiently cold fruit room, these apples cooked with the true acid flavour as if it were January and not May'. Queen Victoria, who was especially partial to baked apples, was said to prefer Dumelow's Seedling above all others. It was also particularly prized as a tart, sauce apple, a use for which Ecklinville Seedling and Warner's King were considered similarly excellent. Herefordshire proudly claimed a Queen of the Sauce apple, and the north of England had its own Yorkshire Goosesauce.

For an apple pie, the requirement was for a sharp, savoury apple that cooked well, but did not collapse into a thin purée, leaving a great void between lid and filling. By the end of the century, there were a number of varieties which fitted the bill, including Bramley's Seedling, which like Dumelow's Seedling was valued for its excellent keeping qualities, and Golden Noble. This beautiful, large, round yellow apple was the one the connoisseur Edward Bunyard considered best of all.

> *The best English apples by long training know how to behave in a pie; they melt but do not squelch; they inform but do not predominate. The early apples, grateful as we are for their reappearance, are not true pie-makers. Early Victoria, Grenadier, Ecklinville, all cook to a white froth, which though permissible in a dumpling is not meet in a pie. We pardon these adolescents, who do the best they can, but we pass on to the later autumn apples to find pie manners at their best. And what should an apple do in a pie? Well I think it should preserve its individuality and form, not go to a pale, mealy squash, but become soft and golden. In flavour it must be sharp or what's the use of your Barbados sugar? It should have some distinct flavour of its own, not merely a general apple flavour.*
>
> *Fulfilling all the conditions comes Golden Noble, golden before and after cooking, transparent in the pie, and in every way delectable. This is a September–October apple; and after this we begin our Bramley's, and can have them until May in most years . . .*
>
> *At the end of the season comes Wellington [Dumelow's Seedling], a fruit which has found a way to keep its acidity till June and has a crisp transparency quite ideal for pies.*[6]

'Apples à la Portugaise'

If a firmer purée was required, or it was necessary for the cooked fruit to retain its shape, then dual-purpose varieties were preferred. For moulded dishes such as Apple Charlotte, cooks asked for the old Golden Reinette, Blenheim Orange, or Mère de Ménage, all of which would make the necessary stiff 'marmalade'. If there was a French chef in the kitchens he would demand his native Calville Blanc d'Hiver, which when cooked retains its shape and has a rich taste, making it especially suitable for open French tarts. Golden Pippin was still used and recommended by the renowned chef

Francatelli for his pyramid of 'Apples and Rice'. The little apples were peeled, cored, poached whole and arranged in tiers on top of a mound of cooked rice, which was heaped in a pastry shell. A preserved cherry and a sliver of angelica was placed in the centre of each apple. In 'Apples à la Portugaise' a pastry case was first filled with a stiff 'marmalade' of apples and then a dome of sliced Golden Pippins would be piled on top. This was then covered with meringue, and after cooking the frosty edifice was decorated with stripes of redcurrant and apple jelly 'arranged to show their colours distinctly which will produce a very pretty effect'. When Golden Pippins became scarce towards the end of the century, one of its seedlings, such as Yellow Ingestrie, made a good substitute.

'Apples and Rice'

WHILE APPLES HELD THEIR OWN on the dining table during the middle years of the 19th century when the craze for hothouse fruits was at its height, the late 1870s saw them rise to new prominence as the supreme English dessert fruit. This was partly a reflection of the fact that exotic fruit had begun to lose the social cachet which had initially made it so attractive. Imported pineapples, whose quality equalled the gardeners' best, had started to come in from the Azores, while excellent grapes could now be cultivated by the commercial glasshouse industry. Now that these luxuries could be bought in the market – albeit at a price – it ceased to be socially and economically worthwhile to maintain acres of glass pavilions of fruit. In addition, a more widespread change in taste was creating a climate in which apples appeared far more seductive than forced strawberries or glasshouse peaches and figs. The relentless march of technology and the changes it brought in its wake were making people nostalgic for the perceived pleasures and skills of a more rural England. One result was the appearance of the Arts and Crafts movement; another was an increase in the gastronomic attention paid to regional and traditional specialities. The combination of wholesome simplicity with epicurean diversity, not found in any other English fruit, now brought apples a level of esteem greater than anything that had gone before.

It was a perception which by now extended far beyond the privileged confines of the aristocratic dining table, and led to a paradox which itself added further emotional piquancy to the apple's position. The second half of the century had seen a rapid expansion in the numbers of more or less affluent, middle-class families who, while they could not – or chose not to – aspire to a country estate and all it entailed, nevertheless aspired towards the quality of life it represented. They made up the growing ranks of customers for whom the commercial markets strove to provide everything from hothouse flowers and grapes to high-quality apples. Vying for their custom, however, were the fruit importers, and although the top-quality market gardener's fruit held its own, the bulk of the English table apple crop was passed over in favour of the bushels of cheap, attractive French, Canadian and 'Yankie' apples which arrived on the quayside and swiftly appeared on sale. Imports from across the Atlantic quadrupled between 1875 and 1879, and the gardening journals of the 1880s began to report that 'go where you will

Plate 16: Cox's Orange Pippin

American apples are in the fruit shop windows . . . The unattractive looking crab-like produce of the home orchards driven out by the cherry bright, clear skins and good looking proportions of the red Baldwins and Northern Spy', not to mention the esteemed Newtown Pippins.

In the face of what seemed like a tidal wave of imports, enthusiasm for English apples acquired the peculiar intensity that arises from a sense of imminent loss. Every corner of the English fruit-growing world shared in the concern. Estate owners might still be able to satisfy their own needs, but as enthusiasts, patriots and farmers they were as concerned as anyone over the threat to their fruit industry. Furthermore, British agriculture had gone into severe depression in the late 1870s, and as in the 17th century, it was to fruit – and apples, in particular – that farmers now looked for salvation.

Connoisseurs, enthusiasts, landowners, head gardeners, farmers, nurserymen and market gardeners all rallied to mount a veritable 'fruit crusade' to modernise old orchards, guide the new fruit growers, beat the 'Yankies' and persuade fruiterers and their customers to buy English apples before it was too late. The crusade opened with the National Apple Congress of 1883 held in the Great Vinery of the Chiswick gardens of the Royal Horticultural Society. The aim was to resolve the confusion of identities and select a good range of apples for growers to concentrate on in terms of appearance, yield and reliability; but it was also a triumphant display of the diversity of apples growing throughout the country. There were 10,150 separate dishes of apples, grouped by county, and representing 236 different exhibitors and over 1500 varieties. When it opened on Thursday, 4 October, the *Journal of Horticulture* reported that the Great Vinery was

> *a spectacle of remarkable beauty, the roof being draped with vines – the abundant pendant golden and black clusters having a charming effect – while beneath are broad tables of brilliantly and diversely tinted Apples, a few graceful plants in the centre of the stands added much to the general beauty.*

It was 'visited every day by about 500 persons from all parts of the country' and had to be kept open for an extra week to accommodate all those who wanted to see it.[7]

In 1890 the Worshipful Company of Fruiterers staged the first of a series of Shows, at the Guildhall, in the heart of the City, which were designed to show 'the citizens of London that the most important hardy fruits can be grown at home as well as abroad'. Although it was a bad year for apples, nearly 5000 dishes were exhibited and some 35,000 people attended.[8] That year, too, the Fruiterers' Master, Sir James Whitehead, insisted that the fruit they presented to the Lord Mayor – a custom which dated back to the time of Queen Elizabeth – should be all home-grown. In 1895, the Royal Horticultural Society held the first of its annual October exhibitions of 'British Grown Fruits' at the Crystal Palace. Here, head gardeners showed just what could be achieved in the much maligned British climate, and editorial writers 'questioned if a finer display of noble fruit had ever been seen in the world.'

The main display in the Great Vinery at the National Apple Congress of 1883. The Vinery housed exhibits from the Home Counties.

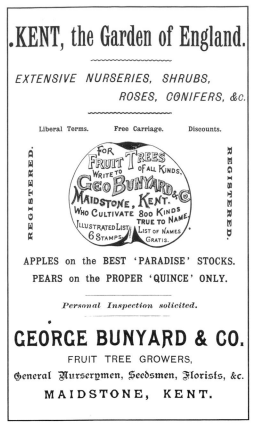

By the end of the 19th century Bunyard's had become one of Britain's leading fruit nurseries. Now all that remains is a sign at Barming Station: 'Alight here for Bunyard's'.

In the last 20 years of the century the English apple was applauded and promoted as never before. In 1896, a splendid exhibit staged for the Royal Horticultural Society show by nurseryman, George Bunyard of Maidstone, featured more than a hundred hardy fruit varieties and had at its centre

> *a square pyramid of finely coloured apples, upon the top of which rested a large crown composed of ornamental crabs and berries with green leaves for a background. The table itself was laden with dishes of apples, pears, grapes, nuts, medlars, plums etc. Some of the apples were arranged in cones, from the centre of which appeared a Croton or Dracaena.*[9]

Estate owners were now said to prefer a visit to the fruit room to a walk around the flower garden, and the appreciation of apples became a serious topic of conversation. Morton Shand, the wine and food writer of the 1920s and '30s, recalled from his childhood the lengthy after-dinner debates on the subject which took place at his home in Kensington. His father was 'a man of fierce Ruskinian ideas' who despised 'style' and considered apples *the* English fruit. Grapes, in his opinion, were mere concessions to the ladies – who did not worry their heads about serious masculine matters such as the quality of the claret or the Blenheims – and he ate apples every day of the year, 'except during the short fretful interval of the red fruits of summer'.[10]

'Ribston and Blenheim were the clarion names' Shand junior recalls, and these 'were seldom uttered without a just perceptible trace of religious awe . . . Old friends quarrelled perennially over which was the nobler, just as they did over whether Château Latour or Château Lafite was the supreme expression of pre-phylloxera Bordeaux.' The fruit was peeled with special bone-bladed knives, so as not to taint the flesh and compromise a delicate flavour. There was high praise for the good old sorts like the 'glorious little Margil'; and the new Cox's Orange Pippin – firmly established by the 1890s – was continually tested against the Ribston. Shand senior, while considering the Cox one of the finest 20 or so eating apples, nevertheless found it

> *a shade too sweet and rather floridly luscious and so lacks the austere aristocratic refinement that Ribston exemplifies transcendentally . . . beautifully soft but somehow insufficiently rounded and hence not quite perfectly balanced.*

Even then, dedicated apple men such as these were mourning the disappearance of varieties such as 'the true old Golden Pippin' they remembered from their youth. Dismissed as too small to be worth growing any more, this had now become something of a rarity; and Shand recalled one family friend having hampers sent up from different parts of the country each autumn in the hope of tracking one down. He and his pomological companions held solemn tasting sessions 'in committee', but rarely encountered the remembered 'tang' of the childhood Golden Pippin.

I F THE APPLE SEEMED TO BRING the values of an older, gentler age to the dining table it fulfilled
a similar role in the garden. Just as 'Capability' Brown's landscape parks had swept away formal
gardens in the 18th century, so the mid-Victorian period had seen parkland replaced by bright
geometric flower gardens, lush shrubberies, mysterious ferneries and collections of everything
from alpines to conifers. By the end of the century, reaction had again set in, and instead of the
brilliant foreign flowers, French-style parterres and Italianate terracing, architects and designers
wanted a return to what they imagined were the 'true' English motifs of Tudor, Elizabethan and
Stuart gardens. In this they were strongly influenced by Francis Bacon's description of his perfect
garden, first published in 1597. The one-time Lord Chancellor and philosopher evoked a
timeless idyll of flowers and fruit trees, in which, for example, July delighted the senses with 'Gilli-
Flowers of all Varieties, Musk-Roses, and the Lime-Tree in blossom, Early Pears and Plumbs in
Fruit, Ginnitings and Quodlings'. The orchard at Penshurst, the Kentish home of Queen
Elizabeth's favourite, Sir Philip Sydney, was considered still to retain the elements of such a scene
and it was a model for the Arts and Crafts gardens. Enclosed by yew hedges, and set to one side
of the formal flower garden, the orchard was divided diagonally by grass paths. On either side of
a central avenue stood a double row of standard apple trees, each of which, a visitor noted in 1881,
was garlanded by a 'stout hop plant amongst which are distributed
vines and clematis; these climb up the stems and through branches
and form natural festoons of varied colours over the walk.' When the
hops and clematis were in bloom this presented 'a most striking and

*This late Victorian illustration reflects the deep vein
of rural nostalgia which helped give apples new
popularity and significance as the 19th century came
to a close.*

BALCASKIE
Birds-eye View

L·ROME·GUTHRIE·

beautiful arrangement especially on either side of a wide border filled with old-fashioned herbaceous and other plants and roses.'[11]

Once again the apple became an indispensable part of the pleasure ground. 'Trained as an espalier [it] makes a beautiful hedge' wrote the garden architect Reginald Blomfield, and 'set as in an orchard it lets the sun play through its leaves and chequer with gold the green velvet of the grass in a way that no other tree will quite allow.' Apple arches, pergolas and tunnels were planted in gardens all over Britain including Frogmore, the Royal supply centre at Windsor; Madresfield Court in Worcestershire; and at Overstrand Hall, a Lutyens house in Norfolk. At Tyninghame in East Lothian, Keswick Codlins were trained over a series of arches to create a blossom walk 140 yards long.

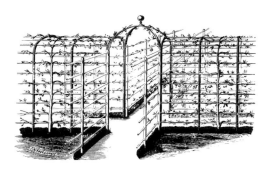

(Above) *Orchards, trained fruit trees and walled enclosures became the height of fashion in the Edwardian pleasure ground as well as in the kitchen garden. Areas would be linked by elaborate tunnels of trained trees and walls bordered by tall espaliers of fruit.*
(Opposite) *When the pleasure ground at Balcaskie in Scotland was redesigned at the end of the 19th century, the aim was to include all these features in a recreation of a 17th-century garden.*

At a more informal level, an appealing combination of beauty and utility could be created by planting a copse of apples in a woodland clearing, or creating an 'artistic' orchard. This idea was inspired by another Elizabethan model, that of William Lawson's *New Orchard and Garden*. 'What can your eye desire to see, your eares to heare, your mouth to taste, or your nose to smell, that is not to be had in an orchard?' he asked. 'What more delightsome than an infinite varietie of sweet smelling flowers? . . . The Rose, . . . faire and sweet scenting Woodbind . . . Cowslips . . . Primrose . . . the Violet . . . And 1000 more will provoke your content.' With broad walks, seats and 'banks of Camamile' the contented owner could view with delight 'fruit trees of all sorts, loaden with sweet blossomes, and fruit of all tastes, operations and colours.' He was recommended to keep a 'Store of Bees' in the orchard and welcome nightingales to 'beare you company night and day . . . and help you to cleanse your trees of caterpillars, and all noysome worms and flyes.'[12] Such an artistic orchard was sought after among others by Miss Gertrude Jekyll who was one of the leading advocates of the Arts and Crafts garden. She encouraged roses to climb up the branches of vigorous apple trees, and naturalised snowdrops, crocuses and narcissi in the grass beneath.

Apple trees also began to appear in the planting repertoire of more formal gardens. The Wiltshire Rector warmly commended an espalier he had seen in Hampshire in 1876. The variety was Emperor Alexander, bearing bright red fruit 'as attractive to the eye as a Paeony in a shrubbery'. According to another clergyman, the Reverend Wilks, secretary of the Royal Horticultural Society, 'a half standard apple with the lower branches encouraged at first so as to sweep down upon the turf is quite as pretty as nine-tenths of the lawn trees one sees'. A prettier spring picture could not be imagined, wrote head gardener Hugh Pettigrew in 1904, than

> a group of pyramidal fruit trees growing in a large shrubbery, with a foreground of the dwarfer evergreens and a background of dark green conifers, with here and there clumps of sloe.

In the autumn, a small orchard of fruit trees made the perfect foil for a herbaceous border:

> the burden of Apples and Plums makes a far from uninteresting background to the broad band of Phloxes, Penstemons, Dahlias, Sunflowers and Michaelmas Daisies below them; in fact the whole forms a natural trophy sufficient to inspire a poet with the theme of Flora half reclining on the turf and Pomona stooping down to kiss her sister goddess on the brow.[13]

With the dual roles of beauty and utility in mind, nurserymen and gardeners began to emphasise the artistic merits of different varieties, in addition to the quality of their fruits. Golden Pippin had 'rich deep pink blossom and a pendulous habit'. Golden Spire was valued for its willowy leaves and appearance, as well as for the golden orb-like fruits that never seemed to fall. More striking was the new Worcester Pearmain, with its scarlet fruit and distinct blossom: 'a delicate silvery white with faint pink veins; salmon pink and yellow on the underside'. Some of the grandest spring displays were to be found on the prolific cookers. Keswick Codlin was strongly recommended together with Lord Suffield, which has immense pale pink flowers. Stirling Castle with 'handsome, long lasting flowers which contrasted beautifully with the bronzy young foliage', and the velvet-like buds of Sandringham also received attention. Vigorous varieties, which made large spreading trees, such as the white blossomed Warner's King, were ideal for planting in ornamental orchards and made a suitable host for a rose. Here also was the place for the deep cerise flowers of Bess Pool and Annie Elizabeth.

IT WAS, HOWEVER, A CRUEL IRONY that even as the English apple enjoyed its heyday, the qualities which made it so special – namely its range and diversity – were under threat from the very quarter that had done so much to highlight them. By the end of the century, not only the United States and Canada, but also Australia and New Zealand and, very soon, South Africa, were sending a large percentage of their export crop to Britain. The English growers' strategy, reasonably enough, was to concentrate their efforts on a handful of good commercial varieties, which was their competitors' strong suit. In this move towards specialisation, the range of apple varieties on sale was inevitably reduced. This trend was reinforced by the fact that the diverse orchards of the market gardener were being swept away by urban development; commercial fruit growing was increasingly passing into the hands of large specialist farmers; and the world of private estates and gardens had also begun to disintegrate as agricultural fortunes fell. On the one hand, many fine flavoured varieties, common enough in the 1890s, ceased to be grown, because their poor or irregular crops, tendency to bruising or disease, or other shortcomings failed to meet the needs of market growers; while on the other, many old and regional specialties simply disappeared from the nurserymen's catalogues as private fruit collections were cut back or abandoned. In addition, increasing social mobility, the rise of restaurant dining at exclusive establishments such as the Ritz and the Carlton, and the increasing popularity of wintering abroad had undermined many of the functions of the country house, including the significance of the fresh fruit dessert. As a result, this became an increasingly modest affair – a fact mirrored in the modification of the term 'dessert' to include the preceding pudding course as well.

As a result, the first half of the 20th century witnessed what, until the recent revival of interest, might have been the swansong of the English apple. That this aspect of the British national heritage could be recovered was due in large part to three men who kept alive the Victorian love affair with the apple long after it had gone into eclipse. They were the nurseryman and connoisseur Edward Bunyard; his protégé J. M. S. Potter, the Royal Horticultural Society's Fruit Officer; and the scholar and linguist Morton Shand.

Bunyard was the eldest son of the Victorian nurseryman George Bunyard, and on his father's death in 1919 he took over as head of the firm. The nursery was famed for its fruit and Edward, like his father, was especially skilled in the science of systematic pomology. He was also a keen student of fruit literature and a food and wine enthusiast. He wrote both for amateurs and professionals on every aspect of fruit and in *The Anatomy of Dessert*, which appeared in 1929, he used his evocative prose to bring alive the disappearing riches and bemoan the sad state of affairs.

How often, after a dinner ordered with intelligence, prepared with art and served with discretion, do we dwindle to a dessert unworthy of its setting? Who has not encountered the [imported American] Jonathan Apple or the Jamaican Banana at a table which would scorn to provide an unacknowledged St Julien or an invalid Port?

He addresses the qualities and range of varieties across the entire spectrum of dessert fruit, but by far the largest chapter is devoted to apples. He pointed out the underlying factors which characterise a distinguished variety. All had their optimum moment of aroma, sweetness and acidity and the aim was 'to catch the volatile ethers at their maximum development, and the acids and sugars at their most grateful balance'. It is 'an incontinent friend [who] eats his Ribston direct from the tree when it should be tested frequently after gathering and will generally be found to be used before Cox comes in November.' He divided the varieties into groups according to flavour and texture and guides his readers through the finest of each season in lyrical prose. After the strawberry flavour of the early Worcester Pearmain, for example, comes the 'melting, almost marrowy flesh, abundant juice, and fragrant aroma' of James Grieve, aniseed-scented Ellison's Orange and the 'very attar of apple' in Gravenstein. In November, Cox was for him the Château Yquem of apples, while in December it was the 'nutty warm aroma' of Blenheim Orange

which is to my taste the real apple gust. The man who cannot appreciate a Blenheim has not come to years of gustatory discretion; he probably drinks sparkling Muscatelle. There is in this noble fruit a mellow austerity as of a great Port in its prime, a reminder of those placid Oxford meadows which gave it birth in the shadow of the great house of Blenheim. Like Oxford, too, it adopts a leisurely pace, refusing to be hurried to maturity or to relinquish its hold on life. An apple of the Augustan Age.

After Christmas came Rosemary Russet, 'an aristocrat in every way', and Claygate Pearmain which was 'fully deserving of a place in the best dozen dessert apples'. Even in March, if properly stored, home-grown English apples would still show 'all the plump turgescence of youth', when King's Acre Pippin would be 'rich and juicy'. Last of all came Allen's Everlasting, which he considered shared the 'dessert honours with Early Rivers Cherry' at the end of June. English culinary apples received a similar analysis in his *Epicure's Companion*, produced in collaboration with his sister Lorna and illustrated by another sister, Frances.

Bunyard also promoted apples through his contacts in the world of London's literary and dining circles, which were associated with the gastronome and writer André Simon. At meetings of the Wine and Food Society and the Saintsbury Club, members would often be treated to a fine selection of apples, which Bunyard would bring up from his Allington nursery for them to sample. He also took advantage of the air waves, joining the Radio Gardener, Mr C. H. Middleton, to promote fine apples, and among the talks he broadcast was 'Apples for the

Epicure'. The varieties he championed included the tiny, yet delicious Pitmaston Pine Apple and Orleans Reinette, which was undoubtedly his favourite. 'Its brown-red flush and glowing gold do very easily suggest that if Rembrandt had painted a fruit piece he would have chosen this apple', he observed, adding that it combined the flavours of Ribston and Blenheim, with 'an admirable balance of sweetness and acidity . . . and for those who incline to the "dry" in food or drink Orleans Reinette is an apple meet for their purpose, rich and mellow, and as a background for an old port it stands solitary and unapproachable.' On the culinary front he singled out the much neglected Thomas Rivers apple, which is almost unique in retaining its aromatic quality even after cooking, and possessed 'a distinct pear flavour with an almost quince like acidity'. So inspired was the popular novelist and playwright, Eden Phillpotts, that he sat down and composed a series of poems entitled *A Dish of Apples*. Illustrated by Arthur Rackham it was a celebration of some of Bunyard's finest offerings, including the Ribston Pippin.

Arthur Rackham's illustrations of c1930 for Eden Phillpotts' poems place apples in a world of enchantment and escape.

A mbrosia, nectar, both; and at his best
A A palimpsest;
For, through the abundance of his native wealth,
Rise magic dreams of eastern fruit by stealth.
Oh more than apple: an elixir too;
Who would not woo
The incomparable mystery he stores
From orient garths and spicy-scented shores?

For all it helped to keep the tradition of apple connoisseurship alive in sympathetic circles, Bunyard's evangelism could not arrest the slide away from diversity as commercial growers sought the holy grail of a few 'perfect' market apples. Fortunately, however, Bunyard was also involved with the setting up in 1922 of the Commercial Fruit Trials at the Royal Horticultural Society's gardens at Wisley in Surrey. The aim was to select out the most promising varieties for the fruit industry, and the trials were jointly sponsored by the Society and the Ministry of Agriculture. Bunyard's presence on the panel ensured that the connoisseurs' interests were not entirely neglected. As a member, and from 1930 also chairman, of the RHS Fruit and Vegetable Committee, he played a key role in helping to assemble the reference fruit collections for the trials. His contacts, knowledge and skills in identifying old varieties ensured that many were saved from oblivion. His partner in this enterprise was J. M. S. Potter who became the society's fruit officer in charge of the trials in 1936.

As soon as Potter began to gather in varieties, he could see that the problem of mistaken identities still persisted, and what was required was a comprehensive reference collection. Fired by the collector's acquisitive passion, he tried to track down as many as he could of the varieties Dr Hogg had described in his *Fruit Manual* of 1884, which included not only the finest for the dining table but many regional apples. As Fruit Committee secretary, Potter had access to all the specimens sent in to the society's shows, which gave him an unrivalled opportunity to spot and secure for the collection any variety not already in the Wisley orchards. As many of the exhibitors were those nurserymen and private gardeners who still maintained

PLATE 17: BRAMLEY'S SEEDLING

significant fruit collections, the potential for adding new varieties was considerable. Fred Streeter, head gardener at Petworth House in Sussex, supplied a number of Sussex apples, while nurserymen Edward Laxton in Bedford, John Allgrove at Langley, near Slough, and Joseph Cheal of Crawley were further sources of old local apples, as were the fruit breeders H. M. Tydeman at East Malling, Kent, G. T. Spinks at Long Ashton, Bristol, and M. B. Crane at John Innes, South London.

It was also the fruit officer's task to identify the fruit sent in every autumn by Fellows, curious as to what they had growing in old orchards. Always alert for the possibility that a 'missing' variety might turn up in one of these parcels, Potter would request some scion wood of any likely candidates, propagate it and hope that when it fruited it did indeed prove to be a new addition. When the Second World War encouraged everyone to produce their own food, many more people wanted to identify the fruit trees in their gardens and the number of 'unknown' fruits and potential finds increased enormously. Potter also brought in for trial new varieties raised abroad, as well as by English breeders, the best of which were added to the collection.[14]

Bunyard himself died in 1939, but his epicurean baton was taken up by the members of the society's Fruit Group, which was founded in 1945 with [Sir] Ronald Hatton, director of the East Mailing Research Station, as its first chairman. One of their aims was to re-examine good, old varieties and among its first members was Morton Shand, the author of one of the most respected books in English on French wines, and an ambassador for modern European architecture. Shand was passionate about the unique qualities of English apples:

> *Our unequalled* aigro-dolce *dessert apples that crunch so crisply . . . are worth all those flaccid-fleshed pumpkins like the Reinette de Canada, that the French rave over.*

He commented in his *Book of Food* in 1927:

> *The French have poor taste in apples, liking them merely soft and sweet. After all, no nation can have an impeccable taste in everything, and the inevitable Achille's heel is just as likely to be found in apples as in drawing room furniture.*

Like Bunyard, he was deeply concerned that the diversity of apples should not disappear. 'Who on earth with a spark of individuality in his make-up wants to eat the self same apple every blessed day in the year', he observed. He spent his spare time during the war years with the Admiralty in Bath tracking down little-known varieties. In a talk broadcast on BBC Radio in 1944 he asked:

> *What would the future be without the sort of apples that are part of the very stuff of the English countryside and its cherished local traditions? The apples men of Norfolk and Devon used to swear by, or that were honoured names to generations of Herefordshire, Somerset, Kent and Sussex yeomen farmers.*

He beguiled his listeners with descriptions of apples such as Ashmead's Kernel, which they might never taste if something was not done to keep them alive in people's minds and gardens.

> *What an apple, what suavity of aroma. Its initial Madeira-like mellowness of flavour overlies a deeper honeyed nuttiness, crisply sweet not sugar sweet, but the succulence of a well-devilled marrow bone. Surely no apple of greater distinction or more perfect balance can ever have been raised anywhere on earth.*

Fruit and information came pouring in from the public. To identify the apples, Shand enlisted the help of Harry Lock, a well-known West Country judge of fruit, who was semi-retired from Long Ashton Research Station, and they were soon joined in the search for 'disappearing and apparently doomed apples' by [Sir] Leslie Martin, the architect, a Miss Holliday who lived in Yorkshire, and the composer and musician Gerald Finzi. He had been so inspired by Shand's broadcast that he had written to offer his help immediately, reputedly throwing himself into the task of collecting old apples with the 'enthusiasm of a Vaughan Williams collecting folksongs'. Whenever possible, scions were obtained of interesting finds and members of the group set about building up their own orchards of forgotten varieties. Finzi's orchard at Ashmansworth in Berkshire increased to about 400 varieties as he propagated the group's discoveries.

As Shand's campaign gathered momentum advertisements were placed in local newspapers and further helpers enrolled. Three years later they had rediscovered nearly 100 'lost' apples, and after a further broadcast in 1947 several hundred more were tracked down. Stoke Edith Pippin, which arose at the Herefordshire mansion of that name, was rescued from obscurity, and the variety that honoured Dr Hogg was tracked down along with Hunt's Duke of Gloucester and Minshull Crab. Red Joanetting or the Margaret apple of the 17th century was found, as well as the richly flavoured Lamb's Abbey Pearmain and Cellini, the London Market gardeners' apple.

Shand's finds proved to be another valuable source for Potter's collection at Wisley. Shand recalled taking 'wood that was frequently not merely dishearteningly unpromising but to all appearances utterly sere and shrivelled' along to Potter, who with a 'wizard's sleight of hand' grafted and propagated new trees. Shand contributed hundreds of new additions altogether, which included regional apples from all over France, and from Australia, Scandinavia, Bulgaria, Hungary, Romania, Italy and Switzerland, and many more were received through his network of friends and contacts.[15] Many of these were esteemed national varieties, which not only enriched the collection considerably, but also served to aid the resolution of the many synonyms which were then in existence. The Bénédictin of Normandy, for instance, proved to be identical with Blenheim Orange.

In 1946, the trials and its fruit collection had become wholly the responsibility of the Ministry of Agriculture and, as the National Fruit Trials, began the move to its present home at Brogdale, near Faversham in Kent in 1952, with Potter as its first director. Appropriately, the new horticultural station lay not only in the centre of the East Kent fruit belt, but close to the village of Teynham, where Henry VIII's fruiterer, Richard Harris, planted the first 'pippins' and other fruits for English growers to try out. It took eight years for the entire collection to be propagated and moved to Brogdale, which became the centre for the trialling of new varieties in northern Europe. As the collections continued to expand through the addition of new varieties from plant breeders and research institutes throughout the world, its international importance as a reference collection increased, and in 1965 it took on responsibility for the testing of new varieties for the award of Plant Variety Rights. Many of the breeders, including W. T. Macoun in Canada, R. Wellington in New York and their successors, and the New Zealander, D. W. McKenzie, sent in local as well as new varieties, and from J. D. G. Lamb came a whole collection of Irish apples. In 1972, Potter's colleague, Muriel Smith, published the *Apple Register*, which was the most comprehensive international directory of apple varieties ever compiled, and finally brought order to the confusion of synonyms.

The National Apple Collection now contains over 2000 varieties and includes almost every apple that has ever been esteemed anywhere in the world. It continues to expand, albeit more gently. It is a gene bank for the future, a source of propagating material for rare varieties, a reference collection for identification purposes, and since 1991 has also been open to the public. Since the closure of the National Fruit Trials in 1989, the Ministry of Agriculture, now the Department for the Environment, Food and Rural Affairs (DEFRA), has retained the Collections and continues to fund their maintenance and development, but their home at Brogdale is now provided by a charitable trust, which has opened up the Collections to the public. Visitors are now guided through these remarkable orchards and can see, buy and sample the apples that were eaten by Renaissance princes, Tudor Kings and Victorian connoisseurs; sustained generations of New England, French, Russian and Antipodean farmers; and made pies and compôtes everywhere from Canada, Scandinavia and Germany, to Turkey and Japan.

Chapter 5
APPLES for the MANY

*A*n *apple a day keeps the Doctor away.*

J. T. STINSON: *Address to the St Louis Exposition,*
Missouri, 1904

T THE END OF THE 18TH CENTURY commercial apple growing was hovering
between the old world and the new. Towns and cities were growing fast
together with the network of roads and canals which linked them, and like
every other aspect of life, the organisation of food production and distribution
was changing. In particular, road improvements meant that areas which had
previously been too far from major towns to take advantage of their markets could now be drawn
into the network of supply. The market gardener, Henry Scott of Weybridge in Surrey, advertised
that he could now supply customers by 'the Chertsey coach which goes every day to London', and
by the 1790s the metropolis and its affluent suburbs could buy table fruits grown as far away as
Hertfordshire, Essex and even Worcestershire.[1] Perishable goods were not usually transported by
canal, but the new network of inland waterways had nevertheless created the opportunity for
Worcestershire also to send barrels of late apples up to the Potteries in Staffordshire, the cotton
towns of Lancashire and to the new industrial centre of Manchester. Demand from these growing
industrial areas also stimulated Kent fruit farmers of the 1790s to send cargoes of apples back on
the returning coal ships, to be off-loaded in East Anglia, the North East and even Scotland.

Forty years later, the completion of the railway line from London to Dover brought the
orchard villages south of Maidstone to within a few miles of a station and the markets of the City.
Soon afterwards new railway connections brought not only the whole of Kent into the railway
network, but encouraged many farmers in the West Midlands who grew apples only for cider to
turn over to fresh fruit, and enabled East Anglia to start commercial fruit growing for the first time.

At the same time the changes taking place in sea transport were beginning to affect British
fruit growing in a different way. As it became possible first for North America, and later for the
countries of the Southern Hemisphere, to send barrel upon barrel of fine apples to England, the
home industry was to face competition on an unprecedented scale. For a variety of reasons,
however, this was a challenge it was ill-equipped to meet.

In Europe, waterways were widely used for transporting fruit and to connect with sea-going vessels bound for Britain and elsewhere. Here, cargoes of apples and pears from Bohemia are being discharged in Berlin.

D ESPITE THE OPPORTUNITIES OFFERED by the new markets, the expansion of English commercial apple growing in the late 18th and early 19th centuries was an erratic affair. One reason for this was that apples were not grown by specialist producers, but remained part of a traditional pattern of mixed farming and market gardening. Historically, farmers only took a keen interest in fruit when grain and livestock prices fell, and since the 1760s, when profits from mainstream agriculture had begun to rise again, farmers had paid less attention to their orchards. Table fruit, in any case, had never promised reliable profits, as English apples always had to contend with French competition. The warmer Continental climate produced earlier and heavier crops, which often took the best prices in the markets. For tenant farmers the long-term investment necessary to plant or renew orchards further discouraged fruit growing, and the fact that beer had finally triumphed over cider for first place in the Englishman's tankard also had an effect. In Kent, Herefordshire and Worcestershire many farmers deserted apples for hops. In 1778, it was said that Rainham, near Sittingbourne in Kent, had 'within living memory great plantations of Cherries and Apples but the greater part of them had been displanted some years since.' In the neighbouring village of Newington, orchards had also fallen into decay: 'the price of hops making them a more advantageous commodity than fruit, most of the orchards in the parish were displanted and hops raised in their stead'.[2] Even market gardeners who had a stronger tradition in quality fruit, tended to regard apples mainly as a cushion against disaster in other areas, and increasingly turned towards high-value crops such as asparagus and, later, hot house fruit and flowers, for the best return on space.

The problems were compounded by the fact that partly as a result of this ambivalent attitude towards apple growing, the producers did not constitute an organised lobby, and there was no central body solely devoted to the needs of the commercial men. The Horticultural Society, which was set up in 1804, aimed to serve both the market producer and private gardener, but its domination by representatives of the country house 'school' of fruit growing meant that it never really came to grips with commercial reality. The society's early work in collecting, identifying and trialling varieties, encouraging fruit breeding, and by drawing attention to good varieties undoubtedly laid the foundations of Victorian commercial fruit growing, but it did little to help growers grapple with the more immediate problems.

Apart from the period of the Napoleonic wars, when French apples were often blockaded and home-grown fruit enjoyed a monopoly, it seemed at times as if everything conspired against English apple producers. In 1838 the implementation of Britain's free trade policy led to the lowering of the duty on imported apples from three shillings a bushel to a purely nominal sum, and with cheap French apples undercutting the English producers at every turn, it was felt that in the 1840s 'orcharding and fruit tree management were at their lowest ebb. The orchards of Kent which had for centuries furnished the supplies of fruit to the great metropolis had become gnarled, cankered and unproductive.'[3] A decade later the Horticultural Society itself had started to ail and, no longer able to undertake the trialling of new varieties, cut back its orchards. The new interest in exotics saw market gardeners and private owners alike turn away from hardy fruits. It was small wonder that with the growing populations of the towns hungry for apples, imported American and Empire fruit found little opposition in the British markets.

THERE COULD NOT HAVE BEEN GREATER CONTRAST between the apple industry of early Victorian England and that of America. On the other side of the Atlantic, growers had unlimited space, a pioneering commitment to apple orchards and geographical advantages which they exploited to the full. They were also single-minded in their pursuit of good commercial varieties. The days of seedling orchards were over, and it was on carefully chosen grafted trees supplied by enterprising nurserymen that the new industry was founded. Serious apple growing west of the Allegheny Mountains had begun after 1796 when Israel and Aaron Putnam, sons of George Washington's second in command in the Revolutionary war, brought 23 varieties of apple scions from their uncle's orchards in Connecticut, and set up a nursery near Marietta in Ohio. The first orchards of grafted trees in the south central states were those of Captain James Stark, who settled in Bourbon County, Kentucky, in 1785 and scions from these trees were taken by his son, James, to Missouri where, in 1815, he founded the famous Stark Nursery, which still exists. In 1847, Iowa's first nurseryman, Henderson Lewelling, set out with his family on the Oregon Trail with 700 grafted fruit trees, and the vision of a commercial apple industry in the west.

PLATE 18: REINETTE ROUGE ÉTOILÉE

Despite Indians, the death of his partner and two oxen, and a water shortage which finished off half the trees, Lewelling and his party made it to the Willamette Valley in Oregon, where with another Iowan, William Meek, they set up the first nursery in the west. A few years later, they had grafted no less than 20,000 apple trees which Seth, Henderson's brother, took to Sacramento and sold for five dollars each. Meek followed with more trees, and in 1854 Henderson Lewelling transferred his nursery from Oregon to California.[4]

In Canada, Empire loyalists who had fled New England and Pennsylvania at Independence set about revitalising the old French orchards along the Saint Lawrence, and founded new fruit areas in Nova Scotia, New Brunswick, Quebec and most notably, Ontario. Quakers from Pennsylvania who settled on the northern shores of Lake Ontario planted orchards as soon as they arrived, and fruit growing spread out so as to stretch from Kingston to Toronto, around to Hamilton, and into the Niagara district.

These pioneering fruit growers chose their varieties carefully and were able to take full advantage of the work of William Coxe and his pomological successors, and of the new horticultural societies, in highlighting the best of the apples raised and selected by the colonial farmers of the eastern seaboard. As a result, many of these including Newtown Pippin, Rhode Island Greening, Esopus Spitzenburg and Roxbury Russet were taken right across America. Newtown Pippins provided the basis not only for the early fruit industry of New York and New England, but under the name Albemarle Pippin, also for that of Virginia, and later Oregon. When a good new seedling arose, growers exploited their advantages as quickly as possible. Baldwin, which arose on a farm in Massachusetts, was swiftly taken up by the New York and New England home and export trade; Grimes Golden, possibly a Johnny Appleseed seedling, helped to set fruit growing on its feet in West Virginia; Rome Beauty, which arose near the town of that name, became the apple of Ohio. Ben Davis, which was possibly a native of Kentucky, became as widely planted in the southern states as Baldwin was in the north. Growers were also adroit at turning the local situation to commercial advantage. In the north, winters were often so severe as to kill apple trees, but the sunny days and cool nights of early autumn also gave an exceptional colour to the skin of varieties such as Fameuse. By concentrating on these, producers made Montreal's and Quebec's fruit famous. Northern growers also set about trialling Russian varieties, and using these to raise crosses which combined hardiness with wonderful colour. In New Brunswick, Canada's first fruit breeder, Francis Peabody Sharp, produced the very hardy Crimson Beauty; and in chilly Minnesota, Peter Gideon raised scarlet Wealthy, which is still a commercial variety, by cross-breeding with the Siberian crab.

Transatlantic apple growers also benefited from the continent's natural geography and the rapid expansion of industrial centres, waterways and railroads. New York, for example, which enjoyed both deep anchorage and a strategic position on the River Hudson, became a major exporting centre. It was able to ship apples grown in New Jersey and Pennsylvania, and stimulated the planting of apples along the Hudson River itself. Its banks boasted not only the largest orchard in the country but also in the world: that of the 'Apple King', Robert Pell, who grew 200 acres of Newtown Pippin at Esopus. With the opening of the Erie Canal linking Albany on the Hudson to Buffalo on Lake Erie in 1825, New York became accessible to goods from the north and Midwest and enabled the apple industry to establish itself around the Great Lakes. Proximity to these gave protection from the late spring frosts, and the first large orchards

The careful packing of the American apple crop into used flour barrels ensured the fruit reached its distant markets in good condition. It would be weeks, if not months, before fruit from these orchards reached Liverpool, London, or other British ports.

of apples and peaches were established on Grand Island, in the Niagara River, by Lewis Allen in 1833. Soon the whole of the Niagara Peninsula was planted with fruit trees, and orchards spread along the shores of Lake Ontario and around the Finger Lakes to form the west New York fruit belt. By 1855 these orchards were producing half a million bushels of tree fruits annually.[5] The growth of Chicago, which through its canal link to the Mississippi waterways became the major distribution centre of the Midwest, was largely responsible for the development of the Michigan fruit belt. This had its origins in the successful, local fruit-growing enterprises started up around Detroit by the early French settlers, but it really began after the Great Chicago fire of 1871. Rebuilding the city called for large amounts of timber, which came from the forests of southern Michigan, and the land was subsequently replanted with fruit trees.[6]

In 1844, Henry Ward Beecher, a clergyman and leading figure in Indiana horticulture, had predicted that 'the apple crop of the United States will surpass the potato crop in value for both man and beast', and with the founding in 1852 of the American Pomological Society, the new industry acquired a great clearing house through which growers, nurserymen and dedicated amateurs could compare fruit and exchange experiences. Among the society's founders were Charles Downing, Robert Manning, and Marshall P. Wilder. Wilder, who was the society's president from 1852 until his death in 1886, was a prosperous businessman. He lived in Dorchester, outside Boston, and his passionate interest in fruit is commemorated by the Wilder Medal, awarded annually for the best new varieties. Manning was Massachusetts' first pomologist, and Downing the author of the massive *Fruits and Fruit Trees of America* which remained the definitive fruit register until the early 1900s. The society's unswerving aim was the selection and promotion of good market apples. In this it succeeded. At their very first meeting members voted Northern Spy, a popular apple in East Bloomfield, New York, the most promising newcomer, and it did indeed go on to become a leading fruit in home and commercial orchards for the next hundred years.

Apple production continued to rise steadily throughout the second half of the century as the area under orchards increased. In particular, the completion of the Union Pacific Railway in 1869 and the coming of the Northern Pacific Railway to Washington and Oregon in 1885 stimulated the planting of thousands of acres of apple trees along the west coast. In Oregon they blossomed in the Willamette Valley and Hood River Valley; in Washington in the Yakima Valley and Wenatchee Valley and further north in British Columbia the Okanagan Valley became the main orchard area.

Eᴺɢʟɪꜱʜ ᴀᴘᴘʟᴇ ɢʀᴏᴡᴇʀꜱ made their first attempt to prepare for the unstoppable tide of American imports in 1854, with the founding of the British Pomological Society, two years after the American one. It was abundantly clear that if the British industry was to stand any chance against the competition, it also needed a central organisation, and a clear strategy for improving the quality and quantity of its fruit. The Horticultural Society was in financial difficulties and the English fruit men seized the initiative to form an alternative group.

The suggestion had first been aired in a letter written to *The Florist, Fruitist and Garden Miscellany* by the nurseryman Thomas Rivers, and John Spencer, head gardener to Lord Lonsdale at Bowood in Wiltshire. The new society's aim was to 'compare and classify the fruits of Great Britain, America and the Continent; and likewise for examining and reporting on newly introduced or seedling varieties.'[7] The garden luminary Joseph Paxton, who was head gardener to the Duke of Devonshire and the designer of the Crystal Palace, agreed to be president, and its secretary was Dr Robert Hogg, who was soon to emerge as not only Britain's leading fruit authority but as the moving force behind the fruit men's response to the American imports.

The society's success in rekindling interest in hardy fruit in turn encouraged the Horticultural Society to revive its own orchard and fruit trials, and in 1858 it set up a Fruit Committee which included all the key members of the Pomological Society as well as other nurserymen, head gardeners and landowners. It held monthly meetings and introduced a system of awards for new varieties: its First Class Certificate (FCC) and Award of Merit (AM), which remained the only formal assessments of new fruits for nearly a century.[8] In January 1860 Hogg accepted the post of secretary to the committee, and immediately set about rebuilding the fruit collection. Later that year, probably at Hogg's instigation, the Horticultural Society's Fruit and Vegetable Committee absorbed the Pomological Society. When the worst threat yet began to hit British apples in the 1870s, it was Hogg who kept not only the fruit committee and at times even the society going, but also prevented the tentative foundations of the English apple industry from collapsing completely.

Hogg's unique strength was his unrivalled knowledge of apple varieties and his position as a pomological colossus with a foot in the worlds of both commerce and the country house. The eldest son of a well-known Scottish nurseryman, he had studied medicine at Edinburgh University, but switched to botany in preference to 'the prospective drudgery of the general practitioner'. Following a university degree, his interest in fruit was awakened in 1836 during a spell of practical work with nurseryman Hugh Ronalds of Brentford, in the Thames Valley, and he had continued his training with the leading authorities in France and Germany. On his return he began to travel around Britain familiarising himself with the local fruits, especially apples, that cottagers and farmers in different areas valued. He examined nurserymen's collections and built up as wide a range of contacts as possible among professionals, amateurs, commercial men and head gardeners alike. In 1844 he married, and his father purchased him a partnership in the famous Brompton Park

Dr Robert Hogg (1818–97). A Hogg Medal for fruit is still awarded by the Royal Horticultural Society in his memory.

Nursery.[9] With its long history as the foremost nursery in England for fruit, it gave Hogg access to unrivalled archive material, a large fruit collection, and the opportunity to produce a new and badly needed reference work on fruit varieties.

In 1851, Hogg published the first and, as it turned out only, volume of *British Pomology*, which was to have been the most ambitious study of fruit over published. By alphabetical good fortune, the opening volume was *The Apple* and within it Hogg documented 401 varieties in minute botanic detail, using the names resolved by 'that patient and indefatigable pomologist Mr Robert Thompson', the Horticultural Society's Fruit Officer, who had 'liberally provided information'. Line drawings of the 'newest, rarest and most esteemed' were given, together with copiously referenced notes on each variety's origins and history. Too scholarly for most people, however, it did not become the popular guide he had intended and plans for further volumes were abandoned.

In 1854 the Brompton Park Nursery closed, and Hogg started work in his father-in-law's business. By now, his own experience with *British Pomology*, combined with the parlous state of the British fruit industry, had made him keenly aware of gardeners' and growers' need for accessible information. Through his friendship with nurseryman Thomas Rivers, he became both closely involved with the setting up of the Pomological Society and, in 1855, joint owner and editor of the future *Journal of Horticulture and Cottage Gardener*, which was one of the leading gardening weeklies providing a much-needed forum for information, and the appraisal of new varieties and techniques. By the time he was appointed secretary of the new Fruit Committee and director of the society's orchards, he had also become a prominent judge at all the major shows, and a highly respected member of the head gardeners' world where fruits were selected according to the criteria of the dessert, and presented in such a manner that 'one could transfer a collection from the exhibition table to that for the "company to dinner" without an off dish or difficulty.'

Hogg was equally concerned to keep abreast of events and opinion in the commercial arena. Nurserymen and growers sent him seedling fruits for comment, and Thomas Rivers often presented him with samples of the new foreign varieties that he was trialling at his Sawbridgeworth nursery. He corresponded with Downing in America, the leading pomologists in Germany, and with his great friend, André Leroy of Angers, who in his *Dictionnaire de Pomologie* was busy doing for France what Hogg was doing for Britain. The commercial fruit growers' most urgent need was guidance as to which sorts were the best to plant, and this Hogg set about helping to provide through his *Fruit Manual*, which appeared in 1860. In it he covered all the fruits grown in Britain from apples to walnuts, and carefully collated all the available information on the different varieties. Conscious of the financial failure of *The Apple* it was not too learned. The first edition was 'a little volume of 300 pages forming a companion which gardeners may carry in their pockets as constantly as a pruning knife' and at 3 shillings and 6 pence the reviewer concluded that it was 'within the reach of all'. Within 18 months an enlarged second edition had appeared and, with new varieties appearing thick and fast, a third and even larger edition followed in 1866. By 1875 it reached encyclopaedic proportions and included much of the information concerning the origins and history of varieties that Hogg had gleaned from poring over the old literature and inviting *Journal* readers' reminiscences. According to a regular correspondent it made 'the history of fruit as interesting as a fairy tale' and was already on his shelf of favourites: 'pencil marked and thumbed and threatens to want binding from constant reading of it'. In the final edition, which appeared in 1884, Hogg documented over 700 apple varieties in comprehensive botanic detail. It remains the

PLATE 19: BLENHEIM ORANGE

standard reference work to old varieties, and Hogg fully deserved the accolade bestowed at an event in Belgium: the Linnaeus of pomology.

Passionately committed to Britain's apple heritage; keenly aware of the strengths, weaknesses and differing needs of both private and commercial fruit growers, and with a national platform, Hogg was uniquely placed to rise to the aid of the emergent English fruit industry.

IN THE MID-1870S BRITISH AGRICULTURE plunged into depression. In addition to apples America was also producing huge quantities of cheap grain, and while continental Europe reacted swiftly by reimposing import duties, the British government continued to maintain its free trade policy in order to encourage manufacturing industry. As grain prices plummeted, the call was for agriculture to turn, as before, to fruit growing as its saviour. In an electioneering speech given at his Hawarden constituency, Gladstone urged farmers to plant fruit instead of corn in order to reduce the five million pounds that had been spent on imported fruit and vegetables in 1879 alone. It was now that the full extent of the weakness of British commercial apple growing was revealed. The Kent fruit farmers' reluctance to modernise and replant meant that their crops were often poor, and the fruit small and blemished; while the market gardeners and smallholders, who had been more receptive to horticultural developments, had ended up with largely haphazard collections, influenced by the purple prose of the nurserymen's catalogues, which were for many the main source of information. As a result, in the markets varieties tended to be bulked up together, and often neither growers nor fruiterers had much idea what apples they were selling. Imported fruit, on the other hand, was not only clean and bright but consisted of single, clearly identifiable varieties such as green Newtowns and red Baldwins.

Hogg and his colleagues knew that to rebuff the competition and the 'Yankies' in particular, a campaign was needed both to persuade commercial growers to concentrate on a few good varieties, and to encourage people to buy them. 'The public will soon learn to discriminate between the brightly coloured but dry and flavourless American kinds and fresh home grown apples' wrote nurseryman George Bunyard in his *Fruit Farming for Profit* in 1881. The problem was, however, to know which these should be. Despite all the work of the last decades it was still the case that 'growers knew little of the varieties they possessed' and the chaos of mistaken identities was being compounded by the stream of new varieties being introduced in response to the quest for novelty in the country house dessert.

The first area to take up the challenge was the county of Herefordshire. This was a cider stronghold, but with the establishment of railway links to the main industrial regions, farmers had begun to find sales of 'pot' fruit more profitable and were eager to invest in high-quality table fruit. The Woolhope Naturalists' Field Club, which was based near Hereford, organised a survey of the country's orchards with the aim of identifying and evaluating all the varieties in order to pinpoint the best. Its members also sought to bring in new ones to see how they fared under local conditions. In 1876 the Woolhope's resident cider expert, the Reverend Charles Bulmer, invited Dr Hogg to attend the club's annual show, and Hogg was so impressed by the quality of the display that he suggested the club produce a *Pomona* to record the county's fruits. He volunteered to be their technical editor. Dr Henry Graves Bull, a local physician and 'the life and soul of the club', took up the challenging position of co-editor and became the 'energising spirit' behind the production.[10]

Over the next few years the project rapidly expanded. Soon its aim was to describe not only the best apples and pears of Herefordshire but also of the country as a whole. Hogg came down to examine the fruits at the annual shows and the hundreds of apples which arrived at Dr Bull's door, as described by one M. Cooke in his lengthy *Ode* to the *Herefordshire Pomona* of 1885:

> *There came apples in hampers and apples in sacks*
> *Apples in boxes and apples in packs . . .*
> *Apples hung over his head all night*
> *They dangled before him in morning light*
> *And filled up his slippers at dawning*
> *They trundled before him all down the stairs*
> *At breakfast they rattled a storm in his ears*
> *And came with the 'Times' in the morning*
> *Throughout the long day they were everywhere*
> *Filling like snowflakes the ambient air.*

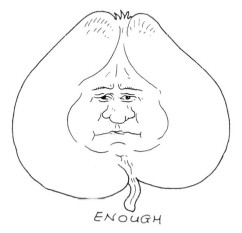

The *Ode, which was recited at a dinner held to celebrate the completion of the* Pomona, *was never published. The handwritten manuscript is illustrated with sketches like this one of 'Dr Pom' – alias Dr Bull.*

The Hereford shows soon became 'the Mecca to which all pomologists make their annual pilgrimage' and the finest examples of each variety would be given to the *Pomona* artists to paint. When Miss Bull, the doctor's sister, and Miss Ellis, a Gold Medallist of the Bloomsbury School of Art, had completed each year's set of water-colours they were sent away to the renowned printing firm of G. Severeyns in Brussels to be made into chromolithographs. With its subtly coloured plates, which were considered to come 'near to natural in general complexion, as showing the leaves and sometimes flowers of the varieties and as representing fair typical examples rather than exceptional fruits abnormally coloured', it was the most accurate pictorial record of the best known varieties of the time, taking its place alongside the earlier masterpieces of Hooker and Ronalds. At the suggestion of Lady Foley of Stoke Edith, whose well-known fruit collection had been made available for the club's use, a special copy was prepared for the Royal library at Windsor. The artists each received a cheque for one hundred guineas, and a miniature portrait painted on ivory of Dr Bull, who had died a few weeks before the last instalment was published.

THE ACTIVITIES AT HEREFORD were successful in resolving many identities and pinpointing some good varieties, but Hogg and his colleagues on the, now, Royal Horticultural Society Fruit and Vegetable Committee knew that more needed to be done at a national level. In 1883 they decided to act. It was clearly going to be an exceptional year with every variety seeming to fruit and a bumper crop in prospect. Now was surely the moment to survey the entire English

PLATE 20: RIBSTON PIPPIN

apple population. If the chaos of mistaken identities could be resolved once and for all, then performances could be compared and it would be possible to draw up some sound recommendations.

It was almost certainly at the fruit committee meeting held on 28 August that the decision was taken to mount the 'Apple Congress' in the society's Great Vinery at Chiswick. The ambivalence with which commercial fruit was still regarded by the horticultural establishment was clear from the somewhat cynical comments made by a Fellow in an editorial in the *Gardener's Chronicle*. This asked whether the society 'was not departing from its dignity in allying itself with an exhibition of Apples'. All fastidious wavering was, however, swept aside in a growing mood of nationalistic and pomological fervour.

A committee of 50 prominent fruit men from all over the country was swiftly convened. They were to drum up support in every county to ensure that as complete a collection of apples as possible went on show in London. With only three weeks to go before the event, circulars were sent out to nurserymen, enthusiasts, head gardeners and to anyone who might help in the search for every possible variety. The response was immediate and gratifying. Fruit and promises came in from as far away as the Channel Islands and Sweden. Archibald Barron, superintendent of the Chiswick garden, and Lewis Killick from near Maidstone in Kent and the only fruit-farming member of the fruit committee, found themselves working all hours to unpack and lay out the specimens which were arriving by every post, and to clear further areas to accommodate the unprecedented demand for space.

The Illustrated London News *was much amused by the spectacle of visitors to the Congress comparing their unnamed apples with the named varieties on display.*

Never before had so many varieties been brought together in one place, and probably never will be again. The committee was forced to work until late into the night to identify, cross check and finally resolve a definitive list of 1545 varieties.[11] Every one which could conceivably be of interest went on display, helping exhibitors and visitors clinch the identity of their fruit and celebrating the unique richness and diversity of British apples. It also marked the formal beginning of the national Fruit Campaign against the 'Yankie' invasion.

THE ORGANISERS HAD TAKEN the important step of asking exhibitors to fill in questionnaires designed to highlight the best varieties in each area, and from the replies a list of the top 60 dessert and top 60 culinary apples was drawn up. The most popular dessert variety proved to be the heavy-cropping King of the Pippins, which was the market gardeners' favourite. The new Cox's Orange Pippin came next, beating its esteemed parent, Ribston Pippin, into third place. Fourth was Kerry Pippin and fifth Blenheim Orange. This also featured in the culinary list, and overall there were more dishes of Blenheim Orange than of any other variety on display. The congress also brought to prominence Worcester Pearmain which, like Cox, became and remains a major commercial variety. Among the cookers, the very early and prolific Lord Suffield topped the poll. Then came Keswick Codlin, third was the generously sized Warner's King, and fourth Dumelow's Seedling, also known as Wellington. Bramley's Seedling, then only known around Nottingham, was shown by two local firms and received the coveted RHS First Class Certificate. As a direct result of the congress, the most popular varieties were widely adopted and planted throughout the apple counties.

The best market varieties, from the growers' point of view, were early, brightly-coloured eating apples, which would have a free run in the markets until imports – which were of late dessert apples – arrived; and culinary apples, which none of the competitors grew. As Southern Hemisphere apples started to arrive in March there was now little to be gained by growing very late dessert varieties that would keep almost 'until apples come again', especially as the quality of the best English late keeper, Sturmer Pippin, was improved by the antipodean sunshine and was rapidly becoming the basis of the Australian and New Zealand export industry. As a result, the late 1890s saw the new Kent apple season open with the cookers Keswick Codlin and Lord Suffield. These were soon followed by the early dessert varieties, scarlet Gladstone, ripe at the end of July, and then maroon Devonshire Quarrenden. King of the Pippins and Worcester Pearmain were harvested in September and always found a good market. Some of the large, heavy cropping cookers, such as Warner's King and Lord Derby, would also be picked before the rush with the hops in September, and the arrival of the new season's 'Yankie' apples in October. The long-standing reputation of Ribston Pippin and Blenheim Orange ensured that they could hold their own and still make high prices up to Christmas, as would Cox's Orange Pippin, which was being grown by farmers in Kent and Essex as well as by market gardeners in London and in the Vale of Evesham by the end of the century. Wellington would last with few problems until March when the English season was over, and the markets turned to forced rhubarb and Australian and New Zealand apples, which were coming in increasing quantities. Tasmania, the chief Australian apple-growing state, had exported its first large consignment from Hobart in 1884, and in the 1890s New Zealand began to ship apples from Hawkes Bay in the North Island.[12]

Producers and wholesalers strove to promote and identify their apples as attractively as possible in the competitive export market.

Breeders and nurserymen also began to concentrate on producing what the market wanted. Beauty of Bath, which is ready to be picked at the beginning of August; Early Victoria, a summer codlin raised in Wisbech; and Newton Wonder, which combined the hardiness of a Blenheim Orange with the keeping qualities of a Wellington, all appeared in the 1890s, secured the Royal Horticultural Society's coveted First Class Certificate and went on to be widely planted. The endorsement by the congress of the high reputation of Cox's Orange Pippin stimulated the breeders' use of it as a parent in order to try to raise more varieties with a similar fine flavour. Inspired by Thomas Andrew Knight's work, Charles Ross, head gardener at Welford Park in Berkshire, began careful cross-pollination experiments and raised a 'larger Cox' and the apple that bears his name. The Laxton brothers, sons of the distinguished nurseryman and breeder of peas and strawberries, turned their attention after his death to apples and began to raise a succession of varieties, starting with Laxton's Advance in August through to Laxton's Superb at Christmas which maintained the sweet, delicate flavour of Cox throughout the season.

The congress also marked a decisive shift in the horticultural establishment towards the interests of the commercial man rather than the country house. In 1888, at the Royal Horticultural Society's Apple and Pear Congress, only those varieties thought worthy of planting for profit were exhibited, and at the associated meeting, marketing and cultural problems were the main issues. Leading nurserymen began to address themselves to the needs of commercial fruit growers rather than those of the private gardener as the former started to become the main customer for their trees. Even head gardeners found they were increasingly called upon to play a commercial role. Tenant farmers on many estates were being encouraged to invest in fruit, and as they usually looked to the head gardener for advice, these men now found themselves directly involved in setting the new industry on its feet. John Robson of the Cornwallis estate at Linton Park near Maidstone in Kent had initiated an annual census of varieties in the *Journal of Horticulture* as early as 1876, and his successors remained closely involved with fruit, which was the main activity of the estate tenants. William Crump, head gardener to Lord Beauchamp at Madresfield Court, Worcestershire, had 230 varieties

The great Victorian fruit exhibits initiated a tradition of grand fruit shows which still continues. This is the Royal Horticultural Society's Autumn Show of 1934 held at the Crystal Palace.

on trial in 1892 and a fruit tree nursery raising some 2000 trees for sending out to tenants. In 1894, the Duke of Bedford went so far as to set up the Woburn Experimental Fruit Farm at Ridgmont on his estate, and employed Lewis Castle, a former head gardener, to manage the fruit trials.

As the Fruit Crusade gathered momentum, similar congresses and exhibitions with the emphasis on the needs of the industry were held throughout the country including Chester, Pershore, Worcester and Exeter. The century came to a close with the great fruit shows of the 1890s which were sponsored by the Worshipful Company of Fruiterers and the RHS. These celebrated British grown fruit, and whereas in the earlier shows the battles for the prizes had been fought over pineapples and grapes, it was apples that were now the stars. Promotion of British fruits to the public was one of their main aims, while nurserymen took the opportunity to show the latest most attractive varieties that might have commercial appeal. In the 1890s, for example, Bunyard's introduced the showy Lady Sudeley apple, proclaiming that its colour 'rivalled the best American fruit . . . and its flavour to some palates excelled the well-known Cox's Orange Pippin'.

REFLECTING THE NEW CONFIDENCE in British fruit, the area under orchards increased by about 3000 acres annually during the 1880s and 1890s.[13] The greatest expansion took place in the traditional apple-growing regions. In the Thames valley, the area of orchards doubled to nearly 5000 acres, and also moved further westwards as the growth of London and the pollution it created drove all but the glasshouse industry further into the country. The orchards lay mainly between Putney and Hampton in the south, and ran from Twickenham through Hounslow to Southall and Uxbridge in the north-west. Holdings ranged from 30 to as large as 150 acres, and fruit bushes and vegetables were grown in between the rows of fruit trees. At Ham Common, in Richmond, Mr Walker, who was famous for his daffodils, had 35 acres of fruit trees with narcissi and paeonies growing beneath, while others grew musk roses.

In Kent the area under fruit increased even further, rising from 10,000 acres in 1873 to 25,000 acres in 1898. A number of farms had now turned over completely to fruit, and orchards extended down over the Weald and into Sussex, while the eastern fruit belt extended beyond Canterbury. As in the Thames Valley, tree and bush fruits were often grown together. One of the largest apple growers was Mr F. Smith, who farmed 200 acres of apples in the village of Loddington outside Maidstone. Here he had an undercrop of gooseberries and cobnuts, with shelter provided by damson hedges.

In the West Midlands, in the area around Evesham, there were over 1000 acres of smallholdings and market gardens, with apples coming second to plums as the favoured fruit. In Herefordshire apples were the main interest, and the new plantations were not of traditional cider apples but of eaters and cookers to supply the Midlands, South Wales and the North. Gloucestershire boasted Lord Sudeley's huge Toddington Orchard Company, which extended over 1000 acres and had its own railway terminus, while Somerset and Devon remained cider counties, but still sent quantities of fruit to market.

In addition, important new fruit areas had developed in eastern England. Cambridgeshire had first entered the world of commercial fruit growing following the purchase of land at Histon

PLATE 21: SPARTAN

Apples were widely promoted as a simply prepared and health-giving food.

by Stephen Chivers in 1851. Apples were grown in the surrounding villages and also further north towards Wisbech. The Essex industry was said to have been launched by the nurseryman William Seabrook, who returned from a visit to the United States in 1887 so impressed by the apples on sale that he determined to supply the English market with equally colourful, good-quality dessert apples. He invested in new land around Colchester and planted orchards of Cox's Orange Pippin.

Even outside the main areas, there was not a county in England which was not producing apples. To the north of London, young orchards at Enfield and Cheshunt helped supply the capital. From Cheshire and Lancashire apples were sent to Liverpool, Manchester and other cotton towns. Many people near Nottingham invested in fruit, encouraged by the nurseries Pearson's and Merryweather's. Around fashionable Bath, market gardeners, who were famed for their strawberries, also turned their attention to apples.

As well as planting up new orchards of the latest recommended varieties, growers also began to give greater attention to the management, picking and storing of the crop. In Kent, for example, it had always been traditional to auction the crop while it was still on the tree and then pack the apples in hay or soft straw in wicker sieves. These were then covered with paper and a final layer of straw, held fast by split branches of ash, hazel or willow pushed under the rim. George Bunyard, among others, recommended that English farmers should 'take a leaf out of the French books and put up our produce in an attractive form'. They should take as their example the gardeners of the Montreuil area to the north-west of Paris who had taken up cordon training on a major scale. Their large, well-coloured apples which were graded and packed in light card boxes were taking over the luxury end of the market. Bunyard also urged growers to think more carefully about storage if they were to keep their apples in good condition and catch the high prices after Christmas. Rather than simply heaping the apples on the bottom floor of the oast, or in a barn, and covering them with straw, they should consider something closer to the gardener's fruit store.

The expansion of the fruit industry was encouraged by – and in turn helped to encourage – Britain's new jamming industry. Because it carried no import duty, sugar was cheaper in England than anywhere else in Europe, and jam, by providing a market for surplus and second-quality fruit, helped to give the developing industry confidence. The Toddington Orchard Company in Gloucestershire had its own jam factory, while the building of John Chivers' factory in 1873 and the establishment of factories at Tiptree Heath and Elsenham Court had done much to encourage fruit growing in East Anglia. While jam did most to stimulate investment in soft fruit for the more expensive end of the preserve market, there was also a huge and growing market for cheap jam, which had become a staple of the working-class diet. This was usually made from mixed fruit, according to whatever was cheapest, and included large quantities of apples. 'Smashers' would go round orchards, particularly in the West Country, buying up fruit by the ton for the factories.

By the end of the century, the consumption and quality of English apples had soared, and compared with 20 years earlier their price had more than doubled, to five shillings a half sieve, by 1899. The new plantations were thriving and 'giving good returns and most excellent samples'. Even so, the struggle against increasing imports continued. Growers in Ontario, who sent five-sixths of their crop to Britain, actually found it profitable to ship boxes of fancy apples to private addresses in the home country, and the beautifully-coloured, large Winter Banana apples from Summerland in British Columbia found a ready market in London.[14] As the 19th century turned into the 20th, the success of the Fruit Campaign was to have established the basis of a modern industry, rather than to have won the war.

THE NEW CENTURY saw fruit production expanding world-wide, and as the orchards of the west coast of North America and Canada came into full production, and South Africa joined the ranks of Empire growers, reasons were sought for the continuing failure of England's new industry to hold its own. During 1908 and 1909 it was feared that it might collapse altogether, and much of the blame was laid at the door of the recommended varieties, such as Cox. These had proved prone to disease and as there was no really effective means of control, the fruit was often too blemished to be saleable. At a fruit growers' conference held in conjunction with the first commercial Kent fruit show in 1911 at the South Eastern Agricultural College, Wye, the Canterbury grower, Spencer Mount, went as far as to say that anyone who planted Cox 'was playing the fool'.

Even more galling was the fact that even good English apples now faced new competitors: the rising imports of reasonably priced exotics from British colonies. West Indian bananas, which arrived in the early summer, brought about the demise of the English dessert gooseberry, while apples were threatened by South African and Palestinian oranges, and canned fruit such as Malayan pineapples. This was a direct result of the government's policy of supporting the Empire at the expense of home production, a strategy which was to continue to undermine the position of English apples for some time to come.

Since the late 19th century Britain's position as the world's leading industrial nation had been challenged, while her agricultural interests were also being damaged by competition, especially from the United States and Germany. In the post-war years of the 1920s and 1930s, the British Empire, then at its largest extent, was seen as the motherland's salvation. The Empire's economic development was encouraged, ties strengthened, and the short-lived but high-profile British Empire Board (1926–33) was founded to promote all Empire goods, from timber and tea to meat and fruit. The British market became a battleground in which American and Continental apples fought it out with those from the Empire, and home-grown apples struggled to stay in the field at all. Such competition had been encouraged by the launching in 1923 of the 'Eat more Fruit' campaign. Organised by the shippers of Liverpool and seized upon by the fruit trade in general, its aim was primarily to boost its own business which stood to gain more from Empire than homegrown fruit. Through a massive advertising campaign the British public – who consumed a third less fruit than their American counterparts – was to be convinced of its health-giving value and induced to buy more. As in the 1890s, shows had been used as an important means of getting the message across. The first Imperial Fruit Show, sponsored by the *Daily Mail* and the Ministry of Agriculture, was held at the Crystal Palace in 1921 and had proved so successful that it was taken over by the fruit trade and shows followed in other major cities.

PLATE 22: PITMASTON PINE APPLE

The fierce competition between exporting countries meant that they vied with each other to produce the best-looking and most attractively presented fruit. The States sent firm red Jonathans, maroon Winesaps, Golden Russets, Grimes Golden and Oregon Newtown Pippins. From Canada came the striking scarlet McIntosh. Spring and summer imports from Australia and New Zealand, which had been of English varieties, now also included the red flushed Rome Beauty and crisp 'Cleos'. England was prevented by climate from growing any of these, and the main English varieties could not compete on grounds of colour or 'finish', which in the market place increasingly seemed to count for more than flavour. Bruised but unbowed, English growers addressed themselves more seriously than ever to the problems of quality and presentation.

The Ministry of Agriculture's National Mark Scheme, which aimed to raise the standard of British produce in general, was applied to apples in 1928 and laid down the criteria growers

(Above) Posters, vans, shops and canteens all exhorted customers to consume fruit in greater quantities. The sun image on the van rotated to drive the message home.

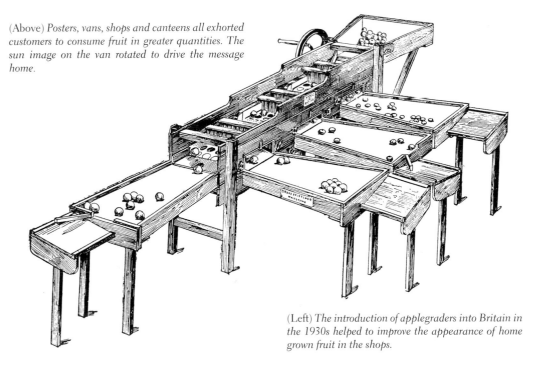

(Left) The introduction of applegraders into Britain in the 1930s helped to improve the appearance of home grown fruit in the shops.

125

Few concessions were made to home producers by the Empire Marketing Board. This advertisement appeared in The Times.

should aim for in terms of a series of carefully defined grades. These ranged from the 'Extra Fancy', luxury product of 'superlative fruit, free from blemish, fully coloured . . . and of first-class flavour'; through the main 'Fancy' grade fruit of good-quality and 'sound value for money'; to the bottom or 'Domestic' grade for 'those of more limited means'. Produce that satisfied the ministry's inspector earned the right to carry the National Mark logo which consisted of a map of Britain and the slogan 'Empire Buying Begins at Home'. By the 1930s English growers were joining the National Mark scheme in increasing numbers. They armed themselves with ministry handbooks and new spraying equipment, invested in American grading machines, built packing sheds and replaced the old wicker sieves with standardised wooden boxes. As part of the policy to encourage Empire produce the old free trade policy was reversed, and in 1932 import duties were levied on non-Empire goods. With the United States and Europe disadvantaged by tax, it was imports of Canadian apples – recognised by their 'Maple leaf' ticket – which became the main competitor for English growers. Southern Hemisphere fruit did not clash with the English season and became no threat.

Despite the fears of many people that the diversity of English apples was disappearing, there was still a wide range on sale. The pattern was much the same as in the 1890s with the emphasis on early eating apples and cookers. Gladstone and Beauty of Bath were first on the market in early August followed by Lady Sudeley, Langley Pippin, Worcester Pearmain, Ellison's Orange, James Grieve, Charles Ross, Rival and King of the Pippins. The main season variety for October up until Christmas was still Cox's Orange Pippin which had been saved by the advent of lime sulphur sprays to control disease. Laxton's Superb was being planted as the late-season variety to carry supplies into the new year. Bramley's Seedling, which had been extensively planted in the earlier years of the century, was firmly established as the best English cooker, but the culinary range was still enormous and faced no competition. The quality of flavour and the freshness of fruit which did not have to travel half-way around the world to reach the market were other strong selling points, and the publicity championed the virtues of 'Apples Grown at Home'.

With the coming of the Second World War, however, the English gains of the 1930s slipped away. Although they had no competition from imports, quality declined. Prices were fixed, labour was short, old orchards were pulled up to make way for basic foodstuffs, while production controls forbade new plantations on arable land. When the war was over, the cycle had to begin again. New orchards were planted during the 1950s, and following the example of the Kent's Marden Show, which had started in 1933, annual shows helped to interest both growers and the public. At first, the main competitor was still Empire fruit which was protected through a system of quotas and licences imposed on other foreign producers. As the Empire began to break up, however, this protection was phased out and the major competitor became European fruit from the new orchards planted since the war in France and Holland. Quality was high and with the

European Common Market in prospect, the government set up new quality controls in 1964 to help raise standards of British fruit. The present grading system was introduced and the relatively greater importance of size and appearance, rather than flavour was established.

By the time Britain became a full member of the European Common Market in 1973, other aspects of the English apple industry were also changing fundamentally. The fruit shippers and auction houses of London and Liverpool had now more or less disappeared with the introduction of containerised transport; and even more important, apples were increasingly being sold through supermarkets. As by far the largest buyers of apples in the market, they now began to set the standards, and by and large these could only be met by overseas growers. In the mid-1970s France seized its advantage with both hands and launched its now notorious campaign to persuade the British public to eat French-grown Golden Delicious. This was originally a West Virginian apple, but it grew well in the Loire and Rhône valleys, where colonials returning from Algeria had been encouraged by the government to invest in fruit. In many ways it was the perfect modern commercial apple. Easy to grow and a heavy cropper, it was also ideally suited to sale through supermarkets, which required a continuous supply of a uniform product. Golden Delicious could be picked and marketed by late September, or kept in store through the winter. It also graded out at the top size, was regularly shaped, and could be attractively packaged.

Golden Delicious presented English growers with a formidable challenge which, once again, they were ill-equipped to face. Neither the English climate, nor the varieties grown suited the supermarket approach and whereas French and Dutch growers packed and sold their fruit through large co-operatives – membership had been obligatory to qualify for replanting grants – most English producers remained fiercely independent. It was clear, however, that to capture the top grade and prices, the industry had to be rapidly restructured. Orchards were modernised, co-operatives set up and growers combined to mount a campaign to promote the superior quality of English apples. This focused on the uniquely rich flavour of Cox's Orange Pippin and the splendid cooking qualities of Bramley's Seedling. It has succeeded in so far as it has established the notion of a special virtue in English apples in the public's imagination, but more recently the English have grown to like or even prefer the modern imports, such as Braeburn, Fuji and Pink Lady, which come in from Europe, North America, the Southern Hemisphere and China. Britain now imports 75 per cent of her apples.

ONE SIDE EFFECT OF THE 20TH-CENTURY expansion of apple growing throughout the world and the exacting standards now set for table fruit has been the development of an apple processing industry to make use of fruit that fails to meet these stringent criteria. Sound fruit of a reasonable size is sliced for pies, freezing, drying, canning or bottling. It may also be steam cooked into apple sauce for baby food or other uses. Medium-quality fruit finds an outlet as apple butter, chips, rings, pie fillings, sauces and wine. Apples of both qualities, as well as small misshapen fruits, are also turned into juice, which is a major English end-product.

The Swiss were the first to start selling non-alcoholic ciders and perries (that is, pasteurised juice) in 1896, and the American industry was launched soon afterwards. Prohibition created a ready market for the juice, which was further encouraged by the exhortations of the followers of the Reverend Sylvester Graham – immortalised by the Graham cracker. He was a former Presbyterian preacher, temperance lecturer, self-styled doctor of medicine and dietetic expert, who had claimed earlier in the century that 'fruits, vegetables and nuts make for temperance' and that 'culinary processes make food less healthful'. As Americans sought to improve their moral and physical health by eating more fruit and vegetables, the foundations of the modern health food movement were being laid. J. T. Stinson, who lived to be 92 and was the first director of the Missouri State Fruit Experimental Station, coined the famous adage that 'an apple a day keeps the doctor away' in 1904, and apples and apple juice have benefited from the association ever since. It was the Nazi government's emphasis on health and fitness which led to the first large-scale production of apple juice in Germany between the wars. In Britain apple juice manufacture commenced in 1936 to mop up the down-graded apples from the National Mark scheme, but it did not become important until the 1970s.

By this time, the techniques that enable juice to be concentrated and reduced in volume ten- to fifteen-fold had become available, allowing large quantities to be stored economically. In the process of concentration, the volatile aromatic compounds are collected and either added back to the concentrate later, or sold on to the cosmetic industry for use in soaps and shampoos. Concentrated apple juice is used to make juice to drink and also, without the addition of the aromatic fraction, on a large scale in the cider industry and for making apple wine and cider vinegar.

World-wide, the main consumers of apple juice are now Eastern Europe and the former Soviet Union where apple juice is the main non-alcoholic drink after water. Europeans also consume a wide range of mixed drinks based on apple, such as strawberry and apple, which are increasingly finding favour in Britain.

THE MODERN APPLE INDUSTRY now bears little resemblance to its counterpart of 200 years ago. No longer is it a sideline for mixed farmers and market gardeners, but a highly specialised enterprise. Intensive orchards with their tightly packed rows of dwarf trees are far removed from the meadow orchards of the past and their management, like the storage and distribution of fruit, has itself become a science.

The first government-funded training and research organisations devoted to fruit were set up around the middle of the nineteenth century in the United States. In 1862, the new Department of Agriculture recognised fruit growing as a separate enterprise within its brief, and in 1886 it set up a Pomological Division which began to establish a network of fruit-research stations in key locations across the continent. Canada's first Agricultural College was founded at Guelph, in the centre of Ontario's fruit region in 1880, and the government established an experimental farm at Ottawa in 1886. In Britain, the National Institute for Cider Research at Long Ashton, near Bristol, was set up in 1903, and the John Innes Horticultural Institute at Merton in South London was established in 1909 with money left by its eponymous benefactor who was a City of London merchant. In 1913, the South Eastern Agricultural College, later to become the University of London's Wye College (now Imperial College at Wye), set up its Fruit

PLATE 23: NEWTON WONDER

Experimental Station which, with the support of Kent fruit growers, became in 1921 the independent East Malling Research Station (now Horticulture Research International).

This new generation of British commercial fruit growers was at pains to distance itself from the world of the Victorian country estate, whose long-standing influence on fruit growing they liked to blame – with some justification – for many of their misfortunes. The promotion of Cox, in particular, was regarded as commercially damaging on a grand scale. Like their counterparts elsewhere in the world, these 20th-century growers aligned their interests with those of the new biological sciences. The new genetics inspired by the rediscovery of Mendel's work gave fruit-breeding programmes fresh impetus and direction, and every aspect of orchard practice from the nutrition of the tree to pollination, pest control, fruit storage and transport came in for reassessment. Fruit breeding and evaluation passed into the hands of the professionals and men such as Edward Bunyard and S. A. Beach in his *Apples of New York* in 1905 were the last representatives of the discursive old style pomology.

One of the first problems to be tackled by the research institutes was the confusion surrounding dwarfing rootstocks. It was now being recognised that these were the key to production of good-quality fruit on a large scale as they were easily managed, gave early returns on investment and made most efficient use of manpower and space. Although there were a number of different kinds available, they were often misidentified, and there was in any case little reliable information about their relative merits. Robert Wellington, the first director of the future East Malling Research Station, began to establish a collection of rootstocks in 1912, and order was finally achieved by his successor [Sir] Ronald Hatton. It was he who, during the 1920s, made the selections that became the basis of the East Malling series of standardised rootstocks, which are now deployed world-wide.

Growers began to adopt semi-vigorous stocks on a large scale in the 1930s but since the 1970s they have been moving over to the dwarfing stock, M9, which will give a tree in full bearing, and no more than eight or ten feet high, within five years compared with the 12 years or more it takes for a standard tree to come into full production. The most intensive of modern orchards may now grow 1300 trees per acre rather than the 48 of a standard orchard, and as a result there have been enormous changes in management. The whole plantation will probably be renewed every ten to 15 years, whereas old orchards lasted 50 years or longer. This ensures that the grower always benefits from the healthy vigour of young trees. Modern orchardists also shape and train their trees, in order to maximise the amount of light the developing fruit receives. The favoured system, known as a centre leader tree, was invented in Germany in the 1930s and subsequently developed further by the Dutch. Resembling an extreme form of the gardener's pyramid, the tree is supported by a tall stake, and the lower branches are encouraged towards the horizontal creating a low framework which bears most of the crop.

There have also been major developments in pest and disease control. The advent of carefully tailored insecticides and fungicides together with improved spray technology now allows growers effectively to control the main problems of aphids, mildew and scab. In recent years the development of pest resistance and concern over safety has also seen the emphasis move towards a more finely tuned approach, known as 'integrated fruit production'. Based upon sophisticated prediction of pest and disease levels, and the use of natural predators where possible, it reduces chemical spraying to a minimum.

A similarly precise approach to picking and storing ensures that the apples reaching the supermarket shelves are as perfect in appearance as the Victorian head gardener's most carefully nurtured specimens. Fruit is chemically analysed before harvest to determine its storage potential, and modern temperature- and atmosphere-controlled stores can maintain it for several months close to the condition in which it was picked. The first experiments in cold storage were probably the barrels of apples which were packed among the blocks of ice being shipped all over the world as part of the Massachusetts ice trade, and by the second half of the 19th century, research into mechanical refrigeration was gathering momentum. In Australia, America and Argentina, particularly, there were huge profits to be made from the movement and export of perishable goods such as fresh fruit and meat, and between 1873 and 1886 Australian growers, in cooperation with the Royal Horticultural Society of Victoria, shipped experimental lots of apple varieties in iced chambers to various exhibitions in Vienna, Florence, Paris, London, Ceylon and India.[15] In the United States, the first refrigerated railcar – the 'Tiffany' – came into use in 1872 to carry fresh produce from New Jersey and Long Island to New York, and in 1899, as part of a strenuous effort to expand their fruit exports, the eastern states sent a refrigerated consignment of choice winter apples to the Paris Exhibition. Packed in barrels, they were shipped in refrigerated compartments to Southampton, and from there taken to a meat storage establishment at Le Havre. No other country, not even France, succeeded in maintaining a display of fresh fruit throughout the entire period of the exhibition, and the success of this experiment stimulated a major research initiative into the handling and storing of fruit in the United States.[16]

As the chemistry of the ripening process was unravelled, it became possible to devise more sophisticated methods of storage which not only allowed fruit to keep in good condition for its natural term, but could also extend its season. Once the development of an apple is complete, there is a sharp rise in the respiration rate and a marked increase in ethylene synthesis, which stimulates the ripening process and, eventually, decay. Once it was realised that cold storage slowed down the production of ethylene, researchers began to look for more efficient ways of doing this. As part of a programme to improve the transport of Empire produce in the 1920s, Drs Kidd and West of the Low Temperature Research Station, Cambridge, found that production of ethylene could more or less be brought to a halt if the cold atmosphere was also low in oxygen and high in carbon dioxide, compared to air. This discovery led to the development of controlled-atmosphere storage, which essentially holds the newly harvested fruit in suspended animation. Initially, it proved particularly useful for cooking apples, whose acidity tends to fall with keeping, and the first gas store was installed by the fruit grower Spencer Mount outside Canterbury in 1929, and filled with Bramley's Seedling.

Further honing of the chemistry and technology at Ditton Laboratory and then at East Malling Research Station refined the possibilities still further, and modern controlled-atmosphere stores allow growers to keep Cox's Orange Pippin beyond the end of its natural season, which is up to early January, right through until the following spring. For this type of long-term storage, however, fruit has to be picked at the correct stage, and once removed from store it tends to mature quite quickly. Traditionalists argue that as a result many apples now on sale are simultaneously unripe and past their best, but modern storage has undoubtedly made crispness an important test of a good apple and done much to remove the concept of seasonality. It is this, among other factors, which enables supermarket chains to demand – and be offered – the same narrow range of varieties all year round.

PLATE 24: CORNISH GILLIFLOWER

The demands of modern growing, storage, and distribution methods have also directly influenced breeders. The ability to store well under modern conditions, withstand the rigours of handling and transport, maintain appearance on a bright supermarket shelf and still be edible after days or weeks in the customer's fruit bowl now number among the qualities sought in a new commercial variety. The new stars of the apple world – Braeburn, Elstar, Fuji, Gala, Jonagold, Pink Lady and Empire – are offsprings of the old major international varieties: the American Delicious, Golden Delicious, Jonathan and Canadian McIntosh. In England, however, in the main, neither these parents nor most of their progeny thrive sufficiently well to be commercially competitive. The stud varieties of English breeders have, therefore, been largely the traditional favourites, and in particular Cox's Orange Pippin. The aim has been to raise varieties with fine flavour but with improved crops and resistance to disease. Fiesta, now called Red Pippin, is the most widely known of these improved varieties, while the most recent introductions are Saturn and Meridian.

IT IS THE COMBINED EFFECT of supermarket criteria, technological sophistication and the sheer scale of the modern apple industry which has both drastically reduced the range of apples grown, and made apples an international, rather than a national commodity. Relatively few varieties meet the demands of modern production and distribution. As every variety has different requirements an individual producer is compelled to concentrate on only one or two in order to reap the benefits of scale. A modern store, for example, holds up to 200 tonnes of apples, which should ideally be of the same variety. Co-operative systems of marketing and distribution also conspire to reduce the number of varieties, by providing centralised facilities in order to produce the standardised product that the market now demands. Even the need for promotion makes it desirable for producers to keep the number of varieties to a minimum. The success of the French campaign to promote Golden Delicious is ample demonstration of the value of establishing a strong brand image. In consequence, the number of varieties that are now grown on a large commercial scale in England is limited. The season begins in August with Discovery, followed by Delbarestival, Worcester Pearmain, Alkmene (Early Windsor) in September. Cox goes on sale in October and continues until the spring, supplemented by Spartan, Royal Gala, Jonagold, Fiesta (Red Pippin), Elstar, Falstaff and Egremont Russet. Bramley's Seedling is available virtually all year round. The majority of English apples are grown in Kent, but apples are also grown commercially in most of the old regions – the West Midlands, East Anglia and Wisbech, the West Country – and also in Hampshire and Sussex. The market gardeners, however, have faded from the picture and the last of the Thames Valley orchards disappeared under Heathrow Airport.

On a global scale, the restriction of fruit grown to a few internationally acceptable varieties means that buyers anywhere in the world can maintain their continuity of supply by buying wherever availability, quality and the state of the currency markets dictate. Britain, for example, buys apples from France, which is still our major supplier, as well as importing apples from Italy, Spain, Germany, Holland and Belgium. The United States and Canada remain major sources and we also have begun to import apples from China. South Africa is the main Southern Hemisphere supplier, followed by New Zealand and Chile. The principal varieties of a decade ago – Golden Delicious, Granny Smith and Delicious in its various forms – are still major international varieties, but they are gradually being eclipsed by Braeburn, Gala, Elstar, Empire, Fuji, Jonagold and Pink Lady. McIntosh and Cox continue to be imported.

Britain is ultimately prevented by climate from competing on anything like equal terms with the major apple-growing countries, and an increasingly competitive market has taken its toll on British fruit growing. Since entry into the European Union the area of commercial orchards has fallen by some 70 per cent. Britain's strength is that the British climate can produce exceptionally good flavour and some of the finest tasting apples in the world. Globally, however, apple production is beginning to exceed demand. As well as increasing output from Eastern Europe – Hungary, Poland, Romania, Czech Republic and Bulgaria – China, the world's largest apple grower, has now entered the export scene, which will ultimately have profound effects on all apple producing countries. Consumers have, however, begun to rebel against mass-produced and characterless food, and there is increasing demand for high-quality, less intensively produced goods, including apples. Alongside the technological efficiency of the modern apple industry it may yet be possible to retain something of the glories of the past.

Apples, apples everywhere and hardly a one to eat. The big red and yellow plastic spheres, waiting in the market for the unsuspecting, are so suspiciously, so blatantly, thick skinned and shiny, it is easy to pass on by. What we must live on is the memory of what a good apple tastes like.

Daily News, NEW YORK, 1977

THE REDUCTION IN THE NUMBER OF APPLE VARIETIES grown and the pursuit of the ideal 'commercial' apple has slowly and insidiously changed our perception of the fruit itself. The notion of a succession of apples, with a series of different qualities ranging from the refreshing summer apples through the richer autumn and winter kinds and finally to those that must be carefully stored until they reach maturity in the new year has almost disappeared. From a hundred years ago when every country and even every locality had its own specialities we have now arrived at a point where the apple has acquired a generic identity largely defined in terms of a narrow range of colour, shape, texture and flavour.

In the last 30 years, however, the movement to try and regain something of the richness of our apple heritage has gathered momentum. In part this reflects a more general realisation that the price of modern production has been a loss of individuality and diversity. There are similar campaigns to recover the qualities of traditional beer, bread and cheese. There is a general recognition too, that traditional, regional food – like vernacular architecture and local accents – is a mirror of history; and that in ironing out the differences we are diminishing our culture.

We have witnessed too, particularly in Britain and the United States, a revival of interest in the quality and nature of food and cooking in general. Even more recently, there has been growing concern about the effects of modern food production on the environment and human health. A concern expressed in the burgeoning demand for organic produce, including fruit, grown without modern pesticides, fungicides and herbicides. The key to successful organic fruit growing lies with varieties that are naturally resistant to disease and can be grown without the protection of chemical sprays. To try to meet this challenge growers are reviving and planting old varieties and fruit breeders throughout the world have turned their attention to raising new varieties that combine good quality with disease resistance.

Fruit Collections are the gene banks for this work. Most of the major apple growing nations have large apple collections, where breeders seek out forgotten varieties that have the resistance they need. At the same time it has been felt that the conservation of fruit varieties is too important a task to leave to government institutes alone. There is a wide movement to rediscover and conserve old varieties by growing them in private gardens, planting them in parks, and by forming regional collections. As a result, nurserymen's lists have expanded, and across the world local groups have formed to study and increase public interest in their fruit heritage. In many countries government and local groups collaborate and, as in England, many of these collections are now open to the public.

In addition, farm shops, which began to appear in the 1960s and 1970s in both Britain and America, have responded to the increased demand for seasonal apples and a more extensive range. Meanwhile, in France and other parts of Europe, local markets and regional agriculture have helped to sustain awareness of local varieties. The popularity of local farmers' markets in America has also spread to England, and created a niche for local produce with the emphasis on organically grown foods.

Since 1990 Britain has celebrated Apple Day on 21 October. Apple Day was launched by the charitable trust, Common Ground, to raise public awareness of local fruits and highlight the importance of orchards in the landscape of fruit growing areas, such as Kent and the West Country. Traditional orchards of standard trees are now on the conservation agenda and fruit trees can be given preservations orders. Conservation and planting grants for fruit trees are available. Sadly, however, neither the interest in the diversity of apples nor of locally grown produce has halted the decline in English fruit growing. The supermarket demand for perfect yet cheap fruit has made it increasingly difficult for the English grower to compete with foreign imports, despite a growing demand for home-grown well-flavoured apples.

We hope that this book will encourage even more people to discover the interest and pleasures of good apples, among both new and old varieties.

> W*ill there be a singer*
> *Whose music will smooth away*
> *The Furrow drawn by Adam's finger*
> *Across the meadow and the wave?*
> *Or a runner who'll outrun*
> *Man's long shadow driving on,*
> *Burst through the gates of history,*
> *And hang the apple on the tree?*
>
> EDWIN MUIR: *Merlin*

135

Plate 25: Foxwhelp, Kingston Black

Chapter 6
The CIDER STORY

In Herefordshire and north of the Severn generally, the cider is light and brisk, though the best is nutty and mellow. Devonshire cider is heavier and sweeter, and often luscious as honey. In Worcestershire it is light yellow in colour, and has a tart and stimulating after taste. Somerset cider is more of the Norman type, full of flavour, but with a pronounced acid tang. Excellent cider is now made in Kent, but it lacks a characteristic savour. Norfolk cider is dark, and rather flat and insipid.

P. MORTON SHAND: 'CIDER AND CIDER-MAKING',
The Listener, 7 October, 1931

JUST AS THE FLAVOUR OF FINE DESSERT APPLES bears comparison with the subtleties of good wine, so the best ciders have always offered connoisseurs similar satisfaction. Depending upon the variety – or combination of apple varieties – from which it is made, the qualities of the local soil and climate, and the techniques used to extract and ferment the juice, cider can be sparkling or still, dry or sweet, smooth as silk or mouth-puckeringly tannic. Cider can be the equal of fine wine – full of fruit, refreshingly astringent, with a strong bouquet – but it can be the roughest, rustic brew imaginable. Although it is now mainly associated with northern Europe and north-west Spain, cider is also produced on a significant scale in Japan, Korea and the USA where the term is applied both to the unfermented juice, dubbed sweet cider, and the alcoholic version, which is known as hard cider. In the past, cider was almost certainly made at some time or another in every apple- and crab apple-growing region of the world.

Its origins are lost in antiquity, but as fermentation occurs naturally whenever yeasts – the ubiquitous micro-organisms which convert sugar into alcohol – come into contact with the juice of organic material, the possibility of making an alcoholic drink from apples must have been discovered and rediscovered many times over. The Celts, Greeks and Romans all seem to have had fermented apple drinks, and cider making of some kind seems to have survived the retreat of the Roman forces from Britain and France.

One of the simplest techniques was to pile the ripe apples into a trough or hollowed-out tree trunk and beat or crush the fruit with sticks to release the juice, which was then left to ferment. Cider was still being made in this way, in England, as late as the 17th century. Another method was to steep pieces of apple in water which made a more dilute brew. Charlemagne's laws included regulations governing the production of this *dépense*, and it seems likely that his decree in the year 800, to plant fruit trees in every town, was partly a device to keep his subjects well – and contentedly – supplied with alcohol.

A horse-drawn cider mill at Tintern in the Wye Valley. This area was renowned for its cider, particularly the strong brew made from the local variety, Forest Styre.

The next development in cider making was to use mills and presses to extract the juice. The Greeks and Romans crushed olives using a heavy upright wheel which ran round inside a circular trough, and they may well have applied the same 'edge runner' technology to the processing of apples. The technique was certainly adopted in the Arab world, and it was from here that the technology was introduced, or reintroduced, into medieval Europe via north Africa and Moorish Spain.

In France, cider making centred first on Normandy. Monasteries and manors invested in acre upon acre of orchards. In 1371 nearly as much cider as wine was sold at Caen, and cider was also shipped up the Seine to Paris.[1] Quality was high. When François I travelled through the region in 1532, he ordered several barrels of cider made from the variety 'Pomme d'Éspice' and drank it all himself. Some 82 varieties of cider apples were listed by Charles IX's Norman physician Julien le Paulmier, in his treatise *De Vino et Pomaceo* of 1588, and such was his praise for cider itself that production spread beyond Normandy to Brittany, Maine, Picardy, l'Île-de-France and Orléans over the next two centuries. The wooden mill trough, or *l'auge*, even gave its name to Normandy's most famous cider region, the Pays d'Auge.

The Normans also seem to have been responsible for stimulating the British enthusiasm for cider. After their arrival in the 11th century it soon became the most popular drink after ale, and was widely used to pay tithes and rents. In 1204, the tenancy of the manor of Runham in Norfolk brought in an annual rent of two *mues* of cider or 'wine made of pearmain' apples, and a century later as many as 74 out of 80 west Sussex parishes paid their church tithes in cider.[2] The most skilled cider makers were the monks, who found it an obvious supplement – particularly in the English climate – to their wine-making activities. Most monasteries planted orchards of at least three acres, which provided not only enough cider for their own needs but a lucrative surplus as well. At Battle Abbey in Sussex the cellarer's accounts record that in 1369–70, he made '55s. from the sale of 3 tuns of cider from the garden, 12 tuns having been deducted as expenses, and 20s, for the purchase of barrels and the collecting of apples'.[3]

Throughout the middle ages cider and ale battled for first place in the English tankard. Both were essentially popular drinks, enjoyed by all but the very rich who preferred foreign wines, and the poorest of the poor. Cider received a temporary setback in the early 16th century when hops – which improved the flavour and keeping qualities of ale – were introduced from Flanders, but a century later the tables were turned when cider, like apples in general, received the powerful boost of Puritan favour. In 1607 John Norden's *The Surveior's Dialogue* proposed that apple and pear trees should be planted in hedgerows, and the following year Arthur

Standish included in his *Commons Complaint* a plea for the planting of 'apples, Wardens, Pears, Walnut Trees and Chestnuts'.[4]

The aims were practical as well as spiritual. Britain was short of wood, but as well as providing much needed timber, apple and pear trees would keep even a small establishment in cider and perry for a year and give them a surplus to sell. Drinking cider instead of beer also helped save wood, which was used as fuel to malt the barley. Further up the social scale, a switch from wine to cider was seen as a desirable step in lessening British dependence on potentially hostile nations abroad.

The campaigners were assisted in changing people's drinking habits by the Civil War. Soldiers camped in the West Midlands developed a taste for the local Herefordshire cider and sang its praises as they moved around the country. The brew also received the royal stamp of approval. The Herefordshire parson and cider expert, John Beale, recalled that 'when the late King (of bleffed memory) came to Hereford, in his Diftrefs, and fuch of the Gentry of Worcestershire as were brought thither as Prifoners; both King, Nobility and Gentry, did prefer it before the beft wines thofe parts afforded'.[5] Furthermore, Cromwell's Navigation Act of 1651 – which was designed to prohibit the use of third country ships between England and foreign ports – had the effect of restricting the import of French and German wine.

Cider making received its greatest boost, however, when mainstream agriculture went into depression in the 1660s. This sent farmers and estate owners searching for profitable supplements, among which apple orchards were the most attractive. With no easy access to large markets in most parts of the country, cider rather than table fruit was the main option, and in any case cider apples were easier to grow than apples for sale. Appearance and size were not important and cider orchards combined very well with other farming activities. Most cider apples were not ready for gathering until late October after the corn harvest was safely in, and would be processed in November and December, which were otherwise slack months. Orchards were also very accommodating. Other crops could be grown between the widely spaced fruit trees, or the land could be grassed over to provide sheltered pasture for cattle or dairy cows. Pomace, the remains of the apples from the press, could also be fed to the stock, bringing cattle's coats to a healthy sheen and, in Beale's opinion, accounting for the fine flavour of Herefordshire's bacon, which he claimed was superior to any other in the kingdom.

By the end of the 17th century, orchards and cider production had become firmly established as part of the farming pattern of the southern half of England, particularly in the West Country, the West Midlands and Shropshire and across the borders into parts of Monmouthshire, Breconshire and Radnorshire. In Devon, the main cider areas were the South Hams district, and the area around the busy port of Exeter. In Somerset, orchards were mainly around Taunton and on the Levels to the east. The mild, moist climate and the rich meadow grass also supported dairy herds, and the West Country became equally famous for its cider apples, clotted cream and Cheddar cheese. It was said that in parts of the West Country the area of orchards had doubled by the end of the 17th century. According to Daniel Defoe, 10–20,000 hogsheads of cider were shipped annually from the Exeter region during the 1720s, and Hereford was exporting cider in bottles to London.[6]

The 17th century saw not only an increase in the amount of cider made and drunk, but also a marked increase in quality. Cider enthusiasts such as John Beale and later John Evelyn and

Systema
Agriculturæ.
Being
The Mystery Of Husbandry
Discovered and
layd Open
by
J W

H Van Houe Scul

John Worlidge had begun to devote considerable attention to improving quality. The ambition was to produce a drink to rival foreign wine, and they soon realised that this depended primarily on the apples used. Cider can be made with any mixture of surplus eating and cooking apples and it is traditional in many counties, including Kent, Sussex and East Anglia, to use table rather than cider apples. It can even, as happens now, be made from concentrated apple juice. But it is the selection and blending of specific *cider* varieties which gives character and savour, and it was this which the 17th-century cider makers set out to achieve.

Cider apples fall into two main groups, the bittersweets and the bittersharps. Both are firm fleshed, high in tannin and readily yield up their juice. Bittersweets, as the term implies, contain the two opposing tastes. The rind and the flesh have a bitter taste, while the juice is sweet and luscious. The juice of bittersharps has less body than the bittersweets but a brisker flavour due to their high acid content. While the sugar in the apple gives the alcohol, it is the tannins that give the cider body, contribute the spicy aroma and 'fine' or clarify the juice; the acids give it balance and zest. During the Protectorate, John Beale, who came from a long line of Herefordshire cider makers, had set about systematically tasting the local brews from different parts of the country and trying to establish the best cider varieties and combinations, the best soils and localities. Later, at the Royal Society, Evelyn and his associates discussed flavours in as much detail as any wine connoisseurs, and Worlidge catalogued every variety that might conceivably be of use in his *Vinetum Brittanicum* of 1676, which became the ciderist's Bible.

The fame of Herefordshire cider lay with the Redstreak apple which could 'excel common cider as the Grape of Frontignac, Canary, or Baccharach, excels the common French grape'. It had been raised from a pip by Charles I's ambassador to France, Lord Scudamore of Holme Lacy. Evelyn recalled that it was following 'the noble example of my Lord Scudamore and some other spirited gentlemen in those parts [that] all Herefordshire is become in a manner but one entire orchard'. According to the cider poet John Philips, the Redstreak was '. . . a pulpous fruit, With Gold irradiate, and Vermilion shines'. In the opinion of Captain Sylas Taylor, another contributor to Evelyn's *Pomona*, it made cider of 'a sparkling yellow, like Canary and full of good body', but it required, said Dr Beale 'Custom and Judgement to understand the Preferency of the Red-ftrake, whofe mordicant Sweetnefs moft agreeably gives the Farewel, endearing the Relifh to all flagrant Palates; which both obliges, whets and fharpens the Stomach with its mafculine and winey Vigor'. Captain Taylor found the taste 'like the Flavour or Perfume of excellent Peaches, very grateful to the Palate and Stomach'.[7]

Another variety that stood on its own was the Foxwhelp, which reputedly derived from the Forest of Dean in Gloucestershire. This produced a full bodied cider and 'a long lasting drink' with a 'vinous, musky, flavour'. The older and more generally planted Genn et Moyle yielded cider of 'smaller body' that appealed to 'tender palates', but those accustomed to the more robust Redstreak considered this 'too effeminate and soft'.

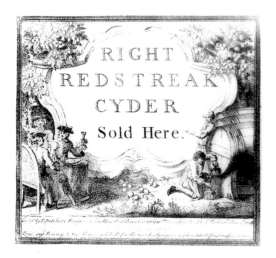

(Above) *The prized cider is being drawn from the barrel and served to tavern customers in fluted glasses.*

(Opposite) *John Worlidge wrote an agricultural manual which was published with this frontispiece showing an English landscape dominated by orchards.*

Redstreak and Foxwhelp were both winter fruits ready for milling in December, as were two other famous varieties, Hagloe Crab and Forest Styre. Both strong bittersharps, they produced ciders with a high acidity, which helped protect against spoilage. Carefully stored, the acidity would mellow in the barrel by the following summer, and in the case of Forest Styre produced a treacherously strong brew.

Softened by age, it youthful Vigour gains,
Fallacious drink! Ye honest men beware!

Around London, and in the southern counties, Golden Pippin was esteemed for making 'the most delicious liquor, most wholesome and most restorative'. The old Pearmain, too, could compete with the best in some people's opinion, but others found it weak 'unless encouraged with some agreeable Pippin to inspirit it.' Pippin cider was, however, generally regarded as inferior to brews made from the celebrated cider varieties, although some authorities, notably Sir Paul Neile, did consider that the best eating apples could produce a pleasant liquor. A summer cider, ready to drink in a month, was widely made from early apples, such as Codlins, Jenettings, Spice Apple and Summer Queening, while Gilliflowers, Marigolds, Golden Reinette, Winter Queening, as well as Pippins and Pearmains all went into winter cider.

As well as experimenting with different varieties, 17th-century enthusiasts examined cider-making techniques. After gathering the fruit was 'tumped' up in the orchard and one of their first recommendations was that the apples should be brought under cover instead, and stored for a month or two before processing so as to give well-ripened fruit with plenty of sugar and flavour. As a result, lofts became a feature of West Country cider houses where the cider was made and stored. The next stage in the process was the crushing of the fruit, either in a large stone mill, or the new, small wooden 'Ingenio'. This was the invention of John Worlidge, and as well as being relatively cheap to buy, portable and efficient, the Ingenio also found some favour among the experts as its rollers, unlike the mill wheels, did not grind up the pips, which could sometimes impart a bitter taste.

The serious cider maker would not press the crushed apples immediately, but allow them to stand for at least two days. This gave 'a more amber bright colour' to the finished cider and extracted maximum flavour and perfume from the skins and flesh of the fruit. Some tannin was also oxidised in the process so the high astringency of the juice was reduced. The crushed apples were then transferred to the press and packed in layers to form a 'cheese'. A press could be a modest affair, with the plate screwed down by hand, or an enormous contraption which took all day to load and exerted its pressure via a system of levers made out of tree trunks. In Herefordshire, each layer was enclosed in an envelope of horsehair cloth while, in Devonshire, it was customary to use straw and reeds. The expressed juice would then be strained and transferred into large barrels or hogsheads, which could hold up to 100 gallons (455 litres).

Nothing else was added. The natural yeasts, picked up from the orchard and accumulated on the presses, could now get to work. The liquor would be left to ferment without any further attention apart from keeping it topped up to prevent air getting in. Keen ciderists, however, began to follow the French example and adopted a more painstaking approach. In order that fermentation should proceed as slowly as possible they would transfer the juice to a vat in a cold cellar. In the vat, the heavier particles would sink to the bottom while the rising bubbles of gas from the fermentation carried pectin and other insoluble substances to the surface where they formed a brown cap. The clear liquor which formed between the upper and lower lees was then racked off into a fresh cask. This step removed not only debris, but also micro-organisms which might spoil the cider. It also depleted the brew of yeasts and yeast nutrients, so that the subsequent fermentation was weak and gentle, and produced a cider with the desired residual sweetness. The process, which was called 'keeving' in England and *défécation* in France, could be refined further by using another Worlidge invention – a glass siphon. This made it easier to carry out the critical process without accidentally exposing the liquor to air.

For best results, the advice was to store the cider in casks that had been previously used for 'Canary, Malaga, or Sherry wines or after Metheglin'. This, according to Worlidge, would 'advance the colour and flavour of cider'. He also recommended an improved regimen of hygiene. Old musty barrels were to be refreshed and scented using a sulphur candle made from a scrap of canvas dipped in a mixture of brimstone, alum and *aqua vitae*, and sprinkled with nutmeg, cloves, coriander and aniseed. This was set alight and wedged in the bung hole.

17th-century cider making. The apples are being crushed in a wooden 'Ingenio' (top), and the resulting pomace is then squeezed in a portable screw-press (below).

Even the experts recognised, however, that things might not always go according to plan. Should the cider fail to clear, Worlidge advised using isinglass, which is a type of gelatin used to 'fine' white wines. Should the cider go flat then a small parcel of ground apples, honey, or sugar might revive the fermentation and introduce some sparkle. Too acidic a cider could be ameliorated by the sweetness of crushed wheat or a pound or so of figs, while the corrective for musty cider was mustard seed.

Most 17th-century ciders were drunk as draught cider from the barrel but this had its problems. It was well known that air was the mortal enemy of cider and once the barrel was tapped there was the ever-present possibility that the brew might turn to vinegar. For the serious cider maker, bottling brought another dimension to his art.

The invention of the tough glass, wine and cider bottles some time during the 1630s is credited to Sir Kenelm Digby,[8] an English gentleman, wit and writer on everything from religion to cookery. Previously, English glass had been of the expensive, fragile, Venetian type, but James I's ban on timber-fired furnaces to save wood provided the incentive for setting up coal furnaces. The higher temperatures of coal furnaces allowed a stronger glass to be made. There is some suggestion that Sir Kenelm's new process was invented at Newland-on-Wye in Gloucestershire, close by the collieries of the Forest of Dean, and at the very heart of the area

143

Lord Scudamore (1601–71) is credited with planting the pip which gave rise to the renowned Redstreak apple; pioneering bottled cider; and setting the fashion for drinking cider from special glasses resembling champagne flutes. His own flutes were engraved with the royal arms on one side and the Scudamore escutcheon on the other.

where the renowned Styre cider was made. It could even have been English cider makers who pioneered the use of glass bottles. Cider was being bottled in Herefordshire by Lord Scudamore during the 1640s but it took until 1709 for Sir Kenelm's process to be adopted in France, at which time bottles were still being described as the 'English fashion'. Sir Kenelm himself was keenly interested in cider, a Fellow of the Royal Society, and among the contributors to Evelyn's *Pomona*, all of whom followed the Scudamore lead.

Bottling not only kept cider sweet but also – and even better – produced a sparkling drink that fizzed in the glass like the newly invented champagne. Although perceptible fermentation had ceased, a minute conversion of sugar into alcohol continued in the bottle. As the bubbles of carbon dioxide produced were trapped, they created an effervescence which was released when the cider was uncorked. The carbon dioxide also acted as a preservative and barrier at the top of the bottle and kept the dreaded vinegar bacteria at bay. It was crucial, however, that bottling took place at just the right time – too early and the bottles would burst – and that the corks were wired on, and the vintage kept cool. A cellar would do, but it was even better if the bottles were buried in sand, suspended in wells, or packed into little vaults cut especially for the purpose in the sides of the well shaft. Best of all was to stand them in a 'cistern of spring water'.

FINE WINE AND GOOD HEALTH

THE RESEARCH AND CARE which went into 17th-century cider production paid off. The finest vintages did appear to equal foreign wines in subtlety and quality, and ciderists vied with each other to achieve the finest vintages. To show off the sprightly, golden liquor to best advantage, hosts served their prized ciders in tall, graceful flute glasses resembling those used for champagne. These were as far removed as possible from the crude pottery mugs from which the farm workers drank their lesser brews. The poet, John Philips, whose family lived in Herefordshire, went so far as to compose 1500 lines of verse in celebration of the cultivation, manufacture and virtues of his native drink.

Nectar! on which always waits
Laughter and Sport, and care beguiling Wit,
And Friendship, chief Delight of Human Life.
What should we wish for more? or why in quest
Of Foreign Vintage, insecure and mixt,
Traverse the extremist World; why tempt the Rage
Of the rough Ocean! when our native Glebes
Imparts from bounteous Womb, annual Recruits
Of wine delectable, that far surmounts
Gallic, or Latin grapes, or those that see
The setting Sun near Calpe's tow'ring Height.[9]

PLATE 26: ROSEMARY RUSSET

In John Worlidge's opinion, too, it was superior to any foreign liquor.

*H*aving tasted a little of those several Dainties that are in most Countries liquidly prepared to please the Palate, I hope every English man; or Native of this Isle on his return hither will conclude with me, that our British Fruits yield us the best beverages, and of these Fruits the Apple the best, which is here called Cider.

Cider rapidly became the national drink, with more cider than ale houses reputedly licensed in London during the reign of Charles II. It was canvassed as a particularly wholesome and health-giving drink with all the benefits of apples as well as some extra ones of its own, including the promotion of longevity. In 1667 the vicar of Dilwyn in Herefordshire reported that parishioners, who lived to ages ranging from 90 to 114, had drunk nothing but cider. Even more remarkable, he claimed that a Morris Dance had been performed by ten people whose united ages amounted to more than a thousand years.[10] Baked apples and sugar had long been prescribed for all sorts of problems, especially 'to abate the tedious cold', but in Worlidge's opinion 'cider is much to be preferred, it being the more pure and active part separated from the impure and seculent and without all peradventure is the most wholesome Drink that is made in Europe for our ordinary use'. To relieve colds it would often be heated up and spiked with ginger and gin. Special copper 'cider shoes' were made, whose 'toes' were pushed into the fire to warm the mixture.[11] Worlidge also held that there was 'Not any drink more effectual against Scurvy, against the stone, disease of spleen . . . and excellent against Melancholy'. Cider was also widely held to be good for rheumatism:

*W*old Zam could never goe vur long
Wi'out his jar of virkin;
A used the Zider zame's twur ile
To keep his jints vrim quirken.

In addition, cider was the 'proper vehicle to transfer the vertues of many aromatic and medicinal drugs, spices, flowers, roots etc into every part of man'. Just two to three spoonfuls of elderberry juice to one quart (1.14 litres) of cider for example would endow it 'with all the medicinal virtues of elderberries'.

Cider apples also found their way into the kitchen. Gennet Moyle was excellent baked, Foxwhelp made apple sauce of an appealing 'rough piquant flavour' and cooks claimed that the Coccagee 'triumphs over all others in sauce, tarts and pies as much as its juice does in cider'. Varieties now known as sharps, such as Tom Putt for example, are lower in tannin than the bittersharps and close to culinary apples, and these could be sold as 'pot fruit' as well as contribute to the cider blend. The pure sweets, such as Morgan Sweet, are also low in tannin, and sweet and juicy enough to eat fresh.

Boot-shaped cider or ale warmers. These were made from copper, brass or tin. Cone-shaped versions were also used.

Y THE END OF THE 17TH CENTURY, cider and cider making had become part of the fabric of English rural life. The finer vintages were the preserve of the well-to-do, but less sophisticated brews and the weaker cider known as water cider, ciderkin, purr or purrkin were the staple drink of all country people. The weaker cider was made by re-working the pomace after it had been pressed. Water was added and the pomace was pressed again and the juice fermented. It had a low alcohol content and could either be used to dilute a strong cider or be drunk on its own as 'common cider'. It was often regarded as a servant's drink, analogous to 'small beer' which was made by a second working of the barley.

In the 18th century, a cider allowance became part of a farm worker's wage and the cider house or 'pound house' in which the cider was made and stored was the focal point of farm life. The usual allowance was two quarts a day for a man and one quart for a boy, but a man mowing with the scythe might consume half a gallon (over two litres) before breakfast. The cider was taken out to the fields in special containers, known as owls, firkins and jars. The ancient 'owl' or 'hedgehog' was a round earthenware pot suspended on a thong. A firkin was a small wooden cask that at its largest held up to half a gallon of cider and at its smallest – called a 'goose-egg'– a mere half a pint (284 ml). These containers would be hung from the horse's collar and tapped as required. In hot weather, stone jars were preferred for keeping the cider cool. Protected by cradles of basketwork, they would be brought out to the field where the two-handled cider mug, called a clome or clombe, would pass from man to man.

> *Oh let the cider flow.*
> *In ploughing and in sowing,*
> *The healthiest drink I know*
> *In reaping and in mowing.*

Wassailing the orchards on the eve of Twelfth Night – the practice of thanking or appeasing the deity of apple trees to ensure next year's crop – became one of the most important events of the year in cider counties. The origins of wassailing are obscure, but it was widely regarded as essential if disaster was not to befall the orchards. The custom survived in places until the beginning of this century and attempts have been made to revive it as recently as the 1970s, when the Taunton Cider Company staged an annual wassailing ceremony.

In 1686 John Aubrey described the Somerset ritual as follows:

> *the ploughmen have their Twelve-cake, and they go into the Oxhouse, to the oxen with the Wassell-bowle and drink to the ox with crumpled horn, that treads out the corne; they have an old concerved rhythme; and afterwards they goe with their Wassell-bowle into the orchard and goe about the trees to blesse them, and putt a piece of tost upon the roots.* [12]

In some areas it was essentially a family occasion with the grandfather or head of the clan leading the procession to the family's orchards and making the offerings. Everyone down to the babies had to be there or ill luck would follow. Elsewhere, groups of neighbours would go from orchard to orchard, performing the ceremony in each one.

In most places, the ritual retained something of the nature of a sacrament despite all the jollifications. Sometimes, as in west Somerset, a jug of cider and pieces of cider-soaked toast, or cake, were placed in the 'vork' of the biggest apple tree as if to honour the gods. In other versions,

Wassailing the apple trees in Devonshire on the eve of Twelfth Night, 1851. The practice of thanking or appeasing the deity of the apple trees was one of the most important events of the year in cider counties.

the trees would be sprinkled with cider in turn. In the eastern part of Cornwall and western Devonshire it was the custom to take a milk-pail full of cider and roasted apples into the centre of the orchard, and each person would dip a cider mug into the pail, toast the trees, and throw the remainder up into the branches. Sometimes the boughs would be lowered to the bucket as if to drink.

The pieces of toast or cake were everywhere regarded as being 'for the robin' or the 'blue tit'. The origin of the connection with apple trees is lost but both birds were once sacred. In a Devonshire version, written down in 1876, it was the custom to hoist a small boy into the tree to perch on a branch. He was 'Tom Tit' and had to cry 'Tit, Tit, more to eat', upon which cider-soaked toast and pieces of cheese were handed up to him. More commonly, the pieces of toast would simply be hung from the branches. It was also important to leave a few small apples on the tree for the 'piskies' (pixies).

In most versions the offerings would be accompanied by a chant or song.

Here's to thee, old apple tree.
May'st thou bud, may'st thou blow,
May'st thou bear apples enow!
Hats full, caps full!
And my pockets full, too! Huzza!

The ceremony would usually conclude with a great deal of noise. Horns would be sounded, kettles, trays and the trees themselves beaten with sticks, and shots fired up into the branches. This was regarded as either frightening away the evil spirits, or making sure the tree god woke up in time for spring. An account in the *Gentlemen's Magazine* of 1791 describes the celebrants as then bowing to the trees three times at the end of the ceremony, each time raising themselves as if with heavy baskets of apples.

THE 17TH CENTURY also saw a short-lived English attempt to popularise distilled cider spirits. The art of distilling was known to the Chinese as far back as 800 BC and also to the Persians, and it seems to have been through contact with the Arab world that the process was introduced into Europe. At first, spirits were intended strictly for medicinal purposes, acting as vehicle for numerous mineral and vegetable infusions, and do not appear to have been established as drinks in their own right until the 16th century. By this time cider was being distilled in Normandy, and the wine growers of Alsace were producing brandy from their excess wine. The Dutch, who were the main wine shippers, quickly realised the commercial advantages of distilled wine and grain spirits (gin), with their combination of high value and small volume, and during the 17th century they persuaded the wine growers of Cognac to distil their wine. It was the import of gin and this newly popular French spirit – brandy was first mentioned in England in 1678 – that probably stimulated experiment with a home product based upon cider.

Richard Haines, author of *Aphorisms on the New Way of Improving Cider*, published in 1684, proposed that the cider be distilled into 'Cider Brandy', which would then be added to fresh cider to give what was in effect a fortified wine. The technique was to:

> take eager, very hard or sowre Cider (for that yields by much the more Spirits) twelve Gallons; distil it as other Spirits are distill'd, in a Copper Body and Head, and a refrigeratory Worm running through a Cask of Cold Water, under whose Beak a Receiver is placed. From which, with a gentle Fire, draw off two Gallons of Cider Brandy, or Spirits, for the Use [of Royal Cider].[15]

Neither cider brandy, nor the fortified 'Royal Cider' became popular, however. Like cider itself in the next century, it lost out to French brandy at the top end of the market and to cheap Dutch gin at the bottom.

THE HARBINGERS OF ENGLISH CIDER'S DECLINE were the cider merchants or middle men who seized on the popularity and profitability of the new national drink as an opportunity to make a quick profit. They bought up unfermented juice straight from the press and, in pursuit of quantity rather than quality, turned out barrel upon barrel of watered down, or otherwise inferior brew which they 'flavoured and fortified to suit in their own estimation the public taste'. Some of this went on sale in London and Bristol, the great centres of the 18th-century trade, while rather more found its way 'to the Continent to return again to this country, in the shape of cheap Hamburg Ports and Sherries; or, more probably, it was manipulated at home for these purposes.' Even their best cider, which they bottled, was held to have served 'to (mis)represent pure wholesome Cider in the home market.'[14]

Cider found itself in the vice of a two-pronged attack on its virtue. Not only did the emergence of these inferior ciders threaten its general reputation, but at the same time it had started to face renewed competition from wine at one end of the social scale and from beer at the other. Whereas in Gloucestershire at the beginning of the century 'the principal gentlemen of the county rivalled each other in their cyders' and 'wine was seldom produced, but at superior tables and then only occasionally', by the end it was reported that 'the case is altered; and cyder and perry are seldom produced but at dinner and then only for a draught, as small beer; after the cloth is taken away you must treat with foreign wines, or incur the imputation of not making your friends welcome'.[15]

Out in the streets, cider fought a losing battle for the poor man's pocket. The wholesomeness of cider in general had been challenged by an outbreak of Devonshire 'colic' which turned out to be lead poisoning. It transpired that its victims were being poisoned by the cider they were drinking, which contained lead leached out from the joints, pipes and taps of the cider-making and dispensing equipment. In addition, cider's appeal was being adversely affected by the fall-out from the propaganda war taking place between gin and beer. This was the time of the emergence of the great beer dynasties, such as Whitbread, Bass and Guinness, and in their battle to win converts from gin – which they portrayed as 'mother's ruin' – they promoted beer as the giver of strength and good cheer, and as a thoroughly wholesome drink. With this positive image, beer triumphed not only over gin, but also won over much of the increasingly shaky ground from cider.

At the same time, changes in the law and in agriculture were conspiring to make cider production itself less attractive to farmers. In 1763 the government decided to treat cider like wine and imposed a tax of four shillings a hogshead on cider and perry. Since virtually every farm in the western counties produced cider the levy gave the excise men the right to gain access to every farmhouse and cottage, and aroused great opposition from both producers and consumers. Lord Bute, the prime minister, was represented as the Scot imposing a yoke on the native English comparable to the 'Norman Yoke' upon the Saxons. His effigy was burnt in market squares and a flurry of broadsheets, demonstrations and processions bore witness to the strength of popular feeling.

Although the resistance was effective and the tax was reduced in 1766, it was for many farmers the last straw, and they gave up. Each successive decade of the 18th century had seen orchards become less profitable. The cost of planting orchards had doubled, while competition from cider merchants and the expanding brewing industry had kept the market price of cider down. Other kinds of land use started to appear a more attractive proposition. With soaring beer consumption, hop yards

The ROASTED EXCISEMAN
or
The JACK BOOT'S EXIT.

Protesters against the cider tax used all available weapons, from satire to pelting politicians with apples, to make their point.

were more lucrative than fruit trees, while the new developments in stock breeding and crop management – the so-called agricultural revolution – also presented a new range of profitable alternatives. Large estates were being built up, land was being enclosed at a furious pace and the possibilities of corn production seemed almost limitless. By the 1790s the Board of Agriculture's surveyor reported that in Herefordshire farmers were neglecting their orchards for the plough and considered 'cider making an intrusion on operations of greater importance'. Since then, the story of English cider, has been one of attempted, but only partly successful, revivals.

EVEN BEFORE THE 18TH CENTURY WAS OUT, the first attempt was being made to resurrect the glories of Herefordshire's cider past. Its moving force was Thomas Andrew Knight, later to become the indefatigable fruit breeder. He had spent all his life among orchards, first at Wormsley Grange in Herefordshire, then at Elton Hall in Shropshire and finally at his brother's mansion, Downton Castle in the Teme Valley. It was through his more sociable brother, the well-known aesthete Richard Payne Knight, that Thomas Andrew was brought into contact with Sir Joseph Banks, then President of the Royal Society, director of George III's botanic gardens at Kew and one of the more active members of the Board of Agriculture. With the advent of the Napoleonic war, the prime minister instructed the board to conduct a survey of the nation's resources as a step towards raising taxes, and Banks invited Knight to help prepare the report on Herefordshire.

Faced with the sad condition of his native orchards, Thomas Andrew Knight made it his personal mission, as well as his considered advice to government, to ensure that every effort was made to restore the quality and prestige of the Herefordshire cider industry. His *Treatise on Cider*, in which he spelled out each stage in its production, was published in 1797, and the *Pomona Herefordiensis* followed in 1811. The *Treatise* contained Knight's theory that the poor condition of the county's orchards was a consequence not only of bad management, but also of a decline in the vigour of the varieties themselves. In a paper read to the Royal Society in 1795, he suggested that each variety had a limited life and that a graft taken from an old variety, even from the celebrated Golden Pippin or Redstreak, would never produce a vigorous, healthy new tree. 'A tree, like an animal', he wrote, 'has its infancy, its flowering spring; its ardent strength, its sober autumn fading into age and its pale concluding winter.' Although his premise was incorrect – the problem was probably due to the build up of virus infection in the parent tree which was passed on in the scion wood – it was this line of thought which led Knight to make his first attempts at controlled fruit breeding.

Knight's *Pomona* was a handsome, illustrated volume compiled at the request of the Herefordshire Agricultural Society. It described all the locally grown varieties of cider apples and perry pears, both old and new, and provided, for the first time, reliable information about the sugar content of each variety. Knight's contacts at the Royal Society had introduced him to the portable glass densimeter and with this he was able to measure the density – and hence deduce the sugar content – of the juice of each variety. From this he was able to predict the eventual alcohol content of the cider. The densimeter later also allowed cider makers to follow the progress of the fermentation more accurately.

Plate 27: Dumelow's Seedling

Knight's initiatives were not, however, sufficient to make any real impact on cider's fortunes. Victorian agriculture prospered, and the years from the 1850s to the 1870s were the heyday of British farming. As a result, farmers took little interest in their orchards, particularly in the rich stock and grain lands of Herefordshire. There was also little demand for cider as fashionable English Society now demanded French wine as well as cuisine. Cider remained, therefore, a drink of indifferent quality; the liquor of the farm labourers – whose wages continued to be paid in cider until 1878 when the practice was made illegal – and of the urban poor who could not afford beer. Almost another century was to pass before a further attempt was made to revive fine English cider. In the meantime, however, the history of cider on the other side of the Atlantic had also been one of early triumph and subsequent decline.

THE PROTESTANT ENTHUSIASM FOR CIDER, like that for apples in general, travelled with the colonists to the New World. By 1775 one in every ten farms in New England possessed a cider mill, and cider had virtually become a form of currency. It was used to pay the cobbler, the tailor, the doctor, the subscription to the preacher, and even for the children's education. 'One half-barrel of cider for Mary's schooling' appears in a New York account book of 1805. It was freely given to travellers, pedlars, business callers, Indians and to all guests.[16]

In America, however, cider was to hold its position for longer. Apart from the fact that cider making accorded with the spiritual philosophy of the settlers, it played a central role in their domestic economy. It was a valuable way of preserving, in a limited space, some of the hundreds of bushels of apples that their orchards were producing. It was widely drunk as an alternative to water which they distrusted, and was the staple beverage, whatever the time of day. One traveller in 1789 recounted that the Virginian plantation owner, after rising at six o'clock and riding around his stock and crops, 'breakfasts about ten o'clock on cold turkey, fried hominy, toast and cyder'. Cider itself was used in the making of apple butter and apple sauce, while cider vinegar, which was considered far superior to imported wine vinegar and also cheaper, was employed for pickling fruit and vegetables.

In addition, cider in America was not subject to the government and local pressures that had damaged English cider making. European vines would not grow in America and the native grapes produced a wine that was deemed barely drinkable. Imported wine was expensive and so the cosmopolitan plantation owners of Virginia and Maryland concentrated on their ciders instead. One enthusiast went so far as to claim that from a Maryland orchard he had made 'an Excellent Cider, not much inferior to that of Herefordshire when kept to a good Age'. The best cider, and the greatest quantities were, however, made in New Jersey, especially in Newark. Fame had come as early as 1682, when Governor Carteret wrote, 'At Newark is made great quantities of cider, exceeding any that we have from New England, Rhode Island or Long Island.' That year a thousand barrels were filled. In 1810 Essex County produced 198,000 barrels and in 1812 one citizen made 200 barrels daily, the products of six mills and 23 presses.[17] The most celebrated of Newark's ciders were made from local varieties, the best known of which was the Harrison apple which produced a cider of 'a high colour, rich, and sweet of great strength, commanding a high price in New York'. Next in reputation was the red Campfield or Newark Sweeting which gave a 'very strongly flavoured cider'. The Graniwinkle was usually mixed with Harrison's for 'cider of superior quality'. For a 'fine early cider', Poveshon, ripe in September, was recommended.

William Coxe's *Fruit Book* of 1817 gave as much attention to cider as to table fruit and producers relied heavily on his advice. Coxe himself had made some of his finest liquor from the Virginian Hewes's crab, with a small portion of Harrison and the dark red Winesap of West Jersey. Like Knight, he advised cider makers to take immense pains in selecting the right varieties to grow, to use ripe fruit and to follow the keeving procedure to produce a fine, sweet cider. His emphasis on hygiene and his improved mill design both became widely adopted.

AMERICAN CIDER BRANDY, or applejack, enjoyed a similarly charmed life throughout the 18th and early 19th centuries. Every well-to-do colonial household owned a copper still for making apple and other fruit brandies and anyone could take advantage of the harsh winter to do a similar job by a different means. If barrels were exposed to frost, the unfrozen centre yielded a concentrated liquor, which if not as strong as distilled cider brandy, was at least twice the normal strength of cider and comparable with that of Madeira wine.

Commercial manufacture of applejack was also centred on New Jersey. In 1810, Essex County produced 307,310 gallons of cider spirits and by 1834, 388 distilleries were operating in New Jersey. Aside from being a fortifying drink, applejack was used to preserve peaches, plums and cherries, and it had many uses in early medicines as an antiseptic, anaesthetic, and as a sedative or stimulant depending on the size of the dose.

BOTH APPLEJACK AND AMERICAN CIDER were, however, to receive a severe blow when the temperance movement began to gather momentum. By 1833 there were over 6000 local groups in the United States dedicated to promoting moderation or abstinence in the consumption of alcoholic liquors, and the first attempts to limit the sale of liquor by law were made in Massachusetts in 1838, and in Maine in 1846. People started drinking the newly carbonated waters instead. More impassioned followers even started to cut down orchards in New England, while the new apple growers began to find the temptations of the rapidly expanding markets in high-quality table fruit far more inviting.

Its fortunes were never to revive and by the turn of the century the word cider had come to mean the unfermented apple juice which was produced from surplus eating apples. In England, however, awareness of the flourishing state of cider and Calvados production in France encouraged another attempt to resurrect a national tradition of fine local ciders.

Celebrating the end of the Normandy apple harvest.

Apples were grown all over France, but the chief cider areas were still Normandy, Brittany and parts of Maine and Picardy. Normandy's combination of dairy farming and the production of cheese, butter and cream alongside cider making had created a way of life which proved remarkably stable, and which was fostered by the close proximity of the Paris markets. The premier cider area was the Pays d'Auge, whose deep, rich soils yielded well-coloured, fine-flavoured fruit. Bittersweet varieties predominated, and made a rich, sweet cider with a pronounced bouquet. In Brittany, on the other hand, bittersharp varieties were favoured and the local cider was much sharper.

Almost every farm producing cider also made cider brandy, *eaux de vie de cidre*. The first documented reference to cider spirits in Normandy dates back to 1553 when Gilles de Gouberville installed an alembic still at his manor in Mesnil-au-Val, on the Cotentin Peninsula. The term 'Calvados' was introduced in the 19th century. It recognised the quality of the cider brandy produced in the *département* of Calvados, and the use of the title is restricted to certain regions of Normandy. The rich Norman heritage of decorated glass bottles, small decanters, pitchers, dropper bottles and so on testifies to the significance of the apple spirit in every aspect of Norman life. Calvados was no less than a universal remedy. A little drop of cider brandy heated over a burner drove off colds, and every newborn, whether a baby or a calf, received a strength-

155

PLATE 28: ASHMEAD'S KERNEL

giving spoonful of Calvados. No serious feast could proceed without the traditional *trou normand*, or small glass of Calvados, which was taken half-way through the long and copious meal, in order to reinvigorate the palate and enable diners to stay the course. Cider brandy was also used by Norman families to make a drink called *pommeau*, in which fresh apple juice was mixed with the spirit in the proportion two to one, and allowed to mature in oak casks for up to two years. *Pommeau* is still made and has become a fashionable *apéritif*.

Calvados maturing in oak casks.

As they do today, cider and Calvados, like butter and cream, featured heavily in the characteristic cookery of the region. Norman chefs will use cider to make the cream sauce for *Sole à la Normande*; Calvados is an essential ingredient of *Tripe à la mode de Caen* and used to finish the famous *Valée d'Auge* chicken. Rabbit is marinated in cider; ham and pork are cooked with cider; and Calvados is routinely added to seafood, for which the region is celebrated. Apples and cider are also the main ingredients of the region's sweet dishes. Apples may be cooked in butter, flambéed with Calvados and served, or used to fill pancakes, and *Tarte aux Pommes* has become a national dish. Another speciality is the *pâté de fruit*, the tiny squares of apple jelly that are reminiscent of 17th-century sweetmeats. The writer Mrs Robert Henrey, who lived near Le Havre, recounted in the 1950s that a rich, dark brown conserve made from a mixture of apples and Calvados often provided something sweet to finish off the meal in Norman households. Like the rest of Europe and the United States, France used the same apple varieties, such as Calville Blanc d'Hiver in the kitchen and dining room and Reine des Reinettes also went to the cider press.

The cows that grazed the celebrated Normandy orchards were haltered in this way to prevent them reaching up and damaging the apple trees.

Outside its own regions, however, French cider still took second place to wine. Then in the second half of the 19th century, a series of disasters struck the French wine industry and directed the attention of the country as a whole to investigating the production of cider as well as wine. First, in 1862, an embarrassingly large number of bottles of wine sold by reputable merchants and delivered to important foreign customers turned out to be vinegary and undrinkable. Louis Pasteur, the French chemist, who had recently proved that the agents of fermentation were yeasts, was called in to help. He identified vinegar bacteria as the cause and developed pasteurisation, a brief heat treatment which did not affect the flavour of the wine, in response. This solved one problem, but worse was to come. In 1863 vines began to die inexplicably. It transpired that phylloxera, which destroys a vine's roots, had inadvertently been introduced from America. By the 1880s this aphid, which had prevented American colonists from growing French grapes, was devastating vineyards all over France. As well as galvanising all the resources of science and horticulture into the services of French viticulture these crises also turned attention to cider.

A study of cider and perry fruits had already begun, earlier in the century, under the auspices of the Horticultural Society of the Seine-Inférieure in Normandy, and in 1862 this was extended to include all types of apples and pears. Two years later the country's pomologists succeeded in persuading the French government to mount a *Congrès pour l'étude des Fruits à Cidre* at Rouen, and meetings were subsequently held throughout the main cider districts. This resulted in the publication of the comprehensive and highly scientific volume, *Le Cidre*, in 1875, which covered every aspect of cider making. Nearly 400 varieties of cider apples from France, England, America and Germany were described, their juice analysed, and the quality of the cider they produced classified. Pasteur's work on fermentation and hygiene was brought to bear on cider production, thousands of new trees were planted, new mills were designed to process them, and each region and even locality was producing its own distinctive type of cider by the end of the century.

THE VICTORIAN REVIVAL

ALL THIS OPTIMISTIC ACTIVITY on the other side of the channel persuaded English cider enthusiasts to make another attempt to revive their favoured brew. English agriculture was booming, orchards lay neglected, and the poor fruit was turned into poor juice or poor cider and sold on to the cider merchants. Even worse was the 'horrible liquor' given to the farm workers. For this 'windfalls are used . . . or he has the second wringing . . . which can only be preserved by the addition of four gallons of hop water to every hogshead, for without this addition it would, from its thinness and inferiority, turn to vinegar'. At one point it was predicted that cider making would die out in Britain altogether. But encouraged by the new industrial markets which could be reached by rail, the members of the Woolhope Naturalists' Field Club did not allow their enthusiasm for table fruit to distract their attention from their native heritage; and the *Herefordshire Pomona* considered cider apples as well as table fruit. In 1886 the subject was covered in greater detail, by Drs Hogg and Bull in a less lavish production, *The Apple and Pear as Vintage Fruits*. The term vintage fruits was their own coinage for particularly good cider varieties, and their book became a textbook for British cider makers. Following the French example, it analysed the juice of each variety, and provided lengthy discourses on the finer points of practical cider making.

The Woolhope Naturalists also set out to discover whether the county's so-called 'Norman' cider apples, such as Hawkin's Norman and Philip's Norman, had indeed originated in France.[18] In 1883 they arranged for a collection of 85 of the best Normandy apples to be sent from France to Hereford. Not one was found to be similar to any Herefordshire fruits. The following year the experiment was repeated at the exhibition of the *Société Centrale d'Horticulture de la Seine-Inférieure* held at Rouen. Drs Hogg and Bull, and George Piper, of Ledbury, took over a collection of fruit which included 18 of the best-known 'Norman' apples of Herefordshire. They compared these with the French fruits on display, but again found them to be quite different. As a result, the Woolhope resolved that the appellation Norman be replaced by Hereford, although even today the word Norman is taken as the sign of a 'traditional' cider apple.

Whatever the origin of the 'Norman' apples, the English exhibit as a whole was considered a great success. The Hereford contingent returned loaded down with awards for their dessert and culinary fruit, their bottles of cider, and their work in compiling the *Herefordshire Pomona*. They also brought back a number of bittersweets to try out in Herefordshire. Several of these, including Médaille d'Or, Michelin and Argile Grise, which was renamed Brown Thorn, were subsequently widely planted, and helped breathe new life into the county's orchards.

Pasteur's work on the chemistry of fermentation was also taken up in England with a view to improving standards and quality. The pioneer was the Somerset landowner, Sir Robert Neville Grenville, who in 1893 set up laboratories at his home Butleigh Court near Glastonbury, which eventually attracted government support and led to the setting up of the National Institute for Cider Research in 1903 in the village of Long Ashton outside Bristol. Collections of cider apples and perry pears were established and evaluated and the expertise of the emerging biological and biochemical sciences was brought to bear on cider production.

It was not, however, the traditional farm cider makers who took advantage of these developments, but the new small cider factories. These bought in fruit and made cider in bulk for the large industrial cities, particularly those of South Wales and Birmingham. Between 1870 and 1900 no fewer than 12 factories opened around Hereford,

Judging cider at the Royal Bath and West Show of 1896. The Royal Bath and West Society was one of Sir Robert Grenville's sponsors before his work was given government support.

including the famous Bulmer firm which was founded in 1887 by the Reverend Bulmer's son, and the still thriving Weston family business.

Nevertheless, as late as the 1930s Edward Bunyard and Morton Shand could still write lyrically of the variations in English cider, which reflected the different apples from which they were made. Devonshire had become known for its sweet ciders as a result of the planting of pure 'sweets', such as Sweet Alford, Sweet Blenheim, Northwood, Slack-my-girdle and Sweet Coppin. Devon cider makers also continued the old process of keeving and produced a cider with a residual sweetness. Bittersweets characterised Somerset's orchards where the 'Jersey' apples such as the Chisel, Royal, Red, Russet and White Jersey were popular. Yarlington Mill and Harry Masters of the Martock area are also of the 'Jersey' type – an old local term for such fruits was 'Jay-see' – which produced full-bodied bittersweet cider. The most famous Somerset cider apple is, however, a mild bittersharp, Kingston Black, which originated in Taunton Vale and is a vintage variety, giving a fine cider with a distinctive flavour. Somerset and Devon also made light ciders from summer apples such as Morgan Sweet and Court Royal, which were ready for drinking by Christmas. Apples were also sold as table fruit to many industrial areas, and in the 1900s Morgan Sweet, especially, was packed off to South Wales where the miners used to enjoy them with Caerphilly cheese.

The large markets of the industrial Midlands and the valleys of south Wales prompted the cider makers of Herefordshire and Worcestershire to invest in dual purpose 'pot' fruit such as Cherry Norman, Gennet Moyle and Tom Putt as well as the French bittersweets, but traditionally the West Midlands favoured bittersharps. Gloucestershire, particularly west of the Severn, was the home of Foxwhelp and Redstreak, and their descendants still provided plenty of acidity in the blend. The Frederick apple of Monmouthshire and parts of Radnorshire was a full sharp, and the Styres of Gloucestershire gave ciders high in acid. The sharp character of these Hereford and border county ciders would, however, have softened by the summer. As cider makers now know, the acid would undergo a secondary fermentation, whereby the malic acid is converted by bacteria into lactic acid and carbon dioxide, to give a cider with half the acidity and a mellow 'nutty' flavour.

159

As well as the good local ciders celebrated by Bunyard and Shand many farmers produced more nondescript brews. Cider makers everywhere consigned surplus apples of any kind to the cider house and in counties producing mainly table fruit, any apples that did not find an outlet in the markets ended up in the cider heap. As they possessed little tannin they produced a pale drink, with a taste closer to that of a 'thin' white wine than a cider. In this they resembled German and Swiss ciders, which were also made from surplus eating and cooking apples. The days of farm cider – good, bad or indifferent – were, however, numbered. The competition from the new factory-based ciders was already making home production increasingly uneconomic, and local brews were steadily being replaced by the standardised factory product bereft of all its regional individuality. From the 1930s onwards, French and English ciders began to follow different paths and to develop into essentially different products.

THE LATE 19TH AND EARLY 20TH CENTURY was a golden age for French cider. The attention of numerous congresses, associations and publications was devoted to every step of production and nuance of quality, and in 1901 a cider research institute was founded at Caen in Normandy. The experience of controlling a complex wine industry was directly transferable to cider, and as a result cider making in France was tightly constrained. These regulations, which were first drawn up in 1905, laid down strict guidelines for each stage of the process and ensured that cider making continued along traditional lines. As is still the case today all ciders must be produced with fresh fruit, natural yeasts and the old processes of clarification.

Orchards and dairy herds both expanded during the 1920s and '30s, and in most years there would be a surplus to sell off to the English cider industry and those of Germany and Switzerland. After the devastation of the war years orchards were replanted, but the 1960s and '70s saw increased demand for Normandy's dairy produce and farmers neglected their apple trees. Now in another reversal, European milk and butter surpluses have diverted attention back to orchards, and cider production figures in the government's scheme for diversifying Norman agriculture. Traditional orchards, with the large standard trees and newer, more intensive plantations are both being established, and the cider growers are making a bold attempt to raise standards.

In 1996, a group of growers in the Pays d'Auge secured the designation *appellation d'origine contrôlée* for their cider, that is, 'true to name'. As with wine, this means that cider bearing the *appellation* is made from particular varieties of cider apple grown in that region. Production is by the old painstaking process and strictly controlled. This is seen as the only way to ensure that home-made cider with its distinctive local character will survive in the face of mass-produced factory brews. The hallmark of a Pays D'Auge cider is its fine 'mousse', or head of bubbles, its amber colour, strong perfume, and complex intense sweet taste. The sweetness comes not only from the trace of sugar remaining, but the inherent 'sweetness' of the bittersweet apples which predominate in the blend. Cornouaille cider of Brittany also secured the designated *appellation d'origine contrôlée* in 1996. Cider makers continue to produce other blends and sweet ciders, which are made by arresting the fermentation and racking off, bottling the cider when the appropriate reading on the densimeter is reached. Visitors to Normandy can follow *La Route du*

Plate 29: Médaille d'Or, Gennet Moyle

Cidre, which winds through the lanes of the Pays d'Auge, past farms where the tasting of ciders, kept at the correct temperature of 12°C (54°F), selection and purchase proceed in the same way as in a Burgundy or Bordeaux vineyard.

Normandy Calvados has been the subject of *appellation contrôlée* since 1942. Several areas qualify, and each is distilled from ciders made from permitted varieties of local apples, and aged in oak casks for two to 20 years. This gives each *appellation* its characteristic quality, and that of the Pays d'Auge is considered the most distinguished.

WHILE FRANCE HAS TRIED TO KEEP CIDER a natural product, the large English cider makers have aimed at a consistency and uniformity. The two products are almost different drinks. French cider is perfumed, very fruity, and tastes strongly of apples, whereas English cider is more alcoholic – strength has become an important quality – often with little or none of the characteristic astringency of cider apples. French cider is sold like a wine, in a champagne-type bottle, while English cider is a robust 'pub' drink which competes with beer.

Apart from the profound differences between English and French drinking cultures, the lack of restrictions on either the varieties or the production techniques of English cider, together with recent agricultural history, are among the main reasons that the situation has developed the way it has. After the last war, subsidised mainstream agriculture, and the profits to be made in dairying, hastened the trend away from orchards among the small mixed farms of the West Country and West Midlands. Factory cider producers made up the shortfall in English apples with French imports, and gave financial incentives to specialist growers to plant intensive orchards of certain selected varieties. Under these circumstances the notion of producing anything but a uniform, blended product receded.

The trend was reinforced by the advent of more stringent controls on the quality of eating and cooking apples. This produced an increase in the quantities of downgraded fruit which was mopped up by the juice industry. With no restrictions over the type of fruit from which cider could be made, the industry began to supplement supplies with juice. The industry now uses fresh and concentrated juice from cider, eating and cooking apples.

Apart from the loss of local cider varieties and blends, the old methods of cider production have also been superseded by the chemist's and the technologist's art. The original process is a seasonal one, best done in the coldest months of the year. Techniques such as keeving were already being passed over in the early years of this century, and became redundant when filtration could be used to clarify the juice, and when apples low in tannin were used. Unlike their French counterparts, English cider makers are allowed to use

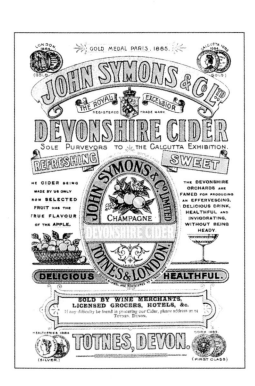

Devon cider makers prided themselves on their characteristically sweet, sparkling cider, made using traditional techniques and varieties.

sulphur dioxide to kill all spoilage organisms, and to replace the wild natural yeasts with a standardised and more reliable yeast culture. Sugar can be added to increase the final alcohol content, which can be as high as eight per cent, and all ciders are fermented to dryness. Sweet cider is made by adding a sweetening agent to dry cider, and an injection of carbon dioxide gives it fizz. French cider, on the other hand, is naturally sweet because of its arrested fermentation, which also makes it much lower in alcohol. It is always sparkling – *cidre bouché* – due to the production of carbon dioxide gas in the bottle.

English factory cider has steadily overwhelmed farmhouse cider, and since the 1980s, cider production and consumption has soared with the introduction of keg cider, which can be bought on draught. There has been little reason to revive farmhouse cider with cider apples barely paying to pick. In addition, government funded cider research at Long Ashton has ceased, thus removing the main resource of expertise and information. Nevertheless, a small number of cider producers remain, and they are bringing about a renaissance in farmhouse English cider. Mills and presses have been bought and restored, networks have been established to encourage small producers, and strenuous efforts are being made to conserve traditional old orchards and to again plant standard trees in meadow grass. Cider routes have appeared; there is a cider guide; and a cider centre outside Lewes in Sussex, at which over a hundred ciders from all over the country are available for tasting. Even the distilling of cider has been revived by Hereford Cider Museum with the launch of 'Hereford Cider Brandy' in 1987, and 'Royal Cider' by Burrow Hill Cider at Martock, Somerset, in 1992. As a result, it is becoming easier to again find local English ciders that evoke the traditional cider apples from which they were made, and to differentiate the regional qualities of the rich Devon cider, the more robust Somerset brew and the 'nuttiness' of a good Herefordshire cider.

*S*weet *cyder is a great thing,*
A great thing to me,
Spinning down to Weymouth town
By Ridgway thirstily
And maid and mistress summoning
Who tend the hostelry:
O cyder is a great thing,
A great thing to me!

THOMAS HARDY: *'Great Things'*

Loading cider apples into a storage loft at Castle Cary near Glastonbury, Somerset.

PLATE 30: NORFOLK BEEFING

DIRECTORY
of APPLE VARIETIES

*N*o other fruit unites the fine qualities of all fruits
as does the apple. For one thing, its skin is so
clean when you touch it that instead of staining the
hands it perfumes them. Its taste is sweet and it is
extremely delightful both to smell and to look at.
Thus by charming all our sense at once, it deserves
the praise that it receives.

PLUTARCH: *'Table Talk V'*

DIRECTORY
PREFACE

TOWARDS
THE
PERFECT
APPLE

VER 2000 DISTINCT VARIETIES OF APPLE grow in the former Ministry of
Agriculture Apple Collection (now Department for Environment, Food and
Rural Affairs (DEFRA)) at Brogdale in Kent, and a further hundred in the
Cider Collection. They offer not only a huge variety of colour, size, shape and
flavours, but also of origins and uses. Even their names are enormously
suggestive, spanning countries, continents and centuries: Kentish Fillbasket, Kerry Pippin,
Boston Russet, Pumpkin Sweet, Wolf River, Ontario, Tasman Pride, Belle-Fille Normande,
Orleans Reinette, Red Astrachan, Groninger Kroon, Tropical Beauty, Fuji . . . In the Directory
which forms the second part of this book, each of these varieties is listed and its characteristics
and history are noted.

Consumers have always been influenced by an apple's **colour**, which can range from the palest
milky yellow to deepest purple, while the possible combinations of flushes and stripes are almost
infinite. Many are also freckled, veined or covered in russet. Those from North America are among
the most vivid that we now possess, and include the deep maroon Baldwin, bright red Esopus
Spitzenburg and crimson Delicious, McIntosh and Spartan. The British, however, have always
held that the more softly tinted red and orange skins, with their touch of russet, of Cox's Orange
Pippin and Ribston Pippin were a sign of fine quality. More exotic examples include the Åkerö
with its pink flush over pale primrose, the almost metallic lustre of Friandise and the blue-black
Violette with its deep bloom like that of a plum. Early season varieties will be fully coloured on the
tree, but late keepers are often dull when picked and slowly develop their wonderful shades in
store; greenish-yellow backgrounds change to gold, and flushes brighten and lighten.

An even **shape** is also regarded as a desirable feature in an apple. Nowadays it aids packaging,
but in the past it was in demand for the dining table and exhibition bench. Adams's Pearmain,
which was a popular Victorian dessert apple, is perfectly conical, but the range of shapes is
almost as wide as that of colours. North American Delicious and its offsprings are heavily ribbed

165

DIRECTORY
PREFACE

TOWARDS
THE
PERFECT
APPLE

and crowned; Kandil Sinap of the Crimea is spire-like; Oaken Pin of Devonshire is oval; and Thorle Pippin and Reinette de Champagne are disc-like. Many, like the Calvilles of France, are markedly angular, while the English Catshead, which is probably a descendant of the medieval Costard, is flat-sided and almost box-like.

A good **size** is now considered essential in an eating apple, but this is a comparatively modern trend. Connoisseurs always believed that the best flavour was to be found in medium-sized fruits, which were also more appropriate for the dessert. After an eight-course Victorian dinner, nothing too substantial was required. As one country house owner explained to his gardener: 'I cannot eat a big Apple or Pear at once after dining and I cannot ask anyone at table to share a piece because it would not be decent.' These inhibitions, however, did not apply in America to fruit 'eaten out of the hand' nor in Japan where great pains are now taken to produce very large apples weighing in at nearly two pounds (a kilogram). The Japanese, like 19th-century gardeners, also 'stencil' patterns and messages on to the skin of the ripening fruit. Fuji, Yoko and Mutsu (Crispin) apples are individually packaged in attractive boxes to form an *o-miyage*, a gift for one's hostess. They also provide the souvenir that every visitor to the apple-growing area around Aomori in the north takes home. At a meal, these apples are served in slices from a plate in the centre of the table.

The nature of the apple's **skin**, whether it is thin or thick, tough or soft, is also important. In Victorian England, thin, melting skins were much preferred to the tough, waxy coats of the imported American Baldwins, and apple lovers on both sides of the Atlantic continue to bemoan the obtrusive 'chewy' skins of, for example, Spartan and Delicious. With some varieties the skin becomes very greasy after picking and this is also an undesirable feature, as it attracts dust which dulls the colour, as well as being unattractive to handle.

The **texture** of the **flesh**, too, can enhance or detract from a variety's appeal. Nowadays, crisp juiciness is the prime requirement, but the apple is capable of a much wider spectrum of texture, ranging from the melting, sorbet-like McIntosh to the crumbling flesh of a Blenheim Orange.

It is **flavour**, however, which has received the most detailed analysis and inspired the most lyrical descriptions. Edward Bunyard wrote that 'apple flavour is as varied as the scents of flowers', and Morton Shand found 'shades of the greatest refinement, subtlety and delicacy'. Flavour in apples depends upon the relationship between sweetness and acidity. When well balanced, these give richness to an apple, and plenty of sugar and acid will provide a good, intense, fruity taste. Too much acid, on the other hand, will make the fruit sour, while too little makes for blandness. The best varieties also contain volatile compounds which are released as the apple is eaten. It is these, combined with the blend of sugar and acid, that give each variety its characteristic flavour. Although there has been no systematic chemical analysis of varieties, almost 200 different volatile substances – ethers, aldehydes, esters and so on – have been identified in just the two varieties, Red and Golden Delicious.

Flavour, size and appearance are all influenced by a number of factors, particularly the weather. A cold, wet season is no good for the choicer varieties, which need a warm summer and good long autumn to build up the sugars, aromas and consequent depth of flavour. The soil affects flavour, too, as does the way that the tree is grown: whether it is an old, large standard in a grassed orchard or an intensively trained cordon, for example. In consequence, the same variety may taste appreciably different from area to area or in successive years. The tasting notes given in the Directory are based upon fruit grown in south-east England, and a number of varieties may not live up to their reputations if grown further north. Similarly, many Continental, American, Japanese and Australian apples only do well in England in a really hot summer, while a few never seem to develop properly at all and have 'green' or 'woody' flavours. In these cases, comments have been included, wherever possible, on the quality of the fruit as grown in its native country.

To appreciate any apple, the most important factor is that it should be **ripe**. Picked too soon it will be hard, taste of acid and starch, and be a disappointment. Later maturing varieties may never achieve their full flavour if harvested too soon but, on the other hand, if left too long on the tree they will not keep in store. The notes given below and in the Directory are based upon well-ripened fruit. Early varieties were eaten from the tree and later ripening fruit stored under simple, natural conditions.

DIRECTORY
PREFACE

TOWARDS
THE
PERFECT
APPLE

APPLE FLAVOURS, although infinitely varied, can nevertheless be grouped into families with a characteristic flavour that reflects a common ancestor. Thus varieties with, for example, Cox's Orange Pippin in their make-up will usually possess elements of Cox flavour. Golden Delicious has imparted a very sweet, honeyed quality to many of its offspring, and most apples with McIntosh in their parentage have its characteristic soft, juicy flesh and taste. Similarly, there are groups of apples which are akin to the much older Golden Pippin, Golden Reinette and Nonpareil.

Flavours, and to some extent colours and textures, also seem to mirror the season in which the variety ripens. Summer apples are light, crisp and refreshing, while as the season advances so flavours become more interesting and skins become bronzed and russeted. About a dozen dimensions can be distinguished, but perception of flavour is very subjective and these comments are offered only as a general guide.

Refreshing and savoury is the defining feature of summer apples. They are crisp and juicy with plenty of sprightly acidity, such as Beauty of Bath, and at the end of August, Miller's Seedling. Into this savoury category might also come the very juicy James Grieve and the old German apple, Gravenstein, which both ripen in September. A number of northern European apples such as Filippa of Denmark, which is a Gravenstein seedling. Åkerö of Sweden and the Red and White Astrachans are all brisk and fruity.

A **strawberry** flavour is found in many red August and September apples which are descended from Worcester Pearmain and also from its probable parent, the ancient Devonshire Quarrenden. These include Ben's Red and Duchess's Favourite, which were once market apples, the modern Discovery, Katy and many more. It is not so much a strawberry but **raspberry** flavour that one detects in the vivid pink Reinette Rouge Étoilée, an old Dutch or Belgian variety, and this flavour has been passed on to the modern Rubens. A number of American apples, such as the crisp-fleshed Northern Spy and Esopus Spitzenburg, also often seem to have a definite raspberry acidity.

Vinous was a term often used of apples in the past and is now usually applied to McIntosh. Early in its season, the flavour seems more like a cross between that of a strawberry and a melon, but with keeping it tends to become 'winey'. McIntosh has produced a whole dynasty of colourful offspring with juicy, melting snow-like flesh and this sweet, perfumed taste. It is seen in the August ripening Melba, and at its sweetest and most decadent in Beverley Hills. Lobo, Niagara and Milton have a similar quality, and Spartan also has a hint of strawberries when properly ripe. In the English climate, however, some McIntosh-type apples can readily tip over into 'green' or 'metallic' flavours as if tainted by aluminium.

Densely fruity is a characteristic feature of a number of small, golden apples, which are probably derived from the old Golden Pippin. This is now an almost forgotten flavour, but one with a brisk tang, and a strong and concentrated taste of fruit. It occurs in Summer Golden Pippin, Downton Pippin and Yellow Ingestrie, and also in Kerry Pippin and Wyken Pippin.

Aromatic, the most prized quality of all, is particularly associated with the best English varieties, and the rich flavour of Cox's Orange Pippin. The ancestor of Cox was probably Ribston Pippin or Margil, which both appeared in the 18th century and seem to have been the first varieties to show this more intense and complex flavour. Ribston and Cox have produced numerous offsprings, which have inherited to varying degrees their aromatic character. The family can, perhaps, be broadly divided into two groups comprising on the one hand the more robustly flavoured, which lean towards Ribston and, on the other, those that are sweeter, more delicate and resemble Cox. Into the Ribston-like group would go varieties such as Holstein, Sunset, Suntan, Tydeman's Late Orange and the more recently introduced Jupiter and Fiesta. Many more, like a number of the Laxton apples – Advance, Epicure and Superb – have less acidity and strength, but the delicacy of Cox.

The desire to create new varieties with this rich aromatic quality is not restricted to Britain and a number of modern German varieties, such as Alkmene and Tumanga, show much of the Cox flavour, while the marriage of Cox with American varieties has added a new dimension, that of an intense aromatic flavour which is also scented. The New Zealand Kidd's Orange Red, produced by crossing Cox with Delicious, tastes, some claim, of Parma violets, and the Jonathan crosses – Lucullus, Prins Bernhard and Prinses Irene – produced by Dutch breeders are all gloriously rich and perfumed. Much, however, will depend upon when they are eaten. Early in their season the taste will be sharper and more intense, but with keeping will sweeten and become more delicate and 'flowery'.

Sweet and scented is another category, and most of these are not of English origin. At its best the German Edelborsdorfer provides the most perfect example, but Edelborsdorfer does not do well in the English climate and after a sunless summer can taste rather bleak. The sweetest apples of all seem to have arisen in the United States and most of these derive from Delicious. At its prime, Delicious is scented, very sweet, and bears little resemblance to the sickly cotton wool-like specimens usually on sale. This perfumed sweetness is further intensified in Sweet Delicious and Redgold, but it needs a good summer for the Delicious family to ripen properly. Edward Bunyard believed that St Martin's, bred by Thomas Rivers, was the sweetest apple of all, but he had not tasted the almost artificially-perfumed sweetness of the Japanese Shin Indo and Sweet Cornelly from Holland, which have almost no acidity and very little apple flavour at all.

Honeyed can describe a good ripe Golden Delicious and this is a quality found in the numerous seedlings that breeders all over the world have raised over the last 50 years. Jonagold produced from a Jonathan cross has a red flush, a more balanced acidity and a more intense flavour than its parent. Elstar has a richer taste and Crispin, at its best, can be very honeyed, although often much less interesting. The most honeyed of all, however, has no connection with Golden Delicious and is perhaps Carswell's Honeydew, a Cox seedling raised in Surrey.

Aniseed is the characteristic flavour which develops in Ellison's Orange after it has been picked, and this reflects high concentrations of a particular substance that flavour chemists have identified as estragol. This aniseed flavour is even more pronounced in its descendant Merton Beauty, which can be gloriously rich, although after storing tends to become overpoweringly medicinal. Cellini, a popular apple with London market gardeners in the last century, also develops an aniseed or balsam-like taste, but it is the Fenouillet apples of France that, as their name implies, should definitively taste of fennel or aniseed. No one has analysed these chemically, but Leroy, the 19th-century pomologist, described Fenouillet Gris as 'endowed with perfume of aniseed and musk'. This is by no means so apparent in fruit grown in England, although all Fenouillet apples do develop a curious sweet and scented flavour.

Russet refers to the appearance of the apple's skin but to many people it also implies the distinctive flavour, usually associated with today's best-known example, Egremont Russet. This flavour is not, however, common to all heavily russeted apples and few, if any, develop it as intensely as Egremont. Edward Bunyard described it as 'remarkably nutty'. Shand, more poetically, defined the flavour of russets as 'nutty or marrowy with a slight suggestion of crushed ferns'. To my mind, Egremont Russet develops a 'dry' taste, which sometimes seems almost tannic, as well as nutty, and is suggestive of smoky bonfires and autumn leaves. King Russet, the russeted form of King of the Pippins, can develop a similar flavour and so does the old Irish apple, Ross Nonpareil, but taste buds and perceptions do not always agree. Morton Shand, who undertook a whole study on the subject of 'russet apples', considered that the Fenouillet apples 'exemplify our English idea of a true russet flavour'. Dr Hogg might have agreed with him. Hogg did not have the opportunity to try Egremont Russet, but in Ross Nonpareil he found 'a flavour which partakes much of that of the varieties known by the name of Fenouillet or Fennel flavoured apples'. Flavour chemists have, in fact, detected trace amounts of the aniseed component in Egremont Russet.

Nutty is the word that springs to mind when eating the crumbling flesh of a Blenheim Orange in November. The taste is very like that of Golden Reinette, which was one of the most widely grown apples of the 18th century and must have produced numerous seedlings, one of which may have been the Blenheim. Another relation of Golden Reinette may be Orleans Reinette, which was often called a 'small Blenheim'. It is similarly nutty yet also aromatic, as are Claygate Pearmain and Adams's Pearmain.

A **pineapple** flavour is evident in Pineapple Russet, Pine Golden Pippin and Lucombe's Pine, but the most strongly flavoured of all is Ananas Reinette, the Pineapple Reinette. This is still popular in Germany and it has passed its pineapple flavour on to Roter Ananas and Freiherr von Berlepsch. The ancient Court Pendu Plat also has an intense pineapple acidity, which the modern Suntan has inherited, but the most famous of the varieties possessing a 'rich pineapple flavour' was the Newtown Pippin of America. Every year, Victorians looked forward to its arrival and the strong brisk flavour inspired many enthusiasts to raise new seedlings with a similar quality, but more likely to 'finish' and ripen properly in England. Lamb's Abbey Pearmain, Sisson's Worksop Newtown and Grange's Pearmain were all raised from Newtown Pippin pips and do possess a distinct tang of pineapple.

169

A definite hint of **fruit** or **acid drops** spikes the sweet-sharp taste associated with the Nonpareil family, which are mostly very late-keeping varieties. At the sharpest end is the old Nonpareil. Its best-known offspring, Ashmead's Kernel, is sweeter, and the family also includes Rosemary Russet, Duke of Devonshire and the latest maturing of all, Sturmer Pippin, which has an underlying steely acidity.

Not surprisingly, after centuries of continual development, many varieties resist all attempts at neat classification and stand out for their uniqueness of flavour. In a perfect summer, American Mother, a native of Massachusetts, can show a 'hot', sweet, perfumed, almost vanilla flavour. In September, the Irish Ard Cairn Russet tastes unmistakeably of banana and so, too, does the American apple, Dodd, while the old Essex variety D'Arcy Spice acquires a spicy, almost nutmeg-like quality in the New Year.

BRITAIN IS THE ONLY COUNTRY to have developed a range of sharp culinary varieties. Elsewhere, the apples used in cooking are sweeter and usually dual purpose. A variety is actually viewed as deficient in America if it is not good both for eating fresh and making pies and sauces. In Britain, however, dual-purpose apples were regarded as 'neither one thing nor the other', lacking sufficient 'bite' to be a good cooker and never developing a fine dessert flavour. As a result English fruit men have never looked with much enthusiasm upon either the sweeter continental cooking apples, the Calvilles and Rambours, or American Baldwins and Jonathans. As pie apples, Edward Bunyard damned these as 'a confection of stewed wood', and for a century Bramley's Seedling has been the standard against which English culinary varieties are measured.

Many of the British cookers still grown in gardens and conserved in the national Apple Collection are, however, by no means as green and sharp as the usual Bramley. There are golden and red flushed cookers, many of which owe a good deal of their reputation to their handsome appearance, large size, regular shape and hence prize-winning place on the exhibition table. In terms of cooking properties, culinary varieties also show diversity and a progression of qualities through the season, and textures and flavours which can range from frothy to firm, and from sharp and strong to sweet and delicate.

Culinary varieties are distinguished from dessert apples by higher levels of acidity, as well as by their larger size. The degree and proportion of acid and sugar still determines taste, while the amount of acid also influences the cooking properties. Cooking liberates the acids, which break down the flesh, and in the process sharpness is reduced. Generally speaking, the more acidic the apple, the more easily it will cook to a purée. Less sharp apples tend to retain their shape during cooking, but if an apple contains too little acid it will cook poorly, failing to soften, and often discolour, turning an unattractive shade of fawn. The structure of the flesh is another factor that influences how an apple will cook. Summer codlins, which 'size up' quickly and have loose textured flesh with a good deal of water, will cook swiftly and often to a froth. Keswick Codlin, Lord Grosvenor and Early Victoria, for example, all bake to a juicy fluff.

As with dessert varieties, the cooking quality of an apple and its taste will be affected by the year and the locality in which the fruit is grown. Many Scottish and northern varieties of high culinary repute often lack acidity in southern gardens and are consequently rather bland. On the other hand, a variety such as James Grieve, which is a valued eating apple when grown in the southern counties, is very sharp if deprived of sun and is usually regarded as a cooking apple in the north.

The way in which an apple cooks also depends on when it is used, and how it is stored. Windfalls and early-picked fruit will be sharper and cook more easily than those harvested in their proper season. After picking, culinary apples, like dessert apples, continue to develop. Acidity will fall, the sugar level rise and the fruit cook more firmly. Many varieties, such as Annie Elizabeth and Newton Wonder, become dual-purpose and even Bramley's Seedling will have mellowed sufficiently by March to serve as a brisk, eating apple or for use in a vegetable salad. A number of old culinary varieties, however, while still sound and fresh-looking in the New Year, can taste flat and insipid when cooked and are often best used early in their season. This problem of fading acidity and flavour has been overcome in modern commercial stores and Bramley's Seedling stored in this way is as sharp in the spring as it was when first picked.

An easy way to discover how any apple will cook and whether it will be suitable for a particular recipe is to test it: bake the slices in a sealed foil dish, for 15 minutes at 200°C (390°F). The performance of culinary and dual purpose varieties under these conditions is included in the Directory entries. Apples that readily break down into a froth will cook quickly and easily. They will be good baked and for making 'Apple Snow' and the first apple sauce. Those that turn into a purée or retain a little firmness will also bake well and are good for making apple pies and sauce, and for most traditional English recipes. Varieties whose slices remain intact are suitable for continental and American recipes.

Although no other country has developed a range of sharp apples specifically for the kitchen, many varieties are considered particularly suitable for certain cooked dishes. Among the early ripening apples used all over Europe and North America are Alexander, Duchess of Oldenburg and White or Yellow Transparent which is ready to pick in time for apple pie on the fourth of July. A major criterion for the success of an American pie seems to be that it contains discernible pieces of apple to spike with the fork, and Baldwin, Northern Spy, Rhode Island Greening, Jonathan and Esopus Spitzenburg were, and still are, used for pies, 'Brown Betty' and numerous other dishes. In France, Calville Blanc d'Hiver was the most esteemed for *Tarte aux Pommes*, although nowadays Reinette du Canada, Reine des Reinettes and Reinette Clochard are used. The cooking apple of Holland is Belle de Boskoop, Germany has the Bohnapfel, Hornburger Pfannkuchen and Glockenapfel, which is still prized for *Apfel Strüdel*. Old specialities are, however, being overtaken by international varieties and Austrian chefs now turn to Idared and Gloster 69.

Cooks everywhere use Golden Delicious and Granny Smith and, in England, Cox's Orange Pippin. These all retain their shape when cooked, but tend to be mildly flavoured by comparison with a true culinary apple. They are best cooked early in their season when the acidity levels are higher as this not only makes them cook more easily but also gives some zest to the flavour.

171

ATTRACTIVE APPLES NOT ONLY make good ornamental trees in a garden – a few left behind after harvest will preserve a tree's decorative quality even after the leaves have fallen – but they can also create spectacular displays inside the house. A sumptuous September basket for a dining or hall table can be made from Discovery combined with a few of the darker-coloured Tydeman's Early Worcester, Red Devil or Jerseymac, highlighted by little Yellow Ingestrie. The deep red of Rubens or a Red Cox, mixed with russet-freckled yellow Ananas Reinette, is also very effective in a large bowl. The combination of Wyken Pippin with maroon Spartan or cinnamon russeted Anisa, or of red flushed Jupiter with orange-yellow Court of Wick, is exciting too. Displays of interestingly shaped apples such as Oaken Pin, or of small varieties – Api, Api Noir, Golden Knob and Fenouillet Gris – are also attractive. The possibilities are endless. On the culinary front, Golden Noble set off by purple damsons makes a superb feature, while a basket filled with Bloody Ploughman or the cider apple Stoke Red can be stunning.

Since the 17th century, a pyramid arrangement has been one of the most dramatic ways of displaying apples. To make a tall spire four or five tiers high you need regularly shaped, preferably round, flattish apples with a good colour. Red Devil and Crimson Newton Wonder are both ideal. The Victorians also garlanded their tables with small apples and sprays of evergreens. To create this effect, simply lay evergreens directly on the table-cloth and place small apples at intervals along their length. Periwinkle, either in the variegated or green form, is long-lasting, and fronds of blackberry foliage also work well. In a similar way, apples can be used to decorate Christmas garlands and swags. They can be attached to branches of yew, box, ivy or other evergreens by pushing a piece of thin wire through the centre of the apple from the stalk end to the apex or eye. Make a little hook at the eye end of the wire to anchor it, and leave enough wire at the stalk end to twist around the branch. The addition of a few sprigs of mistletoe and candles to a circlet of evergreens and apples will create a Kissing Bough.

IN THE DIRECTORY a number of points have also been included on the cultivation of each variety, namely its flowering time; vigour and habit; its cropping quality, resistance or particular susceptibility to disease; any other features of interest; and its season of picking and use.

Flowers (**F**): Varieties with particularly attractive blossom are highlighted (**F***). When making a selection of varieties for their ornamental as well as practical value, characteristics such as the timing of the flowers (early, mid or late) and the reliability or otherwise of the crop (and therefore of the blossom) should be borne in mind as well as the actual colour and size of the flowers.

Apple blossom petals can range from deep maroon through palest pink to snow white. Often the flowers, like those of James Grieve, are scented; and in Gravenstein they are as large as a wild rose. The main flowering period in southern England is early to mid-May, although in recent years mild winters have advanced the period to mid-April. There are varieties which blossom at least two weeks before the majority of apples, and others that do not come into flower until June. Very late flowering varieties are of particular value in frost-prone areas, and many including

PLATE 31: NEWTOWN PIPPIN

Stages in the development from flowers to fruitlets.
1. Green bud. 2. Pink bud. 3. Full flower. 4. Petal fall.
5. Early fruitlet.

Court Pendu Plat, Newton Wonder and Edward VII owe some of their lasting popularity to this property. Feuillemorte always lives up to its name, often seeming as if it will never come into flower.

In Britain, spring temperatures are very variable and often low, which makes pollination a problem. Not only does the process of fertilisation itself require a minimum temperature, but cold air and winds inhibit the activity of bees and other pollinating insects. Some varieties are partially self-fertile, but in the British climate all apple trees produce better crops if they are cross-pollinated by another variety which flowers at the same time. The average, peak flowering time is given in each Directory entry.

Most varieties are diploids and will usually cross-pollinate each other, but some are triploids (trip) and this is indicated in the Directory. Triploids produce very little viable pollen and cannot be used as pollinators. For their own successful pollination and good crops they need two other diploids (to pollinate the triploid and each other). Although they have these extra pollination demands, many of the best-known varieties, such as Ribston Pippin and Gravenstein, are triploids, and this is because triploids tend to produce large fruit and heavy crops. Some varieties have been bred or have arisen as mutations possessing four copies of the genome. These are known as tetraploids (tetrap), and there are a number in the Collection. Tetraploids tend to be self-fertile and produce larger fruit than their parent, which has made them of special value and interest in northern countries, where fruit can often be small due to cold springs, and growing the tetraploid form of a variety can overcome the problem to a degree.

Tree vigour and **habit** has a bearing on the situation and the way in which a variety can be grown, which is of relevance to both amateur and commercial growers. Although the eventual size that a tree attains depends primarily upon the rootstock used, it does also depend upon the vigour of the scion, that is the variety. The majority of varieties are of medium vigour (T^2) , and suitable for all purposes, whether they are going to be grown as standard, bush or centre leader trees or trained as cordons or espaliers. A number are very vigorous (T^3) and will produce larger than average specimens, making them difficult to train as cordons, and producing very large standards. These include many of the famous old varieties, but the heavy crops to be gathered from Baldwins and Bohnapfels were more than compensation for the difficulties of picking from tall trees. If a tree strong enough to support a hammock or make an impressive avenue is required, then a vigorous variety would often be the best choice. Triploids, such as Bramley's Seedling, are very vigorous, but this can be reduced by growing on more dwarfing rootstock. Weak growing varieties (T^1) will often make small neat trees, which are ideal for gardens, although they may not be suitable for grafting on to the most dwarfing rootstocks.

Most varieties will form upright spreading trees, but a number have markedly different habits and tend to form wide spreading trees (sprd) or particularly upright growth (uprt). These are noted in the Directory. Commercial growers like upright, spreading growth to form their centre leader trees and this is also the most accommodating habit for the gardener. On the other hand, very upright growth casts little shade and this was one reason why King of the Pippins appealed to the old market gardeners who grew other crops beneath; for the same reason it can also be of use in a modern garden where apple trees and flowers or vegetables are grown together. Spreading trees can be a problem in a small area and are more difficult to grow as centre leader trees. Some also tend to make too much bare wood to train as an espalier or cordon, but the weeping habit of Golden Spire and Yellow Ingestrie can be very decorative if they are planted as specimen trees.

Another feature which is relevant to the way a variety is grown is its ability to form fruiting spurs. In most varieties, fruit is borne mainly on two-year wood, and on short stubby branches called fruiting spurs. Varieties vary in the extent to which they produce fruit spurs, but those which produce spurs freely make compact trees, lend themselves to training, need less pruning and produce heavy crops. Some varieties produce spurs exceptionally freely and these natural spur types (sp) are noted in the Directory. Calville Blanc d'Hiver, Esopus Spitzenburg and Duchess's Favourite and the modern Kendall, Celt and Tasman Pride are all natural spur types and form neat, prolific trees. 'Sports', or mutations that fruit almost entirely on spurs, have also been found in a number of varieties, including the most extreme example of Wijcik McIntosh, which has been used as the parent of columnar or Ballerina trees. Fruit is borne upon spurs along the main central stem and Ballerina trees need little pruning. There are, however, some varieties which produce few fruiting spurs, and form fruit buds mainly or partly at the tips of new growth. These are known as tip bearers (tip) and this is noted in the Directory entry. Cornish Gilliflower and Irish Peach are examples of tip bearers.

Apple trees are subject to attack by numerous pests and diseases. A few varieties have natural resistance to the major diseases. Discovery, for example, has good resistance to scab and mildew; Lane's Prince Albert and Edward VII are resistant to scab. To identify disease resistance many countries have been looking again at old varieties. In Belgium, systematic screening by the Centre for Agricultural Research at Gembloux has resulted in a number of varieties being recommended again for cultivation and these are now sold by nurserymen with the label RGF (Ressources Génétiques Fruitières); these are noted in the Directory. Fruit breeders have turned to old varieties as sources of disease resistance, and also to apple species – *Malus floribunda* carries the gene for resistance to scab, the V_f gene; *Malus X robusta*, the gene for mildew resistance; *Malus fusca*, resistance to fireblight, a significant problem in many countries, although not in the UK. Prima, Redfree and Liberty from the US carry the V_f gene and resistance to scab; as does Florina from France, while a number of the recent introductions from Germany, such as Rewena, are resistant to scab and mildew. Disease resistance and any other significant characteristics are noted in the Directory.

Crops (C): The production of heavy, regular crops is perhaps the single most important property of a variety. Plentiful crops, which kept well in a cellar right through the winter, have in a large measure been the reason for the lasting popularity of old varieties such as Glockenapfel, and these were among the qualities that first recommended Golden Delicious to its promoters. The hardiness of a variety (hrdy), too, has often ensured widespread popularity and been responsible for McIntosh being taken right across northern Europe as well as Canada and northern America, while Antonovka has been planted in the most northerly regions of the former USSR. Very hardy varieties are marked in the Directory.

An important contribution to an improved crop has been the discovery that apple trees carry viruses, which are not transmitted in seed but passed on through vegetative propagating material – scion wood. Some cause severe problems, while others produce no obvious symptoms, but nevertheless reduce the growth and yield of the tree. Research at East Malling and Long Ashton research stations made it possible to 'clean up' many old garden and commercial varieties, which are conserved as 'EMLA' or virus-free stock lines in the Collection, and are available from nurseries in this form.

The identification of 'sports' bearing more highly coloured fruit has also enabled growers to improve the appearance of the crop and its value. For most of the major commercial varieties, sports with more uniform or brighter colours have been found. These are usually the forms planted commercially and often have markedly different names. Starking and Richard Delicious, for example, are sports of Delicious, and Queen Cox of Cox's Orange Pippin. Many of these sports are growing in the Collection and are included in the Directory.

Picking (P) time will depend upon the variety and it will also vary with the season, the area and the rootstock that is used. The information provided in the Directory is for fruit grown on M9 rootstock in southern England; in warmer conditions the seasons will be advanced by several weeks and in colder climates the season will be correspondingly delayed. In addition, dwarfing rootstocks tend to advance a variety's time of harvest.

Early varieties, which are ready to eat off the tree in August, can often be picked over a period of a few weeks. Although this is now considered a commercial defect, as the grower has to carry out several picking operations on each tree, it was – and still can be – a bonus for the gardener and the cook, as the fruit can be used for a long period from the tree and does not need storing. The English cooker, Arthur Turner, for example, can be picked from August until almost October and similarly Khoroshavka of Kazakhstan. Early varieties, once picked, will not keep for long, and this has been one reason among others for the demise of Gladstone and Beauty of Bath. Early varieties are also often very soft fleshed and easily bruised like Melba and Miller's Seedling, which again creates problems for the commercial grower. Mid-season apples harvested in September will usually store for several weeks, while late varieties are picked before they are fully ripe and develop in store. Exactly how long and how well a variety will keep depends upon the store as well as its natural season. The dates given in the Directory are for simple natural conditions – a cool, dark shed, well-insulated against frost and fluctuating temperatures. Commercial cold-stores or controlled-atmosphere stores will give longer periods. Many of the late season varieties, however, store well in the most basic conditions and owe their standing in part to this property. Dumelow's Seedling, for example, will easily keep until the

PLATE 32: STURMER PIPPIN

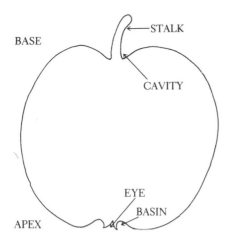

BASE · STALK · CAVITY · EYE · BASIN · APEX

Fruit features. Typical examples of the features listed below appear in brackets:

Eye: *closed (Worcester Pearmain); part open (Cox's Orange Pippin); open (Newton Wonder).*

Basin: *shallow (Charles Ross); medium depth (Cox's Orange Pippin); deep (Blenheim Orange); narrow (Worcester Pearmain); medium width (Cox's Orange Pippin); broad (Bramley's Seedling). Basin may be ribbed, puckered, russeted; beads may be present.*

Skin: *colour may be even, flushed, striped, russeted; lenticels may appear conspicuous as white or russet dots; surface may be dry or greasy, slightly bumpy – hammered; a hair line may be present (Newton Wonder).*

Size: *large (Howgate Wonder); medium (Golden Delicious); small (Golden Knob, Api).*

Shape: *See 'Fruit shapes' diagram, opposite.*

Cavity: *narrow (Newton Wonder); medium width (Worcester Pearmain); broad (Discovery); shallow (Newton Wonder); medium depth (Worcester Pearmain); deep (Discovery). Cavity may be russeted, lined with grey, scarf skin.*

Stalk: *short (Bramley's Seedling); medium (Cox's Orange Pippin); long (Golden Delicious); thick (Bramley's Seedling); thin (Golden Delicious).*

spring, as will Norfolk Beefing, Api, D'Arcy Spice and many of the old export varieties such as York Imperial. Long storage was one of Granny Smith's early claims to fame.

Identification of an Apple Variety: The Directory also gives the information required to identify an unknown apple variety, although this is not easy. The apple's season of ripening and use, whether dessert or culinary, are initial determining features, but identification depends primarily upon a number of botanic features of the fruit (see diagrams, this page and opposite). The most reliable features include: colour of skin; size; shape; nature of the eye (the sepals retained by the fruit); basin (area surrounding the eye); cavity (hollow in which the stalk lies); stalk; and the colour, texture and flavour of the flesh. Botanic descriptions of the fruit (**Frt**) are given in the Directory for varieties likely to be found growing in British gardens and also for the main international varieties; a number are illustrated in the colour plates.

The habit of the tree, appearance of its leaves, fruit buds, and the blossom and flowering time are also factors that will aid identification. All of these features, however, can vary enormously for any variety and be strongly influenced by the weather, pests and diseases, the age of the tree and so on. Frost damage can give russeting, cold springs will result in small fruit as will a heavy crop. The king fruit, the apple that develops from the centre flower of the cluster, will usually be bigger and often atypically shaped as can be seen, for example, in Plate 11 of Mother (American) on page 69. Old trees grown in grass and hence receiving less nitrogen will produce more highly coloured fruit, and the intensity of flavours will depend upon where and how the tree has been grown. Fruit grown in the north of England tends to be longer than that grown in the south, and young trees tend to give larger fruit that will not store so well. The wide variations in colour and shape of a particular variety can be seen in many of the colour plates.

Nevertheless, many varieties do display very characteristic features, such as the wide open eye of a Blenheim Orange, or the tightly closed eyes of Worcester Pearmain and McIntosh. The highly crowned, ribbed shape is a feature of the Delicious and Golden Delicious family, while the aniseed flavour of Ellison's Orange is usually a good pointer, and the bump in the cavity of a Newton Wonder will often clinch its identity.

Detective work can also help pin down the identity of an unknown fruit or of a tree that you have inherited in your garden. It is likely to have been a popular variety at the time that it was planted. If there was a fruit nursery in the locality, then it was probably one of their trees and old catalogues can narrow the possibilities down. There is always

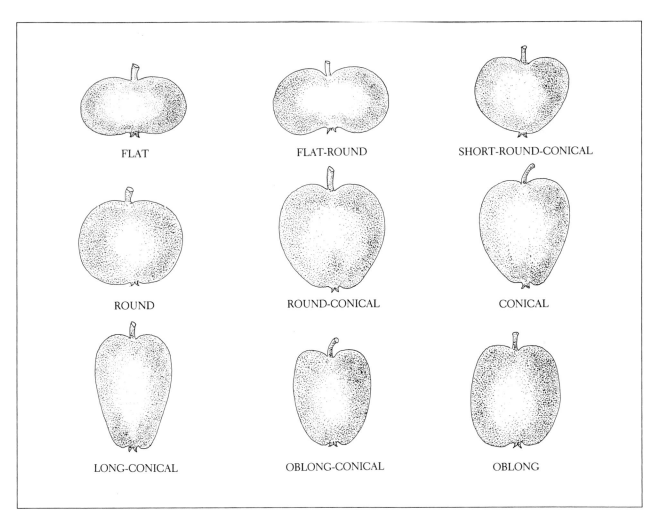

FLAT   FLAT-ROUND   SHORT-ROUND-CONICAL

ROUND   ROUND-CONICAL   CONICAL

LONG-CONICAL   OBLONG-CONICAL   OBLONG

the chance, however, that the graft has for some reason died and it is the rootstock that is growing up and fruiting. Your unknown tree could also be a seedling that has sprung up from a discarded pip, and hence is a completely new variety. This is often the case if it is growing 'out of place' in a hedge or among shrubs or trees.

A brief **history** of each variety is given at the beginning of its entry and this includes its origins, parentage (if known) and any other useful points. The bestowal of an award, for example, such as a Royal Horticultural Society First Class Certificate usually signified the beginning of the variety's rise to popularity and its widespread planting. In addition to the tasting and cultural notes, where possible an indication of a variety's current value is given, that is whether it is still grown in gardens or for market, and any other nuggets of social and horticultural history.

Fruit shapes, as shown in the diagram, are listed with typical examples below:
Flat: *Devonshire Quarrenden, Ben's Red*
Flat-round: *Bramley's Seedling, Dumelow's Seedling, Discovery.*
Short-round-conical: *Epicure, Fortune, Howgate Wonder.*
Round: *Golden Noble, Edward VII.*
Round-conical: *Cox's Orange Pippin, Charles Ross.*
Conical: *Adams's Pearmain, Rosemary Russet.*
Long-conical: *Golden Spire, Mother.*
Oblong-conical: *Cornish Gilliflower, Pitmaston Pine Apple.*
Oblong: *Delicious, Gravenstein, Crispin.*
Many varieties will lie between two fruit shapes: for example, Golden Delicious varies from round-conical to oblong. Shapes can be ribbed and flat sided, and ribs can terminate at the apex to give a 5 crowned appearance as in Delicious and Golden Delicious.

179

KEY TO THE DIRECTORY

THE AIM OF THE DIRECTORY IS to provide a summary of each variety's (cultivar's) characteristics and history.

Each entry begins with the variety name and any common synonym; an indication of its appearance through group numbers 1–8; season of use – early (E), mid (M) and late (L); use – dessert, fresh eating (D), culinary (C), dual purpose (CD). This is followed by its country of origin and brief history, including parentage, if known. Any distinctive features in its appearance not covered by the above grouping are given. Tasting notes are provided, including cooking qualities when relevant. Notes on its social and horticultural history are included where appropriate and also the variety's current value. Each entry concludes with brief details on its cultivation. Entries for varieties likely to be found in gardens in the UK and commercially important varieties also include a description of the fruit to aid identification.

Sports, or mutations, of a variety are under the parent's name and are also given in their alphabetical place in the Directory.

New varieties accessed into the Collection and planted since the publication of the first edition have been added to the Directory. Over the past decade there has been a considerable increase in the number of groups and individuals actively studying their local fruits in Britain and abroad, particularly in France, Belgium and the USA. This has enabled extra details to be added to the Directory to give further information on a number of the accessions.

Study and verification of the Collection is an ongoing process to ensure that the material conserved is true to name. Two types of naming error occur; sometimes a variety is discovered to be held under more than one name in the collection and sometimes material is received under the wrong name. Where synonyms are discovered the duplicates are deaccessed. Accessions that are obviously different to the published descriptions of that variety may be deaccessed or retained under a number, pending the determination of their correct name. These are not included here. For some accessions it is difficult to be certain that they are the true variety described in the literature, either because they almost, but do not quite, match the first published account, or because insufficient information was given when the variety was first recorded. This is indicated by (?) after the variety's name. In a few cases, two varieties have the same name and their source is included to avoid confusion.[1]

The comments, descriptions and information given here are of the variety now growing in the Collection under that name.

Some abbreviations are used in the entries, with more extensive abbreviations in the fruit descriptions and cultivation details.

APPEARANCE CLASSIFICATION

For reasons of space it has not been possible to give a detailed description of each variety's appearance, and instead the classification, through group numbers, developed by John Bultitude has been used.

1 are green and smooth skinned with no or very little russet, and acidic, culinary varieties such as Lord Derby, French Crab.

2 are predominantly green, sometimes with slight brownish orange flush, smooth skinned and sweet, dessert varieties such as Sturmer Pippin, Granny Smith.

3 are flushed and/or striped, smooth skinned, acidic culinary varieties such as Lane's Prince Albert, Newton Wonder.

4 are flushed and/or striped, smooth skinned primarily dessert, but also some culinary and dual-purpose varieties, for example James Grieve, Gravenstein, Cox's Pomona.

5 are predominantly yellow, smooth skinned, both sweet, dessert and acidic, culinary varieties, for example Golden Delicious, Golden Noble, Antonovka.

6 are predominantly red flushed, smooth skinned, sweet, dessert varieties such as Worcester Pearmain, McIntosh, Delicious.

7 are coloured by flush and stripes, with some russet, and usually sweet dessert varieties, such as Cox's Orange Pippin, King of the Pippins, Golden Reinette.

8 are completely or almost entirely covered with russet, and usually sweet, dessert varieties such as Egremont Russet, Ashmead's Kernel, Reinette du Canada.

FRUIT DESCRIPTIONS

Fruit (**Frt**).

Colour (*Col*): red (rd), green (grn), orange (orng), yellow (yell), white (wht); other colours in full.

Size: large (lrg) over 3in (7.5–8cm) diameter; medium (med) 2–3in (5–7cm); small (sml) below 2in (5cm).

Shape: flat (flt); flat-round (flt-rnd); short-round-conical (shrt-rnd-con); round (rnd); round-conical (rnd-con); conical (con); long-conical (lng-con); oblong-conical (oblng-con); oblong (oblng).

Basin: narrow (nrw); medium (med); broad (brd); shallow (shal); medium (med); deep (dp).

Eye: closed (clsd), open, half (hlf) partly (prt); sepals: short (shrt), medium (med), long (lng).

Cavity: narrow (nrw), medium (med), broad (brd); shallow (shal), medium (med), deep (dp); width (wdth), depth (dpth).

Stalk: short (shrt) within cavity; medium (med) just protruding beyond cavity; long (lng) protruding far beyond cavity; thick (thck), medium (med), thin.

Flesh: colour; flavour and texture given in tasting notes.

See also Identification of an Apple Variety, pages 178–9.

CULTIVATION

Flower (**F**); details on blossom, pollination.

F* indicates particularly attractive blossom.

F 1–42 indicates optimum pollination time. For successful pollination the numbers should coincide or overlap by 4 digits before or 3 after. In practice in Kent, F1 corresponds on average to May 1st, F31 to May 31st and F42 to June 11th. A very few varieties flower exceptionally early, and A30 on average corresponds to April 30th. These numbers should not be regarded, however, as dates, which will vary from year to year and locality, depending on weather conditions, but relative positions. Recent mild winters have advanced flowering times but the relative positions remain the same. Early flowering varieties will flower a week or so before mid-season flowering varieties, which in turn will flower a week or more before the late season flowering varieties. For example:

Early: F3 Stark's Earliest; F5 Keswick Codlin
Mid: F11 Cox's Orange Pippin; F12 Golden Delicious
Late: F21 Edward VII; F26 Court Pendu Plat

Some of the new accessions have been growing in the Collection for too short a time to establish their average flowering periods, and in this case an estimate, for example, 'est F10', is given or a more approximate time, such as 'early', 'mid'.

Varieties which are triploid (trip) need two pollinators; tetraploid (tetrap) varieties have no special pollination requirements. It should be noted that parents and progeny may not be good pollinators; for example Cox's Orange Pippin will not pollinate Kidd's Orange Red and vice versa. Biennial bearers are not reliable pollinators.

Tree (**T**); details on vigour, habit, hardiness.

Vigour: weak T^1; medium T^2; vigorous T^3.

Habit: most varieties make upright spreading trees, but those with markedly different growth are noted: spreading (sprd); upright (uprt).

Most varieties fruit on spurs and two-year-old wood and vary in extent to which they produce fruit spurs. Some varieties produce spurs exceptionally freely and these – natural spur types (sp) – are noted. Varieties which produce fruit buds mainly or partly at the tips of new growth – tip bearers (tip). Hardy (hrdy) Any known resistance to, or particular susceptibility to disease or other problems and useful information is included: prone to (prn) mildew, scab, bitter pit; resistant to (res) mildew, scab; partial (prt) resistance.

Crop (**C**) quantity and quality of crop.

Most varieties in the Collection crop reasonably well; additional information is given for a number: heavy (hvy); good (gd); light (lght) Some varieties tend to crop heavily on alternate years: biennial bearers (bien).

Pick (**P**); guide to picking time.

For example early Aug (e-Aug); mid-Sept (m-Sept); late Oct (l-Oct).

Season (**S**); guide to season of use for fruit stored in amateur conditions. Additional later storage dates indicate maximum limits given in literature; i. e. Mar/May. This guide is based upon harvests in south-east England. In warmer climates the season is advanced, and in colder areas delayed.

Additional abbreviations used:

UK counties are standard abbreviations (eg. Beds, Worcs), but note: (Cambridgeshire (Cambs); Cheshire (Ches); Monmouthshire (Mons), Shropshire (Shrops).

RHS, FCC, AM, AGM; Royal Horticultural Society First Class Certificate, Award of Merit, AGM Award of Garden Merit (recommended for garden cultivation). RGF – Ressources Génétiques Fruitières, signifies that after trial at Centre de Recherches Agronomiques de Gembloux, Belgium, under unsprayed conditions it has been reasonably disease free and is recommended for planting.

Accession (acc); conspicuously (conspic); described (desc); dis covered (discv); exhibited (exhib); good (gd); head gardener (HG); introduced (introd); irregularly (irreg); little (ltl); moderate (mod); near (nr); nurseryman (nursmn); partial (prt); possibly (poss); probably (prob); quite (qte); selected (select); susceptible (susc); synonym (syn); tolerant (tol); variable (var); very (v); year (yr); Jan, Feb, Mar, Apr, May, Jun, Jly, Aug, Sept, Oct, Nov, Dec.

Research Institutes:
East Malling Research Station, Kent: EMRS, Kent. Now Horticulture Research International. John Innes Horticultural Institute, Merton, London: JI, Merton. Now associated with University of East Anglia, Norwich, Norfolk. Long Ashton Research Station, Bristol: LARS, Bristol (no longer concerned with fruit research).

Canadian Department Agricultural Research Station, Summerland, British Columbia: CDRS, Summerland, British Columbia. Institut Nationale de la Researche Agronomique: INRA, Angers, France. Instituut voor de Veredeling van Tuinbouwgewassen, Wageningen, Netherlands: IVT, Wageningen. New York State Agricultural Experimental Station, Geneva: NYSAES, Geneva. New Zealand Department Science Industrial Research, Havelock North, Auckland: DSIR Havelock North. Swiss Federal Agricultural Research Station, Wädenswil (Fed. Agric. Res. St. Wädenswil).

Apple: **apfel** (German); **appel** (Dutch); **aeble** (Danish); **äpplen** (Swedish); **alma** (Turkic languages, Hungarian, Kazakhstan); **elma** (Turkish); **maca** (Portuguese); **manzana** (Spanish); **malum** (Latin); **mele** (Italian); **milo** (Greek); **mēlon** (Ancient Greek); **pomme** (French); **ping guo** (Chinese); **ringo** (Japanese); **sagar** (Basque); **sib/seeb** (Persian); **tcfh** (Arabic); **tapuah** (Hebrew); **yablovka** (Russian).

ABBONDANZA 4 L D
Italy; arose near S. Pietro Capofiume, Bologna. Select 1896.
Scarlet flushed. Pronounced raspberry tang, becoming strongly vinous. Sweet, firm white flesh. Formerly grown in Emilia-Romagna; mainly used for cooking.
F 12. T^2. C hvy; frt sml. P m-Oct. S Jan-Mar.

ABONDANCE 5 L D
Received 1947 from France; not French cider apple of this name.
Sharp, fruity.
F 15. T^2. P e-Oct. S Nov-Jan.

ACKLAM RUSSET (?) 8 L D
UK; acc received 1961 from village of Acklam, Yorks, where variety recorded 1768, originated; acc may be variety of Hogg (1884).
Dark brown russet. Rich, sweet-sharp taste.
F 10. T^2. P m-Oct. S Dec-Feb.

ACME 7 L D
UK; raised 1944 by W. P. Seabrook & Sons, Boreham, Essex.
(Worcester Pearmain X Rival) X poss Cox's Orange Pippin.
Resembles Rival in flavour; quite rich, crisp, juicy.
F 8. T^2. C bien. P m-Sept. S Nov-Dec/Jan.

ADAMSAPFEL OSTPREUSSISCHER 6 L D
Germany; received 1951 from Fruit Res St, Jork, nr Hamburg.
Blood red flush, stripes. Very juicy, crisp, sweet, fruity in Dec.
F 15. T^1. P l-Sept. S Oct-Dec.

ADAMS'S PEARMAIN 7 L D
UK; arose Norfolk or poss Herefords where known as Hanging Pearmain. Brought notice c1826 by R. Adams who gave scion wood of variety received from Norfolk to London Hort Soc. Syns many.
Handsome. Rich, aromatic, nutty flavour;

firmly textured. 'Essential' for Victorian and Edwardian dessert. Grown also for market C19th and popular with London fruiterers who always paid 'good price because it was attractive for windows'. Remains valued garden variety.
Frt: Col orng rd flush, rd stripes over grnish yell/gold; russet speckles, veins, patches. *Size* med. *Shape* con. *Basin* brd, shal; sltly ribbed; ltl russet. *Eye* open; sepals brd base, qte lng. *Cavity* med wdth, dpth; sometimes russet bump to side. *Stalk* sht/med; qte thck. *Flesh* crm.
F* 9. T^2. C gd. P e/m-Oct. S Nov-Mar.

ADERSLEBER CALVILLE syn Calville d'Adersleben 7 M D
Germany; raised c1870 by Lichtard at Adersleben, Saxony.
Calville Blanc d'Hiver X Gravenstein.
Resembles Calville Blanc d'Hiver in appearance. Slightly rich, firm flesh. Cooked, retains shape, but low in flavour.
F 8. T^2. P m-Sept. S Sept-Nov.

ADMIRAL 7 L D
UK; raised c1921 by A. K. Watson at Upton, Norfolk. Introd c1937.
Rich, sweet, perfumed at best, but often poor colour and flavour.
F 13. T^2. C bien. P e-Oct. S Nov-Dec.

A. D. W. ATKINS 3 L C
UK; raised c1971 by A. D. W. Atkins, Eymouth, Devon.
Large, grass green, slight flush. Cooks to quite sharp, well-flavoured purée; not as sharp as Bramley.
F 10. T^2. P l-Sept. S Oct-Dec.

ADVANCE 4 E D
UK; raised 1908 by nursmn, Laxton Bros, Bedford. Cox's Orange Pippin X Gladstone. Introd 1929.
Juicy, lightly aromatic; best eaten early to catch some sharpness.
F 11. T^2; sprd; prn mildew, canker. C lght; frt pr col. P eat e/m-Aug.

AÏVANIYA 5 L D
Bulgaria; origin unknown.
Crisp, juicy; tough skin.
Widely grown in central southern Bulgaria in Plovdiv region. Main commerical variety in 1960s, now only of local importance.
F 14. T^2; hrdy; v susc scab. C hvy. P m/l-Oct. S Dec-Mar.

AKANE syn Primrouge 6 E D
Japan; raised 1937 at Morioka Exp St. Jonathan X Worcester Pearmain. Introd 1970.
Crisp, juicy, sweet, slightly chewy flesh; hint strawberry flavour. As Primrouge grown southern France and, as Akane, in Japan, Washington; also US garden apple.
F 14. T^2. C gd. P m/ l-Aug. S Aug-Sept.

ÅKERÖ 6 E D
Sweden (prob). Either seedling arose Åkerö mansion, south of Stockholm, or imported 1759 from Holland by Tessin family, owners of Åkerö estate. Brought notice 1858 by pomologist Olof Eneroth. Mother tree still stands, although only part alive.
Beautiful appearance – pink flush, deep red stripes over pale primrose. Juicy, refreshing with savoury tang; esteemed 'even above Gravenstein' in Sweden. Very hardy, surviving 1939–42 winters when everything else died. There is avenue of Åkerö trees at Bergianska, Stockholm's Botanic gardens, leading from railway station at Fescati to Botanic Institute. Åkerö was also grown in Finland, Norway, Denmark, Northern Germany. Still grown gardens and for sale central Sweden.
F 7. T^3; uprt; v hrdy. C gd. P eat Aug/e-Sept.

ALASTAIR CANNON-WHITE 4 M D
UK; received 1974 from Mrs Margaret Cannon-White, Surrey.
Firm, cream flesh, little richness.
F 13. T^2; tip. P e/m-Sept. S Sept-Oct.

ALBERMARLE PIPPIN syn Newtown Pippin

ALDENHAM BLENHEIM *see* Blenheim Orange

ALDERMAN 5 M C
UK; poss arose Scotland, before 1923.
Large. Cooks to sharp, juicy, cream purée. Held promise as processing apple in 1950s, but not realised.
F 9. T². P e/m-Sept. S Sept-Oct.

ALDWICK BEAUTY 4 E D
UK; arose with Mrs D. M. Alford, Aldwick, Bognor Regis, Sussex.
Pretty, red flush. Becomes sweet by late Aug, but can be hard, astringent.
F 17. T². P eat Aug/e-Sept.

ALEXANDER syns Aport, Emperor Alexander 6 M CD
Ukraine (prob); known 1700s. Introd England 1805 by nursmn James Lee of Hammersmith, London. Sent 1817 by London Hort Soc to Massachusetts Hort Soc. Syns many.
Large, beautiful, rosy red flush. Primarily culinary, but eaten fresh quite sweet, slightly scented, soft, juicy flesh. Cooks to juicy, lemon purée; sweet, pleasant.
Renamed as compliment to Emperor Alexander I by growers around Riga on Baltic, who sent fruit every year to Russian court. Victorian exhibition variety and favoured for large espaliers. Formerly widely grown North America; still esteemed by amateurs there and also in France, Sweden. As Aport widely grown former USSR, especially around Alma Ata, Kazakhstan.
F* 10. T³; hrdy; some res scab, mildew. C gd. P m-Sept. S Sept-l-Oct.

ALEXIS *see* McIntosh

ALFA 68 5 L C
Sweden; raised 1936 at Agric Coll, Alnarp. Belle de Boskoop X Filippa. Introd 1958.
Very large, resembling Belle de Boskoop. Crisp, sharp flesh.
F 11; tetrap. T³; sprd. C slow to bear; frt prn coarse russet. P m-Oct. S Nov-Jan.

ALFRED JOLIBOIS syn Jolibois 7 L D
France; raised by Alfred Jolibois at Gurgy, Yonne (Burgundy); desc 1948 (*Verger Français*).
Deep red flush. Crisp, firm flesh; little flavour, but in France 'sugary and perfumed'.
Well known now Yonne, Seine-et-Marne.
F 12. T²; rep gd res disease. C hvy. P m/l-Oct. S Dec-Mar/May.

ALFRISTON 1 L C
UK; raised late 1700s, by Mr Shepherd at Uckfield, Sussex. Named Shepherd's Pippin; renamed 1819 after sent by Mr Booker of Alfriston, nr Lewes to London

Hort Soc. RHS AM 1920.
Quite sharp. Cooking to lightly flavoured purée in early Nov, but although remaining firm and fresh looking, acidity soon fades. Fishermen took it on sea voyages as it kept so well. Victorian exhibition variety; widely grown in UK; also introduced to Australia. Fruiterers selling it up to 1930s. Still found in Sussex gardens.
Frt: Col bright grn/grnish yell; lenticels conspic as wht dots; scarf skin at base. *Size* lrg/med. *Shape* oblng; flt base, apex; rnded ribs; sltly crowned. *Basin* brd, dp; ribbed, wrinkled. *Eye* lrg, prtly open; sepals brd base. *Cavity* qte brd, qte dp; russet lined. *Stalk* shrt, thck. *Flesh* crm.
F 10. T². C gd, bien. P e/m-Oct. S Nov-Apr.

ALICE 6 E D
Sweden; raised 1943 by H. Jensen, Balsgärd Fruit Breeding Inst.
Ingrid Marie X. Introd 1951.
Named after Alice Wallenberg, wife of patron of Balsgärd.
Pretty, pink flush over cream. Crisp, juicy, refreshing, but soon going soft.
Grown for sale and gardens in Sweden.
F 11. T². C frt cracks, prone brown rot. P eat Aug-e-Sept.

ALKMENE syn Early Windsor 7 L D
Germany; raised 1930s by M. Schmidt, Inst Agric & Hort, Müncheberg-Mark.
Geheimrat Dr Oldenburg, Cox's Orange Pippin cross. Introd 1962. RHS AGM 1998. Renamed Early Windsor 1996 in UK.
Rich, aromatic, honeyed flavour; crisp, juicy flesh.
Grown all over Germany, also commercially in Netherlands, Belgium, England. Ripening before Cox, but shorter season.
Frt Col orng/rd flush, shrt rd stripes over grnsh yell/gold; slght russet patches; lenticels as russet dots. *Size* med. *Shape* rnd-con. *Basin* med wdth; med dpth; sltly ribbed. *Eye* hlf open/open; erect sepals. *Cavity* med dpth; med wdth; lined russet. *Stalk* med lng, med thck. *Flesh* crm.
F 7. T²; res mildew; some res scab. C hvy. P l-Sept. S Oct-Nov.
   CEEVAL; red sport.
   RED ALKMENE; solid red flush; renamed Red Windsor 1998 in UK.

ALLEN'S EVERLASTING 8 L D
Ireland (prob); recorded 1864 by nursmn Thomas Rivers.
Poss Sturmer Pippin seedling; RHS FCC 1899.
Coppery red flush over yellow by spring; markedly flat-round shape. Strong sweet-sharp flavour, reminiscent of Sturmer; firm flesh. Needs long, warm autumn to ripen properly.
F 11. T¹. C gd. P m/l-Oct. S Feb-Apr.

ALLINGTON PIPPIN 7 L DC
UK; raised before 1884 by nursmn, Thomas Laxton, Lincs.
King of the Pippins X Cox's Orange Pippin. Originally South Lincoln Pippin, renamed 1894 by George Bunyard, after his nursery nr Maidstone, Kent. Introd 1896. RHS FCC 1894 as Brown's South Lincoln Beauty.
Mellows to intense fruit drop or pineapple taste, although still fairly sharp by Christmas, but needs good year. Very sharp and bitter-sweet in Nov, but cooks well, keeps shape with good flavour; sweet, not bland.
Widely planted commercially and in gardens in early 1900s, but proved 'cold and sour in Midlands', suffered storage problems and in decline by 1930s. Still often found in gardens.
Frt Col brownish rd flush over pale yell, becoming orng rd over bright yell; lenticels russeted; some russet. *Size* med. *Shape* con. *Basin* brd, med dp, sltly ribbed. *Eye* hlf open or open; sepals lng. *Cavity* brd, qte dp; russet lined. *Stalk* shrt/med, thck. *Flesh* crm.
F 11. T². C gd. P e-Oct. S Nov-Dec.

ALL-RED GRAVENSTEIN *see* Gravenstein

ALNARP'S FAVORIT 3 L CD
Sweden; raised 1945 Agric Coll, Alnarp. McIntosh X Alfa 68. Introd 1967.
Large. Fruity, good in Dec with quite sweet, soft flesh. Cooked, just keeps shape; sweet, pleasant meal.
F 10; trip. T². C frt drops. P m-Oct. S Nov-Feb.

ALTLÄNDER FETTAPFEL 3 M C
Germany; local variety of Jork area of 'Alte Land', fruit region of north west Germany.
Large. Crisp, some richness. Cooks to cream purée; very lightly flavoured.
F 12. T². P m-Sept. S Sept-Oct/Nov.

ALTLÄNDER PFANNKUCHENAPFEL 3 L C
German; chance seedling arose 1840 in Alte Land, nr Hamburg.
'Pancake apple of Alte Land'. Crisp, fruity, sharp. Cooks to lemon purée; good, brisk taste. Now grown mainly eastern Germany; used also for juice.
F 10. T²; res scab; prt res mildew. P m/l-Oct. S Nov/Dec-Jan.

ALTLÄNDER ROSENAPFEL 4 M D
Germany; arose Alte Land, nr Hamburg. Known c1850.
'Rosy apple of Alte Land'; bright red flush. Distinctive brisk, vinous taste; firm, chewy flesh.
F 11. T³. P l-Sept. S Oct-Nov.

ALTON 4 E D
USA; raised 1923 by R. Wellington at NYSAES, Geneva.
Early McIntosh X New York 845 (Red Canada X Yellow Transparent). Introd 1938.
Attractive pinky red flush, stripes over primrose. Soft, melting white flesh, like McIntosh with hint of its flavour.
F 10. T³. P eat Aug.

AMANISHIKI 6 L D
Japan; raised 1936 at Aomori Apple Exp St. Ralls Janet X Indo. Named 1948.
Dark red flushed. Becomes very sweet, slightly scented by Jan, but almost no acidity; yellow flesh. Often fails mature England.
F 12. T². P l-Oct. S Jan-Mar.

AMBASSY see Delcorf

AMBRO 6 M D
Netherlands; unknown parentage; received 1998 from T. Visser, Wageningen.
Crisp, juicy, sweet cream flesh; lightly aromatic.
F est 12. T². P l-Sept. S Oct-Dec.

AMERICAN BEAUTY syns Belle Américaine, Stirling Beauty 6 L D
USA; arose Sterling, Massachusetts. Recorded 1857.
Large. Very sweet, slightly chewy flesh.
F 16. T². P m/l-Oct. S Dec-Feb.

AMERICAN GOLDEN RUSSET 8 L D
USA; arose late 1700s, Burlington County, New Jersey. Recorded 1817 by Coxe as Bullock's Pippin. Renamed Golden Russet, given prefix 1845 by Downing.
Quite small, sugary, lightly aromatic. Esteemed C19th America; widely grown.
F 10. T². P e/m-Oct. S l-Oct-Jan.

AMERICAN GRINDLING 3 E C
Acc received 1943 from Notts; prob variety recorded 1872 by nursmn J. Scott as 'received from north of England' and exhibited from Nottingham at 1883 Congress.
Large. Cooks to pale lemon froth or purée.
F 15. T². C lght. P use Aug.

AMERICAN MOTHER see Mother

AMERICAN SUMMER PEARMAIN 4 E D
USA; described 1817 by Coxe.
Sweet, lightly aromatic flavour; firm becoming soft flesh.
Widely grown, dual purpose home fruit in C19th America and again popular with gardeners.
F 7. T¹. P eat Aug-Sept.

AMES 6 M D
USA; raised by H. L. Lantz, Iowa Ag Exp St, Ames, Iowa.
Allen Choice X Perry Russett. Introd 1921.
Dark red flush. Soft juicy white flesh, vinous flavour.
F 7. T². P e-Sept. S Sept-Oct.

AMOROSA see Aroma

ANANAS REINETTE syn Reinette Ananas 5 L D
Netherlands (prob); recorded 1821 by German pomologist Diel.
Bold russet freckles over gold. Develops definite pineapple flavour by Nov with intense sweet-sharp quality; crisp, juicy.
Very decorative. Formerly much grown Germany, where still valued; also Netherlands, Belgium. Used also for cooking, juice.
F 9. T²; sp; uprt. C gd; frt sml. P e-Oct. S Nov-Jan.

ANANAS ROUGE see Roter Ananas

ANDRÉ BRIOLLAY 5 L D
Received 1947 from France.
Small. Firm, astringent, but sweetening.
F 34. T². P l-Oct. S Dec-Mar.

ANGYAL DEZSÖ 2 L D
Hungary; prob arose early 1900s.
Sweet, firm fleshed, nutty in Jan. Named after Director, Budapest School of Horticulture.
F 15. T². P l-Oct. S Dec-Mar.

ANISA syn Udarria Zagarra 8 L D
France; best known of 'Anise' apples of the Basse Pyrénées; prob dating from C19th.
Pretty red cheek under russet; conical. Sharp, strong aniseed flavour.
Grown in area known as 'Hillsides of Basque Country' and, as Rosalie, sold in markets, mainly Bayonne, Biarritz up to 1960s.
F 24. T²; res codling moth. C gd. P e-Oct. S Oct-Dec.

ANNA BOELENS 7 M D
Netherlands; raised by S. A. Blijstra.
Cox's Orange Pippin X Freiherr von Berlepsch. Introd 1934 by van Rossem Nurseries, Naarden; Cert Merit Netherlands Pom Soc.
Burgundy flush. Rich, quite intense flavour; some aromatic Cox-like quality. Recommended for areas where Cox fails.
F 15. T²; sprd. P l-Sept. S Oct-Nov.

ANNIE ELIZABETH 3 L C
UK; raised c1857 by Samuel Greatorex, magistrates clerk, at Avenue Road, Knighton St Mary, Leicester. Claimed Blenheim Orange seedling. Named after his baby daughter who died 1866, or after daughters of nursmn, Harrison. Introd

c1868 by Harrison's, Leicester; RHS FCC 1868. Original tree still grew in Avenue Road in 1970s
Cooked, keeps shape, quite sweet, light flavour; needs little sugar. Esteemed as stewing apple, also exhibition variety. Attractive and used by Victorian gardeners for large dining and buffet table displays. Also valued in ornamental orchard for blossom. Widely grown in gardens and for market up to 1930s. Remains popular with Midlands, northern gardeners.
Frt Col orng rd flush, many shrt rd stripes over grnish yell/gold; lenticels qte conspic as grey or russet dots; greasy skin. *Size* med/v lrg. *Shape* rnd to rnd oblng; trace rnded ribs; flt base, apex; sltly 5 crowned; hammered surface. *Basin* med wdth dpth; ribbed, puckered; can be beaded. *Eye* clsd or hlf open; sepals, lrg, brd. *Cavity* qte brd, qte dp; russet lined; sometimes lipped at side. *Stalk* shrt, thck. *Flesh* wht, sharp.
F* maroon/dp pink; 16. T³; hrdy. C gd; frt easily blown off. P e/m-Oct. S Nov-Apr.

ANNURCA 6 L D
Italy; old variety.
Small, dark red. Crisp, sweet, fruity.
Grown commercially Campania, southern Italy; popular market apple.
F 11. T². P l Oct. S Nov Jan/Feb.

ANTONOVKA 5 M CD
Russia; arose Kursk. Recorded 1826.
Large, milky white, classic Russian apple. Sharp, refreshing, juicy, crisp flesh in Sept. Cooks to brisk, juicy purée. Mellows and softens; esteemed for 'perfumed, vinous' flavour. Grown all over former Soviet Union; also used as rootstock.
F 8. T³; uprt; v hrdy; res scab; some res mildew. C gd. P e-Sept. S Sept-Nov/Dec.

ANTONOVKA 600-GRAMMOVAYA 5 M D
Russia; found 1888 by I. V. Michurin. Sport of Antonovka Mogilevskaya. Introd 1892.
Larger fruiting, later ripening than Antonovka.
F 9. T³. P m-Sept. S Sept-Dec.

ANTONOVKA-KAMENICHKA 5 L C
Ukraine; arose Chernigov region. Recorded 1889.
Later maturing seedling or form of Antonovka, distinguished by slight pink flush and claimed even hardier.
F 14. T³; v hrdy. P m-Sept. S Oct-Dec.

AORI 6 M D
Japan; raised Aomori Apple Exp St.
Toko X Richared Delicious. Received 1977.
Brilliant scarlet. Very sweet, sugary taste, reminiscent Delicious, but can have metallic undertones in England; tough skin.
F 10. T². P l-Sept. S Oct- Nov.

APEZ ZAGARRA 8 L D
Acc appears identical to Anisa. Apez Zagarra is general name for Anise varieties of south west France, from which Anisa was selected for commercialisation.

API syn Lady Apple 4 L D
France; very old variety. Found in ancient Forest of Api, Brittany, according to French botanist Merlet (1675). Name recorded 1628 by Le Lectier; recorded England 1676 by Worlidge. Syns numerous. Lady Apple of North America. Not reputed Roman variety, Appia of C16th, according to French and Italian authorities.
Small, brightly flushed. Crisp, sweet, fruity until spring.
For centuries decorative feature of dining table and garden, also grown for sale. So pretty that, in C17th, it served 'ladies at their toilets as a pattern to paint by'. In gardens, planted alongside paths, in pots, and as low cordons. Popular for wiring onto evergreens to make garlands and Kissing Boughs and for floating on Wassail cup.
Grown New York in C19th and exported to London. Still grown France, USA.
In 1962, in recognition of its enduring appeal, purse of Api, obtained from New York, was presented to Queen Mother at 300th anniversary of London Royal Society.
Frt *Col* rd flush over yell. *Size* sml. *Shape* flt-rnd; trace rnded ribs. *Basin* brd shal; ribbed. *Eye* sml, clsd to hlf open. *Cavity* med dpth, wdth; ltl russet. *Stalk* lng, thin. *Flesh* wht.
F* 11. T^1; uprt, compact. **C** hvy; hngs late. **P** m/-Oct. **S** Dec-Apr.

API NOIR 6 L D
France; believed to be ancient variety. As syn, de Caluau or Calvau Noire recorded 1608 by Olivier de Serre.
Very dark red flush. Very similar taste to Api; highly decorative. Formerly grown all over France.
F 11. T^1. **C** hvy, hngs late. **P** m/l-Oct. **S** Dec-Apr.

APORT syn Alexander

APPEL VAN PARIS 6 L D
Netherlands; raised 1935 by Prof Sprenger, IVT, Wageningen.
Brabant Bellefleur X Jonathan.
Bright red Jonathan colour. Intense sweet-sharp taste, slightly scented.
F 18. T^2. **P** l-Oct. **S** Dec-Feb/Mar.

APOLLO syn Beauty of Blackmore

ARBROATH PIPPIN syn Oslin

ARD CAIRN RUSSET 8 M D
Ireland; discovered c1890 as unidentified orchard tree by Baylor Hartland of Ard Cairn Nursery, Cork. RHS AM 1910.

Striking, orange red flush under golden russet. Eaten late Sept, tastes just like a banana as Hartland claimed. Sweet, little juice and almost no acidity; firm deep cream flesh. With keeping becomes even drier and cloyingly sweet.
F 12. T^3. **P** l-Sept. **S** l-Sept-Dec.

ARDITI 6 L D
Italy; received 1958 from Univ Turin.
Carmine flushed. Firm, refreshing, quite sweet in Jan.
F 15. T^2. **P** l-Oct. **S** Jan-Mar.

ARKANSAS syns Mammoth Black Twig, Blacktwig 4 L D
USA; believed to be seedling of Winesap raised by John Crawford who settled in Washington County, Arkansas in 1842. Named Mammoth Black Twig because it produced larger, redder, but similar fruit to Blacktwig (syn Winesap). Selected for Arkansas exhibit at New Orleans in 1884 by Col E. F. Babcock, who renamed it Arkansas.
Pleasant sweet, balanced flavour; crisp, cream flesh.
Planted southern, south western States in 1890s, but light crops led to its decline; by 1930s used mainly for processing.
F* 10; trip. T^3. **P** l-Oct. **S** Jan-Mar/May.

ARKCHARM 4 E D
USA; raised Univ Arkansas.
Prima X Hybride 36055. Received 1997.
Crisp to soft juicy flesh; sweet and pleasant.
F est 12. T^2. **P** eat Aug; shrt season.

ARLET SYN HUGO RED 6 L D
Switzerland; raised Fed. Agric. Res. St., Wädenswil.
Golden Delicious X Idared. Received 1995.
Attractive rose pink flush. Crisp, juicy, quite honeyed and lightly perfumed.
Grown Swiss–German border in 1970s, but no longer commercially important.
F mid. T^3; uprt. **P** e-Oct. **S** Nov-Dec/Jan.

AROMA 6 L D
Sweden; raised 1945 at Balsgård Fruit Breeding Inst.
Ingrid Marie X Filippa. Introd 1973.
At best, good rich fruity flavour; melting, very juicy, cream flesh. Can be weakly flavoured.
F 15. T^2. **P** e-Oct. **S** Oct-Nov/Dec.
  AMOROSA; improved colour.

AROMATIC RUSSET (SCOTT) 8 L D
UK; acc received 1976 from J. Scott Nursery, Somerset. May be variety introd c1850 and desc by Hogg, Bunyard.
Cinnamon russeted with slight red flush. Intense, sweet-sharp taste of acid drops, like Nonpareil family with similar firm flesh.
F 8. T^2. **P** m-Oct. **S** Nov-Feb/Mar.

ARTHUR TURNER 5 E C
UK; raised by nursmn Charles Turner, Slough, Bucks. Named 1912 Turner's Prolific. Introd 1915 as Arthur Turner. RHS AM 1912; AM for blossom 1945; AGM 1993.
Cooks to well-flavoured, yellow purée, brisk, but hardly needs extra sugar. Makes early pie, pleasant sauce, good baked. Remains popular garden variety.
Frt *Col* brownish pink flush over pale grn/pale yell; scarf skin at base. *Size* lrg. *Shape* rnd-con, almost oblg; trace rnd ribs; flt base, apex. *Basin* brd, dp; ltl puckered. *Eye* hlf open or clsd; sepals, shrt, downy. *Cavity* brd, dp; lined grey russet. *Stalk* v shrt, thck. *Flesh* pale crm, sharp.
F* 8. T^3; uprt; prn mildew. **C** gd. **P** Sept. **S** Sept-Nov.

ARTHUR W. BARNES 3 M C
UK; raised 1902 by N. F. Barnes, HG to Duke Westminster, Eaton Hall, Chester. Gascoyne's Scarlet X Cox's Orange Pippin. Introd 1928 by Clibrans, Cheshire.
Large, bold red flush, esteemed exhibition variety. Cooks to juicy lemon coloured purée; plenty of bite and flavour. Makes good sauce.
F* 11. T^2; sprd. **P** m-Sept. **S** Sept-Oct.

ASCOT 4 L D
Canada; raised 1898 at Exp Farm Ottawa.
Northern Spy X. Introd 1913.
Resembles large Northern Spy. Sweet, gentle, savoury taste; very juicy, soft flesh.
F 12. T^2. **P** e-Oct. **S** Oct-Nov/Dec.

ASHDOWN SEEDLING 6 E D
UK; arose with Ashdown & General Land Comp, Horsted Keynes, Sussex.
Claimed McIntosh seedling. Received 1966.
Very similar to Discovery, but not as good. Light strawberry taste; sweet, soft flesh.
F 8. T^2. **P** eat m-Aug.

ASHMEAD'S KERNEL 8 L D
UK; according to Hogg raised in C18th by physician Dr Ashmead but local historians now believe this to be an error. It is more likely that William Ashmead (d 1782), Clerk of Gloucester City, raised this variety in garden of house, which became Ashmead House, Gloucester. Ronalds records in 1831 that tree was then 100 years' old. RHS FCC 1981; AGM 1993.
Strong, sweet-sharp intense flavour reminiscent of fruit or acid drops and of Nonpareil, but sweeter than its probable parent; firm, white flesh. Long esteemed by connoisseurs; widely planted from mid-C19th and recently regain popularity in England and North America.
Grown commercially on small scale in England, but dull colour, poor crops weigh against it.

Frt Col grnish yell/ yell, some frt flushed in diffuse brownish rd/orng rd; many russet patches, netting, dots. *Size* med. *Shape* flt-rnd; sltly flt sided; sltly ribbed. *Basin* brd, shal; sltly ribbed; russet lined. *Eye* hlf open; sepals med to lng, v downy. *Cavity* med dpth, wdth; russet lined. *Stalk* shrt, qte thck. *Flesh* wht.
F* 14. **T**^{2}. **C** gd but erratic poss due to cold springs; prn bitter pit. **P** e/m-Oct. **S** Dec-Feb.
PLATE 28

ASTILLISCH 4 E D
Germany; received 1947 from Max-Planck Inst, Cologne.
Red Astrachan X Signe Tillisch.
Dark red with dusky bloom, like Red Astrachan. Rather sharp, soft, juicy, cream flesh.
F 8. **T**^{2}. **P** eat Aug; ripens v irreg.

ASTRACHAN LARGE FRUITED 5 E D
Sweden; found 1850s nr Stockholm by E. Lindgren; introd by him early 1860s.
Believed White Astrachan seedling.
Large, creamy yellow. Sharp. Cooks to quite sweet purée.
F 4. **T**^{2}. **P** use Aug.

ASTRAMEL 6 M D
Germany; raised 1993 at Fruit Res. St. Jork, nr Hamburg.
Red Astrachan X (James Grieve X Melba).
Deep rose pink flush, sweet yet refreshing, melting cream flesh. Name is hybrid of parents, but 'mel' suggests 'mehl' in German mind which means floury and has weighed against popularity.
F v. early. **T**^{2}; sprd. **P** eat Aug.

ATALANTA 6 M D
Netherlands; raised 1935 by Prof Sprenger, IVT, Wageningen. (Not Atalanta raised by Charles Ross; desc Bunyard).
Sternrenette (Reinette Rouge Étoilée) X Cox's Orange Pippin.
Striking, dark red colour and firm flesh of Reinette Rouge Étoilée, but sweeter, richer flavour. Some aromatic Cox-like quality, but only slight raspberry taste of R. R. Étoilée.
F 14. **T**^{3}. **P** l-Sept/e-Oct. **S** Oct-Nov.

ATLAS 4 M D
Canada; raised 1898 at Central Exp Farm, Ottawa.
Winter Saint Lawrence X Duchess Oldenburg. Introd c1924.
Pink flushed, carmine striped. Quite brisk, savoury taste; white, juicy, soft flesh. Often metallic, green taste in England. Dual purpose in Canada.
F 9. **T**^{3}; v hrdy. **P** e-Sept. **S** Sept-Nov/Dec.

AURALIA syn Tumanga 5 M D
Germany; raised 1930s by M. Schmidt, Inst

Agric & Hort, Müncheberg-Mark.
Cox's Orange Pippin X Schöner von Nordhausen. Received 1967.
Resembles Cox's Orange Pippin, slightly aromatic, with quite intense, rich flavour; sweet, crisp, juicy flesh.
Grown all over Germany, mainly for juice manufacture.
F 15. **T**^{2}. **C** hvy; frt bruises. **P** l-Sept. **S** Oct-Nov; cold store-Feb.

AUTUMN ARTIC 4 M D
USA; found 1952 in Barnard, Vermont on farm of Dorothy Thompson growing from pile of chicken house litter.
Prob Artic seedling, possibly crossed with Northern Spy. Introd 1956 by nursmn F. L. Ashworth, Heuvelton, NY.
Maroon flushed. Lightly aromatic, soft, sweet flesh.
F 12. **T**^{2}; hrdy; res scab. **P** m/l-Sept. **S** Sept-Oct.

AUTUMN HARVEST 5 E CD
UK; acc received 1950 from Westmorland; may be variety exhib 1934 from neighbouring Cumberland.
Firm flesh with plain taste of fruit. Cooked, keeps shape but light flavour.
F 11. **T**^{3}. **P** l-Aug. **S** Sept.

AUTUMN PEARMAIN 8 M D
UK; acc is Autumn Pearmain of Bunyard and prob Summer Pearmain of Hogg. Impossible to know if acc is ancient variety recorded late 1500s and cited 1629 by Parkinson.
Prettily flushed and russeted. Quite rich taste; firm, rather dry, cream flesh.
F 13. **T**^{2}. **P** l-Sept. **S** Oct.

AVERDAL *see* Delicous

BAILLEUIL syn Gros Hôpital

BAILLY syn Belle-Fleur de Saint Benoît

BAKER'S DELICIOUS 5/4 E D
UK; found in Wales. Introd 1932 by Bakers of Codsall, Wolverhampton.
Handsome, flushed bright orange red over gold. Rich, juicy, lots of sugar, acidity; quite strong, aromatic flavour; deep cream flesh. Grown small extent for Farm Shops.
F 7. **T**^{2}; prn canker. **C** gd; frt bruises. **P** e-Sept. **S** Sept.

BALDER 4 M D
Netherlands; raised 1950 by A. A. Schaap, IVT, Wageningen.
James Grieve X Irish Peach.
Lightly aromatic, quite good flavour. Similar to James Grieve but tougher, chewy flesh.
F 12. **T**^{2}. **C** pr. **P** m-Sept. **S** Sept-Oct.

BALDWIN 6 L D
USA; found c1740 on farm of John Ball, Wilmington, Massachusetts. Originally Pecker or Woodpecker, because the tree was frequented by woodpeckers. Popularised, renamed early 1800s.
Sweet, fruity, retaining crisp, lively character; but often fails to ripen in England. Highly esteemed and widely grown across northern States in C19th; exported to UK. Also used for pies, cider and processing.
Stone apple on pillar, erected in 1895, marks 'the estate where in 1793 Saml. Thompson Esq, while locating the line of the Middlesex canal discovered the first Pecker Apple tree later named the Baldwin'. Colonel Loammi Baldwin was also an engineer on the Middlesex Canal and his statue at North Woburn is wreathed in apples and inscibed 'Disseminator of the apple in honor of him called the Baldwin apple, which proceeds from a tree growing wild about 2 miles north of this monument'. Overtaken by Jonathan in early 1900s; but still popular north eastern US.
Frt Col dp maroon; pale grnish yell ground col; often ltl russet. *Size* lrge. *Shape* flt-rnd to almost oblng; irregular; ribbed; 5 crowned. *Basin* brd, dp; ribbed. *Eye* qte lrg, pit open, sepals lng. *Cavity* brd, qte dp; russet lined. *Stalk* shrt/med lng, qte thck. *Flesh* wht, tinge grn.
F 12; trip. **T**^{3}. **C** gd, bien. **P** m-Oct. **S** Dec-Apr.
   Double-Red Baldwin; brighter colour. Discv 1924 by E. N. Sawyer in Salisbury, New Hampshire; introd 1927.

BALLARAT SEEDLING 6 L DC
Australia; raised by Mrs Stewart of Soldier's Hill, Ballarat, Victoria. Exhibited at Ballarat show, when seen by Francis Moss of Mossmount Nurseries, who was so impressed by its size and colour that he propagated trees. Named 1870s Stewart's Ballarat Seedling.
Keeps crisp, fruity for months, but tends to be sour grown in England.
Recommended for jelly in Australia.
Grown Australia, New Zealand as culinary apple for export until 1970s.
F 10. **T**^{2}; sp; res mildew. **C** hvy. **P** l-Oct. **S** Jan-Apr.

BALLARD BEAUTY 6 L D
UK; raised 1946 by A. Norman, Bedford.
Cox's Orange Pippin seedling.
At best, delicate Cox-like flavour; melting, juicy flesh. Can be sharp; dull appearance.
F 9. **T**^{2}. **P** e-Oct. **S** l-Oct-Dec.

BALLERINA
UK. Trade name of columnar trees developed by R. Watkins and K. Tobutt, EMRS, Kent, from crosses with Wijcik McIntosh. Trees rarely produce lateral

branches, fruit born on spurs on central stem; need almost no pruning. Wijcik McIntosh discovered 1960s in British Columbia by Mr Wijcik who found branchless sport in his McIntosh orchard and reported it to Summerland Res St. Branchless habit is passed on to 50% of seedlings. This potential was developed in England for intensive commercial orchards, but this was never realised, instead 'Ballerina' trees have found an amateur market. See Charlotte, Maypole, Obalisk (Flamenco), Telamon (Waltz), Trajan (Polka), Tuscan (Bolero).

BALL'S PIPPIN 2/5 L D
UK; introd 1923 by nursmn J. C. Allgrove, Langley, Bucks. Cox's Orange Pippin X Sturmer Pippin. RHS AM 1923.
Good, crisp, solid apple; plenty fruity flavour. Crispness of Sturmer married with some sweetness, aromatic, rich quality of Cox. Often dull appearance.
F 12. T^3. P e/m-Oct. S Nov-Jan.

BALLYFATTEN 3 M C
UK; local to Northern Ireland. Recorded 1802.
Cooks to pleasant, creamy, white purée.
F 17. T^2; tip. C gd. P l-Sept. S Oct-Nov.

BANCROFT 6 L D
Canada; raised Hort Exp Farm, Ottawa. Forest X McIntosh. Introd c1935.
Dark red. Refreshing, fruity; soft juicy flesh. Formerly important Canadian commercial variety.
F 7. T^3; hrdy. P m-Oct. S Nov-Dec; cold store-Apr.

BÁNFFY PÁL 2 L CD
Transylvania (Romania); discv, propagated early C19th by Paul Bánffy, leader of county Kraszna.
Crisp, good, fresh, fruity taste in Jan. Cooked, keeps shape; tastes sweet, light. Formerly popular, valued for heavy crops, good storage.
F 8. T^2. C hvy. P l-Oct. S Dec-Mar/Apr.

BANNS 5 M D
Received from Norfolk 1928.
Large. Sweet, scented flavour, reminiscent elderflowers.
F 14. T^3. P m-Sept. S l-Sept-Oct.

BÄNZIGER 6 L D
USA; introd c1890 to Switzerland as Amerikaner.
Sweet, fruity, soft flesh, light flavour, low acidity; tough skin.
F 10. T^3. P m-Oct. S Nov-Dec.

BARCHARD'S SEEDLING (?) 6 M D
UK; acc may be variety desc by Hogg (1884); raised by Mr Higgs, HG to R. Barchard, Putney, London. Recorded 1853.

Exhib Pom Soc 1856 by market gardener Dancer of Fulham. RHS FCC 1873.
Bright red, in style of Nonsuch. Pleasant light taste; quite soft white flesh. Also used early for cooking. Grown extensively by London market gardeners by 1880s.
F 12. T^2. C gd. P l-Sept. S Oct-Dec.

BARKLEY RED ROME see Rome Beauty

BARNACK BEAUTY 7 L D
UK; raised c1840 at Barnack, Lincs. Introd c1870 by Browns of Stamford. RHS AM 1899; FCC 1909.
Showy; bold red markings. Strong, brisk taste, slight richness and aromatic quality; dense flesh. Also used for cooking.
Ornamental tree, striking in fruit and blossom.
Frt Col bright orng red flush, rd stripes over gold; russet dots, ltl netting. Size med. Shape rnd to oblng-con. Basin brd; shal; ltl russet; sltly puckered or wrinkled. Eye lrg, open; sepals brd based. Cavity nrw, shal; russet lined. Stalk shrt/med; thck. Flesh pale crm.
F* 14. T^2; v sprd. C hvy; frt sml. P e/m-Oct. S Dec-Mar.
> **BARNACK BEAUTY SPORT**; more highly coloured. Discv 1944 by George Lamb at Swanley, Kent.

BARNACK ORANGE 7 L D
UK; raised 1904 by W. H. Divers, HG at Belvoir Castle, Leics.
Barnack Beauty X Cox's Orange Pippin. Intense, rich, aromatic; sweeter and ripens earlier than Barnack Beauty.
F 10. T^2. P e/m-Oct. S Nov-Jan/Feb.

BARNHILL PIPPIN 4 M DC
UK; arose N Ireland. Recorded 1934.
Scarlet striped. Undistinguished flavour.
Formerly common County Armagh, Antrim.
F 10. T^2. P l-Sept. S Oct- Nov/Dec.

BARON DE BERLEPSCH syn Freiherr von Berlepsch

BARON WARD 5 L C
UK; raised 1850 by Samuel Bradley at Elton Manor, Nottingham.
Dumelow's Seedling X. Exhib 1859 at British Pom Soc.
Cooks to lemon coloured slices; not as sharp or flavoursome as Dumelow. Small for cooker.
F 12. T^3. C gd. P m-Oct. S Nov-Jan/May.

BARON WOLSELEY syn Dewdney's Seedling

BARON WOOD 3 M C
Acquired late 1940s.
Very sharp. Cooks to sharp, bright yellow purée. With sugar, strong, full of flavour.

F 9. T^3. C prn bitter pit. P m-Sept. S Sept-Oct.

BARRAUDE 5 L D
France; received 1947.
Markedly flat-round shape. Crisp, juicy, quite sweet, good fruity taste.
F 16. T^2. C hvy. P l-Oct. S Dec-Feb/Apr.

BARRÉ 4 L D
France: known Seine-et-Marne (Ile de France), when desc 1948 (Verger Français). Crisp, juicy, sweet; plenty acid, fruit; snowy white flesh.
Grown now Seine-et-Marne and Yonne (Burgundy); used also for juice.
F 22. T^2. P l-Oct. S Jan-Mar.

BARRY 6 M D
USA; raised NYSAES, Geneva.
McIntosh X Cox's Orange Pippin. Introd 1957.
Deep red McIntosh colour. Juicy, lightly aromatic. Dual purpose US.
F 14. T^3. C gd, frt cracks. P m-Sept. S Sept-Oct/Nov.

BASCOMBE MYSTERY 2 L D
UK; recorded 1831; desc Hogg (1884); known Kent since mid-C19th.
Strong, refreshing taste; sweet yet plenty acid. Dull appearance. Grown for market in Kent up to 1930s; valued for good storage.
F 18. T^2. P m/l-Oct. S Dec-Feb; said keep to June.

BASSARD 1 L C
France; acc is prob variety desc 1948 (Verger Français), when found Yonne (Burgundy). Sharp, crisp. Cooks to quite rich, brisk purée.
Grown now Yonne and Loiret (Val de Loire); used also for juice.
F 24. T^2. P l-Oct. S Jan-Mar/Apr.

BASTIEN 5 L D
France; raised Trellières, nr Nantes, Loire Atlantique. Desc 1948 (Verger Français).
Sharp taste of fruit in Dec; becoming sweeter.
F 19. T^2. P l-Oct. S Dec-Mar.

BATUL-ALMA 5 L D
Transylvania (Romania); arose south east; by early C19th beginning to spread and become one of the most popular of all Transylvanian apples. Name derives from Romanian word 'pátul' which means bottom of haystack, where apples were stored until sold in spring, when hay was finished.
Intense fruity sweetness, crisp, juicy; quite rich flavour in Jan.
Remains popular; in 1950s represented 25–30% of apple crop.
F 7. T^3; sp. C hvy. P m/l-Oct. S Dec-Mar/Apr.

BAUMANN'S REINETTE 6 L D
Belgium; prob raised by pomologist Jean-Baptiste Van Mons.
Dedicated to nursmn, Baumann Bros of Bollwiller, Haut-Rhin, Alsace. Recorded 1811. Syns many. RHS FCC 1878.
Brilliant crimson, remarkable for colour rather than flavour. Brisk, plain taste; sometimes hint strawberry flavour; firm flesh.
Formerly widely grown Europe and UK.
F 10. T². C gd. P l-Sept/e-Oct. S l-Nov-Jan.

BAUNEN *see* King of the Pippins

BAXTER 6 M D
Canada; arose in Brockville, Ontario. Known 1800. Originally Red Pound.
Large; very dark red. Some richness; quite soft and sugary, but rather coarse, deep cream flesh.
Briefly popular in St Lawrence Valley at turn of C19th.
F 9. T¹. P l-Sept. S Oct-Nov.

BAXTER'S PEARMAIN 7 L D
UK; arose Norfolk. Introd 1821 by nursmn G. Lindley, Catton, nr Norwich. Resembles Golden Reinette in appearance and flavour.
F 15. T³. C hvy. P m-Oct. S Nov-Jan/Feb.

BEACON 6 M D
USA; raised 1908 at Univ Minnesota Exp Breeding Farm, Excelsior.
Malinda X Wealthy. Introd 1936.
Large; bright red. Soft, sweet flesh, but often woody, metallic in England. Grown commercially south east Minnesota in Mississipi and St Croix valleys; garden fruit of Northern Great Plains.
F 12. T²; v hrdy; tol drought; prn canker; res scab. P e-Sept. S Sept-Oct; cold store-Dec.

BEAUTY OF BATH 4 E D
UK; arose with market gardeners of Bailbrook, nr Bath, Somerset.
Introd c1864 by nursmn G. Coolings.
Brought prominence after 1888 Apple & Pear Conference. RHS FCC 1887.
Distinctive, fairly acid taste, although fully ripe has plenty of sweetness. Too early, it can be like a green gooseberry, but left too late will taste fermented. Flesh often stained pink under the skin.
Formerly widely grown for market and in gardens. Its very short season and tendency to 'drop' before ripe – farmers used to put straw under the trees to catch the fruit – led to its commercial demise.
Frt *Col* rd flushed, striped, over pale grn/yell; mottled with lrg white lenticels; ltl russet. *Shape* flt-rnd; sltly ribbed. *Size* smll/med. *Basin* brd, dp; sltly ribbed. *Eye* usually clsd; sepals, brd. *Cavity* qte brd, qte dp; scarf skin. *Stalk* shrt, thck. *Flesh* crm.
F 8. T³. C hvy. P eat e-Aug.
PLATE 2

More highly coloured forms:
CRIMSON BEAUTY,
CRIMSON BEAUTY OF BATH,
RED BEAUTY OF BATH

BEAUTY OF BEDFORD 7 M D
UK; raised before 1913 by Laxton Brothers, Nursery, Bedford.
Lady Sudeley X Beauty of Bath.
Sweet, modestly rich, some aromatic character; quite firm cream flesh.
F 11. T³. P m-Sept. S Sept-Oct.

BEAUTY OF BLACKMOOR 4 E D
UK; raised 1947 by G. T. Spinks, LARS Bristol.
(Worcester Pearmain X Beauty of Bath) X Beauty of Bath. Formerly Apollo.
Sharp, crisp, like Beauty of Bath, but can be astringent.
F 15. T². P eat m/l-Aug.

BEAUTY OF HANTS 7 L D
UK; acc prob variety raised before 1850 in garden of Mrs Eyre Crabbe, at Basset, Southampton, Hants; acc is variety of Bultitude (1983). Blenheim Orange X.
Quite sweet, plain taste of fruit; crumbly texture.
F 13. T³. P e/m-Oct. S Nov-Jan/Mar.

BEAUTY OF KENT 4 L C
UK; pres arose Kent. Known to Forsyth 1790s; listed 1820 by Brompton Park Nursery, London. RHS AM 1901.
Large. Cooks to pale yellow purée; delicate flavour, needs no extra sugar. Baked, juicy and creamy. Best used early for cooking as acidity fades, but makes pleasant eating apple by Dec.
Grown in most Victorian and Edwardian country house fruit collections; also by commercial growers.
F 11. T². C gd. P l-Sept/e-Oct. S Oct-Jan.

BEAUTY OF MORAY 1/5 E C
UK; prob arose Scotland. Exhib 1883, when favourite in North Scotland.
Sharp; cooks to strongly flavoured cream purée.
F* 10. T². C gd. P use Aug.

BEAUTY OF STOKE 1 L C
UK; raised by Mr Doe, HG to Lord Saville, Rufford Abbey, Notts. Recorded 1889. RHS FCC 1890.
Quite sweet, firm flesh. Cooks to bright lemon, sweet purée in Dec.
F 16. T². P m/l-Oct. S Dec-Mar.

BÉBÉ ROSE 4/6 L D
Received from France 1947.
Small, closely resembles Api.
Sprightly, fresh and fruity. Ornamental in flower and fruit.
F* 6. T²; uprt. C gd. P m-Oct. S Nov-Jan/Feb.

BEC D'OIE 6 L D
France; acc is variety now known Cher (Val de Loire). Believed to be very old; may be variety quoted 1670 by Claude St-Etienne, but impossible to know.
Shaped like beak – conical, rounded on one side, flat on other with red flush. Sweet, crisp, slight strawberry flavour.
F 12. T². P e-Oct. S Oct-Jan/Apr.

BEDFORD PEARMAIN syn Laxton's Pearmain

BEDFORD RED 4 L D
Received 1972 from former Yugoslavia via Scotland.
Sweet, modestly fruity taste; yellow flesh; tough skin.
F 13. T². P m-Oct. S Nov-Jan.

BEDFORDSHIRE FOUNDLING 5 L C
UK; prob arose c1800 Beds.
Large. Plenty of acidity, sugar, almost rich flavour; deep cream flesh. Cooked, slices keep shape; yellow with sweet, good fruity taste.
F 11. T². C gd. P e-Oct. S Oct-Dec/Mar.

BEDMINSTER PIPPIN 5 M D
UK; received 1952 from 100yr old tree in Gloucs. Name suggests poss arose Bedminster, Somerset.
Small, in style of Golden Pippin. Intense taste of fruit; deep cream flesh. Like strong Downton Pippin.
F 18. T². P m-Sept. S Oct-Nov.

BEDWYN BEAUTY 3/7 L CD
UK; raised c1890 by Mr Stone, Great Bedwyn, nr Malborough, Wilts.
Resembles Golden Reinette. Cooked keeps shape; light flavour. Mellows to quite pleasant eating apple.
F 21. T². C gd. P m-Oct. S Nov-Feb/Apr.

BEELEY PIPPIN 7 M D
UK; raised c1880 by Rev C. Sculthorpe, Beeley, Derbys.
Appealing; dusky pink flush and russet. Rich, intense, quite aromatic flavour; cream crisp yet melting flesh.
F 4. T²; v sprd. P e-Sept. S Sep-Oct.

BEL-EL *see* Elstar

BELCHARD *see* Chantecler

BELLEFLEUR-KITAIKA 4 M D
Russia; raised 1908 by I. V. Michurin at Michurinsk.
Yellow Bellefleur X Kitaika (*Malus prunifolia*). Fruited 1914.
In effort to improve all fruits and extend area of cultivation northwards, Michurin crossed western and old Russian varieties and also used hardy *Malus* species. He received patronage of Lenin and Stalin, a

research institute was established at his home in Kozlov, which was renamed in his honour. However, Michurin believed erroneously that acclimatisation was key to succesful breeding; only a few of his seedlings proved of merit.

Flushed, striped in pink; large, ribbed. Sweet, soft flesh.

F 9. T^2. C hvy. P m-Sept. S Oct-Nov.

BELLEFLEUR KRASNYI 6 L D
Russia; raised 1914 by I. V. Michurin at Michurinsk.

Bellefleur-kitaika X Yakhontovoye (Antonovka X *Malus sieversii ssp niedzwetzkyana*).

Bright red flush. Very scented and sweet; firm, rather dry white flesh, often stained red.

F 3. T^3; res scab; part res mildew. C hvy. P m-Oct. S Oct-Dec.

BELGOLDEN NO 17 *see* Golden Delicious

BELLAQUEENY 6 E D
Received 1948 from EMRS, Kent.

Very large; striking pinky red flush. Quite sharp fruity taste; soft flesh.

F 9. T^2. C hvy. P Aug. S Aug-Oct.

BELLE AMÉRICAINE syn American Beauty

BELLE DE BOSKOOP 7 L CD
Netherlands; discv 1856 at Boskoop, nr Gouda, by K. J. W. Ottolander. Apparently bud sport of Reinette de Montfort. RHS AM 1897; AGM 1993. Originally Schone van Boskoop. Syns many.

Firm, juicy sharp flesh, which mellows to resemble Reinette du Canada, but sharper, not as interesting. Cooked, slices keep shape or make thick gold purée; quite brisk, well-flavoured; hardly needing any sugar. Vit C 10–16mg/100gm.

Often found in English gardens; formerly recommended also as large, spreading, ornamental lawn tree.

Grown commercially Netherlands and all over continent: Germany, Belgium, Switzerland, Poland. In Netherlands usually on sale as Schone van Boskoop or Goudrenet.

Frt *Col* orng rd flush, some rd stripes, over gold; much russeting. *Size* med/lrg. *Shape* rnd-con to oblg; tall; rnded ribs. *Basin* brd, dp; sltly ribbed; some russet. *Eye* lrg, hlf open; sepals, lng, brd. *Cavity* med wdth, dpth; russet lined. *Stalk* med lng, thck. *Flesh* pale crm, tinge grn.

F* 8; trip. T^3. C gd. P e-Oct. S Oct-Apr.
   BIELAAR; darker red sport, but brighter appearance.
   BOTDEN; very dark red sport.
   RED BELLE DE BOSKOOP; higher col. Received 1950 from Denmark.

WILHELM LEY; sport of Roter Boskoop, darker red, less russet. Received 1981 from Wilhelm Ley.

BELLE DE FRANCE 3 L DC
Received from Switzerland 1982.

Large, attractive fruit. Crisp, juicy sharp.

F v late. T^2. C gd. P e-Oct. S Oct-Jan.

BELLE DE LONGUÉ 4 L DC
France; arose with Lefant of Longué, nr Saumur, Maine-et-Loire (Western Loire). Introd 1889 by Leroy Nursery.

Large; lightly flushed in red. Sweet, white flesh; poor culinary flavour.

F 18. T^3. C hvy. P m-Oct. S Nov-Feb/Apr.

BELLE DE MAGNY 5 L D
France; recorded 1895 by nursmn Simon-Louis Frères.

Rich, lots of sugar and acid; almost pineapple flavoured in Nov, Dec; firm, pale yellow flesh.

Grown now Haute Savoie and Switzerland.

F 11. T^2. C hvy. P m-Oct. S Nov-Feb.

BELLE DE NORDHAUSEN syn Schöner von Nordhausen

BELLE DE PONTOISE (LUXEM-BOURG) 6 L CD
France; received 1996 from Luxembourg Gardens, Paris; very similar to Jeanne Hardy (Bunyard) in NFC. Acc may be variety raised 1869 by Rémy of Pontoise, introd 1879; Alexander X.

Large, handsome. Fruity, slightly vinous juicy flesh.

F est 12. T^3. C gd. P m-Oct. S Nov-Jan/Mar.

BELLE DES BUITS syn Pierre 5 L D
France; found at Buits, home of M. Guyot de la Rochère, in Vienne; from here grafts were sent to nursmn Bruant of Poitiers in C19th.

Sharp, juicy, crisp pale cream flesh.

Grown for market around Brussière, Vienne in C19th and now found distributed over Vienne (Poitou), Indre (Val de Loire) and Limousin; used for cooking, juice; vit C 20mg/100g.

F 31. T^2; uprt; reported gd res disease. C hvy. P m-Oct. S Dec-Feb/May.

BELLE DE TOURS (?) 6 L C
France; acc may be Lambron, syn Belle de Tours; arose Indre-et-Loire (Val de Loire) presumably around Tours. Received 1947.

Large; boldly striped, flushed in carmine. Cooked keeps shape; good rich taste.

F 10. T^2. C hvy. P e-Oct. S Oct-Jan/Mar.

BELLEDGE PIPPIN 5 L D
UK; local to Derbys according to Ronalds (1831). First recorded 1818.

Clear yellow, slight blush. Sharp, crisp.

F 12. T^3. C hvy. P m-Oct. S Nov-Jan/Feb.

BELLE ENTE 7 L D
France; received 1948.

Blood red flush. In Dec, sweet, quite rich, scented; crisp white flesh.

F 12. T^2. C bien. P m-Oct. S Dec-Jan/Feb.

BELLE-FILLE NORMANDE 1/5 L C
France; arose Pays de Caux, Normandy. Known late 1700s.

Large, tall, conical. Sweet-sharp quite rich taste in Oct. Cooked, keeps shape or falls into lightly flavoured lemon purée. Flavour fades with keeping. Grown Normandy, Indre et Loire (Val de Loire), Vienne (Poitou). Used also for juice, cider. Vit C 20mgs/100g.

F 10. T^3. P m-Oct. S Oct- Dec/Apr.

BELLE FLAVOISE 4 E C
France; received 1947.

Pretty, pink striped; very flat shape. Brisk taste of fruit in Sept. Used earlier, cooks to lightly flavoured, smooth purée.

F 9. T^2. C hvy, ripens over long period; bien. P eat Aug-Sept.

BELLE FLEUR DE FRANCE 6 L DC
Northern France, Belgium or Netherlands; ancient variety of unknown origin; first described 1758 by Knoop, who believed it to be Dutch. Known under this name Belgium; Belle Fleur Double in Northern France.

Large, red flushed, prominently ribbed. Crisp, juicy, balanced fruity taste; rather bland cooked.

Grown Belgium, northern France for fresh fruit; juice; cooking, especially tarts, compôtes and traditionally to make 'pâtes de pommes' in Bavay (Nord).

F 19. T^3; gd res scab. C hvy. P m-Oct. S Nov-Jan/Apr.

BELLE-FLEUR DE SAINT BENOÎT
5 L C syn Bailly
France; found St Benoît-du-Sault, Indre (Val de Loire).

Pink blushed. Crisp, juicy, quite brisk. Cooks to lightly flavoured purée.

Distributed over Indre; Haute-Vienne, Creuse (Limousin).

F 18. T^2. P m-Oct. S Nov-Jan/Apr.

BELLE FLEUR DOUBLE syn Belle Fleur de France

BELLE-FLEUR LARGE MOUCHE
3 L C
Belgium or France; known Belgium under this name, but Lanscailler in France. Desc 1910 by Carpentier as newly introd in Avesnois (Nord) under name Lancashire.

Large, ribbed with large eye, which accounts for name. Crisp, sharp flesh. Cooks to quite brisk, flavoursome purée.

Widely known Flanders, northern France.
F 13. T³. P m-Oct. S Nov-Jan/Apr.

BELLIDA 6 M D
Netherlands; raised IVT, Wageningen.
Idared X Elstar. Received 1994.
Crisp juicy cream, rather coarse flesh,
sweet, slightly honeyed.
F mid. T². P e-Sept. S Sept- Oct/Nov.

BELVOIR SEEDLING 4 L D
UK; raised by W. H. Divers, prob while HG
at Belvoir Castle, Leics.
Annie Elizabeth X Dumelow's Seedling.
Received 1935.
Large. Cooks to brisk purée.
F 14. T². P e/m-Oct. S Nov-Feb/Apr.

BEN APPLE syn Eustis

BEN DAVIS (?) 6 L D
USA; acc may be variety known since early-
1800s, but confused history.
Deep red flushed; sweet, firm flesh, light
flavour.
Most widely accepted history of Ben Davis
begins in 1799 when William Davis and
John Hills moved from Virginia to Berry's
Lick, Butler County, Kentucky, near to Cpt
Ben Davis (Davis's brother, Hill's brother-
in-law). Later, Hill during trip to Virginia or
Carolina, brought back some young apple
trees. One of these was planted on Cpt
Davis's land. Root shoots from this tree were
used to plant an orchard, which attracted
attention. Later, by root suckers, trees of
Ben Davis were spread throughout
Kentucky, Tennessee. Hill family moved to
Illinois taking trees with them. By 1865 Ben
Davis was widely known and millions of
trees were planted all over US, especially
southern states up to early 1900s.
Ideal market apple – heavy cropping, long
keeping, colourful, firm fleshed, also easily
propagated. Long overtaken by Winesap
and then Delicious.
   BLACK BEN; deeper col. Arose c1880
   on farm M. Black, Washington County,
   Arkansas.
F 15. T². C gd. P m-Oct. S Dec-Mar/May.

BÉNÉDICTIN 7 L CD
French and British pomologists agree
identical to Blenheim Orange, but remains
known in Normandy under this name. In
C19th grown by monks of Jumièges
Benedictine Abbey and exported to UK.
Recommended for 'Tarte tartin' and 'Tarte
aux Pommes'.

BENENDEN EARLY 4 E D
UK; raised c1945 by J. J. Gibbons,
Southampton, Hants.
Saint Edmund's Pippin X Lady Sudeley.
Introd 1952 by Stuart Low Ltd, Benenden,
Kent.
Brilliant red stripes of Lady Sudeley and

some of richness of Saint Edmund's Pippin,
but not nearly as luscious. Can lack flavour.
F 11. T². C frt cracks. P m-Aug. S Aug-Sept.

BENONI 4 E D
USA; raised c1830 on farm of Mr Mason
Richards, Dedham Massachusetts.
Golden yellow, flushed in orange red,
crimson stripes. At best perfumed, creamy
with rich blend of sugar and acid.
Highly esteemed in C19th Massachusetts.
Introd England prob by nursmn, Thomas
Rivers. Grown C19th by London market
gardeners, also Netherlands, where remains
garden apple.
F 9. T²; uprt. C hvy, bien. P e-Sept. S Sept.
   RED BENONI; more highly coloured.

BENSEMAN'S SEEDLING 7 L D
Received 1950 from New Zealand.
Honeyed, very sweet; yellow firm flesh.
F 9. T¹; sp. P m-Oct. S Nov-Feb/Apr.

BEN'S RED 6 E D
UK; raised c1830 by B. Roberts, Trannack,
Cornwall. Roberts sent samples to
nursmn George Bunyard, Maidstone, Kent
who introduced it 1890s.
Devonshire Quarrenden X poss Farleigh
Pippin. RHS AM 1899.
Very like Devonshire Quarrenden, but
ripens later, keeps longer and better
flavoured. Firm often pink tinged flesh;
crisp, sweet with definite strawberry/
raspberry flavour.
Grown and still found all over Cornwall.
Planted Kent in early 1900s instead of
Worcester Pearmain, but with introduction
of lime sulphur sprays in 1920s growers
returned to Worcester. Often sold in
markets as 'Quarrenden'. Roots from
branches, known as 'pitchers', and still
propagated in Cornwall in this way.
Frt *Col* dark maroon flush, faint darker rd
stripes; pale grnish yell/yell background.
Size med. *Shape* flt/flt-rnd. *Basin* brd, shal;
sltly wrinkled. *Eye* clsed; sepals brd, v
downy. *Cavity* brd, dp; lined russet. *Stalk*
shrt, thck. *Flesh* pale crm.
F 6. T¹; sprd; reported res mildew, scab. C
hvy, bien. P e-Sept. S Sept.

BERECZKI MÁTÉ 4 L D
Hungary; received 1948 from Univ Agric,
Budapest.
Named after Bereczki, renowned C19th-
Hungarian pomologist.
Crisp, juicy.
F 8. T². P m-Oct. S Dec-Feb/Apr.

BEREGI SÓVÁRI 5 L D
Hungary; recorded 1900 (Molnar).
Resembles Borsdorfer in appearance.
Slightly sweet.
F 12. T³; sp. C gd. P m-Oct. S Nov-Jan/Feb.

BERLEPSCH *see* Freiherr von Berlepsch

BERNA 6 M D
Belgium. Received 1993.
Poss Belle de Boskoop X.
Large. Deep cream flesh with delicate, light
taste.
F est 15. T³; sprd. P m-Sept. S Sept- Nov.

BERNER ROSEN syn Rose de Berne
6 L D
Switzerland; discv by nursmn Daepp of
Oppligen. Fruited 1888.
Very pretty, flushed in deep red. Sweet,
juicy flesh, but often ripens poorly in
England and tastes metallic.
Switzerland's most popular apple; also
grown Germany; Alsace, Franche Comté,
France.
F 11. T²; sp. C hvy. P e-Oct. S Nov-Jan.

BERTANNE see Golden Delicious.

BESSEMYANKA MICHURINA 4 L D
Russia. Raised 1912 by I. V. Michurin at
Michurinsk.
Skrizhapel X Komsin Bessemyanka.
Refreshing, sweet juicy, very crisp firm
cream flesh, often pink stained.
F 4. T². P e-Oct. S Nov-Jan.

BESS POOL 7 L D
UK; discv 1700s in Notts wood by Bess Pool,
who took apples home to her father, who
kept an inn. Fruit achieved local fame and
was taken up and introduced by nursmn J.
R. Pearson of Chilwell, Notts. 'Best Poll'
known to Forsyth 1802.
Beautiful crimson flush. Rich, almost
aromatic by Jan; sweet rather dry crumbly
flesh. Becomes spongey, but still good to
eat. Also used for cooking in C19th and
decoration to give colour to large fruit
displays on Victorian dining table.
Widely grown in gardens and for market in
C19th; still found growing in North and
Midlands.
Frt *Col* crimson flush, dark rd stripes over
yell; russet patches, dots. *Size* med/lrg.
Shape rnd-con/oblng; often flt sided. *Basin*
qte brd, med dp; 5 beads. *Eye* clsd; sepals v
downy. *Cavity* med wdth, dpth; russet lined;
scarf skin. *Stalk* shrt, thck. *Flesh* wht.
F 22. T³; part tip. C slow to bear, erratic.
P e/-Oct. S Nov-Mar.

BETSEY 8 L D
UK; arose England. Recorded 1842; desc
Hogg (1851).
Good taste of fruit; brisk, plenty of acidity.
F 15. T¹; tip. C frt cracks. P m/-Oct. S Dec-
Feb/Apr.

BETTY GEESON (?) 5 L C
UK; received 1982. Acc may be variety of
Hogg, which arose poss Worcs and sent
1854 to British Pom Soc by Dr Davies of
Pershore.
Plenty of acidity. Cooked, keeps shape;

tastes sweet, quite rich; needs no sugar.
F 11. T^2; sprd. P l-Sept. S Oct-Dec/Jan.

BEURRIÈRE 7 L D
France; received 1948 and variety now known northern France under this name, which may be Picardy variety Beuvriére, recorded 1908 by nursmn Charles Baltet.
Bright flush, fine russet dots, sweet with taste of raspberries. Traditionally used for cooking.
F 25. T^2. C gd. P m-Oct. S Oct-Dec/Feb.

BEVERLY HILLS 4 E D
USA; raised 1939 by Dr W. H. Chandler, Univ California, Los Angeles.
Melba X Early McIntosh. Introd 1945.
At best, heavenly. Very soft, melting juicy, white flesh and sweet, strawberry/winey taste. Very like Melba.
F 15. T^2; sprd. P eat l-Aug-Sept; v shrt season.

BIDE'S WALKING STICK syn Burr Knot

BIELAAR *see* Belle de Boskoop

BIESTERFELDER-RENETTE 7 M D
Germany; arose Castle Biesterfeld, Lippe, Westphalia. Introd 1905.
Good, aromatic flavour; sweet; balanced in Oct; crisp, juicy flesh.
F 9. T^3; sprd. P m-Sept. S Oct-Nov.

BIETIGHEIMER 6 M C
Germany; introd 1870 to Canada; received UK before 1905.
Very large, round; flushed in pink. Cooks to light, sweet purée.
F 6. T^3. P e-Sept. S Sept-Oct/Nov.

BIGGS'S NONSUCH (?) 4 M DC
UK; acc may be variety desc by Lindley (1831), Hogg (1851); raised by gardener Arthur Biggs at Twickenham.
Crimson striped over pale yellow. Brisk taste of fruit. Cooked: attractive gold slices, good, savoury, fruity taste, needs hardly any sugar.
F 9. T^2. P e-Sept. S Sept-Oct.

BISMARCK 3/6 L C
Australia. Believed either originated at German settlement of Bismark, Tasmania, or at Carisbrooke, Victoria and named after German Chancellor, Prince Bismark.
According to 1888 survey of Victoria it was raised by F. Fricke, native of Hannover who settled at Carisbrook in 1855. When tree 25 yrs old, in 1886, fruit shown in Melbourne Museum and in London in 1887. By 1894 widely grown Victoria. RHS FCC 1887.
Rather astringent. Cooks to bright yellow purée, brisk, well-flavoured. Exhibition and popular culinary apple in England up to 1930s, as well as southern hemisphere; also grown Europe.
Frt *Col* bright rd flush, rd stripes over grnsh

yell/ yell; lenticels as russet or grey dots. *Size* med/lrg. *Shape* shrt-rnd-con; rnded ribs; 5 crowned. *Basin* med wdth, qte dp; ribbed. *Eye* clsed; sepals, brd, qte downy. *Cavity* med wdth, dpth; russet lined. *Stalk* shrt/med, qte thin. *Flesh* wht.
F 8. T^2; sprd. C gd. P e-Oct. S Nov-Feb.

BLACK BEN *see* Ben Davis

BLACKJON *see* Jonathan

BLACKMACK *see* McIntosh

BLACK MICKEY *see* McIntosh

BLACKMOOR'S UPRIGHT 5 L C cider
UK: received 1990 from LARS.
Soft, white, sharp flesh; poss sharp cider apple.
F mid. T^2; v uprt. P e-Oct.

BLACK ROME syn Ruby Rome Beauty

BLAHOVA RENETA 7 L D
Czech Republic; received 1974, from Prague.
Quite sweet; some richness in New Year.
F 17. T^2. P m/l-Oct. S Jan-Mar.

BLANC SUR 1/5 L DC
France; acc received as Blanc Dur but is prob Blanc Sur of Sarthe (Western Loire); desc Boré (1997).
Not ancient Blanc Dur syn Blanc-Dureau, Blandurel.
Sharp, some sweetness, crisp. Cooked, keeps shape, sweet, rather bland. Grown Sarthe also for juice and cider.
F 37. T^2. C gd. P m-Oct. S Nov-Mar/Apr.

BLANDURETTE 5/4 L D
France; acc may be Blandurette of Corrèze, but number of varieties known by this name in Limousin and difficult to resolve.
Fruity, sweet; crisp juicy.
Blandurette apples are grown high regions of Limousin and considered fine cider fruit, making long-keeping, good-quality brew.
F 24. T^3. C gd; frt sml. P m-Oct. S Dec-Mar/Apr.

BLAXTAYMAN *see* Stayman Winesap

BLAZE 6 L D
USA; raised 1939 at Illinois Agric Exp St, Urbana.
Collins X Fanny. Introd 1958.
Gorgeous carmine flush and stripes. Sprightly taste; sweet, crisp.
F 13. T^3. P l-Sept. S Oct-Jan.

BLENHEIM ORANGE 7 L CD
UK; discv c1740 by Kempster at Woodstock, Oxon. Exhibit London Hort Soc 1822, awarded Banksian Medal.

Scions sent c1820s to Massachusetts, Canada and the continent; later to Australia. Syns – 67. RHS AGM 1993.
Addictive plain taste flavoured with nuts; quite sweet; crumbly texture; good with cheese. Larger fruits were used for 'Apple Charlotte' as it cooks to stiff purée, or keeps shape. (Vit C 12mg/100gm).
Found growing against boundary wall of Blenheim Park by local cobbler, or tailor, who moved it into his garden and 'thousands thronged from all parts to gaze on its ruddy, ripening orange burden; then gardeners came in the spring-tide to select the much coveted scions' of Kempster's Pippin. After Duke of Malborough's approval, renamed and advertised in 1804 by Worcestershire nurseryman Biggs as 'the new Blenheim Orange'. Widely grown in C19th; more dishes of it were on display at 1883 Congress than of any other variety. Still in markets of 1930s; also imported from Canada and southern hemisphere. Rarely seen for sale now, but remains valued amateur variety; often found in old gardens. Still valued in Germany and as Bénédictin in Normandy.
Makes strong limbed tree with very hard wood which was used to make cog wheels for railways.
Frt *Col* orng rd flush, few rd stripes, over grnsh yell/gold; russet patches, veins. *Size* lrg/med. *Shape* flt-rnd; rnded ribs; sltly crowned. *Basin* brd, dp; ltl russet. *Eye* lrg, open; sepals sep at base. *Cavity* med wdth, dpth; russet lined. *Stalk* shrt, thck. *Flesh* pale crm.
F 12, trip. T^3; part tip; some res mildew. C bien, erratic. P l-Sept/e-Oct. S culinary from l-Sept; dessert, Oct-Dec/Jan.
PLATE 19
ALDENHAM BLENHEIM; discv before 1929 by H. G. Edwin Beckett at Aldenham House, Herts. Said to be more highly coloured, but not noticeably so.
More highly coloured form:
RED BLENHEIM (F. E. Wastie); discv on old tree by F. W. Wastie at Welland, Malvern, Worcs; received 1966.

BLOODY BUTCHER 3 M C
Ireland; desc 1951 (Lamb).
Maroon flushed; remarkable for its colour rather than quality. Soft juicy flesh. Cooked, flat, insipid.
F 12. T^2; sprd. P m-Sept. S Sept-Oct/Nov.

BLOODY PLOUGHMAN 6 M D
UK; old variety of Carse of Gowrie, Scotland; may be variety exhib 1883 Congress.
Blood red, heavily ribbed. When very ripe, flesh becomes stained pink. Sweet, light taste; crisp, juicy.
Reputedly takes its name from ploughman who was caught stealing Megginch estate apples and shot by game-keeper. His wife

got the bag of apples, but she threw them on a rubbish heap and one of seedlings that emerged was rescued by workman and subsequently named.
F 12. T³. P m-Sept. S Sept-Nov.

BLUE PEARMAIN 6 L D
USA (prob); known early 1800s. Introd UK prob late 1800s; RHS AM 1893, FCC 1896.
Large, covered in blue bloom. Delicate aromatic quality in Dec; sweet soft cream flesh.
Widely planted in New England, New York in C19th; now being taken up again by amateurs.
F 11. T²; sprd. P e-Oct. S Dec-Feb/May.

BÖDIKERS GOLD-REINETTE 7 M D
Germany; raised by Bödiker, Meppen, Hannover. Recorded 1875.
Resembles large Golden Reinette. Sharp, little richness; crisp cream flesh.
F 9. T². P m-Sept. S Oct-Nov/Dec.

BODIL NEERGAARD 5 L D
Denmark; found c1850 by nursmn, Andersen at Flintinge, Lolland.
Named after proprietress of Flintage estate where it arose.
Sweet, savoury taste in Nov; soft juicy flesh. Formerly widely grown Denmark, also Sweden.
F 10. T³; sp; weeping; hrdy. C hvy. P m-Oct. S Nov-Jan/Apr.

BOHEMIA see Rubin

BOHNAPFEL 4 L DC
Germany; prob arose Rhineland, late 1700s. Syns many.
Brisk, firm flesh in New Year.
Formerly much grown in Rhine for cider, cooking and fresh fruit. Still grown Germany, chiefly for drying and preserved purée. Continues to be planted in some areas as street tree.
F 11; trip. T³; uprt. C bien. P m-Oct. S Jan-Mar/Jun.

BOÏKEN 5 L DC
Germany; arose Bremen, named after dike-warden Boike. Known 1828.
Bright yellow; pinky orange flush. Sharp, crisp; sweetening, but fresh and fruity to Mar.
Still valued eastern Germany for cooking, drying, juice manufacture as well as fresh fruit; also grown Bas Rhin (Alsace). Grown early 1900s in New York as late market culinary variety.
F 14. T¹; sprd. P l-Oct. S Dec-Mar/Apr.

BOLERO see Tuscan

BONDON 7 L D
France; desc 1948 (*Verger Français*) when

grown Indre (Val de Loire).
Name means logshead plug on account of its shape – long, conical. Brisk, plain, quite good in Nov; firm, rather dry flesh.
Known now Indre, Indre-et-Loire; grown for juice, fresh fruit.
F 13. T³. P m-Oct. S Nov-Jan/Apr.

BONNE HOTTURE 8 L D
France; prob arose around Angers, Maine-et-Loire (Western Loire); long grown there according to Leroy (1873). Recorded 1867 by nursmn Simon-Louis Frères.
Taken to market in baskets or 'hottes'. Small. Sweet rich, quite nutty; cream, green tinged flesh.
Found now in Mayenne, Sarthe, Maine-et-Loire, Indre-et-Loire, Vienne.
F 13. T¹. C hvy. P m-Oct. S Nov-Jan/Apr.

BONNET CARRÉ 2 L D
France; desc 1948 (*Verger Français*). Syn Belle Dunoise; originated region of Dun-le-Palestel, Creuse (Limousin). Bonnet is popular name for Calvilles; Carré refers to square angular shape. Often confused with Calville Blanc d'Hiver.
Quite intense sweet-sharp taste, crisp, juicy. Grown central France, especially Indre, Creuse. Used for cooking, juice (vit C 15mg/100g).
F 23. T¹. C hvy; hangs v late. P l-Oct. S Jan-Mar.

BONNET DE COMTE 5 L CD
France; received 1947.
Some richness by Jan; sweet, firm flesh. Cooked, keeps shape; sweet, light taste.
F 16. T². C hvy. P l-Oct. S Nov-Feb/Apr.

BOROVINKA syn Duchess of Oldenburg

BORSDÖRFER syn Edelborsdorfer

BOSBURY PIPPIN 6 M D
UK; prob variety desc 1920 by Bunyard, which arose west of England.
Resembles Baumann's Reinette in appearance. Quite sweet, light flavour; firm, rather dry cream flesh.
F 12. T³. P e/m-Sept. S Sept-Nov.

BOSKOOP syn Belle de Boskoop

BOSSOM (JOHNSON) 5 L D
UK; received 1991 from Rev D. Johnson, Petworth, Sussex, from tree long known as Bossom in village of Graffham. Acc poss variety desc Hogg (1851); first exhib 1820 and prob raised at Petworth House.
Golden. Yellow tinged flesh with intense sweet, sharp, fruity taste.
F mid. T³. P e-Oct. S Nov-Jan.

BOSTON RUSSET syn Roxbury Russet

BOTDEN see Belle de Boskoop

BOUET DE BONNÉTABLE 2/4 L D
France; desc 1948 (*Verger Français*), when grown Sarthe (Western Loire).
Dusky pink flush over cream by Jan. Sharp, fruity, refreshing, but very bracing and acidic.
F 20. T³; sprd. C gd, frt sml. P l-Oct. S Jan-Mar/Apr.

BOUNTIFUL 4 M C
UK; raised 1964 by Dr F. Alston, EMRS, Kent.
Cox's Orange Pippin X prob Lane's Prince Albert. Introd 1985 by Highfield Nurseries, Gloucs.
Cooked, slices just keep shape; soft, juicy, light, sweet.
F 9. T²; bears early; res mildew. C hvy. P l-Sept. S Sept-Nov.

BOUQUEPREUVE 2/5 L D
France; poss arose Mediterranean region. Recorded 1884.
Sweet, some taste fruit, but rather dry, cream flesh.
Local Marseille fruit in 1940s and still known in area.
F 17. T³. P m-Oct. S Nov-Jan.

BOUSCASSE DE BRÈS 2/5 L D
France; believed raised C19th by M. Brès, who brought seeds back from the Far East and planted them on his property in St-Jean-du-Gard, Lozère.
Sweet, fruity; light, pleasant. Widely grown Cévennes.
F 20. T³. C hvy. P m/l-Oct. S Jan-Mar/Apr.

BOVARDE 3 L DC
Switzerland; desc France 1948 (*Verger Français*).
Large; brightly flushed. Sharp, fruity; crisp, juicy, cream flesh. Cooks to sharp, well-flavoured purée.
Local to Valais and canton of Vaud; used also for juice.
F 11. T³; sp; v hardy; reported res scab. C gd. P l-Oct. S Dec-Mar/May.

BOWHILL PIPPIN 7 L C
UK; raised by A. S. White, Bow Hill, Maidstone, Kent. Introd c1893 by nursmn, G. Bunyard, Maidstone. RHS AM 1893.
Sweet, balanced; plain, slightly nutty taste; firm, deep cream flesh.
F 11. T². C gd. P e-Oct. S Nov-Feb/Apr.

BOX APPLE 5 E D
UK; arose Cornwall. Received 1955.
Tough little apple. Sharp, juicy, lots fruity flavour.
F 10. T². P l-Aug. S Sept-Nov.

BRABANT BELLEFLEUR 3 L C
Netherlands (prob); uncertain origin; brought notice late 1700s. RHS AM 1901.
Syns many; incl Glory of Flanders, Iron

Apple.
Hard, crisp, developing fruity brisk taste with hints of strawberry flavour. Cooked, keeps shape; quite rich flavour in Dec. Vit C 19 mgs/100g.
Widely grown on continent in C19th; also valued England; still esteemed Belgium, Netherlands.
F 23. T³; sprd. C gd. P l-Oct. S Dec-Mar/Apr.

BRADDICK NONPAREIL (?) 7 L D
UK; acc may be variety desc Ronalds, Hogg; raised by John Braddick, Thames Ditton, Surrey; exhib 1818 London Hort Soc.
Characteristic sweet-sharp, fruit drop quality of nonpareil type; rich intense flavour. Finely textured, crisp, cream flesh.
F 13. T². P m-Oct. S Nov-Feb/Apr.

BRAEBURN 4 L D
New Zealand; discovered on property of O. Moran, Waiwhero, Upper Moutere, Nelson.
Grown commercially 1952 by William Bros, Braeburn Orchards.
Believed Lady Hamilton seedling; first known by this name. In 1970s, two forms of Braeburn recognised in New Zealand, one maturing earlier than other; NFC has later form; commercial plants now are of earlier form.
Refreshing, crisp, firm flesh; can be perfumed. In England often fails to mature. Grown commercially New Zealand and all warm apple growing regions: Australia, South Africa, South America, France, Germany, Italy, Canada, US.
F 18. T²; sprd. C hvy. P l-Oct. S Jan-Mar.
   HIDALA, syn Hillwell; discv Hastings, New Zealand, 1981; introd 1990; deeper red; said to ripen 10 days earlier.

BRAINTREE SEEDLING 7 L D
UK; raised 1930 by Mrs Humphreys, Braintree, Essex. Claimed from Gladstone pip, but nothing like parent.
Intense sweet, rich, aromatic flavour with hint of pineapple; quite firm, cream flesh.
F 9. T². C gd. P e/m-Oct. S Oct-Dec/Jan.

BRAMLEY'S SEEDLING 3 L C
UK; raised 1809–1813 by Miss Mary Anne Brailsford and planted in garden Church Street, Southwell, Notts. Introd 1865 by nursmn H. Merryweather; exhib 1876. RHS FCC 1883; AGM 1993.
Cooks to pale cream purée with strong acidity and flavour; rarely overwhelmed by most sugary and highly spiced recipes. With keeping, acidity falls and home stored fruit can serve as sharp eating apples in spring, but commercially stored fruit retains high acidity.
Bramley's Seedling was first admired by Henry Merryweather in about 1857, when tree and cottage belonged to Mr Bramley,

the local butcher. Became widely known after 1883 Congress. Original Bramley tree blew down in storm in early 1900s, but branch grew up from the old trunk and still survives and fruits. First trees planted commercially in Kent were by Mr F. Smith, Loddington, nr Maidstone in 1890, supplied by Merryweather. Extensively planted early 1900s; now almost only cooker grown commercially and available all year round. Remains valued garden variety. Recently introd Japan, to Obuse, Nagano prefecture.
Frt Col grn/grnish yell, brownish orng flush, brd rd stripes; lenticels as russet dots. Size lrg/v lrg Shape flt-rnd; sometimes lopsided; flt sided; ribbed at apex. Basin brd, med dpth; ribbed, puckered. Eye lrg, clsed/part open; sepals brd, downy. Cavity qte brd, dp; lined russet. Stalk shrt, thck. Flesh white, tinge grn.
F* bright pink 12; trip. T³; qte hrdy; prt tip. C hvy; prn bitter pit. P e-Oct. S Nov-Mar; commercially-June.
PLATE 17
   CRIMSON BRAMLEY; arose Southwell, Notts; catalogued 1913.

BRAMSHOTT RECTORY 5/7 E D
UK; prob arose garden Bramshott Rectory, Liphook, Hants. Received c1938.
Good flavour, quite rich, lightly aromatic; crisp, juicy.
F 8. T³. P l-Aug. S Sept-Oct/Nov.

BREEDON PIPPIN 5 M D
UK; raised c1801 from cider pomace by Rev. Dr Symonds Breedon at Bere Court, Pangbourne, Berks.
Resembles Court of Wick in appearance, but firm, rather chewy brisk flesh, and inferior in taste.
F 5. T³. C frt drops, cracks. P e-Sept. S Sep-Oct.

BREITLING 4 L D
Germany (prob); acc is variety of Hogg, popular German apple of C19th; introd c1879 to England by Thomas Rivers.
Large, ribbed; scarlet flushed, striped. Very sweet, delicate flowery taste in early Nov, like rose petal jam.
F 11. T³. C hvy. P m-Oct. S Nov-Jan.

BRENCHLEY PIPPIN 5 L D
UK; raised Brenchley, Kent; desc Hogg (1884).
Like Blenheim Orange in taste. Fruity, sweet, crumbling, rather dry, deep cream flesh.
F 14. T¹. P m/l-Oct. S Nov-Jan/Apr.

BRETTACHER SÄMLING 3 L C
Germany; chance seedling, either discv 1900 by Kutruff or raised by M. Feinauer, at Brettach, Heilbronn, Baden-Württemberg.
Poss Reinette de Champagne, Jacques

Lebel cross.
Quite sharp. Cooks to brisk purée.
Grown Germany in Jork region around Hamburg; also Alsace.
Used also for juice.
F 13; trip. T²; sp. P l-Oct. S Dec-Mar/May.

BREUHAHN see Geheimrat Breuhahn

BRIDGEWATER PIPPIN (CHEAL) 1/5 M C
UK; acc received 1943 from nursmn J. Cheal, Crawley, Sussex. Not Bridgewater Pippin of Hogg, also prob not variety recorded 1665 by Rea.
Large. Cooks to slightly sharp purée; lovely, creamy texture.
F 12. T³. P m-Sept. S Sept-Nov/Dec.

BRIGHTON 4 L D
Received 1950 from New Zealand.
Pinky red flush and stripes with bloom. Quite rich; sweet juicy, firm flesh. Often poor coloured and flavour in England.
F 14. T². C bien; hangs late. P l-Oct. S Jan-Mar/Apr.

BRITEMAC 4/6 M D
USA; raised 1934 by M. A. Blake at New Jersey Agric St, New Brunswick, NJ. (Melba, McIntosh cross) X (Kildare, Langford Beauty cross). Introd 1964.
Brighter maroon than McIntosh, with its flavour; sweet, juicy, white flesh. Raised for New Jersey, Pennsylvania climate.
F 16. T². hrdy. P m-Sept. S Oct-Nov.

BROAD-EYED PIPPIN (BULTITUDE) 1 M C
UK; acc received 1929 from E. Beckett, HG, Aldenham House, Herts; desc 1983 (Bultitude). Acc not Broad-Eyed Pippin of Hogg; prob also not variety recorded late 1600s.
Large open eye set in deep basin accounts for name. Large; cooks to purée or keeps little of form, but flavour light, even in Sept.
F 13. T³. P e-Sept. S Sept-Dec/Jan.

BROOKES'S 7 M D
UK; arose Shrops. Recorded 1820 by London Hort Soc.
Aromatic, rich, sweet, like early Ribston Pippin. Best early, soon becomes dry.
F 11. T²; sp. C gd. P m-Sept. S Sept-Nov.

BROWN CROFTON 8 L D
Ireland; old variety; received 1950.
Slight red flush beneath russet. Intense sweet-sharp, rich flavour. Can be very good.
F 12. T³. C frt cracks. P m/l-Oct. S Dec-Jan/Feb.

BROWN KENTING 5 M D
UK; arose England; recorded 1831; desc Hogg.

Sweet, nutty taste; firm flesh.
F 17. T². P l-Sept. S Oct-Nov/Dec.

BROWNLEES' RUSSET 8 L D
UK; introd c1848 by William Brownlees, Hemel Hempstead, Herts.
Intense sweet-sharp taste of acid/fruit drops, but sweeter than Nonpareil; quite juicy. Victorians prized it for the late dessert and Edwardians thought it almost worth growing for its blossom – deep carmine buds opening to pink.
Frt *Col* grey brown russet, sometimes scaly, silvery sheen, over yellow; sometimes slight brownish rd flush; lenticels as russet dots. *Size* med. *Shape* shrt-rnd-con; rnded ribs. *Basin* med width, shal, slightly puckered, russeted but usually base of sepals grn *Eye* clsd/prt open. *Cavity* med wdth, dpth; russeted, sometimes sml lip one side. *Stalk* shrt, med. *Flesh* crm, tinge grn.
F* 9. T². C gd. P m-Oct. S Dec-Mar.

BROWN'S SEEDLING 3 L CD
UK; raised before 1874 by nursmn Brown at Stamford, Lincs.
Quite sweet, rich, good flavour in Dec. Cooks to sweet bright lemon purée. Dull appearance.
F 16. T². P m-Oct. S Nov-Jan.

BUDAI IGNÁC 5 L D
Hungary; received 1948 from Agric Univ, Budapest.
Sparkling, refreshing taste of fruit, not aromatic or rich but good.
F 15. T¹. P l-Oct. S Dec-Feb.

BUDIMKA 5 L C
Serbia; received 1936.
Fruity, quite sharp. Cooked, keeps shape, light taste.
Winter apple of Belgrade in 1930s.
F 8. T². P m-Oct. S Nov/Dec-Feb/Apr.

BUKHOVITSA 3 L C
Bulgaria; received 1957 from Bulgarian State Res St, Kustendil.
Bright red striped. Sharp, firm flesh. Cooked, keeps shape, brisk savoury taste.
F 10. T³; uprt. P m-Oct. S Nov-Feb/Apr.

BURCHARDT'S REINETTE 8/7 L D
Russia; raised by von Hartwiss, director royal garden at Nikita, Crimea; recorded 1863. Named after pomologist O. Burchardt of Landsberg.
Sweet yet sprightly with refreshing savoury taste. Soft, white, juicy flesh. Grown Crimea, Transcaucasia.
F 13. T²; sprd; needs warm summer. P e/m-Oct. S Nov-Dec/Jan.

BURGESS SEEDLING 8 L D
UK; received 1948 from EMRS.
Brisk, fruity, nutty; crisp, cream flesh.
F* 11. T³. P m-Oct. S Nov-Feb/Apr.

BURGUNDY 6 M D
USA; raised 1953 by R. C. Lamb, NYSAES Geneva.
Monroe X (Macoun X Antonovka). Introd 1974.
Stunning colour – dark burgundy over bright pink. Cream flesh heavily stained with pink. Sweet, light savoury taste; crisp flesh.
F 15. T². P l-Sept. S Oct-Nov.

BURN'S SEEDLING (?) 7 L D
UK; acc may be variety of Hogg (1851); raised by Henry Burn, HG to Marquis of Aylesbury, Tottenham Park, nr Malborough, Wilts. Recorded 1831. Acc close to Hogg's desc, but not culinary, as he states.
Very sweet, scented, firm flesh; almost like sweetmeat by Nov.
F 16. T². P e-Oct. S Nov-Dec.

BURR KNOT 5 M C
UK; accs Howard, Lascelles; neither agrees completely with Hogg, but prob many forms existed in past. Lascelles acc desc Bultitude (1983). Burr Knot recorded 1818, England. Bears burrs at base of branches which will root. Syn, Bide's Walking-Stick, derives from Mr Bide, who in c1848 cut a branch from tree in Cheshire as a walking stick. He stuck it in the ground in his Hertford garden, forgot about it and it rooted. Burr Knot type apples were widely grown by cottagers and farmers particularly in South Wales in C19th and known as 'Pitchers', a name given to any variety that could be propagated by cuttings. Burr Knot was used as rootstock and useful kitchen apple.
Cooks to yellow purée, sweet, juicy, pleasant.
F 9. T². P m-Sept. S Sept-Oct/Nov.

BUSHEY GROVE 3 M C
UK; raised by Mrs Good, Bushey Grove, Herts.
Believed Queen X Bismark. Introd 1926 by G. C. King, Sidcup, Kent.
Colourful red flush like Bismark. Cooks to sharp fruity purée.
F*, dp pink; 9. T². C prn scab. P m-Sept. S l-Sept-Nov/Dec.

BUZÁVAL ÉRÖ ALMA 6 M D
Hungary; received 1948 from Univ Agric, Budapest.
Oval, almost egg-shaped; maroon flush. Brisk fruity taste; soft, juicy flesh.
F 7. T². P m-Sept. S Sept-Oct/Nov.

BUZZAN syn Rosemary Russet

BYELOBORODOVKA 4 E C
Russia; recorded 1842.
Style of Duchess of Oldenburg, but not as highly coloured. Juicy, sharp. Cooks to dark cream purée or froth, juicy, quite brisk.
F 10. T¹; uprt. P m-Aug. S Aug-Sept.

BYFLEET SEEDLING 3 L C
UK; raised 1915 by George Carpenter, HG, West Hall, Byfleet, Surrey.
Bramley's Seedling X Lane's Prince Albert. Very large, exhibition apple. Crisp, sharp flesh. Cooks to lightly flavoured purée.
F 10. T³. P e Oct. S Oct Jan/Mar.

BYFORD WONDER 1/5 L C
UK; arose England. Recorded 1893; AM RHS 1893. Introd 1894 by Cranstons Nursery, Hereford.
Large. Sweet-sharp, crisp, juicy flesh. Cooked, keeps shape, yellow; sweet, delicate, sometimes slightly pear-like in Nov. Taste easily overwhelmed; becomes even lighter with keeping.
F* 10. T³. P e-Oct. S Nov-Dec/Jan.

CABUSSE 5 L DC
France; old variety, arose Lozère (Languedoc-Roussillon).
Brisk, strong, refreshing taste; plain but pleasant. Crisp, juicy to spring.
Formerly widely grown Lozère and all over Cévennes, even up to 950m. Also used for cider and cooking; in past baked in bread ovens as they cooled.
F 17. T³ hrdy; reported gd res pests, diseases. C gd. P l-Oct. S Jan-Mar/May.

ČAČANSKA POZNA 6 L D
Serbia; raised 1959 at Inst for Fruit Res, Čačak.
Starking X Jonathan. Introd 1971.
Strong, robust flavour in Nov; quite soft, deep cream flesh.
F 13. T³. C gd. P l-Oct. S Nov-Mar/May.

CAGARLAOU 6 L DC
France; arose Lozère (Languedoc-Roussillon); desc 1948 (*Verger Français*), but much older. In cévenol patois 'cagarlaou' means snail, which local people say love to climb onto the leaves of this tree.
Pretty; bright red flush, stripes. Sweet, rather insipid, firm becoming crumbly, cream flesh, but said to be 'perfumed and cooks well'.
Well-known in Cévennes and traditional apple of Vallée Française in cantons Barre, St Germain-de-Calberte.
F 20. T². C hvy, bien; hangs late. P l-Oct. S Dec-Feb/May.

CALVILLE
Name given in France to a group of varieties, which are characterised by heavily ribbed, crowned shape.

CALVILLE BLANC D'HIVER 5 L CD
Europe, prob France/Germany. Believed to be very old variety.
Recorded 1598 by Jean Bauhin under syn,

Blanche de Zurich. Recorded 1628 as Calville Blanc by Le Lectier. Named poss after commune of Calleville, Eure, Normandy. Syns – 151; internationally distributed.

Cooked, keeps shape, pretty yellow and good strong taste. Ideal for 'Tarte aux Pommes'. Mellows to intense, rich, sweet-sharp flavour; deep cream flesh; (vit C 35mg/100g).

Best-known of Calville varieties and, although now rarely seen on sale in France, its name still evokes fond memories of fine tarts and conserves that it made. Baked, also made 'perfect sweetmeat'.

In late C19th/early C20th grown as luxury fruit around Paris on espaliers, occupying over 500 hectares. Needs hot summer to ripen properly in England; Victorians grew it against wall or in pots under glass for French chefs who held it in high regard.

Frt *Col* grnish yell/yell; slight orng flush; russet dots, patches. *Size* med/lrg. *Shape* oblng; irregular; flt base, apex; prom ribbed; 5 crowned. *Basin* med wdth, dpth; ribbed; russet. *Eye* hlf-open; sepals, med to lng. *Cavity* brd, dp; russet lined. *Stalk* shrt/med; qte thin. *Flesh* crm.

F 12. **T**[1]; sp. **C** gd. **P** m/l-Oct. **S** cul, Nov-Dec; des, Dec/Jan-Mar.

CALVILLE D'ADERSLEBEN syn Adersleber Calville

CALVILLE D'ANGLETERRE syn Cornish Gilliflower

CALVILLE D'AOUT syn Six Côtes du Boulonnais E 5 DC
France; chance seedling obtained 1925; introd 1935 by Ets Delaunay of Angers.
Markedly angular, hence syn. Brisk white flesh. Cooks to well-flavoured purée; plenty of acidity, with added sugar rich and savoury.
Found as standards on farms of Boulonnais (Picardy).
F est 2. **T**[2]. **C** gd. **P** Aug-e-Sept.

CALVILLE DE DANTZIG syn Danziger Kantapfel

CALVILLE DE DOUÉ 5 L D
France; received 1947.
Golden, slightly ribbed. Brisk, juicy, crisp with plenty of fruit in Nov.
F 16. **T**[3]; sp. **C** hvy. **P** m-Oct. **S** Nov-Feb.

CALVILLE DE SAINT-SAUVEUR 1/5 L C
France; shoot of rootstock found by M. Despréaux, Saint-Sauveur, Oise (Picardy); distributed by 1836 by nursmn Jamin.
Ribbed, oblong.
Cooks to lemon coloured purée, or keeps shape; good, rich taste.

F 16. **T**[2]. **P** m/l-Oct. **S** Nov-Feb/Apr.

CALVILLE DES FEMMES 1/5 L C
France; found 1850 in garden of Comice Horticole, Angers, Maine-et-Loire.
Large, Calville shape. Crisp, juicy, sharp in Nov. Cooks to pretty gold slices; sweet but can be rather flat.
F 15. **T**[3]. **P** m/l-Oct. **S** Nov-Feb/Jun.

CALVILLE DES PRAIRIES 5 L DC
Belgium (prob); acc is variety recorded by Belgian pomologist A. Bivort (1853–60); poss variety now recognised in Belgium as Cwastresse-Simple (syn Calville des Prairies).
Markedly Calville shaped. Cooked, retains shape; sweet, quite rich taste.
F 18. **T**[2]; tip. **P** e-Oct. **S** Nov-Jan/Feb.

CALVILLE D'OULLINS 6 L DC
France; found c1850 by nursmn Armand Jaboulay at Ouillins, nr Lyon.
Conical rather than Calville shape. Crisp, sharp, pale cream flesh. Cooked, keeps shape, sharp with little flavour. Still grown France.
F est 12. **T**[3]; claimed gd res disease, pests. **P** e-Oct. **S** Oct-Jan/Mar.

CALVILLE D'ULZEN (?)1 L C
Received from France 1947. Acc may be Apfel von Ulzen raised early 1800s at Ulzen, Hannover, Germany by Hoefft.
Large; grass green; Calville shaped. Considered first quality cooker by continental writers. Cooked, tends to retain shape; quite rich taste.
F 17. **T**[2]. **C** hvy. **P** m-Oct. **S** Nov-Feb/May.

CALVILLE DUQUESNE 3 M C
Belgium; raised before 1895 by nursmn Dusquesne at Mons-Pont-Canal.
Sharp astringent. Cooked, keeps shape; sweet, pleasant flavour.
F 8. **T**[1]. **P** l-Sept/e-Oct. **S** Oct-Dec/Jan.

CALVILLE MALINGRE (BUNYARD) 6 L C
Acc is variety desc 1920 by Bunyard; not variety of Hogg, *Herefordshire Pomona*; prob not variety known to the monks of Chartreux, Paris in 1775.
Cooks to lemon coloured purée, sharp, quite strong, good flavour; needs sugar. Still quite sharp in Feb.
F 10. **T**[2]; sp. **C** gd. **P** m-Oct. **S** Nov/Dec-Mar/Apr.

CALVILLE ROUGE D'AUTOMNE (BARNES) 6 M DC
Received 1949 from W. Barnes, Sussex. Acc may be Calville d'Automne now known France and possibly dating from C17th, but confused French and English literature.
Large. Sweet, slight strawberry flavour; soft,

juicy flesh. Cooked, keeps form; smooth, creamy texture; sweet, very lightly flavoured.
Calville d'Automne, formerly highly esteemed on continent and still grown France.
F 14. **T**[3]. **P** l-Sept. **S** Oct-Nov/Dec.

CALVILLE ROUGE D'HIVER 6 L CD
France; acc is variety of 19th pomologists, which was believed to be very ancient; poss arose Brittany. As syn, Calville Rouge listed 1628 by Le Lectier; Calville Rouge d'Hiver listed 1690 by la Quintinyne. Syns numerous.
Tall, Calville shape. Sweet, fruity with little richness, almost aromatic; soft, white flesh. Cooks to lemon purée; sweet, good in Dec, but fades with keeping.
Formerly widely known Europe, but never esteemed UK. Still grown French gardens,.
F 16. **T**[2]. **P** l-Oct. **S** Dec-Feb/Apr.

CALVILLE ROUGE DU MONT D'OR 6 M DC
France; found Champagne-au-Mont-d'Or, nr Lyon; propagated by M. Laperriére. Desc 1948 (Vercier).
Bright scarlet; large; Calville shape. Sharp, juicy. Cooked, slices keep shape; quite sweet, rich taste. Quite widely grown in France.
F 10. **T**[2]. **C** gd. **P** e-Sept. **S** Sept-Nov/Dec.

CAMBUSNETHAN PIPPIN 4 M DC
UK; raised either c1750 by HG, Paton at Cambusnethan House, or arose much earlier at Cambusnethan Monastery, Stirlingshire, Scotland.
Sweet, light flavour; firm chewy flesh. Cooked, rather empty, but would have more acidity in north. Still highly regarded Scotland.
F 15. **T**[1]; sprd; part tip. **P** m-Sept. **S** Oct-Nov/Dec.

CAMELOT 3 L C
UK; received 1937 from orchard of Major Willetts, Yeovil, Somerset; exhib 1934. Acc prob not cider variety of this name, desc Copas (2001).
Crisp, sharp, cream flesh. Cooks to sharp, fruity, gold purée in Dec.
F 10. **T**[3]. **P** m-Oct. **S** Dec- Mar.

CAMPANINO 4 L D
Italy; old variety known since late C19th; prob arose Modena, Emilio Romagna.
Sweet yet sharp, pleasant, but not intense taste in Jan; firm flesh.
F 16. **T**[2]; uprt. **P** l-Oct. **S** Jan-Mar.

CANADA BLANC DE LA CREUSE 5 L D
France; arose pres Creuse (Limousin); desc 1948 (*Verger Français*).
Clear yellow skin and pale yellow, firm

flesh. Crisp, fruity in Jan, underlying astringency. Grown Creuse, Indre. Used also for juice; vit C 25mgs/100gm.
F 35. **T**^{3}; **C** claimed gd res disease. **P** l-Oct. **S** Dec-Feb/Mar.

CANVADA 5 M D
Received 1957 from South Africa, but prob did not arise there. Described c1926.
Attractive. Very juicy, sweet, little scented, but low acidity; soon becomes winey.
F 10. **T**^{1}. **P** e-Sept. **S** Sept-Nov.

CAPTAIN KIDD *see* Kidd's Orange Red

CARAVEL 4 E D
Canada; raised at Central Exp Farm, Ottawa.
Melba X Crimson Beauty (Early Red Bird). Introd 1964.
Resembling Melba, but ripening earlier, firmer fleshed. Snow white, juicy, scented flesh; characteristic taste, almost of sweet almonds or coconut. Of its type can be very good.
Grown commercially small extent eastern Canada.
F 7. **T**^{2}; hrdy. **C** hvy, bien. **P** eat Aug.

CARDINAL 6 E D
USA; found 1948 as chance seedling on home of M. E. Race, New Jersey; introd France 1951.
Large. Strong, brisk, fruity flavour; plenty sweetness; juicy, crisp yet melting flesh. Ready July in South France, where recommended for early markets in 1970s.
F 7. **T**^{2}. **P** m-Aug. **S** Aug-Sept.

CARLISLE CODLIN (BULTITUDE)
5 M C
UK; acc is variety of Bultitude (1983); prob not variety of C19th, presumed arose Carlisle, Cumberland.
Readily cooks to white, juicy purée; plenty of tart, fruity flavour. Baked, bursts into juicy fluff.
F 7. **T**^{2}; hrdy. **C** gd. **P** e-Sept. **S** Sept-Oct.

CARLTON 6 E D
USA; raised 1912 at NYSAES, Geneva.
Montgomery X Red Astrachan. Introd 1923.
Dark red flushed; prominently ribbed, crowned. Slight strawberry flavour; quite sweet, white, firm flesh.
F 6. **T**^{2}. **P** l-Aug. **S** Aug-Sept/Oct.

CARMIGNOLLE 8 L D
France; desc 1948 (*Verger Français*) when grown Drôme, Rhône Alpes.
Fenouillet type flavour – sweet, almost sickly; hint of aniseed.
F 14. **T**^{2}. **C** gd. **P** m-Oct. **S** Nov-Jan/Mar.

CARNET 6 M D
Received from Denmark 1950.
Light flavour; juicy crisp, sweet, slightly

perfumed, can be insipid.
F 11. **T**^{1}. **C** prn canker. **P** m-Sept. **S** Oct-Nov.

CAROLA syn Kalco

CAROLI D'ITALIE 2/5 L D
Italy; arose Piedmont, where much grown 1876, when recorded by French nursmn, Simon-Louis Frères; prob much older.
Markedly conical; ripening to clear yellow. Sweet, juicy, can be tannic.
F 14. **T**^{2}. **P** m-Oct. **S** Nov-Jan/Feb.

CAROLINE 4 L D
UK; arose before 1822 in Lord Suffield's garden, Blickling, Norfolk. Named after his wife.
Sweet, some richness; crisp, firm flesh.
F 13. **T**^{2}. **C** pr col. **P** m-Sept. **S** Oct-Dec/Jan.

CAROLINE HOPKINS 5 L C
South Africa; found at Worcester, Cape of Good Hope. Received 1956.
Needs very little winter chilling and in England after mild winter can flower by late March. Burrs form at base of shoots; cuttings will root. Fruit poor quality; fruits twice a year in South Africa.
F A28. **T**^{2}. **P** m Oct. **S** Nov-Dec/Jan.

CARRARA BRUSCA 2/5 L D
Italy; received 1958; poss arose near city of Carrara, North Italy.
Markedly flat-round shape. Sharp, crisp, juicy. Cooked, keeps shape, sweet, light, little richness.
F 13. **T**^{3}; uprt. **P** l-Oct. **S** Jan-Mar.

CARRATA 2/5 L D
Italy; received 1958; syn Callera Dolce.
Very sweet, little acidity, light flavour; firm flesh.
F 13. **T**^{3}; uprt. **P** m-Oct. **S** Nov-Jan.

CARREY 5 L D
France; arose Béarn (Pyrénées-Atlantique); desc 1948 (*Verger Français*).
Finely textured, cream flesh, quite sweet but little flavour in England.
Grown Béarn; traditionally baked before open fire.
F 19. **T**^{3}; uprt. **P** m-Oct. **S** Dec-Feb.

CARSWELL'S HONEYDEW 4 M D
UK; raised 1939 by J. W. Carswell, Ashtead, Surrey.
Cox's Orange Pippin X.
At best very sweet and honeyed, yet balanced, not at all sickly. Crisp, juicy, deep cream flesh. Can be over sweet; dull appearance.
F 20. **T**^{1}; sprd. **C** hvy. **P** m/l-Sept. **S** l-Sept-Nov.

CARSWELL'S ORANGE 7 M D
UK; raised 1938 by J. W. Carswell, Ashtead

Surrey.
Cox's Orange Pippin X.
Handsome. Aromatic, sweet, balanced; good flavour; almost yellow flesh.
F 14. **T**^{2}; sprd. **P** m/l-Sept. **S** l-Sept-Oct.

CARTAUT 5 L D
France; arose Puy de Dôme (Auvergne). Desc 1948 (Vercier).
Large. Refreshing, crisp, quite juicy, savoury in Jan.
Formerly grown higher areas Puy de Dôme.
F 17. **T**^{1}. **P** l-Oct. **S** Dec-Mar/Jun.

CARTER'S BLUE (?) 4 L D
USA; acc may be variety raised 1840s by Col Carter, Mount Meigs Depot, nr Montgomery, Alabama. Desc 1869 by Downing.
Large with blue bloom. Sweet, very fragrant flavour, like rose-water. Quite soft, juicy cream flesh.
Rediscovered by US enthusiasts; grown again gardens.
F 7. **T**^{3}. **P** e-Oct. **S** Oct-Feb/Jun.

CARTER'S PEARMAIN 7/8 L D
UK; received 1948 from EMRS Kent; prob variety exhibited 1934; desc Taylor (1946.
Resembles Claygate Pearmain. Quite rich, aromatic; sweet, plenty of acid; fairly crumbly texture.
F 12. **T**^{2}. **P** m-Oct. **S** Nov-Feb/Apr.

CASE WEALTHY *see* Wealthy

CASSOU syn De Casse

CASTLE MAJOR 5/3 M C
UK; arose Kent. Desc Hogg (1875).
Large, ribbed. Sharp, firm flesh. Cooks to pretty, soft, melting slices; light flavour.
Grown Kent for London markets in C19th. Valued now for its ornamental blossom.
F*, dp carmine, 11. **T**^{3}. **P** m-Sept. **S** Sept-Oct/Nov.

CATHERINE 5 L C
UK; received 1977 from tree over 100 yrs old outside public house in Combs, Stowmarket, Suffolk.
Cooked, keeps shape, sweet, lightly flavoured.
F 23. **T**^{2}. **P** m-Oct. **S** Dec-Feb; said to keep for a year.

CATSHEAD 1 L C
UK; arose England. Acc is variety of Hogg (1884), which he believed very old; cited 1629 by Parkinson.
Viewed in profile, lives up to its name. Cooks to quite sharp, firm purée.
Well known by C18th and conveniently box-shaped for parcelling up into dumplings for farmers to take out to fields. Widely grown C19th; still found in old orchards.

197

Frt *Col* grn/pale yell; many russet dots. *Size* lrg/v lrg. *Shape* oblng-con to oblng; tall, flt sided; ribbed; crowned; irregular. *Basin* qte brd, qte dp; ribbed, puckered; ltl russet. *Eye* lrg, open; sepals sep at base. *Cavity* qte brd, dp; lined scaley russet. *Stalk* shrt, qte thck. *Flesh* wht.
F 11. T^2; sprd. **C** gd. **P** e-Oct. **S** Oct-Jan.

CAUDAL MARKET 3 L C
UK; raised by F. W. Wastie at Eynsham, Oxford.
Lane's Prince Albert X Hambeldon Deux Ans. Received 1953.
Quite juicy, sweet, crisp. Cooked, keeps shape, but poor flavour.
F 18. T^2. **P** m-Oct. **S** Nov-Dec/Mar.

CAVALLOTTA 4 L D
Italy; received 1958 from Univ Turin.
Quite fruity but often sour and tannic in England; crisp, green tinged flesh.
F 12. T^3; sp. **C** gd. **P** m-Oct. **S** Dec-Feb/Apr.

CEEVAL *see* Alkmene

CELLINI 6 M CD
UK; raised by nursmn Leonard Phillips, Vauxhall, London.
Believed Nonesuch seedling. Introd c1828.
Showy, bright red apple. Curious 'balsamatic' or aniseed taste; quite brisk, crisp, juicy white flesh. Cooks to cream, rather bland, purée.
Popular with London market gardeners; sixth in culinary poll held at 1883 Congress, but overtaken by early 1900s. Continues to be valued in northern France, especially for juice.
F 12. T^2; sprd. **C** hvy. **P** m-Sept. **S** Sept-Nov.

CELT 7 L D
UK; raised 1943 by David Harris, Melksham, Wilts.
Handsome. At best, very sweet, honeyed; not intensely aromatic but rich, almost scented flavour. Juicy, bright cream flesh. Can be only moderately flavoured.
F 8. T^2; sp. **C** hvy. **P** e-Oct. **S** l-Oct-Nov/Dec.

CHAD'S FAVOURITE 6 M D
UK; raised by M. B. Crane, JI, Merton, London.
Northern Greening X Cox's Orange Pippin. Named 1952.
Named after Sir Chad Woodward, who liked it. Handsome, quite large. Rich, quite intense aromatic flavour in Nov; plenty sprightly acidity.
F 10. T^2. **C** bien. **P** e-Oct. **S** l-Oct-Nov/Dec.

CHAMP-GAILLARD 2/5 L D
France; desc *Pomologie de la France* (1863–73); arose Alpes-de-Hte-Province.

Quite large. Fruity, fresh in New Year, but quite sharp; firm, white, green tinged flesh. Grown Hautes and Basses-Alpes in 1950s.
F 12. T^3. **P** m/l-Oct. **S** Dec-Feb/Apr.

CHANNEL BEAUTY 5 L D
UK; raised by C. H. Evans, Blossoms, Mumbles, Swansea, South Wales.
Cox's Orange Pippin seedling. Received 1922.
Good savoury taste; crisp, juicy; can be too sharp.
Popular South Wales 1920–30s.
F 8. T^3. **P** e-Oct. **S** Nov-Jan.

CHANTECLER 5 L D
France. Raised 1958 INRA, Angers.
Golden Delicious X Reinette Clochard. Introd 1977. Trade name Belchard.
Rich and honeyed, crisp, juicy, cream flesh. Grown commercially southwest France.
F mid. T^3. **C** hvy. **P** m-Oct. **S** Nov-Jan.
   CHANTEGRISE; handsome, russeted form.

CHARLAMOWSKY syn Duchess of Oldenburg

CHARDEN 5 L D
France; raised 1959 at INRA, Angers.
Golden Delicious X Reinette Clochard. Introd c1975.
Sweet and honeyed; more intensely flavoured and more acidity than Golden Delicious. Grown France for late markets.
F 15; trip. T^3. **C** hvy. **P** m-Oct. **S** Nov-Jan; commercially-May.

CHARLES EYRE 1 M C
UK; raised by Charles Ross, HG at Welford Park, Newbury, Berks. Introd c1911 by nursmn William Pope, Wokingham, Berks. RHS AM 1911.
Large, exhibition apple. Cooked, keeps shape, but very light flavour.
F 14. T^1; tip. **P** m-Sept. **S** Sept-Nov.

CHARLES ROSS 4 M CD
UK; raised by Charles Ross, HG at Welford Park, Newbury, Berks.
Peasgood Nonsuch X Cox's Orange Pippin. Exhib 1890 as Thomas Andrew Knight; renamed 1899 at request either of employer, or Ross's friend, nursmn William Pope. RHS AM 1899, FCC 1899; AGM 1993.
Lightly aromatic; quite juicy, firm flesh. Cooked, tends keep shape; sweet, slightly pear-like flavour. Best used early for cooking.
Handsome, with even conical shape; still prized exhibition variety. Grown for market 1930s; remains valued garden apple, especially in Scotland.
Frt *Col* orng rd flush, broken rd stripes, over grnsh yell/yell; some russet patches. *Size* med/lrg. *Shape* mid-con to con, regular.

Basin brd, shal, v sltly ribbed. *Eye* hlf or fully open; sepals, med/lng, downy. *Cavity* brd, qte dp; russet lined. *Stalk* shrt, thck. *Flesh* crm.
F 11. T^2; hrdy. **C** gd. **P** m-Sept. **S** l-Sept-Dec.
   RED CHARLES ROSS; received 1948 from H. Merifield, Staines, Middx.

CHARLOT 2/4 L D
France; received 1947.
Firm fleshed, crisp, juicy sharp.
F 10. T^2. **C** hvy, bien; frt sml; hangs late.
P l-Oct. **S** Jan-Mar/May.

CHARLOTTE syn Irishspire (US) 3 L C
UK; raised EMRS, Kent.
Wijcik McIntosh X Greensleeves; introd 1988. *See also* Ballerina.
Sharp; cooks to a purée, but not as brisk as Bramley's Seedling.
F 10 T^2. **P** e-Oct. **S** Oct- Dec.

CHÂTAIGNE DE LÉMAN syn Franc Roseau

CHÂTAIGNIER 6 L D
France; acc is prob variety of Leroy (1873), which arose Normandy and which he believed to be very ancient, but impossible to know. As syn, 'De Castegnier', recorded Rouen market 1370–1423; as 'De Chastignier' listed 1540 by Charles Estienne.
Small, prettily flushed and striped. Crisp, sweet, fruity; fresh and crisp in Feb.
Still grown Normandy, Seine et Marne (Ile de France); used also for cider.
F 23. T^2; sp. **P** l-Oct. **S** Nov-Feb/Mar.

CHAUX 5 L D
France; prob arose south west Haute Vienne (Limousin). Desc 1948.
Good, intense, rich flavour; quite aromatic in Dec, but rather dry.
Found now, especially, in Saint-Yrieix (Haute Vienne) and also north east Corrèze (Limousin).
F 13. T^2; uprt. **P** l-Oct. **S** Nov-Feb/Mar.

CHAXHILL RED 4 M cider
UK; raised by Mr Bennett of Chaxhill, Westbury on Severn, Gloucs. Recorded 1873. FCC 1873 for cider.
Sharp, firm flesh.
F17. T^3.

CHEDDAR CROSS 4 E D
UK; raised 1916 by G. T. Spinks, LARS, Bristol.
Allington Pippin X Star of Devon. Introd 1949.
Dusky pink over cream. Savoury with mellow acidity; crisp, juicy. Can develop trace of fennel after picking, but can be merely sharp. Finely textured, almost yellow flesh.

F 8. T^2. P eat l-Aug/Sept.

CHEHALIS 5 M D
USA; discv 1955 nr Chehalis river, Oakville, Washington by Lloyd Lonborg. Prob Golden Delicious seedling. Introd 1965.
Resembles Golden Delicious, but ripens earlier. Sweet, slightly honeyed; juicy, crisp yet melting pale cream flesh.
F 12. T^2; res scab. C gd. P m-Sept. S Oct-Nov/Dec.

CHELMSFORD WONDER 3 L C
UK; raised c1870 by mechanic nr Chelmsford, Essex. Introd 1890 by nursmn William Saltmarsh. RHS FCC, AM 1891.
Cooks to sharp, well-flavoured cream purée; making good sauce. Baked, soft juicy like Dumelow's Seedling, but not as sharp. By Jan, sweeter, cooking more firmly, but still good. Formerly quite widely planted; still found in old Essex orchards. Also introduced to France, Poland.
Frt *Col* diffuse pinky orng flush, faint rd stripes, over yell. *Size* lrg. *Shape* rnd; flttened; bit irregular; sometimes hair line present. *Basin* brd, dp; ltl puckered. *Eye* hlf open; sepals brd based. *Cavity* brd, qte dp, russet lined. *Stalk* shrt, thck. *Flesh* pale cream.
F 12. T^2. C gd. P e-Oct. S Nov-Mar.

CHEMISETTE BLANCHE syn Weisses Seidenhemdchen

CHENANGO STRAWBERRY 6 E D
USA; arose either Lebanon, Madison County, New York, or brought to Chenango County from Connecticut. Known 1850.
Beautifully coloured, red striped over pale yellow, and in America 'finely flavoured, juicy, tender flesh with singularly powerful aroma, which fills the room with its scent'. In England faint strawberry taste, but inclined to be astringent.
F 13. T^2. P e-Sept. S Sept.

CHERRY COX see Cox's Orange Pippin

CHIEFTAIN 6 L D
USA; raised 1917 by S. A. Beach, Iowa Exp St, Ames.
Jonathan X Delicious. Introd 1967.
Dark red flushed. Quite rich, sweet, honeyed taste; crisp, juicy. More sprightly than Delicious, but not strong acidity of Jonathan.
Grown commercially Iowa.
F* 12. T^2; res scab, fireblight. C gd. P m/l-Oct. S Nov-Jan/Feb; commercially-June.

CHIPS 6 E DC
UK; raised c1867 by Mr Sturgeon prob in Kent. Received 1970.

Crisp, brisk, firm. Cooked, keeps shape with some astringency, but often dull.
F 8. T^2. C frt pr col. P use Aug-Sept.

CHIVERS DELIGHT 4 L D
UK; raised c1920 by Mr Chivers, Chivers Farms, Histon, Cambs. Introd 1966.
Golden, flushed in brownish red. Often honeyed flavour; sweet, well balanced. Crisp, juicy to the end.
F 14. T^2. C gd; prn canker. P m-Oct. S Nov-Jan, cold store-Mar.

CHORISTER BOY 6 L D
UK; found Wilts garden. Introd by Keynes, Williams & Co, Salisbury; recorded 1890.
Shiny red flushed, striped. Crisp, juicy, fruity, hint of strawberry flavour. Can be over sharp, taste green, woody.
F* 12. T^2; sprd; part tip; res mildew, red spider. C hvy; frt sml. P e/m-Oct. S Nov-Dec.

CHRISTIE MANSON 1 M C
Received 1905.
Large. Cooked keeps shape; light, sweet taste.
F 14. T^3. C prn bitter pit. P m-Sept. S Sept-Oct/Nov.

CHRISTMAS PEARMAIN 7 L D
UK; raised by Mr Manser. Introd by nursmn G Bunyard, Maidstone, Kent; recorded 1893.
Russeted, pretty apple. Sweet, quite richly flavoured, not really aromatic but good; reminiscent of Mabbot's Pearmain.
F 5. T^2. C gd. P e-Oct. S Nov-Jan.

CHÜSENRAINER 3 L C
Switzerland; found c1852 by John Waller growing amongst some brambles in Chüsenrain-Sempach. Named 1891.
Sharp, juicy, tough flesh. Cooked, keeps shape, but in England low in flavour. Said to make good cider.
F 19. T^2. P m-Oct. S Nov-Jan/Apr, even Sept.

CIGÁNY ALMA 6 L D
Germany (poss); received 1871 by Hungarian pomologist, Bereczki from Germany.
Gypsy apple. Dark red flush. Quite good fruity taste.
F 8. T^3. P m/l-Oct. S Jan-Mar/Apr.

CINQ CÔTES 1 L CD
France; desc 1948 (*Verger Français*).
Markedly angular, five sided. Mellowing to quite intense sweet-sharp taste.
Grown Corrèze, Dordogne, Lot (south west France). Used early for cooking – said to make excellent jam – later for eating fresh.
F 16. T^3. C hvy. P l-Oct. S Dec/Jan-Mar/Jun.

CIODO 2 L D
Italy; received 1958 from Univ Turin.
Sweet-sharp, quite rich and intense, firm; tough skin.
F 14. T^2. P l-Oct. S Jan-Mar/Jun.

CISSY syn Tamplin 6 E D
UK; raised 1750–1800 by Mr Tamplin, Malpas, Mons. Distributed by his sister Cissy. RHS AM 1902.
Crimson flushed. Sweet, quite rich, scented, firm flesh.
Popular in Monmouth market in C19th.
F 10. T^2. C hvy; prn bitter pit. P e-Sept. S Sept.

ČISTECKÉ LAHŮDKOVÉ 4 M D
Czech Republic; arose with Dr Hales c1860 at Cistaa u Rakovnika as sport of Princess Louise (Canadian variety).
Snow white flesh; soft, very juicy; scented with hint of strawberry/raspberry flavour. Delightful surprise in dull looking apple.
Grown commercially in west and central Bohemia until 1960s; now declined; but remains popular garden apple.
F 10. T^2; hrdy. C hvy; prn bitter pit; bruises. P e-Oct. S Oct-Nov/Dec.

CITRON D'HIVER 5 L D
Switzerland; from Valais region. Desc1925.
Not ancient French, Citron d'Hiver.
Quite rich taste.
F 15. T^2. C hvy. P m-Oct. S Nov-Jan/Feb.

CLARINETTE 5 E D
UK; received from Axbridge, Somerset 1950.
Strong fruity taste, on sharp side; crisp.
F 10. T^3. P e-Sept. S Sept-Oct.

CLAYGATE PEARMAIN 7 L D
UK; discv by John Braddick at Claygate, Surrey, growing in hedge of Firs estate. Exhib London Hort Soc 1821. RHS FCC, 1921; AGM 1993.
Strong nutty taste, rich, aromatic. Firm, slightly crumbling flesh, like Blenheim, but some of aromatic quality of Ribston.
Grown in most Victorian and Edwardian country house gardens and one of Edward Bunyard's 'indispensible dozen'. Needs good year; can be poorly coloured.
Frt *Col* sltly or more prom flushed orng rd, few rd stripes, over grnish yell/yell; much russet dots, patches, netting. *Size* med/lrg. *Shape* oblng-con, trace rnded ribs. *Basin* brd, med dpth; sltly ribbed; puckered; some russet. *Eye* lrg, open; sepals sep at base. *Cavity* brd wdth, dpth; russet lined. *Stalk* shrt, thck. *Flesh* pale crm.
F 12. T^2. C gd. P e/m-Oct. S Dec-Feb.

CLEARHEART 4 M C
Ireland; desc 1951 (Lamb), when grown Piltown, County Kilkenny.
Probably used for cider; astringent, juicy.
F 22. T^2.

CLEEVE 5 E D

UK; raised before 1930 by Sir Stanley Machin from seed of imported Canadian apple at Cleeve, Weybridge, Surrey.

Quite savoury, crisp. By late Aug sweet, honeyed; soft, juicy.

F 3. T^2; res rosy apple aphid; prn mildew. C gd; easily bruised. P eat Aug.

CLEMENS 2/5 L D

Belgium, received 1948.

Very sweet, but not much taste of fruit; crisp, juicy cream flesh.

F 11. T^3. P e-Oct. S Nov-Jan/Apr.

CLEOPATRA syn Ortley

CLIMAX 5/4 E D

Acquired 1921 from M. B. Crane, JI, Merton, London.

Soft, juicy, pale cream flesh, but often metallic taste.

F 11. T^2. P e-Sept. S Sept-Oct.

CLIVIA 6 M D

Germany; raised 1930s by M. Schmidt at Inst Agric & Hort Müncheburg-Mark. Geheimrat Dr Oldenburg X Cox's Orange Pippin. Introd 1961.

Bright orange red flush, like Clivia flower. Scented, reminiscent poss of elderflowers, but soon fades; crisp, juicy, cream flesh.

F 6. T^2. P l-Sept. S Oct-Nov.

CLOCHARD syn Reinette Clochard

CLODEN 5 L D

France; raised INRA, Angers.

Golden Delcious X Reinette Clochard. Introd 1977.

Crisp, deep cream flesh; good taste of fruit, but not as flavoursome as Chantecler.

F est 9. T^3. C hvy. P e-Oct. S Dec-Feb; commercially-May.

CLOPTON RED 6 M D

UK; raised 1946 by Justin Brooke, Clopton Hall, Wickambrook, Suffolk.

Cox's Orange Pippin X. Introd 1961.

Bright red flush. Delicate, aromatic flavour; sweet, juicy.

F 18. T^2. P l-Sept. S Oct-Nov.

CLOSE 6 E D

USA; raised by C. P. Close, USDA, Arlington Exp Farm, Virginia. Introd 1938.

Large; beautiful bright, pinky red flush. Soft, juicy, white flesh with lots of acidity. Grown northern US.

F 6; trip. T^3; uprt; v hrdy. P eat e-Aug.

CLUDIUS HERBSTAPFEL 5 M D

Germany; raised by Supt Cludius of Hildesheim, Hannover. Recorded 1850. Syns many; widely distributed Europe.

Savoury almost cidery flavour; crisp, juicy, cream flesh.

Remains valued eastern Germany.

F 10. T^2; uprt. P m-Sept. S l-Sept-Oct/Nov.

CLYDESIDE 4 M C

UK; old variety; received 1947 from Scotland.

Cooks to juicy, sharp, fruity purée.

F 6. T^3. P m-Sept. S Sept-Oct.

COCKETT'S RED 6 L D

UK; received 1929 from Wisbech, Cambs. Bright red; hard, sharp flesh.

F 12. T^1. C hvy; frt sml. P e/m-Oct. S Nov-Jan/Mar.

COCKLE PIPPIN 8/5 L D

UK; raised c1800 by Mr Cockle, Godstone, Surrey.

Quite rich, firm; some aromatic quality in good year.

Widely popular in C19th; grown gardens and for market

F 11. T^2; uprt. C hvy. P m-Oct. S Nov/Dec-Mar.

COCKPIT 1 L C

UK; prob arose Yorks; rec 1831 by Ronalds. Quite sharp, soft, juicy flesh. Cooks to sweet purée, some bite; pleasant flavour. Grown in north would have more acidity.

Highly esteemed C19th Yorkshire apple, which is still valued.

F 11. T2; sprd. P e-Oct. S Oct-Dec.

COEUR DE BOEUF 6 L DC

France; acc prob variety of Leroy (1873) and Coeur de Boeuf now known in Brittany. Poss variety known 1200s, but impossible to know. As syn 'De Rouviau', featured in chants of Paris fruit sellers in C13th. Leroy believed Coeur de Boeuf of C19th France was prob 'De Rouviau'.

Large, blood red flushed. Fruity, quite sweet in Nov. Cooks to sweet, lemon coloured purée. Keeps well.

F* 11. T^3; sprd. C gd. P m/l-Oct. S Nov-Mar/May.

COLA 5 L D

Italy; known early 1900s. Received 1958 from Catania, Sicily.

Pale yellow; markedly, oblong-conical shape. Sweet, scented; very tough skin.

F 15. T^2; sp. C gd. P l-Oct. S Dec-Mar/Apr.

COLA GELATA 5 M D

Italy; arose early 1900s in Giarrita-Petralia region of Catania, Sicily.

Cola, Gelata cross.

Long, conical shape; clear yellow. Very sweet, firm flesh.

F 13. T^2. P e-Oct. S Oct-Dec.

COLAPUY 6 L D

Believed brought from Crimea to France by soldier, Nicholas Puy in mid-1800s.

Dark red flushed, striped. Strong taste of

fruit; scented quality; firm, rather dry.

Grown Somme, Oise (Picardy), Nord and Pas de Calais. Used also for juice, cider. Disease resistance has led to use in French breeding programmes.

F 15. T^3; gd res scab, mildew. P m-Oct. S Nov-Jan/May.

COLLINS 4 L D

USA; arose c1865 at Fayetteville, Arkansas. Syn Collins Red; known Europe as Commercio, Pomme de Commerce.

Some richness, hard, rather dry flesh.

Widely planted south western States and Europe in early 1900s, but long surpassed.

F 12. T^2. C hvy. P l-Oct. S Jan-Mar.

COLLOGGETT PIPPIN 3/4 M C

UK; arose Tamar Valley, Cornwall, pres at Collaggett Farm, nr Botus Fleming. Desc 1920 (Bunyard). Syn Lawry's Cornish Giant. Pronounced 'Cloget' in Cornwall; syn honours James Lawry, distinguished C19th horticulturalist.

Large, red striped. Astringent, soft flesh. Cooks to brisk, gold purée; makes sharp baked apple.

Local story recounts that one year when all other apples failed they made cider using only 'Cloget' Pippin, which produced a sparkling, dry, light champagne cider. Still found Botus Fleming, St Dominic in Tamar Valley.

F15. T^2. P m-Sept. S Sept-Nov.

COLONEL VAUGHAN syn Kentish Pippin 4 L D

UK; arose Kent. Kentish Pippin of Hogg, Bunyard; believed to be very old; may be variety of late 1600s.

Small, pink flushed over cream. Pleasant, fruity taste in Oct; juicy, sweet, plenty of acidity; quite crisp.

Kentish Pippin was recommended for pippin sweetmeats and tarts in C17th and also for cider. Grown for London markets for centuries, but overtaken by C19th.

F* 11. T^2. C hvy; frt sml. P e-Oct. S Oct-Dec.

COLONEL YATE 3 L C

UK; raised c1905, prob by nursmn W. H. Divers, Surbition, Surrey.

Lane's Prince Albert X Peasgood Nonsuch. Exhib 1918.

Large, handsome. Cooked, little flavour. Eaten fresh, sweet, light.

F 12. T^2. P m-Oct. S Nov-Jan.

COLWALL QUOINING 4/6 M D

UK; received 1949 from Tenbury Wells, Worcs; Colwall is village nr Malvern.

Prominently ribbed or 'quoined'; often almost completely red flushed and striped. Faint Worcester flavour; soft, juicy flesh.

F 15. T^2. P l-Sept. S Oct-Nov.

COMRADE 4 M D
UK; prob raised by George Carpenter, HG, West Hall, Byfleet, Surrey. Received in 1932.
Rich, quite aromatic; fairly brisk in Sept; firm, crisp flesh.
F 9. T^2; sp. **P** c-Sept. **S** Sept-Oct.

CONTESSA 3 M C
Italy; received 1958 from Univ Turin.
Large. Cooks to pale lemon purée, sweet, pleasant in Oct.
F 11. T^2. **P** e-Sept. **S** Sept-Oct.

COOPER'S SEEDLING 3 M C
UK; received 1982 from Maidstone, Kent. Acc may be variety exhib 1934, when grown Kent.
Cooks to sharp, soft purée in late Sept.
F 5. T^2. **P** e-Sept. **S** Sept-Oct.

COO'S RIVER BEAUTY 6/4 L D
UK; received from EMRS, Kent. Accs may be variety exhib 1934.
Very large. Sweet, lightly flavoured, some savoury taste; soft, loosely textured flesh.
F 11. T^2. **P** e-Oct. **S** Oct-Dec/Jan.

COQUETTE 2/5 L D
France; prob arose Roussillon, eastern Pyrénées; desc 1948 (*Verger Français*).
Sweet, firm flesh but in France 'good, perfumed flavour'.
Renowned in Aude (Languedoc-Roussillon), also grown Midi-Pyrénées where formerly market apple.
F 14. T^3. **P** l-Oct. **S** Dec-Mar/Apr.

COQUETTE D'AUVERGNE 6 L D
France; arose Auvergne. Received 1947.
Brilliant scarlet flush. Fresh fruity taste in Jan; crisp, cream flesh, but often woody, green quality.
Grown Auvergne; also used for juice. Highly decorative.
F 16. T^1. C gd. **P** m-Oct. **S** Dec-Mar.

CORNISH AROMATIC 7 L D
UK; Cornwall; brought notice 1813 by Sir Chistopher Hawkins, who sent fruit to London Hort Soc, but prob much older.
Very handsome and at best, firm fleshed with sweet-sharp, pear drop quality yet almost spicy too. Can be merely good looking, flavourless, chewy.
Frt *Col* bright rd flush, faint rd stripes over gold; much russet netting, dots, patches. *Size* med/lrg. *Shape* con to oblng-con; ribbed; 5 crowned. *Basin* med wdth, dpth; ribbed, puckered; ltl russet. *Eye* clsed or prt open; sepals, lng. *Cavity* brd, med; russet lined. *Stalk* med lngth, med thck. *Flesh* crm.
F 15. T^3. C lght. **P** m-Oct. **S** Dec-Feb.

CORNISH GIANT syn Collogett Pippin

CORNISH GILLIFLOWER 7 L D
UK; found in cottage garden nr Truro, Cornwall c1800. Introduced 1813 to London Hort Soc by Sir Christopher Hawkins. Prob not Gilliflower mentioned by Evelyn, Switzer. Syn, continent, Calville d'Angleterre.
Knobbly exterior conceals yellow, perfumed flesh. In late Oct, intensely flavoured, rich, aromatic, in style of Ribston. Becoming more delicate and developing its characteristic flowery quality. Prized by Victorians; remains popular garden variety.
Frt *Col* dark rd flush, rd stripes over gold; lenticels as fine, grey russet dots; some russet, as patches, veins. *Size* med. *Shape* oblng-con; ribbed; 5 crowned; often irregular. *Basin* nrw, qte shal; ribbed; puckered; ltl russet. *Eye* clsd, pinched; sepals brd, sltly downy. *Cavity* nrw, qte shal; some russet. *Stalk* med/lng, qte thin. *Flesh* dp crm, usually tinged grn around core.
F* mauve buds; 12. T^2; tip. C lght; needs warm spot. **P** m-Oct. **S** Nov-Jan/Feb.
PLATE 24

CORNISH HONEYPIN 5 M D
UK; received 1955 from Cornwall.
Very sweet, soft, honeyed in late Sept; almost no acidity.
F 10. T^3. **P** e-Sept. **S** Sept-Oct.

CORNISH PINE 7 L D
UK; raised Exminster, Devon. Believed Cornish Gilliflower seedling. Desc 1920.
Very good flavour, rich, aromatic, quite intense, yet has delicate Cornish Gilliflower quality of flowers or rose petals. Quite soft flesh, sweet, well balanced.
Grown Devon and Cornwall in 1930s.
F 12. T^3; sprd. C gd. **P** e-Oct. **S** Oct-Dec.

COROLCA 6 L D
Received from Romania 1948.
Bright red flushed. Quite intense sweet-sharp taste; firm flesh; tough skin.
F 12. T^3. **P** l-Sept. **S** Oct-Dec/Mar.

CORONATION 4 M D
UK; raised at Buxted Park, Sussex by H. C. Princep. Introd nursmn Pyne, Topsham, Devon. Recorded 1902. RHS AM as Edward's Coronation 1902.
Large, handsome, exhibition apple. Sweet, light taste.
F 13. T^3; prt tip; res red spider. **P** m-Sept. **S** Sept-Nov.

CORRY'S WONDER 3 L C
UK; raised 1917 by E. Corry Hanks, Eynsham, Oxford.
Beauty of Stoke X poss Lane's Prince Albert. Crisp, sharp, fruity. Cooks to cream purée, quite brisk, well flavoured in Dec.
F 14. T^2. C bien. **P** l-Oct. **S** Dec-Feb.

CORTLAND 6 L D
USA; raised 1898 by S. A. Beach, NYSAES, Geneva.
Ben Davis X McIntosh. Introd 1915. Silver Wilder medal 1923.
Striped, flushed in red; McIntosh type. Very sweet, soft melting white flesh, vinous flavour; can be very good.
Leading commercial variety New York State, Canada; also grown France, Poland. Recommended for freezing in US.
F 9. T^2; uprt; v hardy; v prn canker. C hvy. **P** e-Oct. **S** Oct-Jan/Mar; commercially May.

COSTARD
Medieval Costard was 'rarely met with' by C19th. Gloucestershire and Herefordshire Costards grew locally in 1870s, but Hogg observed that Catshead was supplanting them. Costard was large, ribbed, greenish yellow cooker. Catshead is prob nearest we have to it now. Accs received at NFC up to present have not proved to be Costard described in literature.

COTTENHAM SEEDLING 3 L C
UK; raised by Robert Norman, Cottenham, Cambs.
Dumelow's Seedling X. Introd by H. J. Gautry. Received 1924.
Attractive light flush over gold. Cooks to bright lemon purée; juicy, sharp, savoury, similar to Dumelow's Seedling but not as sharp. Good sauce apple.
F* 20. T^2; prt tip. **P** m-Oct. **S** Nov-Mar.

COUL BLUSH 5 M D
UK; raised by Sir George Mackenzie, Coul, Ross-shire, Scotland. First fruited 1827.
Delicately flushed over gold. Scented, sweet, lightly flavoured, soft cream flesh. Used early and sharper, cooks to quite brisk, lemon froth or purée. Grown Scotland would have more acidity and reputed to be even finer than Hawthornden.
F 7. T^3; sprd. **P** e-Sept. **S** Sept-Oct/Nov.

COULON REINETTE syn Reinette Coulon

COURT OF WICK 7 L D
UK; arose at Court of Wick, Claverham, nr Yatton, Somerset. Believed from Golden Pippin seed. Introd 1790 by Wood of Huntingdon; also known as Wood's Huntingdon. Prob arose much earlier, as then well known locally.
Red flushed, russet freckled over gold. Intensely fruity flavour in style of Golden Pippin; plenty of sugar, acidity; almost yellow flesh.
Widely grown in C19th gardens and, in west country, said to withstand 'most severe blasts from Welsh mountains'; also used for cider. By 1920s, 'good old sort, now little grown'.
F 15. T^3. C gd. **P** l-Sept. **S** Oct-Dec.

COURT-PENDU DE TOURNAY syn Reinette de France

COURT PENDU PLAT or **ROUGE** 7 L D

Europe; variety now known believed to date back to C17th or earlier. Mentioned c1613 by J. Bauhin, who found it growing among Gallic Roman ruins at Mandeure in Doubs, Franche Comté.

Known Court Pendu Rouge in France. History much confused with Court Pendu Gris. Syns numerous; include Wise Apple because flowers late and escapes spring frosts.

Claimed Roman origin, but Leroy (1873) considered Court Pendu Gris had more ancient associations.

'Capendu', syn Court Pendu Gris, appeared in accounts of Normandy abbeys in C15th; in 1420 it was on sale Rouen market; before 1483 grown by Benoist, gardener to Louis XI in Plessis-lez-Tours. By C16th grown all over France, Italy, Switzerland. Some claimed it was Roman and Cestiana of Pliny. Estienne named it Court Pendu in 1540, because it was 'short hanged', ie short stalk. 'Capenda' known to Parkinson in 1629; Court Pendu Plat known England by C18th.

Court Pendu Plat is rich, fruity, with strong pineapple-like acidity; mellows to become sweet, scented, yet still intensely flavoured in Feb.

For centuries valued for dessert and among top ten Victorian dessert apples, prized also for its vermillion flush. Valued too as decorative tree to border walk or grown in pots to set about formal garden.

Both Court Pendu Gris and Court Pendu Plat grown in France, Belgium. Both are also used for cooking, dried and made into 'pâtes des fruits'. Vit C 15–20mgs/100g.

Frt *Col* flushed sltly or more prom in orng rd, rd stripes over grnish yell/bright yell; lenticels conspic as russet dots, some netting russet nr base. *Size* med/sml. *Shape* flt-rnd; sltly crowned. *Basin* brd, dp; sltly ribbed; puckered, sometimes russet rings. *Eye* lrg, open; sepals sep at base; qte downy. *Cavity* med wdth, dpth; russet lined. *Stalk* v shrt, qte thck. *Flesh* pale cream.

F* 26. T^2; uprt; gd res scab, mildew. C gd, hangs late. P m/l-Oct. S Dec/Jan-Apr.

COURTAGOLD *see* Golden Delicious

COX'S EARLY EXPORT 7 E D

New Zealand; raised by Ivory's Ltd at Rangiora.

Believed Cox's Orange Pippin X King of the Pippins. Received 1950.

Pretty red flush, speckled with russet. Lightly aromatic, firm crisp,

F 10. T^2. C lght; frt cracks. P eat Aug-Sept.

COX'S ORANGE PIPPIN 7 L D

UK; raised c1825 by Richard Cox, retired brewer, at Colnbrook Lawn, Slough, Bucks. Believed from pip Ribston Pippin. Introd c1850, prob by Smales and Son, Colnbrook. Prob first exibited 1856 at British Pom Soc by W. Ingram, HG to Queen Victoria. Syns, many European. RHS FCC 1962.

Perfectly ripe, deliciously sweet and enticing with rich, intense aromatic flavour; deep cream flesh. Described as spicy, honeyed, nutty, pear-like, but subtle blend of great complexity. Not as strong or sharp as Ribston Pippin, softer fleshed, more regularly shaped and smaller; ideal for C19th dessert. Voted best dessert apple of south at 1883 Congress; in 1895 RHS Fruit Committee found it best apple from Oct to Jan.

First grown commercially prob 1862 by nursmn Thomas Rivers in Herts; taken up by London and Vale of Evesham market gardeners and by 1890s being planted by Kent farmers. Problems with disease led to its rejection in early 1900s, but regained commercial popularity with introduction of lime sulphur sprays in1920s. Since 1970s main English apple; also grown Holland, Germany, Belgium, northern France, New Zealand.

Frt *Col* orng rd flush, rd stripes, over grnsh yell turning gold; ltl russet as dots, patches. *Size* med. *Shape* rnd-con. *Basin* med wdth, dpth; sltly ribbed; usually russet present. *Eye* sml, hlf-open; sepals med lng, nrw. *Cavity* med brd, qte dp; ltl russet. *Stalk* med lngth, qte thin. *Flesh* crm.

F 12. T^2; prn mildew, scab, canker; not suitable north, wet areas. C gd. P l-Sept/e-Oct. S m-Oct-Dec/e-Jan; commercially-Apr.
PLATE 16

Several coloured clones exist including:

CHERRY COX; red flush. Discv c1942 by Mr Roed in Kjer's Orchard, Linved Hjallese, Denmark; introd England 1951.

COX RED SPORT (Potter); brightest red sport. Arose with Mr C. A. Potter, Howlett's Farm, Molash, Canterbury, Kent; received 1957.

COX ROUGE DES FLANDRES; red sport, flatter shape, grown northern France, Belgium. Discv by Dr Heem, director Flemish Fruit Centre.

COX'S ORANGE PIPPIN (Fryde-land); received from Sweden.

KORTEGÅARD COX (Alnarp) syn Blangsted Cox; unstriped sport. Discovered 1914 at Kortegåard, Denmark by Otto Stowo.

QUEEN COX; more even flush; selection of Queen Cox made by H. Ermen 1982.

QUEEN COX (Maclean); more even flush. Discv by Maclean, Appleby Fruit Farm, Kingston Bagpuize, Berks; received 1953. Original Queen Cox.

COX'S POMONA 3/4 M CD

UK; raised c1825 by Richard Cox, Colnbrook Lawn, Slough, Bucks. Believed pip of Ribston Pippin. Introd c1850.

Crisp, brisk eaten fresh. Cooks to bright yellow, quite sweet purée; makes good baked apple.

Formerly grown small extent for market. Introduced to Europe; still valued Sweden.

Frt *Col* rd flush, many broken rd stripes over grnsh yell/pale yell; skin greasy. *Size* lrg. *Shape* flt-rnd, irregular; flt base; rnded ribs; 5 crowned. *Basin* brd, dp; ribbed, puckered. *Eye* lrg, open; sepals sml, qte downy. *Cavity* brd, dp; russet lined. *Stalk* shrt, thck. *Flesh* wht.

F* 14. T^2; hrdy. C gd. P m/l-Sept. S l-Sept-Dec.

COXSTONE 6 M D

Canada; prob raised by W. J. Macoun, Central Exp Farm, Ottawa.

Stone X prob Cox's Orange Pippin.

Handsome. Sweet, aromatic, reminiscent Cox's Orange Pippin.

F 11. T^2; tip. C prn bitter pit. P m/l-Sept. S Sept-Oct/Nov.

CRĂCIUNEŞTI 5 L D

Romania; well known 1960s in districts Muscel, Tîrgoviste.

Often prominently pear-shaped, narrower at base than apex; king fruits have fleshy base to stalk which curls like hook. Tasty, sweet; often slightly astringent, coarsely textured.

F 5. T^2. P l-Sept. S Oct-Dec/Feb.

CRAGGY'S SEEDLING 4 M C

UK; acc received 1992 from Kirby Steven, Yorks; may be variety recorded 1946 by Taylor as grown in West Riding, Yorks.

Large, orange red flushed. Cream, brisk, succulent flesh. Cooks to quite sharp, deep gold purée; well flavoured.

F est 12. T^2. P l-Sept. S Oct-Nov.

CRÂVERT 2 L D

France; found c1870 either in field of Dépigny family, or arose earlier in forest of Allogny in Cher (Val de Loire).

Firm, quite sharp. Cooked, keeps shape. Chefs suggest it for 'Tarte Tartin'. Grown now Cher.

F 28; trip. T^2. C bien. P l-Oct. S Dec/Jan-Mar/Apr.

CRÂVERT ROUGE; more highly coloured; arose Cher.

CRAWLEY BEAUTY 3 L CD

UK; found c1870 in garden at Tilgate, nr Crawley, Sussex. Introd 1906 by nursmn J. Cheal. RHS AM 1912. Acc is variety of Bultitude (1983).

Cooks to lightly flavoured purée early in

season. With storing loses acidity to become preferable fresh.

Popular Sussex and elsewhere as garden variety; especially in frost prone areas.

Frt Col brownish rd flush, many broken rd stripes over pale grn/grnish yell; greasy skin. *Size* med. *Shape* flt-rnd to rnd; flt base, apex. *Basin* brd, qte dp; sltly puckered. *Eye* open; sepals sep at base; qte downy. *Cavity* med wdth, dpth; ltl russet. *Stalk* med, thck. *Flesh* wht, tinge grn.

F 29. **T**^{2}; sp; sprd. **C** hvy. **P** m-Oct. **S** Nov-Feb.

CRAWLEY REINETTE 7 L D
UK; introd by J. Cheal & Sons, Crawley, Sussex. Recorded 1902.
Quite sharp; little richness developing by New Year; firm, cream flesh.
F 15. **T**^{2}. **P** m-Oct. **S** Dec-Feb.

CREȚESC 3 L C
Romania; received 1957 from Agric Inst Bucharest.
Crețesc means reinette. Very tall, ribbed. Cooked, retains shape; good fruity taste.
F 10. **T**^{2}. **P** l-Oct. **S** Dec-Feb.

CREȚESC DE BREAZA 5/4 L D
Romania; received 1957 from Agric Inst Bucharest.
Light flavour; strange mealy texture; juicy, pale cream tinged green flesh.
F 23. **T**^{2}; uprt. **C** hvy. **P** l-Oct. **S** Dec-Feb.

CRIMSON BEAUTY (CANADA) syn Early Red Bird 4 E D
Canada; raised 1850–80s by Francis Peabody Sharp, Canada's first fruit breeder, at Upper Woodstock, St John's Valley, New Brunswick.
Believed Fameuse X New Brunswick; parents most often used by Sharp. New Brunswick, reputedly superior form/seedling of Duchess of Oldenburg.
Crimson flushed, striped; ripe very early. Brisk lemony acidity yet plenty of sweetness; often definite raspberry flavour; juicy quite soft flesh.
Formerly grown commercially New Brunswick.
F 4. **T**^{3}; v hrdy. **P** eat e-Aug.

CRIMSON BEAUTY OF BATH see Beauty of Bath

CRIMSON BRAMLEY see Bramley's Seedling

CRIMSON COX see Cox's Orange Pippin

CRIMSON KING 4 M D Cider
UK; exhib from Somerset 1934. Recorded 1895.
Red striped, flushed over yellow. Very sweet, juicy; slight astringency.

F 14. **T**^{2}. **P** e-Sept. **S** Sept-Oct.

CRIMSON NEWTON see Newton Wonder

CRIMSON PEASGOOD see Peasgood Nonsuch

CRIMSON QUEENING 6 M D
UK; acc is variety of Bultitude, Hereford Queening of *Herefordshire Pomona* and Crimson Quoining of Hogg. Recorded 1831, but prob much older.
Prominently ribbed or quoined, that is angular; dark red flushed. Sweet, flowery, scented quality; soft pale cream flesh.
F 14. **T**^{1}. **P** e-Sept. **S** Sept-Oct.

CRIMSON SPY see Northern Spy

CRIMSON SUPERB see Laxton's Superb

CRIPPS PINK syn Pink Lady

CRISPIN syn Mutsu

CRITERION 4 L D
USA; found 1968 by Francis Crites on his Parker Heights Nursery, nr Wapato, Washington.
Poss Golden, Red Delicious cross. Introd 1972.
Known as 'Candy' apple. Quite honeyed, fruity; firm rather chewy flesh; can have underlying green taste in England.
Grown Washington; exported to UK.
F 18. **T**^{2}. **C** hvy. **P** m-Oct. **S** Nov-Jan; cold store-May.

CROWN GOLD see Jonagold

CSÍKOS ÓRIÁS HALASI 4 M D
Hungary; received 1948 from Univ Agric, Budapest.
Quite sweet, juicy; crisp to melting flesh; low in flavour.
F 9. **T**^{2}. **P** e-Sept. **S** Sept-Oct/Nov.

CURÉ 3 M C
France; prob arose Pays de Bray, Normandy. Desc 1948 (*Verger Français*).
Large; blood red flush, stripes. Sharp, firm flesh. Cooked, keeps shape; rather bleak taste.
Grown Normandy. Used also juice; vit C 20mg/100gms.
F 9. **T**^{3}. **P** m-Sept. **S** Oct-Nov.

CURL TAIL 5 M CD
UK; believed arose Woking, Surrey; recorded 1872 by nursmn J. Scott (Somerset).
King fruits have fleshy base to stalk which curls round like tail. Firm fleshed, brisk. Cooks to sweet, juicy pleasant purée.
F17. **T**^{3}; prt tip. **P** m-Sept. **S** Sept-Nov/Dec.

CUTLER GRIEVE 6 M D
UK; raised prob by James Grieve, Red Braes Nursery, Edinburgh, Scotland and among batch seedlings sent to nursmn David Storrie, Glencarse, who introd c1912.
Pretty, cherry red flush. Plain taste; hint of strawberry flavour; firm, white chewy flesh.
F 12. **T**^{1}. **P** m-Sept. **S** Oct-Nov.

CWASTRESSE DOUBLE syn Reinette des Vergers

DAKOTA 6 M D
USA; raised by W. A. Oitto, Northern Great Plains Field St, Mandan, N Dakota.
Wealthy X Whitney Crab. Introd 1965.
Sweet, soft, juicy cream flesh; light strawberry taste; very tough skin. Can taste sickly and metallic in England.
F* 11. **T**^{2}; v hrdy. **C** gd. **P** e-Sept. **S** Sept-Oct.

DALICE 7 L D
UK; raised 1933–37 by A. C. Nash, Scutes Farm, Hastings, Sussex, from pip of New Zealand Cox's Orange Pippin.
Rich, delicate aromatic flavour; reminiscent but not as intense as Cox. Can be less interesting; poor colour.
F 12. **T**^{2}; sprd. **P** e-Oct. **S** Nov-Jan/Feb.

DALIEST see Elstar

DALILI see Delcorf

DAMJANICH 5 M D
Hungary; origin unknown; prob named in honour of C19th General, János Damjanich.
Crisp, juicy, sweet; often empty in England.
F 11. **T**^{2}; uprt. **P** m-Sept. **S** Oct-Nov/Dec.

DÁNIEL-FÉLE RENET 7 L D
Hungary; received 1948 from Univ Agric, Budapest.
Sweet, rich, nutty flavour; crisp, cream flesh; similar to Blenheim Orange but smaller and sharper.
F 10. **T**^{2}. **C** gd. **P** e-Oct. **S** l-Oct-Jan/Mar.

DANZIGER KANTAPFEL syn Calville de Danzig 6 M DC
Poland; arose pres Danzig (Gdansk). Cited 1760 under syn by Dutch botanist Knoop. Syns numerous.
Large, Calville shape; deep red. Sharp in Sept, but quite intense rich, flavour; crisp, firm, juicy almost yellow flesh. Develops hints of aromatic quality. Cooked, keeps shape, good, quite brisk taste.
Widely grown northern, central Europe in C19th.
F 11. **T**^{1}. **P** e-Sept. **S** Sept-Oct/Nov; becomes v greasy.

D'ARCY SPICE 8 L D
UK; found c1785 in garden of The Hall, Tolleshunt D'Arcy, Essex; poss older. Introd 1848 by nursmn John Harris of Broomfield nr Chelmsford as Baddow Pippin.
Hot, spicy, almost nutmeg-like flavour by New Year; fairly sharp but sweetens. Firm white flesh, becomes rather spongy by spring but flavour remains. Needs plenty autumn sunshine or can be tannic, musty. Traditionally, picked Guy Fawkes Day (Nov 5), or Lord Mayor's Day and stored in sacks hung on the trees or packed away in trunks. Remains popular Essex variety; on sale locally.
Frt: Col: bright grn/yellish grn, sometimes trace purplish brown flush; much dark ochre russet patches, netting. Becomes dark brick rd flush over gold. *Size:* med. *Shape:* oblng, ribbed; 5 crowned. *Basin:* med wdth, dpth, ribbed, russeted. *Eye:* clsd/prt open; sepals lrg, brd base. *Cavity* nrw, med dpth, russet lined. *Stalk:* shrt, thck. *Flesh:* wht tinge grn.
F 14. T². C erratic. P l-Oct. S Jan-May.

DARK RED STAYMARED *see* Stayman's Winesap

DARU SÓVÁRI 4 L D
Hungary; old variety, first desc 1875.
Tall, conical. Modestly rich, fruity, good flavour; quite sweet, crisp, firm flesh.
Widely planted villages on eastern side of River Tisza in C19th; popular market apple.
F 14. T². P m/l-Oct. S Jan-May.

DAVEY 6 L D
USA; discv 1928 by S. Lothrop Davenport, North Grafton, Massachusetts.
McIntosh X. Introd 1950. FCC of Mass Hort Soc 1945.
Good brisk taste, hint strawberry flavour, crisp, juicy. Grown commercially small extent US.
F 7. T¹; res scab. C gd. P l-Sept. S l-Oct-Jan.

DAWN 6 L D
UK; prob raised by R. Staward, Ware Park Gardens, Herts. Exhib 1940.
Raspberry flavoured, sweet, balanced, juicy, soft flesh; can be over-acidic.
F 11. T². P l-Oct. S Nov-Jan/Mar.

DE BONDE 1/5 L C
France; prob arose early C20th; well known 1925 Indre-et-Loire, Indre (Val de Loire).
Cooked, keeps its shape; sugary, light taste. Formerly widely grown Indre-et-Loire; famed for its productivity. Used for juice (vit C 15mg/100g), tarts, *pommes tapées* (dried apples).
F17. T²; sp. C gd. P m-Oct. S Nov-Jan/Mar.

DE CASSE syn Cassou 2/4 L D
France; discv early 1900s by nursmn in Bonnut, Béarn (Pyrénées-Atlantiques).

Large. Crisp, juicy, brisk. Cooked, keeps shape; quite sweet, pleasant.
Distributed throughout Pyrénées-Atlantiques and Landes (Aquitaine).
F* rd; 13. T². P m-Oct. S Dec-Mar.

DECIO 2/4 L D
Italy; believed to be ancient variety, but impossible to know if acc and variety known in Italy in C20th, dates from C16th, or is Roman.
Small, flushed in brownish pink, over green, but ripening to red and cream. Quite sweet, fruity, fresh taste in Feb.
In C16th Francesco Bocchi, noted that 'melo d'Ezio' or 'd'Ezio' was taken from Rome by general Ezio to Adria, nr Padua where he fought Attila in cAD450. Decio apples were recorded in 1529 and 1536 when served at banquet given by Duke of Ferrara. Decio may be one of unnamed apples painted by Bimbi for Cosimo III de'Medici (1642–1723), but Decio not mentioned again until 1823 when French horticulturalist living in Venice included 'melo d'Ezio' in list of worthy varieties. Decio well known around Venice in 1939.
F 16. T¹; uprt. P l-Oct. S Jan-Mar/Apr.

DECOSTA *see* Jonagold

DE DEMOISELLE syn Weisses Seidenhemdchen

DE FLANDRE 4 L D
France; acc is prob variety desc 1948 (*Verger Français*) when grown Aube (Champagne).
Maroon flushed; small. Fruity, refreshing in Jan; crisp cream flesh.
F 22. T². P l-Oct. S Dec/Jan-Mar.

DE JAUNE, syn Reinette du Mans 5 L DC
France; prob arose c1700 around Le Mans, Sarthe (Western Loire); syns include Sarthe's Yellow.
Develops good rich taste; succulent, firm in Mar.
Known generally as Reinette du Mans; highly regarded C19th to early C20th; formerly exported to England. Still grown commercially Sarthe, Indre-et-Loire. Favoured for apple dumplings, tarts; also juice (vit C 25mg/100g), cider.
F 20. T²; reported res scab. C hvy. P l-Oct. S Jan-Apr.

DEKKER'S GLORY 4 M D
Netherlands; selected 1930 by C. Dekker, Blokker.
Soft, sweet flesh, winey taste; but low acidity, bland. Dutch market apple 1960s.
F19. T³; v sprd. P m-Sept. S Oct-Nov.

DE LA ROUAIRIE syn Fenouillet de Ribours

DELBARESTIVAL *see* Delcorf

DELBLUSH syn Tentation 6 L D
Not in NFC.
France; seedling raised by nursmn G. Delbard, Malicorne, nr Commentry, Allier (Auvergne). Trade name Tentation. Crisp, firm, honeyed flesh. Grown France; exported to UK.

DELCORF 4 M D
France; raised 1956 by nursmn G. Delbard, Malicorn, nr Commentry, Allier (Auvergne).
Golden Delicious X Stark Jonagrimes. Trade name Delbarestival. RHS AGM 1998.
Very sweet, honeyed, deep cream, crisp flesh, with hints of aniseed flavour. As Delbarestival grown commercially France, also Belgium, Italy, Germany, small extent England. Red sport Monidel is also known as Delbarestival and also grown commercially.
F 11. T². C hvy. P e-Sept. S Sept-Oct; cold store-Jan.
   DALILI; more highly coloured form; trade name Ambassy.

DELECTABLE 6 M D
UK; arose with Mr Trayhorne, Berks. Introduced 1952 by Pope Bros, Wokingham, Berks.
Delicate, aromatic quality, sweet, very juicy, fine texture. Dull appearance.
F 8. T²; prn canker. C lght. P m/l-Sept. S Oct.

DE L'ESTRE syn Saint-Germaine 2 L D
France; arose prob Corrèze (Limousin); popularised late 1700s by M. Turgot of Limoges and lawyer Cabanis. Now exists in number of forms, widely distributed over Limousin and Massif Central.
Some richness and intensity of flavour; quite crisp; good in Jan.
Name derives from l'estro in Limousin dialect, meaning window. Turgot, while visiting Corrèze in May 1770, sheltered from a storm in a farm house, nr Saint-Germaine-les-Vergnes. Some bread and an apple, taken from a window ledge, were given to him. Later, Turgot wished to propagate the apple he had enjoyed, but could only remember it as apple of the window. Formerly most widely grown apple of Massif Central; as St Germaine grown for Paris markets in C19th; again planted Limousin and further south Midi-Pyrénées. Valued for flavour and long keeping qualities; used also for cooking and cider.
F 19. T¹. P l-Oct. S Jan-Apr.

DELGOLLUNE 4/6 M D
France; raised 1960s by nursmn G. Delbard, Malicorn.
Golden Delicious X Lundbytorp. Trade

name Delbard Jubilé; syn Jubilé.
Bright red flush; quite large. Delicate aromatic flavour, hint of raspberries or flowers in Nov, but can be merely sweet; crisp juicy; tough skin. Grown commercially France; also popular garden apple.
F 7. T^2. C hvy. P m/l-Oct. S Nov-Feb.

DELGRARED 6 L D
France. Raised by G. Delbard, Malicorne.
Jenimota X Akane. Introd early 1990s.
Deep scarlet. Firm to rather coarse white flesh, sweet, lightly scented.
F mid. T^2. P e-Oct. S l-Oct-Dec.

DELICIOUS L D
USA; developed c1870s from seedling rootstock after scion has broken off, according to W. A. Taylor (assist pomologist USDA,1893). Arose on farm of Jesse Hiatt, Peru, Iowa. First named Hawkeye. Renamed, introd 1895 by Stark Bros, Missouri; introd UK c1912.
At best, densely sweet, not sickly, lightly aromatic, glistening cream flesh. But fruit on sale is sweet, soft, tough skinned. Needs hot summers and grown England, often lapses into metallic woody taste.
Brought to prominence after winning prize at Stark Brothers show in 1893; when C. M. Stark bit into apple he exclaimed 'My that's delicious – and that's the name for it'. He spent three quarters of a million dollars advertising what proved to be ideal commercial variety, producing heavy crops of sweet apples that remained shiny, bright red, no matter how long they stood out on display. Planted throughout the warmer apple regions and formerly the world's most widely grown variety, now being supplanted by newer varieties. Mostly grown for fresh fruit, also used for juice.
Original tree was almost killed in winter of 1940, but shoot grew up from root, fruited and still stands protected by a fence. Also commemorated by monument in Wintersct Park, Iowa.
Over 100 sports identified in US.
Frt *Col*: rd flush, shrt rd stripes; over yellish grn/yell background. *Size*: med. *Shape*: oblng-con, ribbed, flt sided and prominently 5 crowned. *Basin*: brd, dp; ribbed; puckered. *Eye*: hlf-open, sepals qte downy. *Cavity*: brd, dp; ltl russet. *Stalk*: med lngth, v thck. *Flesh* crm, tinge grn.
F12. T^2. C hvy. P m-Oct. S. Dec-Mar/Jun.
   AVERDAL; sport Red King Delicious.
   HARED; sport of Oregon Spur Delicious.
   IDAHO DELICIOUS; tetrap; received 1965.
   LANCRAIG; sport Topred.
   OREGONSPUR
   RICHARED DELICIOUS; more highly coloured. Discv c1915 by J. L. Richardson, Monitor, Washington; introd 1927.

STARKING; improved colour. Discv 1921 by Lewis Mood at Monroeville, New Jersey; introd 1924.
STARK'S LATE DELICIOUS received from Scotland 1967.
STARKRIMSON DELICIOUS; coloured sport of Starking; spur type. Discv c1953 by Roy A Bisbee, Hood River, Oregon; introd by Stark Bros, 1956.
WELLSPUR DELICIOUS; spur form of Starking. Discv 1952 by L. Green & C. Nelson, Azwell Orchard of Wells & Wade Fruit Co, Wenatchee, Washington; introd 1958.

DELJENI 4/5 M D
France; raised 1960s by nursmn G. Delbard, Malicorn. Syn Primgold.
Jenimota X Stark Jonagrimes.
Resembles Golden Delicious, but matures earlier. Sweet, slightly honeyed flavour; crisp, cream flesh.
F 13. T^2. C hvy. P m-Sept. S l-Sept-Nov.

DELJORAM 6 L D
France. Raised by G. Delbard, Malicorne.
Jonalicious X LawRed Rome. Trade name Delbartardive. Introd early 1990s.
Deep scarlet. Sweet, delicate, quite scented flavour.
F late. T^2. C hvy. P l-Oct. S Nov-Mars.

DELKISTAR 6 L D
France. Raised 1978 by G. Delbard, Malicorne.
Kidds Orange Red X Starkrimson Delicious. Trade name Régali. Introd 1990s.
Handsome deep red colour of Starkrimson. Deep cream flesh, crisp juicy, sweet with hints of richness of Kidd's Orange Red.
F est 10. T^3. C hvy. P e-Oct. S l-Oct-Jan; comercially-Apr.

DELNIMB 5 M D
France; raised 1960s by nursmn G. Delbard, Malicorn.
Maigold X Grive Rouge. Introd 1980s.
Rich, intense flavour; scented, flowery. Plenty juice, sugar, acidity. Rather coarse texture; tough skin.
F 5. T^1. P l-Sept. S Oct-Nov.

DELORGUE 6 M D
France; raised 1982 by G. Delbard, Malicorn.
Delcorf X Akane. Trade name Festival. Introd late 1990s.
Crisp, pale cream flesh; sweet juicy; sometimes scented with hints of aniseed.
F mid. T^2; sprd. P e-Sept. S Sept.

DELORINA 6 L D
France; raised 1985 by nursmn G. Delbard, Malicorn.
Bushing Golden X Querina Delbard. Trade

name Harmonie. Introd mid 1990s.
Crisp, rather chewy, pale yellow flesh; sweet, lightly flavoured, but Delbard claims 'scented banana and nuts'.
F mid. T^2; sprd; res scab. P m-Oct. S Nov-Feb.

DELPRIM 6 E D
France; raised 1960s by nursmn G. Delbard, Malicorn.
Delcorf X Akane. Intro mid 1980s.
Quite rich flavour; plenty sugar, acidity; hard, crisp, juicy. Keeps well into Sept, becoming sweeter, deliciously scented.
F 5. T^3; sprd. C hvy; frt sml. P m-Aug. S Aug-Sept, cold store-Oct.

DELPRIVALE 6 E D
France; raised 1960s by nursmn G. Delbard, Malicorn.
Delcorf X Akane. Introd late 1990s.
Crisp, good texture juicy; sweet, slightly honeyed but plenty of acidity.
F mid. T^2. P and eat Aug.

DELVALE 4 L D
France. Raised by nursmn G. Delbard, Malicorn.
Goldenspur X Jonagrimes. Introd 1990s.
Crisp, open textured flesh; sweet, juicy, with attractive spicy, aniseed taste.
F mid. T^2; sprd. P e-Oct. S Nov-Dec.

DEMOCRAT 6 L D
Australia; found c1900 nr orchard of J. D. Duffy, Glenlusk, nr Hobart, Tasmania. Believed Hoover seedling. Named Duffy's Seedling; renamed Democrat, then Tasma to avoid confusion with US Democrat. Known Tasma New Zealand.
Dark red flushed. Sweet, juicy, crisp in New Year. Grown Tasmania, New Zealand.
F 8. T^2; res mildew. P l-Oct. S Jan-Mar/Jun or later.

DE QUINT 4 L D
France; received 1948 from Drôme, Rhône, where then grown.
Markedly oblong. Quite rich; plenty of fresh, fruity flavour; sweet juicy.
F 13. T^2. P m-Oct. S Nov-Jan.

DERMEN McINTOSH *see* McIntosh

DERMEN WINESAP *see* Winesap

DESSE DE BUFF 3 M C
UK; received from Wrest Park, Beds; England.
Exhib 1900 by Laxton Bros, Bedford as local apple – Dessin de Boeuf.
Sharp, firm. Cooked keeps shape; little flavour.
F19. T^2; sprd. P m-Sept. S Sept-Oct.

DESSEFFY ARISZTID 4 L D
Hungary; received 1948 from Univ Agric,

Budapest.
Attractive, red flush over gold. Light savoury flavour; coarse flesh.
F 12. **T**^{2}. **C** hvy. **P** e-Oct. **S** Nov-Dec/Jan.

DE VENDUE L'ÉVÊQUE 1/5 L C
France; desc 1948 (*Verger Français*), when grown Aube (Champagne).
Sharp, crisp. Used for cider, cooking.
F21. **T**^{2}. **C** hvy. **P** l-Oct. **S** Jan-Mar.

DEVON CRIMSON QUEEN 6 M D
UK; received 1953 from Launceston Cornwall; syn Queenie.
Brilliant red flush, with bloom. Some richness, quite sharp; firm, deep cream flesh; becomes softer, sweeter.
Former favourite of Tamar Valley, Cornwall.
F 7. **T**^{2}; sp; sprd. **C** hvy. **P** l-Aug. **S** Sept-Oct.

DEVONSHIRE BUCKLAND 3/4 L C
UK; recorded 1831; desc Hogg (1851). As syn Dredge's White Lily mentioned 1810 by Forsyth; William Dredge was nursmn of Wishford, Salisbury.
Sharp and fruity. Cooks to lemon coloured purée; sweet good flavour.
F14. **T**^{3}. **C** bien. **P** l-Oct. **S** Dec-Feb/Mar.

DEVONSHIRE QUARRENDEN 6 E D
UK; old variety; arose poss Devon, or introd from France. Poss took name from Carentan, Normandy. Name recorded 1676 by Worlidge. 'Quarendouns' mentioned in Middle English alliterative poem on plant names, of west country provenance, dated c1450.
Distinctive flavour – of strawberries or winey to loganberry taste; sweet, plenty of acidity. Not easy to catch in prime; soon goes soft once picked. Valued for C19th summer dessert and 'grown from Devonshire to the Moray Firth'. Sold in markets as 'Quarantine' and 'aways made a high price'. In decline by 1890s, but still often found in old gardens.
Frt Col: dark, bright crimson flush, few stripes; pale yell background. *Size*: med/sml. *Shape*: flt; rnded. *Basin*: brd, shl; puckered. *Eye*: clsd, sepals sml, v downy. *Cavity*: brd, dp; ltl russet. *Stalk*: med lng, thin. *Flesh*: wht, tinge grn.
F7. **T**^{1}; sprd. **C** bien; frt drops. **P** m-Aug. **S** Aug.
PLATE 5.

DEWDNEY'S SEEDLING syn Baron Wolseley 1 M C
UK; raised c1850 by Mr Dewdney, Barrowby, nr Grantham, Lincs.
Large. Cooks to sharp froth or pale creamy purée. Mellows later.
F14. **T**^{3}. **C** gd. **P** e-Sept. **S** Sept-Nov.

DEWDULIP SEEDLING 1 M C
UK; received 1946 from Tenbury Wells,

Worcs.
Quite rich, firm, cream flesh. Cooked keeps its shape; sweet, very light.
F15. **T**^{2}. **P** e-Sept. **S** Sept-Nov.

D'EYLAU 4 L D
Claimed brought c1812 from Eylau, Russia to France by a soldier of Napoleon's Grand Army.
Markedly oblong; dark red flushed. Very sweet, firm flesh; almost no acidity. Grown now Savoie and Isère (Rhône Alpes).
F 13; trip. **T**^{2}. **P** m-Oct. **S** Nov/Dec-Feb.

DIAMOND JUBILEE 5 L D
UK; raised c1889 poss by Mr Thomas, Rainham, Kent; RHS AM 1901.
Crisp, plenty good brisk flavour in Jan. Earlier cooks to sharp fruity purée.
F10. **T**^{3}; uprt. **P** m-Oct. **S** Dec/Jan-Mar.

DIANA 4 L D
Received 1967 from former Yugoslavia via Scotland.
Bright red flushed; pink stained flesh. Reminiscent McIntosh in taste; sweet, balanced; good fruity flavour; soft, very juicy flesh.
F 10. **T**^{2}. **C** prn canker. **P** e-Oct. **S**-Nov/Dec.

DIANASPUR *see* Golden Delicious

DICK'S FAVOURITE 4 L C
UK; raised prob late 1800s by Carless, foreman at Rowe's Nurseries, Worcester.
Cooked keeps shape; bright lemon; quite well flavoured. Eaten fresh – crisp, brisk.
F 14. **T**^{3}. **P** m-Oct. **S** Nov/Dec-Jan.

DILLINGTON BEAUTY 4 L D
Received 1953 from New Zealand.
Sprightly taste of fruit in Dec; crisp, firm.
F 15. **T**^{2}. **C** prn bitter pit. **P** l-Oct. **S** Dec/Jan-Apr.

DIRECTEUR LESAGE 4 E D
Belgium prob; known northern France as Précoce de Wirwignes.
Pretty, pink flushed, red striped. Light, refreshing; soft, juicy, white flesh. Grown small extent Belgium, France.
F 6. **T**^{2}. **C** hvy, bien. **P** l-Aug. **S** Sept-Oct.

DIRECTEUR VAN DE PLASSCHE 7 L D
Netherlands; raised 1935 at IVT, Wageningen.
Cox's Orange Pippin X Jonathan.
Rich, aromatic. Sharper, more robust flavour than Cox, but same fine textured cream flesh.
F 10. **T**^{2}. **P** e-Oct. **S** Nov-Jan/Mar.

DISCOVERY 6 E D
UK; raised c1949 by Mr Dummer, Blacksmith's Corner, Langham, Essex.

Worcester Pearmain X poss Beauty of Bath. Named Thurston August; renamed 1962 and introd by nursmn J. Matthews, Thurston, Suffolk. Mother tree stills grows in garden at Langham. RHS AGM 1993.
Well ripened on the tree, bright red, with crisp, juicy, often pink stained flesh and hint of strawberry flavour.
Dummer, workman on Essex fruit farm, had raised a number of seedlings from Worcester Pearmain pips and decided to plant the best one in front garden. He only had one arm and so his wife's help was needed, but she slipped, broke her ankle and the tree had to remain, despite frosts, covered by a sack for some weeks before it was planted. Nevertheless it survived, produced colourful heavy crops, which unlike most earlies never seemed to drop. Came to notice of Matthews, who bought grafts and every year held a party under the tree to popularise his new variety.
Main early commercial variety by 1980s; widely grown gardens.
Frt Col: bright rd flush; grnish yell/yell background; flecked fine dots pale russet. *Size*: med. *Shape*: rnd, sltly flattened. *Basin*: brd, med dpth; sltly ribbed *Eye*: clsd; sepals, shrt, v downy. *Cavity*: brd, dp; russet lined. *Stalk*: med lngth; thck. *Flesh* crm.
F11. **T**^{2}; prt tip; res scab, mildew. **C** gd; slow to bear. **P** m/lAug. **S** m/l-Aug-Sept.
PLATE 3

DISCOVERY; spur type.

DOCKNEY (?) 3 E C
Ireland; acc may be variety desc 1951 (Lamb), but incorrect season: early rather than late.
Cooks to a sharp froth.
F15. **T**^{3}. **P** l-Aug. **S** Aug-Sept/Oct.

DOCTOR CLIFFORD 3 E C
UK; arose c1898 with nursmn W. Ingall, Grimoldby, Lincs. Introd 1911.
Larger than most early cookers. Cooks to froth, light, brisk.
F 9. **T**^{2}. **C** hvy. **P** Aug. **S** Aug-Sept/Oct.

DOCTOR HARVEY syn Harvey 5 M C
UK; Harvey apple recorded 1629 by Parkinson; impossible to know if this is variety known C19th when, as Doctor Harvey, desc by Hogg (1884). Acc matches Hogg's desc. Prob arose East Anglia.
Cooks to well-flavoured, sweet purée, with mild acidity.
According to Norwich, nursmn George Lindley, 'When baked in an oven which is not too hot . . . they become sugary and will keep a week or ten days furnishing for the dessert a highly flavoured sweetmeat'.
Long popular Cambs and, according to Hogg, named after local benefactor, Dr Gabriel Harvey, Master of Trinity Hall, Cambridge University, who 'in about 1630

left by will an estate to mend the road from Cambridge towards London, six miles to Fulmer'. One of best Norfolk varieties in 1822. Seen in Norwich and London markets in C19th and on sale locally until 1960s; quite widely popular garden variety. Many old trees remain in Norfolk.

Frt *Col* pale grn/gold; slight pinky brown flush; many russet dots. *Size* lrg. *Shape* oblng-con; rnded ribs; sltly flt sided; sltly 5 crowned. *Basin* brd, dp; ribbed. *Eye* lrg, open; sepals shrt. *Cavity* nrw, dp; russet lined. *Stalk* shrt, qte thin. *Flesh* pale crm.
F* 14. **T**^{2}. **C** gd. **P** m-Sept. **S** l-Sept-Nov/ Dec.

DOCTOR HOGG 5 M C
UK; raised by HG, Mr S. Ford, Leonardslee, Horsham, Sussex.
Believed Calville Blanc seedling. Introd c1880 by W. Paul & Son, Herts; RHS FCC 1878.
Very large. Cooks to quite brisk, good purée. Named in honour of the esteemed Victorian pomologist.
F* 9. **T**^{3}. **P** e-Sept. **S** Sept-Oct.

DOCTOR RAMBURG 3 L C
Received 1967 from former Yugoslavia via Scotland.
Large, handsome. Rich, quite aromatic, cream flesh. Cooked, keeps shape; very light, sweet taste.
F 11. **T**^{2}. **P** m-Oct. **S** Nov-Dec.

DODD 4 M C
USA; arose on lane nr Tolman Sweet orchard of H. Dodd, de Peyster Township, St Lawrence County, NY.
Tolman Sweet, poss Fameuse cross. Introd 1962 by nursmn F. L. Ashworth, Heuvelton, NY.
Tastes like an unripe banana in Sept. Very sweet crisp white flesh; almost no acidity, which fades altogether with keeping.
F 9. **T**^{2}; prt tip. **P** e-Sept. **S** Sept-Oct.

DOMINO 1 E C
UK; prob arose Midlands. Exhib 1883 Congress from Southwell, Notts; desc Hogg (1884).
Codlin type; cooks to pale cream froth or purée, juicy, sharp, good taste.
Grown around Southwell in C19th and in Cambs 1930s.
F 9. **T**^{2}; sp; sprd. **C** gd. **P** Aug. **S** Aug-Sept.

DOMNESC 4 L D
Romania; believed from Iaşi, Moldavia. Recorded 1831.
Quite large. Savoury, refreshing taste; crisp, juicy. Often underlying green, woody taste.
Widely grown Moldavia, Serbia, Bulgaria in C19th; still considered good kitchen, processing apple in 1960s.
F 5. **T**^{3}. **P** m-Oct. **S** Nov-Jan/Mar.

DOMNICELE 5 L C
Romania; received 1948 from Agric Acad Sc, Bucharest.
Crisp, juicy, some sugar in Feb, but sharp, lacking flavour in England.
F 11. **T**^{2}; sp; uprt. **C** gd. **P** m/l-Oct. **S** Jan-Mar.

DORÉE DE TOURNAI 5 L D
Belgium; raised 1817 by Joseph de Gaest de Braffe at Tournai.
Russet freckled over gold. Intense, sweet-sharp taste, spiked with acid drop flavour; like Nonpareil family. Firm, rather leathery, deep cream flesh.
F 9. **T**^{2}. **P** l-Sept. **S** Oct-Dec.

DORSETT GOLDEN 4 L D
Bahamas; arose 1953 at home of Mrs Irene Dorsett, Nassua, New Providence Islands.
Believed Golden Delicious seedling. Introd 1964.
Sweet, lightly aromatic in Dec. Fruits in tropics, where no winter chilling occurs.
F 19. **T**^{2}. **P** m-Oct. **S** Dec-Jan/Feb.

DOUBLE RED BALDWIN see Baldwin

DOUBLE RED NORTHERN SPY see Northern Spy

DOUBLE RED ROME BEAUTY see Rome Beauty

DOUBLE RED WEALTHY see Wealthy

DOUBLE ROSE 2/4 L D
France; acc prob not Double Rose of Aveyron desc 1948, but Double Rose de l'Ain (Rhône Alps); widespread in this area.
Prettily flushed pink over yellow. Sharp, fresh taste to April; firm, crisp flesh.
F 12. **T**^{2}; spr; uprt. **C** gd. **P** l-Oct. **S** Jan-Apr.

DOUD GOLDEN DELICIOUS see Golden Delicious

DOWNTON PIPPIN 5 M D
UK; raised by Thomas Andrew Knight, Elton Manor, Ludlow, Shrops.
Orange Pippin X Golden Pippin. Exhib 1806.
Almost cidery tang; intense flavour of fruit; brisk, crisp, juicy. Knight, claimed also well-suited to the kitchen – cooked keeps shape; strong taste, sweet but not flat.
Popular Victorian variety. Named after Downton Castle, built by Knight's brother in Shrops. Still found Shrops in old orchards of Teme Valley.
Frt *Col*: grnish yell turning gold, lenticels as russet dots. *Size*: sml. *Shape*: rnd or rnd-con. *Basin*: brd, v shal. *Eye*: open. *Cavity*: brd, shal; dark grn. *Stalk*: shrt, thck. *Flesh* dp crm.
F 12. **T**^{2}. **C** gd. **P** m-Sept. **S** Sept-Oct/Nov.

DRAP D'OR 7 L D
France; acc is prob Drap d'Or syn Chailleux of *Pomologie de France* (1863–73). Very old, may be variety listed by Le Lectier in 1628, but impossible to know.
Like gilded brocade – bright flush over gold, veined with russet. Sweet, juicy, crisp, plenty of fruit, acidity; slightly vinous in Nov.
F 9. **T**^{3}. **P** m-Oct. **S** Nov-Jan/Mar.

DREDGE'S FAME 4 L D
UK; introd by nursmn William Dredge, Wishford, Salisbury. Recorded 1802.
Bright red flush. Quite rich, plenty of fruit, sugar and acid
F 9. **T**^{2}. **C** v prn bitter pit. **P** e/m-Oct. **S** Nov-Dec/Mar.

DUBBELLE BELLE FLEUR
Netherlands; received 1983 from Wilhelminadorp Fruit Res St.
Used Netherlands as interstock to give better unions between Cox's Orange Pippin scion and M9 rootstock. Produces better feathered Cox maidens; res collar rot.

DUBBELE ZOETE AAGT
Netherlands; received 1983 from Wilhelminadorp Fruit Res St.
Used Netherlands as interstock to give better unions between Cox's Orange Pippin scion and M9 rootstock. Produces better feathered Cox maidens; res collar rot.

DUCHESS OF BEDFORD 4 E D
UK; raised by Laxton Brothers Nursery, Bedford.
Cellini X Beauty of Bath. Exhib 1918.
Strong aniseed flavour; more balsamatic than Cellini, sweet, soft, white flesh.
F 10. **T**^{2}. **P** Aug. **S** Aug-Sept.

DUCHESS OF OLDENBURG 4 E CD
Russia; known Russia early 1700s. Poss arose Tula area. Originally Borovinka. Spread to Germany, Sweden. Sent c1817–24 from St Petersburg and Sweden to England; renamed. Sent c1835 from London Hort Soc to Massachusetts Hort Soc. Syns numerous, include Oldenburg, Duchess Charlamowsky.
Very beautiful, boldly striped and mottled in red over pale yellow with bloom. Savoury, quite brisk, juicy, soft, deep cream flesh. Cooks to very lightly flavoured purée. Popular Victorian garden apple; also grown for market. Formerly widely planted US. Still grown all over former USSR, N Europe; as Charlamowsky especially valued in Sweden.
F 4. **T**^{2}. **C** gd; v hrdy. **P** m-Aug. **S** Aug-Sept.

DUCHESS'S FAVOURITE 6 E D
UK; raised late 1700s/early 1800s by nursmn Cree, Addlestone, Surrey.

Showy apple, so named because admired by Duchess of York, who lived at Oatlands Park, nr Cree's nursery. Can develop intense flavour of strawberries; firm white flesh, often stained red. Very like Bens' Red in flavour or a sharp Worcester Pearmain. Popular Victorian garden apple; in 1890s also recommended for planting in shrubberies, lawns. Grown C19th by market gardeners around London and Vale of Evesham; also planted Kent early 1900s instead of Worcester.

Frt *Col*: bright rd flush; faint rd stripes; pale grnish yell background; lenticels, russet dots. *Size*: sml/med. *Shape*: rnd to flt-rnd; sltly ribbed. *Basin*: brd, shal; sltly puckered. *Eye*: hlf open, sepals sml, downy. *Cavity*: nrw, dp; russet lined. *Stalk*: shrt, thin. *Flesh* wht.
F 10. **T**^{2}; sp. **C** hvy. **P** e-Sept. **S** Sept.

DUCK'S BILL 4 L D
UK; received 1937 from Fred Streeter, HG, Petworth House, Sussex. Poss not the ancient Duck's Bill.
Ribbed, flat sided – hence its name. Quite rich, sweet but plenty acidity.
F 16. **T**^{2}. **P** e-Oct. **S** Nov-Dec.

DUÈGNE 4 E D
France; received 1947.
Juicy, melting flesh; fruity, pleasant, but rather sharp.
F 3. **T**^{2}. **P** eat Aug-e-Sept.

DUGAMEL *see* Melrose

DUKÁT 4 L D
Czech Republic. Prob raised Exp Botany St, Strizovice.
Golden Delcious X Cox's Orange Pippin. Received 1974.
Dark red flushed. Sweet, scented, slightly honeyed flavour; coarse texture, nearly yellow flesh.
F 10. **T**^{2}. **P** m/l-Oct. **S** Dec-Jan.
DUKÁT SPUR TYPE; also larger fruit.

DUKE OF DEVONSHIRE 8 L D
UK; raised 1835 by Wilson, HG to Duke of Devonshire at Holker Hall, Cumbria. Introd c1875.
Intense, sweet-sharp, fruit drop flavour of Nonpareil but sweeter, yet not as rich as Ashmead's Kernel.
Popular Edwardian apple, 'quite indispensable for the late dessert'. Also appeared in markets of 1930s; remains valued garden apple.
Frt *Col*: grn/gold; sometime brownish gold flush; russet patches, netting, dots. *Size*: med. *Shape*: flt-rnd to shrt-rnd-con. *Basin*: brd, shal; sltly ribbed; puckered, russeted. *Eye*: clsd or hlf-open; sepals shrt, brd based. *Cavity*: nrw, shal; russet lined. *Stalk* v shrt, thck. *Flesh* crm.
F 11. **T**^{2}; sprd; prn mildew. **C** lght. **P** e/m-Oct. **S** Jan-Mar.

DÜLMENER ROSENAPFEL 4 M DC
Germany; raised by schoolmaster Jäger, at Dülmen, Westphalia, c1870.
Gravenstein seedling.
Resembles Gravenstein, but not as well flavoured. Sweet, juicy cream flesh. Cooks to sweet purée. Grown eastern Germany; also used for juice, drying.
F 8; frost res. **T**^{2}; res mildew. **P** m-Sept. **S** Oct.

DUMELOW'S SEEDLING, syn Wellington 3 L C
UK; raised late 1700s by Richard Dummeller, farmer at Shakerstone, nr Ashby-de-la-Zouche, Leics. Believed from Northern Greening pip. Sent from Gopsall Hall, Earl Howe's estate to Richard Williams, Turnham Green, who exhibited 1818 at the London Hort Soc as Dumelow's Crab. Renamed Wellington 1819 or '20. Also known Normanton Wonder in northern England. RHS AGM 1993.
Cooks to sharp, strongly flavoured pale cream purée, similar to Bramley but creamier texture. Splendid baked, smooth and juicy. Keeps acidity and flavour to spring.
Premier Victorian culinary apple; widely grown in gardens and for market. Recommended for making mincemeat. May Day apple in northern England. Overtaken by Bramley in early 1900s, but many old trees remain especially in the west Midlands.
Frt *Col*: pale grnish white/pale yell, sometimes mottled with bright apricot, few stripes; lenticels as purplish dots; becomes greasy. *Size*: med/lrg. *Shape*: flt-rnd, sometimes hair-line rnd apex. *Basin*: brd, shal, sltly ribbed. *Eye*: lrg, open; sepals lng, brd at base. *Cavity*: nrw, qte shal; russet lined. *Stalk*: shrt, thck. *Flesh* wht.
F* 15. **T**^{3}; sprd. **C** gd. **P** e/m-Oct. **S** Nov-Apr.
PLATE 27

DUNNING 6 E D
USA; raised 1923 by R. Wellington, NYSAES Geneva.
Early McIntosh X Cox's Orange Pippin. Introd 1938.
Very handsome, dark red, deep bloom. Very sweet, little acidity, almost like eating candyfloss. White, crisp flesh, often stained pink.
F 10. **T**^{2}; sprd. **P** l-Aug. **S** Aug-Sept.

DUNN'S SEEDLING syn Ohinemuri 5 L D
Australia; believed raised by Mr Condor at Kew, Melbourne. Recorded UK 1890. Known South Africa also as Monroe's Seedling after its propagator.
Good, fresh taste; sweet, almost scented. Formerly grown Australia, New Zealand, South Africa for export.
F 6. **T**^{2}. **P** m-Oct. **S** Dec-Feb/Mar.

DURELLO 4 L D
Italy; desc 1949, but old variety.
Good, brisk taste of fruit in the New Year; some sweetness; hard flesh.
Formerly grown Emilia, Lombardi, Veneto; used for cooking.
F 13. **T**^{2}; uprt. **C** bien. **P** m/l-Oct. **S** Jan-Mar/Jun.

DUTCH CODLIN 1/5 L D
Netherlands (prob). Brought notice 1783; known England also as Chalmers' Large.
Large, ribbed, codlin shaped. Slightly sharp. Cooks to gold, rather flavourless purée.
F 20. **T**^{2}. **P** e-Sept. **S** Sept-Oct.

DUTCH MIGNONNE 4 L D
Netherlands (prob). Introd England c1771 by Thomas Harvey of Catton, Norwich, then named. Syns many, include Reinette de Caux.
Brisk, plain, quite like Blenheim Orange; deep cream flesh. Cooked keeps shape or makes a stiff purée, light, sweet taste. Victorian cooks used it for Apple Charlotte and late in season considered fit for dessert. Widely grown gardens in C19th.
F 9. **T**^{2}; sprd. **P** e-Oct. **S** Nov-Feb/Mar.

DYER *see* Pomme Royale

EADY'S MAGNUM 5 L C
UK; raised c1908 by Miss D. A. Eady at Wellingborough, Northants.
Very large. Sharp, cream flesh. Cooks to very juicy, lemon purée, brisk, intensely fruity. Makes good sauce.
F 15. **T**^{3}; prt tip. **P** m-Oct. **S** Dec-Mar.

EARL COWPER 1 M C
UK; received 1949 from Wrest Park, Beds, but prob old variety.
Name suggests raised between 1859, when 6th Earl Cowper married Lady Earl de Grey, owner of Wrest Park, and 1905 when 7th Earl Cowper died without an heir.
Large, greenish yellow. Firm sharp flesh. Cooks to juicy, white purée; sweet, light.
F 5. **T**^{2}. **P** m-Sept. **S** Sept-Oct/Nov.

EARLY JULYAN Syn Tam Montgomery 1/5 E CD
UK; believed arose Scotland. Known before 1800. Introd southern England early 1800s by nursmn Hugh Ronalds, Brentford, Middx.
Ready to use as early as July; straw yellow, when fully ripe. Good, refreshing taste; firm, brisk, cream flesh. Cooks to bright yellow purée; light, fruity.
Grown for market, popular garden apple in C19th.
F 9. **T**^{3}. **C** gd. **P** use Aug.

EARLY McINTOSH 6 E D
USA; raised 1909 by R. Wellington, NYSAES, Geneva.
White Transparent X McIntosh. Introd 1923.
Very like Melba and McIntosh. Light strawberry flavour; juicy, soft, white flesh. Grown commercially small extent North America, Belgium, France.
F 9. T^3. C hvy, bien. P Aug. S Aug-Sept/Oct.

EARLY RED BIRD see Crimson Beauty (Canada)

EARLY QUEEN see Jonagold

EARLY VICTORIA syn Emneth Early

EARLY WINDSOR syn Alkmene

EASTBOURNE PIPPIN 5 E D
UK; raised 1930 by E. A. Lindley, Eastbourne, Sussex.
Believed Newtown Pippin seedling.
Sweet, tasty; firm cream flesh. Can lack interest.
F 7. T^2. C hvy. P l-Aug. S Sept-Oct.

EASTER ORANGE 7 L D
UK; introd 1897 by Hillier & Sons, Winchester, Hants. RHS AM 1897.
Handsome. Intensely flavoured, rich, aromatic, deep cream flesh; rather chewy texture.
F 12. T^1. C gd; prn bitter pit. P e-Oct. S Nov/Dec-Feb/Apr.

EAST LOTHIAN PIPPIN 1 E C
UK; acc received from Tyninghame Gardens, East Lothian, Scotland; gardens from which first recorded fruit was sent in 1883 to National Apple Congress. Impossible to know if same, as descs very brief, but given source likely to be true.
Cooks to juicy purée; good, quite sweet taste.
F 11. T^2. P m-Aug. S Aug-Sept.

ECCLESTON PIPPIN 4 M DC
UK; found or raised by Mr N. F. Barnes, HG to Duke of Westminster, Eaton Hall, Cheshire. Exhib 1883 Congress.
Showy red flush, stripes; quite large. Cooked, keeps shape; bright lemon; sweet but lacks acidity.
F 11. T^2. P m-Sept. S Sept-Oct.

ECKLINVILLE 1 M C
UK; believed raised by gardener Logan at Ecklinville, Portaferry, nr Belfast, N Ireland. Known 1800.
Cooks to well-flavoured, brisk, soft, juicy, purée, not as sharp as Golden Noble. Esteemed by Victorians as sauce apple.
Widely grown in gardens; also

recommended in 1890s for 'artistic' orchard. Brought prominence as market apple by Dancer, London market gardener; became particularly popular in Worcs. No longer planted by 1930s, as fruit easily bruised.
Frt Pea grn turning yell; many russet dots; becomes v greasy. *Size* lrg. *Shape* flt-rnd. *Eye* hlf open; sepals, shrt. *Basin* brd, dp. *Cavity* med wdth, dp; russet lined. *Stalk* shrt, qte thin. *Flesh* wht.
F* 13. T^3. C gd. P e-Sept. S Sept-Nov.

EDELBORSDORFER syn Borsdorfer 5/4 L D
Germany; arose village of Borsdorf in Meissen and prob very old. 'Borsdorf' was cited as grown 'Misnia' by Valerius Cordus (died 1544). Introd England, listed 1785 by Brompton Park Nursery, London. Syns numerous.
Attractive, flushed in pinky orange over yellow – according to an old German adage 'Her cheeks are as red as a Borsdorfer apple'. Very sweet, scented, as if with elderflowers; firm, white flesh. Needs warm summer for good flavour; can be metallic, flat.
In C18th, 'Pride of the Germans'; so esteemed that Queen Charlotte of Mecklenburg, wife of George III, had consignments sent every year to London.
Formerly widely grown Europe, but never popular England.
F 14. T^7; hrdy. C hvy. P m-Oct. S Nov-Jan.

EDEN 6 L D
UK; arose 1948 with E. J. Ingleby, Forest and Orchard Nurseries, Falfield, Gloucs.
John Standish X Cox's Orange Pippin. Sister seedling of Fon's Spring, not syn. Introd 1957 by nursmn J. Matthews, Suffolk.
Bright red. Crisp, firm flesh, slightly aromatic; earlier season than Fon's Spring.
F 11. T^2. C hvy. P m-Sept. S Oct-Dec.

EDGAR 6 L D
Canada; arose at Central Exp Farm, Ottawa.
McIntosh X Forest. Select 1929.
Brilliant red with bloom, like McIntosh. Pink stained, sweet, juicy, soft melting flesh.
F 15. T^2; uprt. C hvy. P e/m-Oct. S Nov-Dec.

ED GOULD GOLDEN see Golden Delicious

EDITH HOPWOOD 5 M D
UK; raised by F. W. Thorrington, retired London Customs Officer, Hornchurch, Essex.
Cox's Orange Pippin X. Received 1925.
Bright yellow. Sweet, lightly aromatic, juicy, cream flesh. Tends drop before fully ripe, but too sharp if picked early.
F 13. T^2. C gd. P e-Sept. S Sept-Oct.

EDMUND JUPP 5 M DC
UK; arose nr Horsham, Sussex. Recorded 1862.
Russet freckled over gold. Quite brisk, savoury; crisp to soft, white flesh. Cooks to gold purée; very mild taste.
F 11. T^2. C gd; frt sml. P m- Sept. S Sept-Oct/Nov.

EDWARD VII 1 L C
UK; believed Blenheim Orange X Golden Noble. Recorded 1902. Introd 1908 by Rowe, Barbourne Nurseries, Worcester. RHS AM 1903; awarded 1st prize 1909 by RHS for best new culinary variety; AGM 1993.
Cooks to well-flavoured, translucent, cream purée in Nov-Dec; rather like Golden Noble, but not as acidic as Bramley. Becomes sweeter, cooking more firmly and makes pleasant, brisk, eating apple.
Popular garden apple, especially in frost prone and wet areas; grown for sale small extent.
Frt *Col* grn/yell; lenticels as faint grey or russet dots; scarf skin at base. *Size* med/lrg. *Shape* flt-rnd to rnd. *Basin* qte brd, shal; sltly ribbed, puckered. *Eye* qte lrg, open; sepals lng, brd, qte shal. *Cavity* med wdth, qte shal; ltl russet. *Stalk* shrt/med, qte thck. *Flesh* crm.
F* dp cerise/pale pink; 21. T^2; hrdy; uprt; res scab; mod res mildew. C gd. P mid-Oct. S Dec-Apr.

EDWARDS 5 L D
USA; believed arose Chatham County, North Carolina. Desc 1869.
Seedling of Hall.
Quite large. Savoury, brisk flavour in Dec; loosely textured deep cream flesh.
F 11. T^3. P m-Oct. S Dec-Mar; claimed-Sept.

EGREMONT RUSSET 8 M D
UK; prob arose England. Recorded 1872 by nursmn J. Scott, Somerset. Exhib 1883 Congress from Sherborne Castle, Dorset. RHS AGM 1993.
Very distinctive flavour; often described as nutty; Morton Shand said it recalled scent of crushed ferns. Almost smoky, tannic quality develops after keeping and flesh becomes drier.
Name suggests it arose on estate of Lord Egremont, Petworth, Sussex. Gardens were famed for fruit, particularly in early 1800s, when number of new varieties were raised, but none bearing this name. Nevertheless, head gardener, Fred Streeter always maintained it was raised at Petworth.
Popularity came early 1900s, after nursmn George Bunyard praised it as one of 'richest late autumn fruits . . . pretty colour for dessert'. Commercially season clashed with Cox, but following demand for 'Russet' apples in 1960s it was planted to small

extent and 'Russet' now seen on sale.
Frt Col ochre russet, slight orng flush; gold ground col. *Size* med. *Shape* flt-rnd. *Basin* brd, qte dp. *Eye* lrg, open; sepals, brd based; qte downy. *Cavity* nrw, shal; lined russet. *Stalk* v shrt, qte thin. *Flesh* crm.
F 7. T^2; uprt; hrdy; res scab; prn bitter pit. C gd. P l-Sept/e-Oct. S Oct-Dec.
PLATE 12

EIGHT SQUARE 5 E D
Ireland; received 1950, when grown County Monaghan.
Pleasant, but lightly flavoured; juicy, quite sweet.
F 15. T^2. P l-Aug. S Aug-Sept.

ELAN 4 M D
Netherlands; raised at IVT, Wageningen. Introd 1984.
Golden Delicious X James Grieve.
Attractive. At best rich, intensely flavoured with sugary acidity. Can be less exciting, merely crisp, sweet.
F 17. T^2; sprd; tip. C lght. P e-Oct. S l-Oct-Dec.

ELBEE *see* Golden Delicious

ELEKTRA 7 L D
Germany; raised 1930s by M. Schmidt, Inst Agric & Hort, Müncheberg-Mark.
Cox's Orange Pippin, Geheimrat Dr. Oldenburg cross.
In late Nov, very good; rich, aromatic, firm, deep cream flesh. Cox-like flavour, but longer season.
Grown for market in eastern Germany.
F 7. T^2; prt tip. P e-Oct. S Nov/Dec-Jan/Feb.

ELISE RATHKE 4 L D
Poland; either raised by Rathke of Praust near Danzig (Gdansk), or arose with A. Dühring of Elbing (Elblag); named, distributed by nursmn Franz Rathke. Recorded 1884.
Crisp, strong, sharp cidery taste in late Nov. Very greasy skin.
F 14. T^2; weeping. C hvy. P m-Oct. S Nov-Jan/Mar.

ELLISON'S ORANGE 7 M D
UK; raised by Rev. C. C. Ellison, Bracebridge Manse and Mr Wipf, HG to Ellison's brother-in-law at Hartsholme Hall, Lincs.
Cox's Orange Pippin X Calville Blanc (prob d'Été). Recorded 1904. Introd 1911 by Pennells of Lincoln. RHS AM 1911; FCC 1917; AGM 1993.
At best glorious, with intense rich, aromatic quality and taste of aniseed, which develops after picking; crisp yet soft, melting juicy flesh. After storing, can become rather medicinal.
Takes its name from well-known rosarian

and keen fruit man, who in 1900 possessed collection of 1,500 fruit trees. Remains popular garden variety, especially in north of England. Planted commercially 1920–30s but aniseed flavour proved not to everyone's taste.
Frt Col rd flush, stripes over grnish yell/yell; greasy skin. *Size* med. *Shape* rnd-con; trace rnded ribs. *Eye* lrg, clsd/hlf-open; sepals qte downy. *Cavity* med dpth wdth; ltl russet. *Stalk* lng, qte thin. *Flesh* pale crm.
F* 13. T^2; hrdy. C gd. P m/l-Sept. S l-Sept-l-Oct.
PLATE 9
   RED ELLISON; even red flush. Found 1948 in H. C. Selby's orchard Walpole St. Peter, Wisbech, Cambs.

ELMORE PIPPIN 5 L D
UK; received 1949.
Freckled with star-shaped russet dots over gold. Develops rich, intense sweet-sharp, flavour; yellow flesh.
F 16. T^2. P m-Oct. S Dec-Feb/Mar.

ELSTAR 7 M D
Netherlands; raised 1955 by T. Visser, IVT, Wageningen.
Ingrid Marie X Golden Delicious. Introd 1972. RHS AGM 1993.
Intensely flavoured, very honeyed, sweet, crisp, juicy flesh. (Vit C 8–13mg/100gm).
Planted throughout Europe: Germany, Holland, Belgium, France, Italy; also grown Boltzano, Italy, Washington, US, Canada, Chile; and small extent England.
F15. T^2. C gd. P e-Oct. S l-Oct-Dec.
Red sports include:
   BEL-EL
   DALIEST
   ELNICA
   ELSHOF
   RED ELSTAR

ELTON BEAUTY 4 M D
UK; raised by N. W. Barritt, Ince Orchards, Chester.
James Grieve X Worcester Pearmain. Introd by 1952.
Bright red flush. Fully ripe, has Worcester sweetness with slight strawberry flavour and juicy, soft white flesh of James Grieve.
Failed commercially because not ripening early enough for Aug markets and too sharp if picked early.
F 7. T^1. C hvy. P e-Sept. S Sept-Nov/Dec.

EMILIA 4 M D
Canada; raised 1915 at Central Exp Farm Ottawa by T. W. Macoun.
Northern Spy X.
Scented, quite rich; juicy, crisp, fine texture. Can be metallic. Often poor colour in England.
F 15. T^3. C hvy. P m-Sept. S Sept-Oct/Nov.

EMNETH EARLY syn Early Victoria
1/5 E C
UK; raised by W. Lynn, Emneth, Wisbech, Cambs.
Lord Grosvenor X Keswick Codlin. Introd by Messers Cross of Wisbech; recorded 1899. RHS FCC 1899; AGM 1993.
Codlin type, baking like a soufflé, rising up in juicy fluff. Well-flavoured, hardly needing any sugar. Good for summer dishes like 'Apple Snow'.
In early 1900s became early market apple, displacing both its parents, but now rarely seen on sale. Can be small unless thinned and strongly biennial, which led to its commercial demise and fall in popularity as garden apple, but remains widely grown.
Frt Col pale grn/pale yell; lenticels as pale grey spots. *Size* med/sml. *Shape* rnd-con; ribbed; often hair line. *Basin* sml, shal; puckered; beaded. *Eye* sml, clsd; sepals, lng, pinched, downy. *Cavity* brd, shal. *Stalk* shrt, thck. *Flesh* wht.
F* 10. T^2; some res scab, mildew. C hvy, bien; frt sml, unless thinned. P Aug. S Aug-Sept/Oct.
PLATE 8

EMPEROR ALEXANDER syn Alexander

EMPIRE 6 L D
USA; raised 1945 by R. D. Way, NYSAES, Geneva.
McIntosh X Delicious. Introd 1966.
Striking, bright red flush with waxy bloom. Crisp clean taste of fruit; sweet; with hint of McIntosh flavour, quite scented; tough skin. Grown Canada, Michigan, New York; exported to UK. Resists bruising, stores better than McIntosh.
F 9. T^2. C hvy. P e/m-Oct. S Nov-Dec/Jan.

ENCORE 1/3 L C
UK; raised by Charles Ross, HG, Welford Park, Newbury, Berks.
Believed Warner's King X Northern Greening. Recorded 1906. Introd nursmn J. Cheal, Crawley, Sussex. RHS AM 1906; FCC 1908.
Large. Used early, strong acidity. Later makes creamy baked apple; quite rich flavour; hardly needs sugar.
Formerly, popular garden variety.
F* lrg, dp pink; 14. T^2; res scab. C gd. P e/m-Oct. S Dec-Apr.

ENDSLEIGH BEAUTY 7 M DC
UK; arose with H. Whiteley, Endsleigh Place, Torquay, Devon. Received 1906.
Very like Blenheim Orange in appearance and flavour; brisk, fruity.
F 11. T^3. P l- Sept. S Oct-Dec.

ENTZ-ROSMARIN syn Husvéti Rozmarin
5 L D
Hungary; arose with pomologist Dr F. Entz. In 1860 he sent it to pomologist Bereckzi,

who named it. Tall, oblong. Fruity, quite sweet, juicy, crisp flesh.
Grown Hungary 1960s.
F 12. T³. P m-Oct. S Nov-Jan.

EPICURE 7 E D
UK; raised 1909 by Laxton's Brothers Nursery, Bedford.
Wealthy X Cox's Orange Pippin. RHS AM 1931; Bunyard Cup 1929, 1932; AGM 1993.
Delicately aromatic, sweet, juicy; it combines some of Cox flavour with brighter colour of Wealthy.
Popular garden variety; also grown for Farm Shop trade.
Frt Col Dark orng rd flush, thck rd stripes over grnsh yell. Size med/sml. Shape rnd-con or con; regular. Basin brd, shal; sltly ribbed or puckered. Eye clsd; sepals brd base. Cavity brd dp. Stalk lng, qte thck. Flesh crm.
F* 10. T². C hvy; frt sml, unless thinned; prn bitter pit. P l-Aug. S Aug-l-Sept.
   EPICUREAN; more highly coloured. Arose with G. A. Maclean, Kingston Bagpuize, Berks. Received 1953.

ERICH NEUMANNS ROTER JAMES GRIEVE see James Grieve

ERI ZACARRA 1/5 L C
France; old variety of French Basque country.
Strong acidity, some sweetness and flavour.
Cooked, keeps shape; quite rich taste.
Grown Basque country.
F 16. T³. P m/l-Oct. S Dec-Feb/Apr.

ERLIJON 6 E D
USA; arose with R. Banta, Carroll township, Arkansas. Introd 1968 by M. J. Lucas, Green Forest Nursery, Arkansas.
Ripening earlier than Jonathan; dark maroon flush. Soft white flesh, winey flavour; tough skin.
F 11. T²; uprt. P e-Sept. S Sept-Oct.

ERNST BOSCH 5 E D
Germany; raised by Diedrich Uhlhorn jnr, engineer and fruit breeder, in Grevenbroich, Rhineland.
Ananas Reinette X Eva (syn Mank's Codlin). Recorded 1908.
Named after musician from Dusseldorf, who was great friend of Uhlhorn.
Small, golden. Savoury, refreshing, crisp flesh. Reported high vit C content.
F 6. T¹; sp. C hvy, bien. P e-Sept. S Sept-Oct.

EROS 4 E D
UK; raised 1947 by nursmn W. P. Seabrook, Boreham, Essex.
Worcester Pearmain X Charles Ross.
Bright red flushed; large. Quite sharp, fruity, firm, yellow flesh.
F 11. T². P e-Sept. S Sept-Oct.

ERWIN BAUR 4 L D
Germany; raised 1928 by M. Schmidt, Inst Agric & Hort, Müncheberg-Mark.
Geheimrat Doktor Oldenburg cross. Named 1955 after founder of Institute.
Flushed, striped in red-brown over deep yellow. Crisp, juicy, good brisk taste with a distinctive pineapple acidity and some aromatic quality.
Grown eastern Germany.
F 6. T². P e-Oct. S Oct-Jan.
   ROBA; more highly coloured.

ESOPUS SPITZENBURG 6 L D
USA; arose with 'low Dutch' at Esopus, Ulster County, New York. Known before 1790. Sent 1824 by Michael Floy of New York Hort Soc to London Hort Soc.
Most beautiful, bright red. Rich, fruity flavour; lively acidity; lots of character; hard, very crisp, almost yellow flesh. Still fruity and good in Mar.
Widely planted North America both in home and commercial orchards in C19th. Dual purpose: 'Spitzenbergs' were said to make finest pies and perhaps inspired creation of Waldorf Salad. Displaced by Jonathan, then Red Delicious, any remaining orchards supply canning industry for pie fillings. Still esteemed by American fruit enthusiasts.
Grown in England often ripens poorly, but in good year full of flavour.
Frt Col bright rd flush; yell ground colour; numerous russet dots. Size med. Shape con to rnd-con; sltly ribbed, sltly crowned. Basin med wdth, dpth; ribbed, puckered. Eye clsd; sepals shrt. Cavity nrw, dp; russet lined. Stalk med, qte thin. Flesh dp crm.
F 12. T²; sp. C hvy. P e/m-Oct. S Nov-Mar.

ESSCHING 1 L C
Belgium; local to West Flanders. Received 1948.
Cooks to cream purée; brisk, well-flavoured.
F 7. T³. C bien. P l-Oct. S Dec-Feb/Mar.

ESTIVA 6 M D
France; raised 1957 at INRA, Angers.
Usta Gorria (Basque variety) X Red McIntosh. Received 1978.
Dark McIntosh red flush. Scented, snow white, juicy flesh; can be woody, unattractive in England.
F 13. T¹; sprd. P l-Sept. S Oct-Nov, cold store-e-Feb.

ETLINS REINETTE syn Reinette d'Etlin 7 L D
Germany; found on Landenberg estate of pomologist Dr. Etlin. New 1866.
Rich, aromatic, sweet, well balanced. Reminiscent Ribston Pippin with similar robust flavour.
F 6. T³; sprd. P e-Oct. S Nov-Jan/Mar.

EUSTIS (?) syn Ben Apple 7 L D
USA; acc is old American variety, but may not be variety believed arose late 1600s in Massachusetts, either with Benjamin Smith at Wakefield, or at South Ready, Essex County.
Handsome. Honeyed, quite rich, firm, deep cream flesh.
F 7. T². C v prn bitter pit. P l-Sept. S Oct/Nov-claimed Mar.

EVAGIL 5 E CD
Belgium; found at Tielt by Dr van der Espt. Introd 1863 by nursmn L. Van Houtte & Sons.
Juicy, tender pale cream flesh; attractive, brisk perfumed taste. Cooked, keeps shape; pretty gold colour; sweet delicate flavour.
RGF variety, recommended for cultivation by Agric Res Centre, Gembloux because of disease resistance; recently revived as market apple.
F 14. T²; sp; res scab, prt res mildew. C gd. P l-Aug. S Sept-Oct/Nov.

EXCELSIOR 3 M C
UK; raised by nursmn W. P. Seabrook & Sons, Boreham, Essex. Introd 1921.
Bright red flush. Cooks to juicy, brisk, deep cream purée; makes quite strongly flavoured sauce.
Grown for market Essex, Cambs 1930s, but displaced by Bramley.
F 11. T³. P l-Aug. S Sept-Oct.

EXETER CROSS 4 E D
UK; raised 1924 by G. T. Spinks at LARS, Bristol.
Worcester Pearmain X Beauty of Bath.
Ripe early Aug, but can be quite astringent.
F 11. T². C gd; frt cracks; ripens over v long period. P eat Aug.

EXQUISITE 7 E D
UK; raised 1902 by Laxton Brothers Nursery, Bedford.
Cellini X Cox's Orange Pippin. RHS AM 1926.
Flavoured with aniseed, like Cellini or soft Ellison's Orange; quite sweet; can be over-sharp.
F 8. T³. C lght; bien. P l-Aug. S Sept- Oct.

EXTRAORDINAIRE 7 L D
Received from France in 1947.
Quite rich taste by Dec. Sweet, slightly nutty, firm flesh but can be tough, poorly flavoured.
F 9. T³. C hvy. P m-Oct. S Dec-Mar.

EYNSHAM CHALLENGER 5 L C
UK; raised 1935 by F. W. Wastie, Eynsham, Oxford.
Blenheim Orange X Lord Derby.
Large. Cooks to cream purée, lightly flavoured, quite brisk.
F 12. T². C hvy. P e-Oct. S Oct-Dec.

EYNSHAM DUMPLING 3 L C
UK; raised by F. W. Wastie, Eynsham, Oxford.
Blenheim Orange X Sandringham. Received 1960.
Large. Cooks to lemon purée, but little flavour.
F* 13. T². P e-Oct. S Oct-Dec.

FAIRIE QUEEN 7 M D
UK; arose with R. Staward, Ware Park Gardens, Herts. Recorded 1937.
Sweet, rich, aromatic. At best, recalls Cox's Orange Pippin but can be less exciting; coarsely textured flesh.
F 11. T¹. C frt cracks. P e/m-Sept. S Sept-Oct.

FAIRY 6 M D
Received 1967 from former Yugoslavia via Scotland. Not variety of Hogg.
Bright red flushed, striped. Sweet, soft, juicy flesh, plenty acidity, hint strawberry flavour.
F 12. T². P e/m-Sept. S Oct.

FALL HARVEY 5 M D
USA; arose Essex County, Mass. Recorded 1838.
Sweet, light taste, soft, juicy flesh, but 'high flavour' grown in USA.
Recommended for Northern New England by early writers.
F 12. T³; hrdy. P m-Sept. S Sept-Oct.

FALL PIPPIN 5 M DC
USA; recorded 1806. Poss seedling Holland Pippin or White Spanish Reinette. Introd UK c1820s by William Cobbett.
Quite briskly flavoured; some richness. Cooks to purée, but rather bland.
Popular market and dual purpose variety for home orchard in most apple growing areas of C19th America.
F 12. T². P m-Sept. S Sept-Oct.

FALL RUSSET 8 M D
USA; original tree arose c1875. Rediscovered 1950s in old orchard in Franklin, Michigan by nursmn R. A. Nitschke, Birmingham, Michigan.
Attractive golden russet. Intense flavour, rather like 'early' Ribston Pippin, rich and good. In US recommended for drying.
F 5. T²; sprd. C hvy; frt sml. P e/m-Sept. S Sept-Oct.

FALSTAFF 4 L D
UK; raised by Dr F. Alston, EMRS, Kent. James Grieve X Golden Delicious. Introd 1986. RHS AGM 1993.
Pretty red flush, stripes. Fruity, well balanced; crisp, juicy.
Planted commercially Kent.
Frt *Col* orng/rd flush, shrt rd stripes over grnsh yell/gold; becomes greasy. *Size* med. *Shape* rnd-con. *Basin* qt nar, med dp; sltly ribbed. *Eye* clsd/hlf open; sepals, lng. *Cavity* med wdth; med dpth; slght russet. *Stalk* long, med thck. *Flesh* crm.
F 11; frost res. T²; some res scab. C hvy. P e-Oct. S Oct-Dec.
    RED FALSTAFF; even red.

FAMEUSE syn Snow Apple 6 L D
Canada (prob). Either originated as scions, or more prob as seed, brought from France. Trees recorded as planted c1730 on eastern shores Lake Champlain at Chimney Point, from here scions widely distributed to early French settlements.
Deep red flush with pink undertones and snow white flesh. Sweet, crisp, juicy yet soft and melting; light strawberry flavour.
Formerly widely grown, especially around Montreal. Displaced by McIntosh, but still grown.
F 7. T²; v hrdy; res mildew. C hvy; frt often sml. P l-Sept/e-Oct. S Oct-Dec.

FANTAZJA (Phantasy) 6 L D
Poland; raised 1944 by A. Rejman, Warsaw Univ.
McIntosh X Linda. Introd 1960.
Brighter red than McIntosh. Similar sweet, juicy, soft flesh; light strawberry flavour; lovely of its kind.
Grown for market in Poland.
F 7. T²; hrdy; sprd. C hvy. P e-Oct. S Nov-Dec.

FĂRĂ NUME 5 L D
Romania; known north Moldavia, where in 1960s recommended for higher ground.
At best fruity, sprightly; rather tough flesh.
F 14. T²; hrdy; prn canker; claimed gd res pest, disease. C hvy. P m/l-Oct. S Nov-Feb/Apr.

FAROS syn Faro 6 L D
France; acc is variety now known Seine-et-Marne; found particularly around Lagny, Brie and Les Yvelines; prob not Gros Faro, syn Faro of C19th.
Deep red flushed. Crisp, sweet, quite pleasant; can have underlying woody, empty taste in England.
Popular Paris markets in 1950s; also used for juice, cider.
F 19; trip. T². P l-Oct. S Dec-Mar/Apr.

FAVERSHAM CREEK 3 M C
UK; found 1970s by H. Ermen, growing by Faversham Creek, Kent.
Cooks to gold purée, brisk, savoury.
F 13. T²; part tip; res salt water. P m-Sept. S Oct.

FEARN'S PIPPIN 6 L D
UK; raised before 1780 in Mr Bagley's garden, Fulham, London.
Brilliant scarlet. Crisp, juicy, sweet-sharp, almost lemon quality; mellows to hint raspberry flavour. Prized for dining table; among Victorian top dozen varieties.
Widely grown in gardens, by London market gardeners and in Kent, but popularity waned by 1890s.
F 9. T²; uprt. C gd. P e-Oct. S Oct-Jan/Feb.

FEKETE TÁNYÉRALMA 6 L D
Transylvania (Romania); desc 1909 (Molnar).
Dark maroon; large; flattened; its name means black plate apple. Quite rich, sweet, nutty, good flavour.
Grown particularly Besztercze (Bistrita) in early 1900s.
F 9. T². C hvy; prn bitter pit. P e-Oct. S Dec-Jan/Mar.

FELTHAM BEAUTY 6 E D
UK; raised at Veitch Nursery, Langley, Bucks.
Cox's Orange Pippin, Gladstone cross. Described 1908. RHS AM 1908.
Dark red flush of Gladstone. Very sweet, soft flesh; hint aniseed flavour.
F 13. T². P eat e-Aug.

FENOUILLET DE RIBOURS syn de la Rouairie 8 L D
France; found in garden of la Rouairie, nr Angers, Maine-et-Loire. Fruited 1840. Distributed by M. de la Perraudière.
By Jan, scented, sweet, plenty acidity; rather tough, cream flesh. In England, never seems to develop the 'light perfume of aniseed' which characterised Fenouillet apples.
F 10. T³. P m/l-Oct. S Jan-Mar.

FENOUILLET GRIS 8 L D
France; acc presumed to be old variety, which prob arose Anjou. Name Fenouillet recorded 1628 by Le Lectier. Fenouillet Gris recorded late C17th as grown Anjou and sent to Paris markets. Leroy considered it to be Espice d'Hiver listed 1608 by Olivier de Serre. Syns numerous; incl Caraway Russet in UK.
Striking appearance, flushed in very dark red and laced with grey russet; small. Characteristic flavour, scented, very sweet, little acidity, firm, rather dry flesh, but in England not 'endowed with perfume of aniseed and musk, as delicate as well as pronounced' as Leroy described.
Long esteemed in France and among 'seven principal' apples grown by La Quintinye at Versailles.
In England recommended as decorative tree in early C18th to plant alongside paths, grow in pots to ornament formal gardens, but never highly valued as dessert fruit.
F 10. T². C gd. P m-Oct. S Nov-Mar.

FENOUILLET ROUGE (?) 8 L D
France prob; acc may be variety desc by Leroy (1873) which he believed to be variety recorded 1667 under syn Bardin.
Red flush under russet. Curious scented quality, but not obviously fennel; sweet, firm, white, green tinged flesh.
F 12. T². P m-Oct. S Nov-Mar.

FESTIVAL *see* Delorgue

FEUILLEMORTE 6 L D
France; desc 1948 (*Verger Français*) when found Meaux, Seine-et-Marne.
Very late coming into blossom, hence its name. Savoury, brisk, sometimes hint of raspberry flavour; crisp, quite juicy.
Grown Seine-et-Marne; used also for juice.
F 40. T¹. P l-Oct. S Dec-Mar.

FIESSERS ERSTLING 3 M C
Germany; raised before 1904 by gardener Fiesser of Baden Baden, from Bismark seed.
Often very large. Cooks to strong, sharp fruity purée; needs sugar.
F 5. T². P m-Sept. S Oct-Nov.

FIESTA 7 M D
UK; raised 1972 by Dr F. Alston, EMRS, Kent.
Cox's Orange Pippin X Idared. RHS AM 1987; AGM 1993. Renamed Red Pippin 1996 in UK.
Cox-like flavour – rich, aromatic, sweet, plenty balancing acidity, crisp, juicy; rather tough skin. With keeping can betray some of emptiness of Idared, but many cultural improvements on Cox.
Grown commercially Kent, Holland, Italy, New Zealand; popular UK garden apple.
Frt *Col* orng/rd flush, shrt rd stripes over grnsh yell/gold. *Size* med. *Shape* rnd-con. *Basin* shal; med wdth; sltly ribbed. *Eye* sml, clsd; sepals lng. *Cavity* qt shal; med wdth; slght russet lined. *Stalk* lng, qt thck. *Flesh* crm.
F 14 frost res. T², less pm disease than Cox. C hvy. P l-Sept. S Oct-Dec/Jan; comm-May.

FILIPPA 5 M D
Denmark; sown c1880 in pot by Filippa Johannsen in Hundstrup, South Fyn Island.
Poss Gravenstein seedling. Danish Hort Soc FCC, 1893.
Very scented, soft, juicy, white flesh, but in England often tastes metallic.
Formerly very popular in Denmark; also grown Sweden, but now in decline.
F 14. T³; tip. P m-Sept. S Oct-Dec.

FIL JAUNE 5 L D
France; desc 1948 (*Verger Français*); when known Morbihan, Brittany.
Sweet, slightly honeyed, quite juicy, pale yellow flesh.
Grown Morbihan; used also for juice.
F 15. T²; res mildew. P l-Oct. S Dec-Mar.

FILLINGHAM PIPPIN 4 M D
UK; believed raised C19th by Mr Fillingham, a carter between Swanland and Hull, Yorks. He received seed or cuttings from his brother in South America.
Sharp, quite fruity, juicy, firm white flesh. Popular in East Riding of Yorkshire in 1940s. Can be propagated from cuttings; suggested it was variety Darwin found growing on Island of Chiloé in 1835, where orchards made by planting branches in ground.
F* pink; 10. T². P e-Oct. S Oct-Dec.

FINASSO 6/4 L D
France; received 1949 from Lozère.
Small, bright red flushed, striped. Fruity in Jan; sweet, sprightly acidity; crisp.
F 23. T². P l-Oct. S Jan-Mar.

FINKENWERDER PRINZ 7 L D
Germany; arose c1860 on island of Finkenwerder in Elbe nr Hamburg.
Develops distinctive flavour in home territory. Formerly widely grown around Hamburg in Alte Land. Still sought by devotées and recently revived; recommended for juice. Also planted along Baltic coast towards Poland; Netherlands.
F 12. T²; res scab. P l-Oct. S Dec-Mar.

FIRESIDE 4 L D
USA; raised by C. Haralson, Minnesota Agric Exp St, Excelsior.
McIntosh X Longfield; introd 1943.
Sweet, quite juicy; rather tough flesh. Bred for severe North American winters.
F 11. T²; v hardy. P e-Oct. S Dec-Feb/Apr.

FIRMGOLD 2/5 L D
USA; found among Starkspur Golden Delicious and Starkcrimson Red Delicious trees in Zillah, Washington. Introd 1977.
Similar Golden Delicious. Crisp, intensely sweet, honeyed, yellow flesh.
F 11; frost res. T²; res mildew. C hvy. P l-Oct. S Nov-Jan; cold store-Mar.

FISHER FORTUNE *see* Fortune

FLAME 7 E D
UK; raised 1925 by nursmn W. P. Seabrook & Sons, Boreham, Essex.
Worcester Pearmain X Rival.
Pretty pink and red flush. Lightly aromatic, hint of aniseed; sweet, plenty acidity; crisp, juicy. Can be only fair flavour.
F 10. T¹. C lght. P eat l-Aug-Sept.

FLAMENCO syn Obelisk

FLETCHER'S PROLIFIC 7 L D
UK; received 1945 from EMRS, Kent and variety sold 1940 by Barnham Nursery.
Brisk, refreshing, strong acidity, but plenty sweetness in Dec; firm, cream flesh.
F 19. T³; uprt. C hvy. P l-Oct. S Dec-Feb/Mar.

FLEURITARD 7 L D
France; desc 1948 (*Verger Français*), when grown Meuse (Lorraine Vosges).
Bright red flush. Brisk, fruity; firm, crisp flesh.
F 29. T³. P l-Oct. S Dec-Mar.

FLEURITARD ROUGE 6 L D
France; received 1947 from Seine et Marne.
Sharp, juicy, green tinged flesh.
F 37. T¹. C hvy; frt sml. P l-Oct. S Dec-Mar/Apr.

FLORENCE BENNETT 1/5 M CD
UK; raised by Mrs F. M. Bennett, Crosby Green, West Derby, Liverpool.
Poss pip Bramley's Seedling; arose from cores thrown onto garden rubbish heap. Fruited 1960. Received 1971.
Early in season, cooks to bright lemon quite rich purée; needs little sugar. When golden and fully ripe makes savoury, soft eater.
F 11. T². C hvy. P e-/m-Sept. S Sept-Nov.

FLORIANER ROSENAPFEL syn Rose de Saint-Florian 7 L D
Germany; described 1776. Given this name 1859. Syns many.
Prettily red striped. Quite rich, honeyed taste; soft, deep cream flesh.
Widely distributed Europe in C19th.
F 10. T². P m-Oct. S Nov-Dec/Jan.

FLORINA 6 L D
France; raised INRA, Angers.
Complex parentage involving scab resistant seedling. Trade name Querina. Introd 1977.
Handsome, red flushed. Sweet, quite crisp, pale yellow flesh; modest richness.
Planted commercially in France.
F est 13. T ²; tip; res scab; mod res mildew, fire blight; tolerant rosy apple aphid. C hvy.
P e-Oct. S Oct-Dec.

FLOWER OF KENT *see* Isaac Newton

FOLKESTONE 6 E D
UK; received 1964 from Chandler & Dunn, Ash, Kent.
Believed Tydeman's Early Worcester X. Bright red Worcester colour; similar firm, chewy flesh; slight strawberry flavour.
F 15. T¹. P l-Aug. S Aug-Sept.

FON'S SPRING 4 L D
UK; arose 1948 with E. J. Ingleby, Forest and Orchard Nurseries (FON), Falfield, Gloucs.
John Standish X Cox's Orange Pippin; raised from same cross as Eden.
Bright red flush. Pale cream tinged green flesh, crisp, brisk, sharp, like John Standish, but plenty of fruit, some sugar. Firmer fleshed than Eden, later season.
F 19. T². P e-Oct. S Nov-Jan/Mar.

FORESTER 1/5 L C
UK; origin unknown; exhib 1883 Congress from Royal Gardens, Frogmore; desc Hogg.
Sweet, nutty, fruity, plain; recalls Blenheim Orange. Cooks to bright yellow purée; rich, quite sweet; not very juicy.
Grown West Midlands C19th.
F 16. T^2. P m-Oct. S Nov-Jan/Apr.

FORGE syn Sussex Forge 4 M DC
UK; believed arose either at Forge Farm or nr old iron forges around East Grinstead, Sussex. Desc 1851 (Hogg), but then well known.
Early in season, sharp, tannic, but mellowing to be almost rich, lightly aromatic. Cooks to soft, lemon coloured purée, quite brisk, fruity; makes good sauce. Formerly also used for cider.
'Cottagers' apple', widely grown north Sussex and Surrey in C19th; still found around East Grinstead.
Frt Col pale grn/yell, light orng flush; ltl rd. *Size* med. *Shape* con to rnd-con; crowned. *Basin* brd, qte dp; ribbed. *Eye* clsd; sepals, shrt. *Cavity* nrw, shal; ltl russet. *Stalk* shrt, thck. *Flesh* wht.
F 15. T^2. C gd. P l-Sept. S Sept-Dec.

FORPEAR 4 E D
Canada; received 1928 from Ottawa Central Exp Farm.
Sweet but low acidity; firm, rather chewy flesh.
F 11. T^3. P l-Aug. S Aug-Oct.

FORTOSH 6 L D
Canada; raised Ottawa Central Exp Farm. McIntosh X Forest. Introd 1926.
Sweet, strongly scented, like rose water; quite fruity; slightly crumbling pale yellow flesh. Can be less interesting.
F 8. T^3. P m/l-Oct. S Dec-Feb/Mar.

FORTUNE 7 E D
UK; raised 1904 by Laxton Bros Nursery, Bedford. Cox's Orange Pippin X Wealthy. Introd 1931. RHS FCC 1948; AGM 1993.
At best, sweet, rich, lightly aromatic and juicy. Less well ripened, can have little intensity, tasting woody, empty. Best left on tree as long as wasps allow, to colour well and develop good flavour.
Remains valued garden fruit; formerly grown commercially, especially in Worcs, but now confined to Farm Shops.
Frt Col rd flush, shrt rd stripes over grnsh yell/pale yell; some russet patches. *Size* med. *Shape* shrt-rnd-con. *Basin* qte nar, dp; ribbed, sometimes beaded. *Eye* clsd; sepals, lng. *Cavity* med wdth, dpth; russet lined. *Stalk* med lngth, qte thck. *Flesh* crm.
F 9. T^2; hrdy. C gd; bien; frt bruises, often pr col. P e-Sept. S Sept-Oct.
  FISHER FORTUNE; more highly coloured. Discv c1960 by A. M. Fisher Holmlea Farms Ltd., Elverland,

Faversham, Kent.
  RED FORTUNE; even red flush.

FORTY SHILLING 7 M D
UK; received 1950 from tree planted c1800 at Thursby, nr Carlisle, Cumberland.
Savoury, sharp, almost rich quality; very soft, juicy flesh.
F 12. T^3. C hvy. P e-Sept. S Sept-Oct.

FOSTER'S SEEDLING 4 L C
UK; introd c1893 by nursmn G. Bunyard, Maidstone, Kent.
Deep red flushed. Very acidic, juicy, soft flesh. Cooks to sharp purée.
F 17. T^1; sprd. C hvy; hangs late. P m-Oct. S Nov- Mar.

FOULKES' FOREMOST 4 M D
UK; raised 1938 by F. Foulkes at Headington, Oxford from pip of 'foreign' apple.
Attractive, scarlet striped over yellow. Rather bland sweet taste, little acidity; firm, chewy flesh.
F 7. T^2. C lght. P e-Sept. S Sept-Oct/Nov.

FRAISE DE BUHLER 4 L D
Received 1947 from Switzerland; resembles Queen.
Carmine striped, flushed. Quite sweet, hint strawberry flavour; crisp, juicy, white flesh.
F 11; trip. T^3. P m-Oct. S Nov-Jan/Mar.

FRAMBOISE 6 L D
Netherlands prob; first desc 1771 by Dutch pomologist Knoop.
Flushed almost completely in deep maroon. Soft melting flesh, sweet juicy with distinct flavour of possibly raspberries as name suggests.
F mid. T^2. P l-Sept. S Oct -Dec.

FRANC-BON-POMMIER (MOSELLE) 7 L D
France; received from Moselle 1950; acc is not Belle Fleur de France syn Franc-Bon-Pommier.
Crisp, fresh in Feb; sweet, fruity.
F 20. T^2. P l-Oct. S Dec-Apr.

FRANCE DELIQUET 4 L D
Received 1958 from Angers, France.
Crisp, sharp in Feb; little sugar or flavour develops in England.
F 20. T^2; sp. C hvy, bien. P l-Oct. S Jan-Mar/Apr.

FRANCIS 7 L D
UK; raised by F. W. Thorrington, retired London customs officer, Hornchurch, Essex; named after his son.
Cox's Orange Pippin X. Received 1925.
Sweet, lightly aromatic, like Laxton's Superb. Firm, rather dry, cream flesh.
F 12. T^2. C hvy; frt sml/unless thinned. P m-Oct. S Nov/Dec-Jan.

FRANÇOIS JOSEPH syn Kaiser Franz Joseph

FRANC ROSEAU 4 L D
Switzerland or France; known before 1850, and as Châtaigne du Léman in Switzerland.
Good fresh taste of fruit in New Year; firm, sharp flesh.
Principal variety of Valais (Switzerland) and Loire Valley in C19th; still grown Switzerland; Isère, Savoir, Haute Savoie (Rhône Alpes).
F 19. T^2. P l-Oct. S Jan-Mar/May.

FRANKETTU (?) 7/8 L D
France; acc may be Frankettu of Yonne (Burgundy); desc 1948 (*Verger Français*). Acc not Francatu, desc Leroy (1873) and so prob not ancient 'Franceturum' cited 1536 by Ruel.
Ripens to red flush over greenish yellow with much russet. Brisk taste of fruit, some sweetness; quite soft, cream green tinged flesh.
F 13. T^2. C gd. P m-Oct. S Dec-Feb/Mar.

FRAU MARGARETE VON STOSCH 5 E D
Germany; raised 1895 at Fruit, Vine Hort St, Geisenheim, Rhine; first rec 1914/15.
Minister von Hammerstein X White Winter Calville.
Bright gold, Calville shape. Sharp, juicy; intense taste of fruit; almost yellow flesh.
F 7. T^2. P l-Aug. S Sept-Oct.

FRED WEBB 4 M D
UK; raised c1935 by F. W. Webb, while working at nursery of G. Bunyard & Co, Maidstone, Kent.
Very pretty, scarlet flushed over cream. Sweet, slight strawberry flavour; crisp, juicy.
F 18. T^2. P m-Sept. S l-Sept-Oct.

FREEDOM 6 L D
US; raised 1958 by R. C. Lamb, NYSAES, Geneva, NY.
Complex parentage involving Macoun, Antonovka, Golden Delicious, Rome Beauty, *Malus floribunda* (carrying Vf, scab resistant gene). Introd 1983.
Named for its freedom from disease.
McIntosh type; sweet and strawberry flavoured, juicy melting white flesh.
Popular amateurs' variety in US; used also for cooking, cider.
F 7. T^3; sprd; res scab, mildew, cedar apple rust, fireblight. C hvy. P l-Sept. S Oct-Dec; cold store-Jan.

FREIHERR VON BERLEPSCH syns Baron de Berlepsch, Berlepsch 6 L D
Germany; raised c1880 by Diedrich Uhlhorn jnr, engineer and fruit breeder, at Grevenbroich, Rhineland.
Ananas Reinettte X Ribston Pippin. Named in honour of Hans Hermann Freiherr von

Berlepsch, president of Rhineland 1884–90, and then Imperial Germany's Minister for Trade and Industry.
Intense pineapple flavour; sweet, plenty of acidity; crisp, juicy. Also used for cooking; vit C 25–35mg/100gm.
Remains valued; grown commercially Rhine, Belgium.
F 11; trip. **T**³. **C** gd, bien. **P** e-Oct. **S** Nov-Dec/Jan.

FRÉMY 8 L D
France; raised c1830–40 by farmer Frémy of Cherré, Maine-et-Loire (Western Loire). Rich, plenty sugar, acidity; finely textured, firm, deep cream flesh.
F 11. **T**³. **P** m/l-Oct. **S** Jan-Apr.

FRENCH CRAB 1 L C
France (prob); introd England late 1700s. Syns many; Hogg's Winter Greening; Ironsides, Two Years Apple.
Dark green, freckled with fine russet dots. Hard, sharp, little juice; white, tinged green flesh. Cooks to sharp purée in Dec but by New Year acidity fallen, rather bland.
'Old Ironsides' was famed for keeping qualities. Widely grown on continent and UK in C19th; also Australia.
F 15. **T**². **C** gd. **P** m-Oct. **S** Dec-Jun.

FREYBERG 5 L D
New Zealand; raised 1934 by J. H. Kidd at Greytown, Wairarapa Valley.
Golden Delicious X Cox's Orange Pippin. On his death, Kidd's seedlings passed to DSIR, Havelock North; select by Dr. W. McKenzie for amateurs. Introd 1958.
Named after military hero Lord Freyberg, Governor General New Zealand, 1946–52. Amazingly strongly scented flavour; very sweet, honeyed. Can be low in juice, acidity, little taste fruit.
F 11. **T**¹. **P** m-Oct. **S** Nov-Dec/Jan.

FRIANDISE 8/6 L D
Netherlands; desc 1760 by botanist H. Knoop.
Striking – claret flush, metallic lustre, much russeting, long, oval shape. Quite intense flavour, sweet, taste of strawberries, but can be less interesting. Highly decorative.
F 11. **T**². **P** m-Oct. **S** Dec-Mar.

FRIEDRICH DER GROSSE 5 L D
Received 1947 from France.
Quite rich, sweet, plenty of fruit, acidity; tastes very like Golden Reinette.
F 14. **T**². **P** e/m-Oct. **S** Nov-Jan.

FROGMORE PROLIFIC 3 M C
UK; raised before 1865 by Royal, HG, William Ingram at Frogmore, Windsor.
Large; dusted with red. Cooking to white froth or creamy purée; sweet, pleasantly flavoured. Baked becomes juicy soufflé, but

best early as flavour soon fades.
Grown gardens and by London market gardeners in C19th, but by 1920s 'little known'.
F 13. **T**¹. **P** e-Sept. **S** Sept-Nov.

FROMME SCAB RESISTANT 2 L D
Received 1953 from Virginia Ag St, USA. Seedling showing scab resistance.
F 21. **T**³. **C** bien. **P** e-Oct. **S** Dec-Feb.

FUJI 4 L D
Japan; raised 1939, by H. Niitsi, Hort Res St, Morioka.
Ralls Janet X Delicious. Named 1962.
Attractive, orange red flushed. Honeyed sweetness of Delicious; crisp, firm, quite juicy flesh; tough skin. Often fails ripen fully in England.
Grown Japan and China, where main variety and grown to great size. Half of China's 2.6 million hectares are planted with Fuji. Grown also all warmer apple regions of world: southern hemisphere, US, Italy, France.
F 12. **T**². **C** hvy. **P** l-Oct. **S** Dec/Jan-Mar.
  NEW FUJI; improved form.

FUKUNISHIKI 6 L D
Japan; raised 1933 at Aomori Apple Exp St. Ralls Janet X Delicious. Named 1949.
Sweet, quite honeyed like Delicious, but crisp, juicy flesh of Ralls Janet.
F 12; trip. **T**³. **C** pr col Eng. **P** m/l-Oct. **S** Jan-Mar.

FUKUTAMI 6 L D
Japan; raised 1933 at Aomori Exp St. Jonathan X Ralls Janet. Named 1948.
Red Jonathan colour. Crisp, sharp, but little flavour in England.
F 11. **T**². **P** l-Oct. **S** Jan-Apr.

GABIOLA 1 L C
Italy; received 1958 from Univ Turin. Sharp, firm. Cooked, keeps shape, quite sharp, some richness.
F 22. **T**². **P** l-Oct. **S** Jan-Mar.

GAILLARDE 2 L DC
France; known since late C19th; formerly widely distributed orchards northern France; desc Stievenard (1996).
Cooks to sharp, rather dry, yellow purée.
F 13. **T**³. **C** v hvy. **P** l-Oct. **S** Dec-Mar.

GALA 6 L D
New Zealand; raised c1934 by J. Hutton Kidd, Greytown, Wairarapa Valley.
Kidd's Orange Red X Golden Delicious. Select, named 1965 by Dr. W. M. McKenzie, DSIR, Havelock North.
Rich, honeyed, juicy with some of perfumed quality of Kidd's Orange Red.

Grown New Zealand and all major apple regions: France, Italy, Germany, Spain, Australia, South Africa, South America, Canada; also grown UK. Usually more highly coloured sports, such as Royal Gala are planted. Royal Gala named by McKenzie following visit by Queen Elizabeth II, who was so impressed by box of Gala that she requested more. Further royal approval was bestowed in 1971 by Queen of Netherlands, when she visited W. ten Hove's orchards where sport discovered.
Frt *Col* bright orng rd flush, rd stripes over gold; inconspic fine russet dots. *Size* med. *Shape* oblng-con; ribbed; crowned. *Basin* lrg, brd, dp; ribbed. *Eye* clsd; sepals, lng, brd base. *Cavity* med, dp; russet lined. *Stalk* lng, thin. *Flesh* cream.
F* 14. **T**²; prn canker, scab. **C** gd. **P** e-Oct. **S** Oct-Jan/Mar.
  GALAXY; discv 1985 by K. W. Kiddle, Hawkes Bay, New Zealand; introd 1988. Darker red flush, stripes.
  IMPERIAL GALA; syn Mondial Gala, Mitch Gala; darker colour than Tenroy. Most widely planted form in UK.
  PRINCE GALA syn Regal Prince, Gala Must; unstriped, solid red flush.
  ROYAL GALA see Tenroy
  TENROY syn Royal Gala; darker red flush, stripes. Found by Mr W. ten Hove. RHS AM 1988; AGM 1993.

GALANTINE (?) 5 L D
France; acc may be variety desc Vercier (1948), when grown Savoie, Haute-Savoie (Rhone Alpes).
Large. Crisp, juicy, fairly sharp in Feb, but little richness.
F 14; trip. **T**³. **P** m-Oct. **S** Jan-Mar.

GALAXY see Gala

GALEUSE 7 L D
France; originated North Finistère (Brittany) and known around Montbéliard since early 1900s.
Sweet, modest richness, slight perfume. Used also for cooking and as rootstock for standard trees.
F 14. **T**². **P** e/m-Oct. **S** Nov-Feb/Mar.

GALLOWAY PIPPIN 5 L CD
UK; arose Wigtown, Galloway, Scotland. Brought notice 1871, then said to be old. RHS FCC 1871.
Large, russet freckled over yellow. Cooked, keeps shape; quite rich flavour. Best used early for cooking as flavour, acidity fade. Makes sharp eater, but russeted sport, Siddington Russet more attractive with good, plain taste of fruit.
F* 14. **T**²; sprd; hrdy. **C** gd. **P** l-Sept/e-Oct. **S** Nov-Feb.
  SIDDINGTON RUSSET; arose 1923 with John Jeffries & Sons Ltd; Siddington, Gloucs.

GALTON 4 M D
Canada; raised by W. T. Macoun, Central
Exp Farm Ottawa.
Northern Spy X. Introd 1915.
Sweet, soft, tender flesh.
F 15. T[1]. P l-Sept. S Oct-Nov.

GAMBAFINA 4 L D
Italy; believed arose c1900 nr Carraglio,
Cúneo, Piedmont.
Sweet, soft, juicy, greenish white flesh;
often woody undertones in England.
F 18. T[2]. P l-Oct. S Dec-Apr.

GARDEN ROYAL 4 E D
USA; arose on farm of Mr Bowker,
Sudbury, Massachusetts. Desc 1847.
Pretty, flushed dark brick red. Light
aromatic quality; firm, quite juicy, flesh.
F 9. T[3]. P l-Aug. S Aug-Oct.

GARNET 6 M D
UK; raised by nursmn W. P. Seabrook,
Boreham, Essex.
Rival, Worcester Pearmain cross. Received
1936.
Flushed in bright Worcester red. Brisk,
crisp to chewy flesh.
F 11. T[1]. C frt cracks. P e-Sept. S Sept-Oct.

GASCOYNE'S SCARLET 6 M D
UK; arose with Mr Gascoyne, Bapchild
Court, Sittingbourne, Kent. Introd 1871 by
nursmn G. Bunyard, Maidstone, Kent.
RHS FCC 1887.
Sharp, hard, white often pink stained flesh
in Oct, but mellowing to become very sweet
with distinctive taste, which Bunyard
described as balsamatic.
Highly decorative, valued exhibition variety;
formerly widely grown and still found in old
orchards. Introd 1883 to Germany; became
popular in Brandenburg, Saxony and
Thuringia and still grown for its beauty and
'delicate flavour'. As Cramoisie de
Gascogne grown France in Franche-
Comté, Indre-et-Loire, Normandy.
Frt Col bright rd flush, indistinct stripes
over pale grnish cream turning cream;
lenticels prom as russet/grn dots. *Size* lrg.
Shape flt-rnd or rnd. *Basin* lrg, dp; ribbed,
puckered. *Eye* lrg, prt open; sepals, brd, lng.
Cavity nrw, dp, russet lined. *Stalk* med,
thck. *Flesh* white.
F 19. T[3]; res sawfly. C gd. P l-Sept. S Oct-
Dec/Jan.

GAVIN 6/7 L D
UK; raised 1956 by Gavin Brown, JI,
Bayfordbury, Herts.
Merton Worcester X scab res seedling
(Jonathan X *Malus floribunda* X Rome
Beauty).
Lightly aromatic, sweet, quite rich.
F* 13. T[2]; res scab. C lght. P e-Oct. S Oct-
Dec.

GAZERAU 1 L C
France; acc prob Pomme Gazerand
cultivated in 1920s in Neauphe le Château
and Montfort l'Amaury (Yvelines, Ile-de-
France).
Cooked, very light flavour.
F 15. T[2]. P m-Oct. S Dec-Mar.

GEEVESTON FANNY 6 M D
Australia; arose prob at Geeveston, nr Port
Huon, Tasmania, where in 1880 tree grew
in orchard of James Evans.
Bright red flushed. Sweet, but tough, often
bleak grown in England.
Formerly Tasmanian export variety.
F 9. T[2]; res mildew. P l-Sept. S Oct-
Nov/Dec.
RED GEEVESTON FANNY

GEHEIMRAT BREUHAHN syn
Breuhahn 4 L D
Germany; raised 1895 at Fruit, Vine, Hort
Res St, Geisenheim, Rhine.
Halberstädter Jungfernapfel X. Introd 1934.
Sweet, light, pleasant flavour; juicy, finely
textured almost yellow flesh.
Grown Germany mainly for juice
manufacture.
F 10. T[2]. P e-Oct. S l-Oct-Dec/Mar.

**GEHEIMRAT DOKTOR OLDEN-
BURG** syn Oldenburg 4 M D
Germany; raised 1897 at Fruit, Vine Hort
St, Geisenheim, Rhine.
Minister von Hammerstein X Baumann's
Reinette.
Lightly aromatic, good, rich flavour; soft,
juicy, deep cream flesh.
Grown Germany; also used for juice
manufacture; being overtaken by Alkmene.
F 7. T[1]. P e-Sept. S Sept-Nov/Dec.

GELBER EDELAPFEL syn Golden
Noble

**GELBER MUNSTERLANDER BORS-
DORFER** 5 M D
Germany; arose Münsterland, Westphalia.
Received 1951.
Similar appearance to Edelborsdorfer, but
only little of its sweetness; firm flesh.
F 19. T[2]. P e-Oct. S Oct-Dec.

GELBER TRIERER WEINAPFEL syn
Trierer Weinapfel 5 L C
Germany; recorded 1895. Prob arose Trier,
on Moselle.
Very sharp, astringent early in season.
Cooked, keeps shape; remains very sharp.
F 19. T[2]. P m-Oct. S Nov-Jan.

GENE PITNEY 5 L D
New Zealand; raised DSIR, Havelock
North.
Cox's Orange Pippin X Golden Delicious.
Select by Dr. D. W. McKenzie. Introd
1964.

Large. Sweet, juicy, firm flesh; whiff of
aniseed flavour, which is pronounced
grown in New Zealand.
Developed at request of Gene Pitney fan,
who wished to mark visit of her musical
hero with apple from Hawkes Bay area.
F 17. T[2]. P m-Oct. S Nov-Dec.

GENETING syn Joaneting

GENNET MOYLE 5 M C cider
UK; acc prob variety of C19th.
Slight orange red flush. Cooks to bright
lemon purée, sharp, strongly flavoured;
needing little sugar.
Gennet Moyle, known since C17th,
produced cider which was pleasant, if mild
and 'fit only to be drunk by ladies in
summer' but Philips, cider poet, found it
'Of sweetest honied taste'. Still much grown
Herefordshire in C19th; also sold for jam
making. Roots from cuttings; formerly used
as rootstock.
F 5; trip. T[3]; sprd; tip. P e-Sept. S Sept-Oct.

GENEVA ONTARIO *see* Ontario

GEORGE CARPENTER 7 M D
UK; raised 1902 by George Carpenter, HG,
West Hall Gardens, Byfleet, Surrey.
Blenheim Orange, King of the Pippins
cross.
Handsome. Quite rich, aromatic flavour;
rather like modest Ribston Pippin.
F 12. T[3]. P m-Sept. S Oct-Dec.

GEORGE CAVE 4 E D
UK; raised 1923 by George Cave,
Dovercourt, Essex. Acquired by Seabrook &
Sons, Boreham, Essex. Named 1945.
Strong, sweet-sharp taste, more interesting
than most summer apples; reminiscent of
Nonpareil family.
Formerly market apple following Beauty of
Bath in the shops, but now only grown for
Farm Shop trade.
Frt Col rd flush, carmine stripes over pale
grnish yell; lenticels prom as russet dots.
Size med/sml. *Shape* rnd-con; regular; sltly
ribbed. *Basin* brd, shal; sltly puckered. *Eye*
hlf open; sepals, lng, brd, downy. *Cavity*
med, shal. *Stalk* qte lng, thck. *Flesh* wht.
F 8. T[2]. C gd; frt often pr col, cracks. P eat
e/m-Aug.

GEORGE NEAL 3 E CD
UK; raised 1904 by Mrs Reeves, Otford,
Sevenoaks, Kent. Named after, introd by
nursmn, George Neal, Wandsworth,
London. RHS AM 1923; AGM 1993.
Refreshing, brisk eating apple. Cooked,
slices keep shape; sweet delicate flavour;
needs no extra sugar.
Remains popular garden variety.
Frt Col grnish yell turning pale yell,
brownish rd flush, rd stripes; lenticels qte
prom as grey dots. *Size* lrg. *Shape* flt-rnd.

Basin brd, dp; sltly ribbed. *Eye* lrg, open/hlf open; sepals, shrt. *Cavity* nrw, dp; russet lined. *Stalk* med, qte thck. *Flesh* pale crm. F* bright pink; 7. T². C gd. P e/m-Aug. S Aug-Sept.

GEWÜRZLUIKEN 4 L D
Germany; unknown origin; old, local variety of Württemberg.
May be seedling of Luikenapfel.
'Gewürz' means tasty. Crisp, juicy, sweet, plenty acidity, with spicy almost peppery taste. Popular in Baden, Württemberg; also used for processing.
F 17. T²; res scab, mildew. C gd. P m-Oct. S Dec-Mar.

GIAMBUN 4 L D
Italy; received 1958 from Univ Turin.
Markedly conical; bright red flush, stripes. Crisp, firm, sharp.
F 20. T². C hvy. P l-Oct. S Jan-Mar/May.

GIAN ANDRÉ 4 L D
Italy; received 1951 from Univ Turin.
Scarlet flushed. Sweet, quite rich taste in Dec; firm flesh.
F 21. T¹. P l-Oct. S Dec-Mar/Apr.

GIANT GENITON 6/7 L D
USA; acc prob variety raised by J. W. Johnson, Lewis County, Missouri; introd 1920s by Stark Bros Nursery, Missouri.
Resembles large Ralls Janet or Geniton. Sweet, lightly aromatic, juicy, crisp, deep cream flesh; can be weak, watery.
F 10. T². C hvy; hangs v late. P l-Oct. S Jan-Mar/May.

GIBBON'S RUSSET 8 M D
Ireland; arose Cork; recorded 1897.
Cidery taste; firm, sharp flesh.
Popular Cork early 1900s; also used for cider.
F 12. T³. C bien. P l-Aug. S Sept-Oct.

GILLIFLOWER OF GLOUCESTER 4 M D
UK; received 1952, Saul village, Gloucs.
Similar shape to Cornish Gilliflower. Slightly scented, sweet, rather dry, firm, white flesh.
F 9. T². P e-Sept. S Sept.

GIPSY KING (EMRS) 8 L D
UK; received 1948 from EMRS, Kent; acc prob not Gipsy King of Hogg and Bunyard. Brilliant red cheek under russet. Quite intensely flavoured, sweet-sharp quality of Nonpareil type.
F 17. T². P m-Oct. S Nov-Jan/Mar.

GLADSTONE 6 E D
UK; found in field by William Jackson of Blakedown Nursery, nr Kidderminster, Worcs. Introd 1868 as Jackson's Seedling; renamed 1883, when mother tree estimated over 100 yrs old. RHS FCC 1883.

At best, shows definite raspberry to redcurrant flavour; sweet, refreshing acidity, crisp yet soft, melting flesh.
After 1883 Congress, was strongly recommended for profit and widely planted commercially and in gardens; early market apple until 1960s.
Frt Col dark rd flush, stripes; grnish yell background; lenticels conspic as darker dots. *Size* med. *Shape* rnd-con; irregular; ribbed, often 5 crowned. *Basin* brd dp; ribbed. *Eye* clsd or prt open; sepals, brd; qte downy. *Cavity* nrw, v shal. *Stalk* med/lng, thck. *Flesh* wht, tinge grn.
F 13. T². C hvy; frt cracks. P eat l-July/e-August; v shrt season.

GLASBURY 3 M C
UK; received 1979 from collection of O. D. Knight, Norfolk; obtained from Hanniford Nursery, Paignton, Devon in c1918. Impossible to say if same as variety recorded 1872 by Scott as no desc given.
Large, pink flushed. Cooked, keeps shape yet soft, juicy; good flavour. Soon fades, becomes insipid.
F 14. T². P e-Sept. S Sept-Oct.

GLASS APPLE 5 M D
UK; raised by Mr Snell, market gardener, Radland Mill, St Dominick, Tamar Valley, Cornwall. Known 1934; syn Snell's White.
Large, pale gold. Sweet, slightly honeyed.
F 11. T³. P l-Aug. S Sept-Oct/Nov.

GLEBE GOLD 5 M D
UK; raised by James Walker, Ham, Surrey. Yellow Ingestrie X Cellini. Received 1945.
Small, bright gold. Brisk, savoury taste; quite soft, juicy.
F 12. T². C hvy, bien. P l-Aug. S Sept.

GLENGYLE RED *see* Rome Beauty

GLENTON 6 M D
Canada; raised Ottawa Central Exp Farm, Ontario.
Northern Spy X. Named 1911; introd 1946. Bright red Northern Spy flush. Sweet, juicy, soft cream flesh; low acidity, often metallic undertones in England; tough skin. Used also cooking US.
F 18. T². P m-Sept. S Sept-Oct.

GLOCKENAPFEL 1/5 L C
Central Europe; ancient variety; unknown origin; widely distributed by C19th. Syns Weisser Winterglockenapfle, Pomme Cloche.
Bell-shaped as name suggests. Cooked, keeps shape with translucent quality; quite sweet, yet plenty of acidity.
Formerly widely grown Europe; still prized in Switzerland for 'Apple Strudel'; also valued Alsace, Netherlands.
F 8. T². C hvy. P m-Oct. S Dec-Apr; commercially-Jly.

GLOGEROVKA 5 L D
Lithuania or Russia; desc 1961 (Simirenko), but old variety of late C19th.
Savoury taste, definite acidity; crisp to soft, juicy white flesh.
Widely distributed over Baltic States.
F 4. T²; sprd. P m-Sept. S Oct-Nov/Dec.

GLORIA MUNDI 1 L C
Germany or poss USA. Confused history. Recorded USA in 1804; introd 1817 to England. Syns numerous, include Monstrous Pippin, Ox Apple. Poss same as Baltimore Monstrous Pippin, arose c1780 Baltimore Maryland.
Very large. Cooks to dark gold, sweet purée, some flavour; acidity fades with keeping.
Formerly 'show' apple for country fairs in US; in 1860 specimen recorded weighing three and half pounds.
F 13. T²; uprt. P e-Oct. S Oct- Dec/Mar.

GLORIE VAN HOLLAND syn Glory of Holland 5/4 L D
Netherlands; raised or found c1890 by Haselbach at Rockanje.
Bright red flushed over gold. Very sweet, almost like sweetmeat; crisp quite juicy flesh, but low in fruit, acidity. Still grown Netherlands.
F 9. T²; uprt. P e Oct. S l-Oct-Dec.

GLORY OF HOLLAND syn Glorie van Holland

GLOSTER 69 6 L D
Germany; raised 1951 at Fruit Res St, Jork, nr Hamburg.
Weisser Winterglockenapfel (Glockenapfel) X Richared Delicious. Introd 1967.
Attractive, dark red flushed. Sweetness of Red Delicious, some balancing acidity; good flavour; crisp flesh. Dual purpose on continent. (Vit C 4–8mg/100gm).
Widely planted over northern Europe, although now declined, but remains important variety in Germany, Belgium, Austria, northern Italy, Switzerland, Poland.
F 17. T³. C hvy; prn develop mouldy core in store. P m-Oct. S Nov-Feb/Mar.

GLOUCESTER CROSS 7 M D
UK; raised 1913 by G. T. Spinks, LARS, Bristol.
Cox's Orange Pippin X.
Quite intense, rich, aromatic flavour, sweet, crisp, juicy, deep cream flesh; like early, light Cox.
F 12. T². P e-Sept. S Sept-Oct/Nov.

GLOUCESTER ROYAL 6 L D
UK; raised c1930 by J. W. Thornhill, Dursley, Gloucs. McIntosh X. Introd 1956 by nursmn J. Matthews, Thurston, Suffolk. McIntosh rcd. Very sweet, but low acidity, often woody, green taste; tough skin.
F 16. T². P m-Oct. S Nov-Jan/Feb.

GOLCO 6 L D
Begium; received from nursmn J. Nicolaï of Sint-Truiden.
Parentage unknown.
Attractive appearance and flavour; sweet, juicy, rich, lightly aromatic.
F est 12. **T**[1]; uprt. **P** e-Oct. **S** Oct-Dec.

GOLDEN BITTERSWEET 5 E D cider
UK; cider apple sent to Hogg by Mr Rendall of Netherton Manor, Devon; desc 1884.
Frecked in russet. Quite sharp, very soft, juicy flesh.
F[*] 13. **T**[3]. **P** e-Sept. **S** Sept-Oct.

GOLDEN BOUNTY 5 M D
UK; raised c1940 by A. C. Nash, Scutes Farm, Hastings, Sussex.
Cox's Orange Pippin X poss Early Victoria.
Crisp, sharply aromatic; like brisk, early Cox; can be very good,
F 13. **T**[2]. **C** hvy; frt cracks. **P** m-Sept. **S** Sept-Oct.

GOLDEN DELICIOUS 5 L D
USA; arose c1890 with A. H. Mullins, Clay County, West Virginia. Poss Grimes Golden seedling. Originally Mullins' Yellow Seedling. Introd 1916 and renamed by Stark Bros, Missouri. Wilder Medal, 1921. Introd England 1926 by E. Bunyard; exhib 1934. RHS AGM 1993.
At best, honeyed with crisp, juicy, almost yellow flesh; Bunyard found it 'very sweet with rich perfume'. But little acidity and often tastes flat, cloying; especially if picked early.
Needs warmer climate than England to be successful commercially, but often grown as orchard pollinator and popular garden apple. If left on tree to become golden with slight blush and build up sugars and flavour, it can be good. Dual purpose in many countries; used by baby food processors. Cooked, keeps shape, sweet, but very light flavour. (Vit C 5–17mg/100gm).
Began rise to fame after Mullins sent fruit to Stark Brother's Nursery in April 1914, commending to them its excellent keeping qualities and heavy crops. Paul Stark was sufficiently impressed to come and inspect the tree which he bought for $5,000, erected a cage around it to prevent anyone else taking grafts and paid $100 a year for its maintenance. Tree survived until 1958, and just before it was honoured with photograph on centennial issue of Clay County Bank cheques. Extensively planted in USA by 1920s, planted on few Kent farms in 1930s, but came to European prominence in 1960s. French colonials returning from Algeria and investing in fruit growing in Loire valley were recommended to grow Golden Delicious, which began to be exported in 1970s, primarily to Britain. Widely planted in all apple growing countries and continents, although being overtaken by newer varieties. Its one failing is tendency to develop unsightly russet, but many less prone sports identified.
Frt *Col* grnish yell turning gold, slght orng flush; russet freckles. *Size* med. *Shape* rnd-con to oblng; ribbed, 5 crowned. *Basin* med wdth, dpth; ribbed; trace russet. *Eye* clsd/prt open; sepals qte lng, nrw. *Cavity* nrw, dp; russet lined. *Stalk* lng, thin. *Flesh* dp crm.
F[*] 12; gd pollinator. **T**[2]; sprd. **C** hvy. **P** l-Oct. **S** Nov-Feb; cold store- Jun.

> **BELGOLDEN NO** 17; less prone russet.
> **BERTANNE**; entirely russeted; more interesting flavour. Discv south west France by M. Ainée, named after his grandchildren. Grown small extent France. Not in NFC.
> **CLEAR GOLD**; russet free sport. Discv 1962 by O. Stauffer, Myerstown, Pennsylvania; introd 1965 by Worley's Nursery, York Springs, Pennsylvania.
> **COURTAGOLD**; spur type.
> **DOUD GOLDEN DELICIOUS**; tetrap; discv by L. J. Doud in Wabash, Indiana.
> **ED GOULD GOLDEN**; less prone russeting.
> **ELBEE**, partial red, less russet sport. Discv by Leo Basser, Untervaz, Switzerland.
> **GOLDEN DELICIOUS: HORST NO 2**
> **GOLDEN DELICIOUS**; LARS selection.
> **GOLDENSHEEN**; less prone russeting. Trade name Belgolden No 1; discv by M. Ancian, Agen, France.
> **GOLDEN DELICIOUS RUSSET FORM**
> **GOLDEN DELICIOUS**; (4E-26-18-N); tetrap.
> **GOLDEN MORSPUR**; spur type.
> **GOLDENSPUR**; spur type.
> **LYS GOLDEN** (Goldenir); less prone to russeting. Arose 1961 INRA, Angers, France.
> **PENCO**; light, brownish orange flush. Found by Maurice Penxten, Belgium; introd by N. Nicoläi.
> **SMOOTHEE**; larger, firmer, less prone russeting. Discv 1958 in orchard of C. Gibson, Evans City, Pennsylvania. Introd 1957.
> **STARKSPUR GOLDEN DELICIOUS**; spur type. Discv 1959 by P. J. Jenkins; introd 1961 by Stark Bros, Missouri.
> **TESTERSPUR GOLDEN DELICIOUS**; spur type.
> **YELLOWSPUR**; spur type. Discv 1959 by O. Thornton and R. Thompson, Oroville, Washington; introd 1962.

GOLDEN HARVEY 7/8 L D cider
UK; old variety; prob arose Herefords; may be Round Russet Harvey of early 1600s. Syns incl Brandy Apple.
Small, golden, russeted. Intensely flavoured, sweet-sharp, quite aromatic.
Popular for Victorian dessert, but cast aside in 1890s as too small. Called Brandy Apple, because of high specific gravity of juice, hence strong cider.
F 13. **T**[3]. **C** gd. **P** mid-Oct. **S** Dec-Mar.

GOLDEN KNOB 8 L D
UK; arose at Enmore Castle, Somerset. Known since late 1700s.
Small, very like Golden Harvey, but more russet. Similar, intense sweet-sharp taste, but sharper, not as rich.
Grown in Kent and by London market gardeners in C19th; sold as 'little Golden Knobs'. Still remembered in Kent.
F 11. **T**[2]. **P** m-Oct. **S** Dec-Mar/May.

GOLDEN MELON 5 L D
Japan; raised 1931, Aomori Apple Exp St. Golden Delicious X Indo. Named 1948.
Very sweet, quite melon-like flavour, crisp, but almost no acidity.
F 11. **T**[2]. **C** lght, bien. **P** e-Oct. **S** Nov-Feb.

GOLDEN MORSPUR *see* Golden Delicious

GOLDEN NOBLE 5 M C
UK; discv in old orchard at Downham, Norfolk by Patrick Flanagan, HG to Sir Thomas Hare, Stowe Hall, Norfolk. Flanagan exhib London Hort Soc, 1820. But may not have been original tree as 'Golden Noble' listed 1769 by nursmn William Perfect, Pontefract, Yorks. Many syns; distributed over Northern Europe. RHS AGM 1993.
Cooked, keeps little of its form; sharp, well flavoured, but not as acidic as Bramley. Ideally suited to pies; baked has creamy texture; needs only little sugar. In Spring can serve as brisk, almost rich eating apple. Prized Victorian and Edwardian cooker; widely grown in gardens; also by London market gardeners. On sale up to 1930s; remains valued garden variety. Valued also Germany as Gelber Edelapfel.
Frt *Col* pale grn turning gold, sometimes slight flush; faint russet dots. *Size* lrg. *Shape* rnd to rnd-con; v regular. *Eye* hlf open; sepals, brd base, shrt. *Basin* brd, shal; sltly puckered. *Cavity* brd/med wdth, med dpth; russet lined. *Stalk* shrt, thck. *Flesh* crm.
F[*] 16. **T**[2]; prt tip; res scab, mildew. **C** gd. **P** e-Oct. **S** Oct- Dec; keeps-Mar.
PLATE 15

GOLDEN NUGGET 7 L D
Canada; raised 1932 by C. J. Bishop, Kentville Agric St, Nova Scotia.
Golden Russet X Cox's Orange Pippin. Introd 1964.

Small but very attractive. Intense, aromatic flavour; sweet, rich, almost pineapple acidity.
F* 10. **T²**. **C** gd. **P** l-Sept. **S** Oct-Nov.

GOLDEN PIPPIN (?) 5 L D
UK; prob arose England, but origin uncertain. Acc is close to C19th descs, but impossible to be certain if variety is true. 'Golden Pippin' recorded 1629 by Parkinson; well known by late 1600s. Claimed arose Parham Park, nr Arundel, Sussex. Many syns.
Intense, brisk taste, almost lemon tang; plenty sugar, acidity; strong, fruity quality. Cooked keeps shape; sweet, quite rich taste. Small fruit, poached whole, make pretty sweetmeats, like old Golden Pippin, which was used for 'Pippin', jelly, tarts and cider.
Golden Pippin was widely grown in gardens and for market in C18th. Also said to flourish in Oporto, Portugal. Especially ordered by George Washington for his garden at Mount Vernon, Virginia, but did not thrive in USA. English reputation suffered set back in 1790s when Knight erroneously concluded that variety had reached its declining years. By 1890s it was damned as too small. Used by Knight in first breeding programmes; Downton Pippin claimed closely resembles its parent.
Frt *Col* gold; many bold russet dots; some russet at base. *Size* sml/med. *Shape* rnd-con to oblng. *Basin* lrg, brd, shal; sltly ribbed. *Eye* open. *Cavity* brd, shal; russet lined *Stalk* shrt, thck. *Flesh* pale yellow.
F 11. **T²**. **P** e-Oct. **S** Oct/Nov-Jan/Mar.

GOLDEN REINETTE 7 L D
Europe (prob); confused origins; acc is variety of Hogg (1884). Name recorded England mid 1600s. Syns numerous.
Handsome with good, plain taste of fruit; crisp to crumbling deep cream flesh. Very similar to Blenheim Orange with same addictive quality. Cooked, keeps shape or forms stiff purée; good for 'Apple Charlotte', 'Tarte aux Pommes' etc.
Old authorities claimed it attained highest perfection in Herts, but widely grown in C18th as table and cider apple. Continued to be grown for market around London in C19th, 'always commanded a good price'. Gardeners voted it amongst top dozen varieties in 1883, but by 1920s considered 'hardly of first class quality', although still grown Europe.
Frt *Col* orng rd flush over gold; numerous russet dots; some russet netting. *Size* med. *Shape* flt-rnd to oblng; trace ribs, sltly crowned. *Basin* brd, shl; ribbed, trace russet. *Eye* lrg, open (like Blenheim); sepals smll. *Cavity* nrw, dp; russet lined. *Stalk* med, thck. *Flesh* crm.
F 12. **T²**. **C** gd. **P** e-Oct. **S** Oct-Jan.

GOLDEN RESISTA syn Resista 4 L D

Czech; raised Inst for Fruit Growing & Propagation, Holovoúsy.
Prima X NJ56. Introd 1998.
Bright red flush over gold. Crisp firm white flesh; savoury, brisk taste mellowing to slightly perfumed quality; good in Jan.
F mid. **T³**; res scab; v sus mildew. **C** hvy. **P** m-Oct. **S** Dec-Jan; commercially Mar/Apr.

GOLDEN ROYAL 5 E D
Ireland; received 1950, when grown County Monaghan.
Sharp, fruity; crisp, juicy, cream flesh.
F 10. **T³**; sprd. **P** use Aug.

GOLDEN RUSSET OF WESTERN NEW YORK 8 L D
USA; in C19th often confused with English Golden Russet, but Beach (1905) considered variety then widely grown in West New York to be distinct.
Attractive, covered in golden russet. Sugary, honeyed taste, firmly textured cream flesh. Exported to UK up to 1930s.
F 7. **T²**. **P** e/m-Oct. **S** Dec-Feb/Mar.

GOLDEN SPIRE 5 M C cider
UK; found c1850 in Lancs; introd by nursmn Richard Smith of Worcester. Known as Tom Matthews, cider variety in Gloucs.
Golden, tall, oblong. Quite intense, almost cidery flavour; sharp, juicy, deep cream flesh. Cooks to well-flavoured, slightly brisk, yellow purée.
Quite widely grown late C19th; also recommended in 1890s as decorative tree for planting in shrubbery, alongside paths.
F* 5. **T¹**; weeping. **C** gd; hangs late. **P** e-Sept. **S** Sept-Oct.

GOLDENSHEEN see Golden Delicious

GOLDENSPUR see Golden Delicious

GOLDJON 4 L D
Italy; raised by Prof R. Carlone, Turin Univ. Golden Delicious X Jonathan. Introd 1965. Bright red flush of Jonathan with Golden Delicious flavour. Sweet, crisp, honeyed, pale yellow flesh.
F 14. **T¹**. **P** e-Oct. **S** Nov-Jan, cold store-Mar.

GOLD MEDAL 5 E DC
UK; raised by Troughton, nursmn at Preston, Lancs. Introd c1882 as Ryland Surprise.
Large. Savoury, brisk taste; soft flesh. Cooks to juicy purée, some sharpness, pleasant.
F 7. **T²**. **P** l-Aug. **S** Aug-Sept.

GOLD-N-ROSE syn Maigold

GOLDPARMÄNE see King of Pippins

GOLD REINETTE 7 L D

Received 1952 from Estonia via Denmark. Sweet, scented, cream flesh.
F 5. **T²**. **P** e-Oct. **S** l-Oct-Dec.

GOMBA KÁROLY DR 5 L D
Hungary; received 1948 from Univ Agric, Budapest.
Sharp, firm, chewy in England.
F 11. **T²**. **P** e-Oct. **S** l-Oct-Dec.

GOOSEBERRY 1 L C
UK; prob arose Kent. Recorded 1831; desc Hogg (1851).
Hard; as green and sour as gooseberry. Very acidic, cooks to sharp purée.
Widely grown in Kent for London markets in C19th.
F 15. **T³**; sprd. **C** gd. **P** l-Oct. **S** Dec-Mar/Jun.

GORO 5 M D
Switzerland; raised 1951 at Swiss Fed Res St, Wädenswill.
Golden Delicious X Swiss Orange (Ontario X Cox's Orange Pippin). Introd 1973.
Quite large; slight blush over gold. Crisp, fruity, quite sharp; mellowing and softening with keeping. Recommended commercial planting in Switzerland.
F 17. **T³**; sprd; some res scab. **P** l-Sept/e-Oct. **S** Oct-Nov; cold store Dec.

GRAHAM see Royal Jubilee

GRAND'MÈRE 5 L C
France; acc is not variety of Vercier (1948), but prob one of the many Grand'mère varieties that exist in France.
Large. Slightly scented, quite sweet. Cooked, keeps shape, sweet, light.
F 6. **T¹**; sprd. **P** l-Oct. **S** Nov/Dec-Jan.

GRANGE'S PEARMAIN 3 L CD
UK; raised before 1829 by James Grange, market gardener, Kingsland, Middx. Introd by Dickson's Chester.
Quite rich, brisk, crisp flesh, mellowing to perhaps hint of pineapple flavour. Claimed by Bunyard to be 'nearest we have to Newtown Pippin'. Cooked, keeps shape, quite sweet.
F 12. **T³**. **P** e/m-Oct. **S** Nov/Dec-Mar.

GRANNY GIFFARD 4 L DC
UK; exhib 1858 at British Pom Soc by Mr. Swinherd, HG to John Swinford, Esq, Minster, nr Margate, Kent. Desc Hogg (1884).
Sharp, quite fruity, soft flesh; said to 'bake well'.
F 25. **T²**. **C** prn bitter pit. **P** e-Oct. **S** Nov-Dec/Feb.

GRANNY SMITH 2 L D
Australia; arose with Mrs Anne Smith, Ryde, New South Wales. Believed from pip of French Crab; fruiting by 1868.

Hard; can be crisp, refreshing but often very acid, with no distinctive taste, firm greenish white flesh; tough skin. Dual purpose in many countries.

Mrs Smith was born in Peasmarsh, Sussex in 1800 and emigrated to Australia in 1838. In 1860s she found seedling tree growing in creek where she had tipped out last of some apples brought back from Sydney. She used its fruit for cooking, but boy claimed they were good to eat fresh. Tree was propagated and later family increased their orchards and marketed fruit in Sydney, where proved popular and ideal for export market. Exported from Australia to Britain by 1930s. Grown in all warmer fruit regions: southern hemisphere, Spain, Southern France, Italy, Washington, US. English, northern European climate not warm enough. Popularity is waning, but remains major international variety. To commemorate apple, site near Mrs Smith's home was named 'Granny Smith Memorial Park' in 1950 and monument is also planned.

Frt Col bright grn, turning grnish yell; slight brownish flush; lenticels conspic as grey dots. *Size* med. *Shape* rnd-con, sltly 5 crowned. *Basin* med wdth, dpth; ribbed, puckered. *Eye* clsd; sepals, downy. *Cavity* nrw, dp. *Stalk* med lng, qte thin. *Flesh* grnish wht.

F 10. **T**^{2}. **C** hvy; ripening prly Eng. **P** m/-Oct. **S** Jan- Apr.

GRANNY SMITH SPUR TYPE

GRANTONIAN 5 L C
UK; introd by nursmn J. R. Pearson of Nottingham. Recorded 1883. RHS 2nd class cert 1883.
Large. Cooks to lemon coloured purée; fruity, quite sweet in Nov.
F 14. **T**^{3}; uprt. **P** e-Oct. **S** Nov-Jan/Mar.

GRAUE HERBSTRENETTE 8 L D
Europe or England. Syn of, and acc agrees with, Reinette Gris d'Automne of Leroy (1873), which is syn of Reinette Gris d'Angleterre, first listed 1670 by Claude St Etienne. Widely distributed Europe by C19th.
Large; cinnamon russet over yellow. Sweet, fruity, firm flesh.
F 8. **T**^{2}. **P** e-Oct. **S** Oct-Dec/Jan.

GRAVENSTEIN 4 M D
Europe; very old, widely distributed. Claimed arose 1600s in gardens of Duke Augustenberg, Castle Graefenstein, Schleswig-Holstein. Others say it was Ville Blanc, which arose in Italy or South Tyrol and sent to Schleswig-Holstein, or scions from Italy were sent home by brother of Count Ahlefeldt of Graasten Castle, South Jutland. Believed arrived Denmark c1669. Introd England 1820s; from London Hort Soc sent to Massachusetts, Nova Scotia. Introd to California by 1820s by Russian

settlers. Syns numerous.
Savoury, crisp, yet melting flesh which glistens with juice. Cooked, keeps shape; sweet, light taste.
Very popular around Hamburg, Germany in early C19th, became grown throughout northern Europe. Grown English gardens, but too large, angular and brisk for dessert, yet not sharp enough to be a 'good cooker'. Became major Canadian market apple; exported to Britain.
Still grown Canada and California where dual purpose, also used for pies, sauce, juice. Continues to be valued northern Europe; Norway's most popular apple, also grown South Tyrol. Commercially highly coloured sports usually grown.
Frt Col lght pinky orng flush, rd stripes over grnish yell/yell. *Size* med/lrg. *Shape* oblng; irregular, often lop-sided; tends to be ribbed, flt sided; 5 crowned. *Basin* brd, dp; ribbed, russet present. *Eye* lrg, open; sepals, small, downy. *Cavity* med wdth, dp; russet lined. *Stalk* shrt, qte thck. *Flesh* crm.
F* wht, lrg; 3; trip. **T**^{3}; v hrdy; res red spider. **C** gd; slow to bear; frt drops. **P** e-Sept. **S** Sept-Oct.
PLATE 1 (blossom), 10

   ALL-RED GRAVENSTEIN; received 1960 from New Zealand.
   MORKROD; red flushed. Arose Denmark or Sweden; received 1960.

GREASY PIPPIN 5 M D
UK; desc 1951 (Lamb), when common Ballinamallord, County Fermanagh, also found Co Tyrone, N Ireland.
Very greasy skin. Brisk, refreshing taste; juicy, crisp, cream flesh.
Formerly used as pollinator in Irish Bramley orchards.
F 16. **T**^{2}; tip. **P** l-Sept. **S** Oct-Nov.

GREAT EXPECTATIONS 7 M D
UK; raised c1945 by Mr. A. Nobbs, retired gardener. Propagated by Hammond Stock Nursery, Bearsted, Kent.
Red Cox X Egremont Russet.
Handsome; red flush under russet. Flavour recalls Egremont Russet, but sweeter, weaker; low acidity; firm, quite dry, white flesh.
F 19. **T**^{1}; sprd. **P** m-Sept. **S** Oct-Nov.

GREEN BALSAM (HOLLIDAY)
Acc received 1946 from Yorks. Acc not Green Balsam of Hogg, which arose North Yorks, or Rymer with which it was often confused, but it does show curious 'balsamatic' flavour; also very attractive blossom.
PLATE 1 (blossom)

GREEN CUSTARD 5 M D
UK; received 1941 from J. Cheal Nursery, Crawley, Sussex. Costard type; prob 'misprint' for Green Costard.

Very similar appearance to Catshead, but maturing earlier with sharp, plain taste.
F 11. **T**^{2}. **P** m-Sept. **S** Oct-Nov.

GREEN HARVEY 1/5 L CD
UK; received 1930 from W. G. Kent, Wisbech, Cambs. Prob variety exhib 1904, 1934, but impossible to know if Green Harvey listed 1813 as no published desc.
Moderately rich eating apple. Cooked, keeps shape; quite sweet.
F 14. **T**^{2}. **P** l-Oct. **S** Dec-Feb/Apr.

GREEN KILPANDY PIPPIN 1 E C
UK; received 1949 from Tyninghame Gardens, East Lothian, Scotland.
Grass green, covered in scarf skin which gives grey bloom. Greenish white flesh; cooks to brisk purée.
F 19. **T**^{3}. **C** bien. **P** use l-Aug-Sept.

GREEN PURNELL 7 L D
UK; arose Worcs. Received 1945.
Quite sweet; some richness.
F 12. **T**^{2}. **C** prn bitter pit. **P** m-Oct. **S** Nov-Feb.

GREEN ROLAND 2 M D
UK; arose Norfolk. Desc 1946 (Taylor).
Grass green with green tinged flesh. Good brisk taste of fruit, yet quite sugary.
Popular East Anglia during last war as culinary apple, when sugar in short supply.
F 14. **T**^{2}. **P** l-Sept. **S** Oct-Nov.

GREEN ROLLAND syn Greenup's Pippin

GREENSLEEVES 5 M D
UK; raised 1966 by Dr F. Alston, EMRS, Kent.
James Grieve X Golden Delicious. Introd 1977. RHS AM 1981; AGM 1993.
Sweetness and honeyed taste of Golden Delicious but balanced by acidity of James Grieve; firm, juicy, deep cream flesh. Early in season hard, sharp; becoming sweeter, softer.
Grown for Farm Shop trade; popular garden apple.
Frt Col pale grnish yell/yell, sometimes faint pinky orng flush; lenticels as russet dots. *Size* med. *Shape* oblng or rnd-con; sltly 5 crowned. *Basin* brd dp; sltly ribbed. *Eye* hlf open; sepals, lng. *Cavity* med dpth, wdth; russet lined. *Stalk* lng, thin. *Flesh* crm.
F* 7; some res frost. **T**^{1}; hrdy. **C** hvy; early bearing. **P** m-Sept, can be eaten from tree into Oct. **S** Oct.

GREENUP'S PIPPIN syn Green Rolland 5/4 M DC
UK; found in garden of shoemaker, Greenup in Keswick, Cumberland. Introd late 1700s by Clarke & Atkinson Nursery, Keswick. Syns many, incl Yorkshire Beauty,

Cumberland Favourite, Red Hawthornden. Sharp, savoury, quite acidic; soft, juicy, white flesh. Cooks to creamy juicy froth or purée, brisk, well-flavoured.
Grown 'throughout Border Counties' in C19th and 'valuable where choicer varieties do not attain perfection.'
F 15. T². P m-Sept. S Sept-Nov.

GRENADIER 1 E C
UK; origin unknown; recorded 1862. Exhib by nursmn Charles Turner of Slough at 1863 Crystal Palace Show; RHS FCC 1883; AGM 1993.
Cooks to pale cream, sharp purée, but not frothy, sweeter quality of codlins.
Brought to prominence at 1883 Congress by George Bunyard, and soon promoted as 'improvement on Keswick Codlin', as it produced much larger fruit. Early market fruit up to 1980s; but now overtaken by early picked Bramley's Seedling. Remains valued Belgium, where RGF variety, recommended for cultivation by Agric Res Centre, Gembloux.
Frt Col grn/yell; scarf skin at base. Size lrge. Shape rnd con; qte irregular; ribbed, flt sided. Basin nrw, qte shal; ribbed, puckered. Eye sml, clsd. Cavity brd, qte dp. Stalk shrt, thck. Flesh wht.
F 11. T²; gd res disease. C hvy. P m-Aug. S Aug-Sept.

GREY PIPPIN (TANN) 5 L D
UK; local Essex variety found c1980 by J. Tann, fruit grower of Aldham. Impossible to know if variety exhib 1883 by nursmn Saltmarsh of Essex, as only brief desc published.
Golden, heavily strewn with russet dots. Quite sweet, fruity and pleasant.
F est 9. T². P e-Oct. S Nov-Jan.

GRIGIA DI TORRIANA syn Renetta Grigia di Torriana

GRIMES GOLDEN 8 L D
USA; discv 1832 on farm of Thomas Grimes, Fowlersville, nr Wellsburg, West Virginia.
At best some complexity of flavour; sweet, honeyed, crisp, juicy, almost yellow flesh; recalls its offspring Golden Delicious, but better flavour, not as cloying.
Tradition holds that it arose from pip planted by Johnny Appleseed. Founded early fruit industry of West Virginia and became planted throughout southern states and west coast in C19th. Exported to Britain up to 1930s, but now replaced by Golden Delicious and only used as orchard pollinator.
Frt Col grnish yell/yell; netted fine russet; lenticels as grey dots. Size med. Shape oblng; flt base, apex; rnded ribs; 5 crowned. Basin med brd dp; sltly ribbed, fine russet. Eye lrg, open; sepals shrt, sep at base, qte

downy. Cavity med brd, dp; russet lined. Stalk shrt/med, qte thin. Flesh pale crm.
F 10; gd pollinator. T². C gd. P m-Oct. S Nov-Feb.

GRIMOLDBY GOLDEN 5 E C
UK; received from Lincs 1993; believed raised by William Ingall, Grimoldby, Lincs prob early 1900s.
Large golden, similar to Ingall's Dr Clifford.
F est 12. T². P use Aug-e-Sept.

GRIS BRAIBANT syn Reinette de Mâcon

GRONINGER KROON 4 M D
Netherlands; found c1875 by S. H. Brouwer, Noordhoek, Groningen.
Large, handsomely striped, flushed. Some aromatic quality, quite sweet firm flesh, but often unattractive underlying acidity. Remains popular Dutch apple.
F 12. T². C prn mildew. P l-Sept. S Oct-Nov/Dec.

GRONSVELDER KLUMPKE 4 M D
Netherlands; sport of Eisendener Klumpke, discv Gronsveld, Limburg; desc 1955.
Large, oval; carmine flushed; more highly coloured than parent. Quite rich, but rather sharp. Formerly widely grown Netherlands, Belgium; popular standard tree.
F 18. T³. P e-Oct. S Oct-Dec/Mar.

GROS API 6 L D
France (prob); acc prob Gros Api of Leroy (1873), which he believed to be ancient variety. Syn 'Dieu' cited 1628 by Le Lectier. Later claimed arose ancient forest of Api in Brittany. Syns many.
Larger version of Api; bright pink flushed, striped. In New Year sweet, slightly perfumed, good taste of fruit; crisp flesh. Among apples admired by Louis XVI's gardener, la Quintinye. Remains known all over France.
F 13. T³; sp. C gd. P l-Oct. S Jan-Apr.

GROS CROQUET 6 L D
France; acc is variety desc 1948 (Verger Français).
Bright red flushed. Sweet, quite pleasant; crisp, juicy.
F 21; trip. T². P m-Oct. S Nov-Jan.

GROSEILLE 3 L CD
France; grown Indre-et-Loire, Sarthe (Western and Val de Loire), when desc 1948 (Verger Français).
Brisk, juicy savoury in Jan.
Found now Normany, Sarthe, Indre-et-Loire; used for cooking, pâtisserie.
F 21. T³. C hvy. P m-Oct. S Dec-Mar.

GROS FENOUILLET (SHAND) 8 L D
France; acc is variety known by this name in France 1947, when received by NFC; prob not Gros Fenouillet Gris of Leroy (1873),

nor any connection with name recorded 1667.
Covered in cinnamon russet; larger than Fenouillet Gris. Develops characteristic, sweet, scented quality of Fenouillet apples; firm, rubbery flesh.
F 13. T². P m-Oct. S Nov-Jan/Feb.

GROS-HÔPITAL syns Bailleul, Reinette de Bailleul 3 L C
France; arose nr Yvetot, Seine-Maritime (Normandy). Recorded 1864, but prob arose much earlier.
Large, red flushed. Sharp, yet quite rich. Cooked, keeps shape; sweet, good taste. Grown Normandy, used for tarts, fritters; juice (vit C 15mg/100g), cider.
F 14; trip. T³. P l-Oct. S Nov-Jan/Jun.

GROS LOCARD 2 L DC
France; local to Sablé, Sarthe, (Western Loire) according to Leroy (1873); introd 1849 by Leroy.
Crisp, firm, sharp in Feb. Cooked, keeps shape, quite sharp, pleasant.
Several types of Locard apples now exist in France; Gros Locard found in many areas; on sale up to 1950s. Used for cooking, juice.
F 17; trip. T³. C gd. P l-Oct. S Dec-Mar.

GROSSE DE SAINT-CLEMENT 1/5 M DC
France or Belgium; received from France; recorded 1895 by French nursmn Simon-Louis Frères.
Large, slight flush. Little flavour, low sugar, acid, rather hard in England.
Formerly widely grown in Belgium.
F 9. T². P e-Sept. S Sept-Oct.

GROSSE MIGNONNETTE D'HER-BASSY 4 L D
France; desc 1948 (Vercier).
Quite brisk, fruity; soft white flesh.
As Mignonette d'Herbassy much grown in Isère (Rhone) in 1940s.
F 17. T². P l-Oct. S Jan-Mar.

GROSSE ROUGE 4 L D
France; acc prob variety desc 1948 (Verger Français) under this name. Known also as Denfer and grown Lot (Midi-Pyrénées). Sweet, juicy, slightly perfumed.
F 18. T². P l-Oct. S Nov-Mar/June.

GROSSE SAULETTE 1/3 L C
France; sport or seedling of Saulette, variety desc 1948 (Verger Français) when grown Yonne (Burgundy).
Crisp, sharp. Cooks to sharp fruity purée in early Dec.
F 17. T². C hvy; frt sml, hangs late. P l-Oct. S Dec-Mar.

GRÜNER CALVINER 4 M C
Received 1951 from Fruit Res St, Jork, Hamburg, Germany.

Large, Calville shape. Quite sharp, crisp. Cooks to quite sweet, pale cream purée; very light flavour.
F 12. T². P e-Sept. S Sept-Oct/Nov.

GRVENA LEPOGVETKA 6 L D
Serbia; received 1975.
Dark maroon flush; recalls Red Delicious. Intensely flavoured, honeyed with pineapple-like acidity.
F 12. T². P l-Oct. S Dec-Jan/Feb.

GUELPH 6 M D
UK; raised by W. Pope, Welford Gardens, Newbury, Berks.
Charles Ross X Rival. Recorded 1912. RHS AM 1912; FCC 1913.
Soft, juicy, sweet, balanced; light pleasant flavour.
Named after town of Guelph and site of Ontario Agric Coll in Canada.
F 11. T². P e-Sept. S Sept-Nov/Dec.

GULDBORG 5/4 E D
Denmark; known 1870s on Lolland Island; named after Guldborg village.
Long, oblong; pink flush, stripes over light yellow. Good flavour; sweet, refreshing acidity, juicy, white flesh.
Widely distributed throughout Denmark by 1920s, also southern Sweden.
F 7. T². P eat l-Aug-Sept.

GUSTAVS DAUERAPFEL 6 L D
Switzerland; raised by Löbner, Wädenswil. First desc 1899.
Dark orange red flush, stripes. Lightly perfumed, aromatic quality; sweet, slightly crumbling texture.
F 11. T². P m-Oct. S Nov-Mar.

GYLLENKROKS ASTRAKAN 4 E D
Sweden; arose with Gyllenkroks family at one of their mansions, either in Skane or more likely in Regnaholms, Ostergötland. Prob known early 1800s. Introd 1863 by E. Lindgren. Poss original tree survives, now protected in garden, which belonged to Gyllenkroks family 1759–1859.
Bright red flush, stripes. Quite sharp, soft, juicy flesh; quickly goes over.
Most valued of Scandinavian summer apples, which became spread all over Sweden, into Norway, Finland. Still grown Swedish gardens.
F 3. T²; v hrdy. C gd. P eat e-Aug.

GYÓGYI PIROS 6 L D
Transylvania (Romania); recorded 1860. Form of Gyógyi alma, which arose Alsó-Fehér county.
Sweet yet brisk, refreshing taste.
Gyógyi alma is one of oldest, most popular Transylvanian apples; its seedlings often used as rootstock.
F 7. T². C hvy. P e-Oct. S Nov-Dec/Jan.

HALBERSTÄDTER JUNGFERNAPFEL 4 M D
Germany; arose Harz area. Recorded 1885.
Very attractive, bright scarlet flush. Modestly flavoured. Juicy, crisp, firm flesh.
F 11. T². P m-Sept. S Sept-Nov/Dec.

HAMBLEDON DEUX ANS 4 L C
UK; arose c1750 in Hambledon, Hants. Syns many, incl Green Blenheim; Grahams around Maidstone, Kent.
Large; dark maroon flushed, tough apple. Some richness, deep cream rather dry flesh, but mainly kitchen apple.
Cooked, keeps shape or makes stiff, gold purée; sweet, quite rich taste, early in season. With keeping, flavour and acidity fade.
In C19th reputedly every Hants garden had tree of 'Deusans'. Also grown countrywide in gardens; for market in Kent and favourite with Sheffield fruiterers. Still found in old Hants, Sussex gardens.
F* dp pink; 10. T³; sprd; v lng lived tree. C gd; prn bitter pit. P e/m-Oct. S Oct-Mar/May.

HAMBLING'S SEEDLING 1/5 L C
UK; raised by Major W. J. Hambling, Dunstable, Beds. Introd 1894 by G. Bunyard, Maidstone, Kent. RHS FCC 1893.
Large, perfectly round. Some richness, plenty of acidity. Cooks to well-flavoured sweet, gold purée early in season.
F 12. T³. P m-Oct. S Nov-Jan/Mar.

HAMMERSTEIN syn Minister von Hammerstein

HAMVAS ALMA 4 L D
Germany, poss; known since 1880s by this name in Hungary, but poss German variety, Weisser Matapel, desc 1821 by Diel.
Sweet, plenty of fruity acidity; soft flesh.
F 14. T². P m-Oct. S Nov-Jan/Mar.

HANNAN SEEDLING 3 M CD
UK; raised 1928 by Mrs I. Hannan, Walton-on-Thames, Surrey, from seed of Australian apple.
Large; bright red flush, stripes. Crisp, sharp, some sweetness. Cooked, very lightly flavoured.
F 8. T². P e-Sept. S Sept-Oct.

HANWELL SOURING 3 L C
UK; believed raised at Hanwell, nr Banbury, Oxon. Recorded 1820.
Named for its strong acidity. Large, cooking to fruity, sharp purée.
Formerly widely planted; still grown commercially West Midlands in 1930s, but surpassed by Bramley.
F12. T³; sprd. P m-Oct. S Nov-Jan/Apr.

HAPSBURG 5 E C
Received 1948 from EMRS, Kent.
Balsamatic taste, sharp, juicy. Cooks to quite sharp purée.
F 11. T³. C bien. P use Aug-Sept.

HARALSON 6 L D
USA; raised Univ Minnesota Fruit Breeding Farm, Excelsior, Minnesota. Malinda X. Introd 1923.
Named after C. Haralson, Excelsior's fruit breeder. Boldly striped, flushed. Crisp, juicy, firm; can be rather thin, acid in England with curious taste, almost of oranges.
Dual purpose in USA; also recommended for cider; grown commercially Minnesota, North Dakota.
F 12. T¹; v hrdy. C gd. P m-Oct. S Nov-Jan/April.

HARANG ALMA 6 L D
Hungary prob; arose prob C19th.
'Bell apple'. Sweet, lightly flavoured. Grown C19th in Pest county as high quality market apple.
F 15. T². P e-Oct. S Oct-Jan/Apr.

HARBERTS REINETTE syn Reinette Harbert 7 L D
Germany; recorded 1828 by A. Diel, when sent to him by pomologist Harbert from Westphalia. Diel thought poss arose in old monastery. Syns many.
Large; handsome. Sweet, plenty of fruit and acidity, recalls Golden Reinette.
Esteemed in Germany, grown also Alsace, Netherlands; also used for juice, drying and cooking, especially to top 'Apple Cake'. Formerly used for street planting in Saxony.
F* 13; trip. T³; sprd. C gd; prn bitter pit. P e-Oct. S Oct-Jan/Mar.

HARED see Delicious

HARLING HERO 6 M D
UK; raised by Frank Claxton, game dealer, gardener of East Harling, Norfolk; trees on sale 1930s in Norfolk.
Very large; deep red flush. Sprightly, white, melting, juicy flesh.
F mid/late. T³. P l-Sept. S Sept-Nov.

HARMONIE see Delorina

HARRY PRING 4 L D
UK; raised by Harry Pring. Introd by W. Peters, Leatherhead, Surrey. RHS AM 1914.
In Nov savoury, brisk, fruity, soft flesh; like James Grieve.
F* 12. T². P m-Oct. S Nov-Jan/Mar.

HARVEST FESTIVAL 1/5 M C
UK; raised c1949 by Mrs Helen Lloyd, Lytham, Lancs. Brought local prominence 1963–4 when raiser was Lady Mayoress.

Received 1968.
Large. Quite sharp, soft flesh. Cooked, keeps shape; can be bland.
F 13. T². P e-Sept. S Sept; shrt season.

HARVEY syn Dr Harvey

HATSUAKI 6 L D
Japan; raised 1939 at Morioka Branch of Fruit Tree Res St.
Jonathan X Golden Delicious. Select 1967; introd 1976.
Crisp, juicy, plenty fruit but can lack interest.
F est 12. T³; sprd. P m-Oct. S Nov-Dec/Jan.

HAUGHTY'S RED 4 M D
UK; received 1946 from Tenbury Wells, Worcs.
Crimson stripes over flush. Sweet, lightly aromatic. Soft, juicy flesh.
F 12. T². P m-Sept. S Sept-Oct.

HAWTHORNDEN 5 M C
UK; arose Scotland; first catalogued c1780.
Named after birthplace of C16th poet Drummond, nr Edinburgh. Primrose yellow, lightly flushed with pinky red. Cooks to creamy, brisk purée. Makes well-flavoured sauce and claimed best baked 'when its white creamy flesh breaks through the embrowned skin'. Popular Victorian garden apple. Grown by London market gardeners, also in Kent; but easily bruised; popularity waning by 1890s.
F 13. T²; sprd; hrdy. C gd. P m-Sept. S l-Sept-Nov/Dec.

HAZEN 6 E D
USA; raised at Great Plains Exp St, Mandan, North Dakota.
Duchess Oldenburg, Starking Delicious cross. Introd 1980.
Named after nearby town and Arlon Hazen, director of Exp St. Striking, large, deep red flush, with bloom. Very sweet, some perfume; recalls Delicious; yellow flesh. Dual purpose in US.
F 12. T²; v hrdy. P l-Aug. S Aug-Sept.

HECTOR MACDONALD 3 M C
UK; raised by Charles Ross, HG at Welford Park, Newbury, Berks. Recorded 1904. RHS AM 1904. Introd nursmn Pearson's Nottingham 1906.
Named after Ross's childhood friend and hero, 'Fighting Mac' popular soldier figure of late Victorian era. Resembles small Lane's Prince Albert. Cooks to pale cream purée, but low acidity; can be very mildly flavoured.
F 6. T²; sp. C hvy; frt sml, unless thinned. P m/l-Sept. S Sept-Dec.

HEJÖCSABAI SÁRGA 5 L D
Hungary; received 1948 from Agric Univ, Budapest.

Slightly scented, sweet in Nov; snow white flesh.
F 14. T². P m-Oct. S Nov-Dec.

HELIOS 4 E D
Germany; raised by M. Schmidt and H. Muraski at Inst for Agric & Hort, Münchenberg-Mark. Oldenburg X. Introd 1969.
Red flecks over pale primrose. Sweet, savoury, quite rich.
F est 12. T²; rcs mildew. P eat m-Aug.

HENRY CLAY 2/5 E D
USA; introd by Stark Bros, Missouri; received 1951.
Primrose yellow, orange pink flush. Strong, sweet-sharp taste. Soft, juicy flesh.
F 4. T². P e-Aug. S Aug-Sept.

HERCEG BATTHYÁNYI ALMA 2/5 L D
Hungary; recorded France 1876 by nursmn Simon-Louis Frères.
Small; russet freckles over gold. Brisk, densely fruity taste; recalls Golden Pippin type.
F 14. T¹. P m-Oct. S Nov-Feb/Apr.

HEREFORD CROSS 6 E D
UK; raised 1913 by G. T. Spinks, LARS, Bristol.
Cox's Orange Pippin X.
Dark maroon flushed, streaked. Quite rich, but rather sharp; crisp, cream flesh.
F 10. T². P l-Aug. S Aug-Sept.

HEREFORD QUEENING syn Crimson Queening

HEREFORDSHIRE BEEFING 6 L C
UK; arose Herefords. Known late 1700s.
Deep red flush, stripes. Sharp, firm, not very juicy flesh. Cooked, keeps shape; pretty yellow; quite rich flavour. Suitable for open tarts; best early in season, with keeping juice, acidity falls. In C19th recommended for drying.
F* 17. T²; uprt; tip. C frt drops. P e-Oct. S Oct-Dec.

HERMA 6 L D
Germany; raised 1930s by M. Schmidt at Inst for Agric & Hort, Münchenberg-Mark. Jonathan seedling. Received 1977.
Red Jonathan colour. Crisp, juicy, but in England underlying astringency; tough skin.
F 9. T². P l-Oct. S Jan-Mar.

HERRNHUT syn Schöner aus Herrnhut

HERRING'S PIPPIN 6 M D
UK; prob raised by Mr Herring of Lincoln. Recorded 1908. Introd by Pearson's Nursery of Nottingham.
Very striking; large; bright red. Crisp juicy; rather coarse texture with distinct flavour of aniseed.

F 15. T³. P e-Sept. S Sept-Oct.

HEUSGEN'S GOLDEN REINETTE 7 L D
Germany; raised by Pastor Konrad Henzen, Elsen, Rhine. Introd c1877.
Named after pomologist Peter Heusgen. Bright red flushed. Savoury, quite strong taste of fruit and acid in Nov, but mellowing; firm, juicy, finely textured, yellow flesh. Popularised England 1920s by Bunyard's Nursery.
F 17. T². C gd. P m-Oct. S Dec-Mar.

HIBBS SEEDLING 4 L D
UK; found by A. Moody, Southampton, Hants. Received 1961.
In good year, delicate aromatic flavour, scented, reminiscent of flowers; sweet, melting, juicy flesh. Often flavourless, dull appearance.
F 12. T². P e/m-Oct. S Nov-Dec.

HIBERNAL 3 M C
Russia, or arose USA and sent to Russia; known 1880.
Large. Cooks to sharp, bright lemon purée. Favoured for northern US states; still grown North Dakota; also used as rootstock.
F 6; trip. T²; v hrdy; res scab, mildew. P e-Sept. S Sept-Oct.

HIDALA see Braeburn

HIGH VIEW PIPPIN 7/8 L D
UK; raised 1911 by F. Fitzwater, Ernest Hill, Weybridge, Surrey.
Sturmer Pippin X Cox's Orange Pippin. RHS AM 1928.
Sweet, quite intensely flavoured, aromatic; firm, rather dry flesh; can have strange underlying almost tarry taste.
F 12. T². C gd; frt often sml. P m-Oct. S Nov-Mar.

HIGHWOOD see Jonagold

HILLWELLSYN HIDALA

HISTON FAVOURITE 5 M D
UK; raised by John Chivers, Histon, Cambs. Exhib 1883.
Good, brisk taste of fruit. Cooks well. 'Much grown in Cambs' in late C19th.
F 10. T². C gd. P l-Sept. S Oct-Dec.

HOARY MORNING 4 L CD
UK; believed arose Somerset. Recorded 1819. Syns many.
Large, bold red stripes over greenish yellow, with deep bloom, like hoarfrost. Plenty of acidity, some sweetness; firm flesh. Cooked, keeps shape; sweet, quite rich taste. Formerly favourite West Country apple; also used for cider.
F 12. T². C gd. P e-Oct. S Oct-Jan/Apr.

223

HOCKING'S GREEN 5 L DC
UK; raised c1860 by Mr Hocking, Illand Farm, Coads Green, nr Callington, Cornwall.
Dull green, turning yellow. Sharp, firm flesh. Cooked keeps shape; sweet, but can be insipid. Serves as plain eater by Christmas.
Formerly common in south east Cornwall and Tamar Valley.
F 14. T^2; thrives mild, high rainfall areas; res scab, canker. P m-Oct. S Nov-Jan.

HODGE'S SEEDLING 3 E C
UK; sent 1876 to Hogg by J. Vivian Esq, Hayle, Cornwall.
Boldly striped. Sharp, some sweetness, greenish white flesh. Cooked keeps shape; mildy sharp taste.
F 13. T^2. C hvy. P l-Aug. S Aug-Oct-Sept.

HOE 4 L D
South Korea; received 1967 via F. P. Matthews Nursery, Tenbury Wells, Worcs.
Curious taste, very sweet, but not sickly; yellow, firm, coarsely textured flesh.
F 18. T^2; sprd. C gd, hangs late. P e-Oct. S Nov-Dec.

HOG'S SNOUT 8 L D
UK; received 1947.
Conical, ribbed, with flat apex, like snout. Cinnamon russet over gold. Sweet sharp, Nonpareil type flavour by Jan; firm flesh.
F 7. T^2; sp. P m-Oct. S Nov-Jan/Feb.

HOHENZOLLERN 2/5 L D
Received from France 1947.
Large. Sweet, juicy, crisp cream flesh; light flavour.
F 14. T^3. C hvy. P l-Oct. S Jan-Mar.

HOLIDAY 6 L D
USA; raised 1940 by F. S. Howlett, Ohio Agric Exp St, Wooster, Ohio.
Macoun X Jonathan. Introd 1964.
Ripe during Christmas holidays. Deep red with waxy bloom, like Macoun. Sweet, crisp, juicy, slight vinous or strawberry flavour; tough skin. 'High quality' in USA, but often green, empty in England.
F 10. T^2; uprt. P m-Oct. S Nov/Dec-Jan.

HOLLANDBURY 6 M C
UK; first mentioned 1799 by Forsyth as Kirke's Scarlet Admirable, but recorded 1831 as Hollandbury by Ronalds.
Large, scarlet flushed. Valued by Victorians as much for its showy appearance as use. Cooks to brisk, creamy purée; good flavour.
F 7. T^2. P e-Sept. S Sept-Oct.

HOLLAND PIPPIN 1/5 L CD
UK; arose Holland, Lincs. Recorded 1729. Not Holland Pippin of USA.
Plain, fruity, plenty of acidity. Cooked

keeps shape; quite sharp, good flavour.
F 14. T^2; sprd. P m-Oct. S Nov-Mar.

HOLSTEIN 7 L D
Germany; raised or discv c1918 by Vahldik, teacher in Eutin, Holstein.
Poss Cox's Orange Pippin seedling.
Intense rich, aromatic flavour. Like excellent Ribston Pippin in taste and large Cox in appearance. Stronger, sharper than Cox; more coarsely textured.
Grown small scale for Farm Shop Trade.
Frt Col orng rd flush, brd rd stripes over gold; russet dots, some patches, veins. *Size* med/lrg. *Shape* rnd-con; flt base; sltly lop-sided; sometimes 5 crowned. *Basin* brd, qte shal; ribbed, russeted. *Eye* lrg, open; sepals lng, brd base. *Cavity* med/nrw, shal. *Stalk* shrt, thck. *Flesh* dp crm.
F 10; trip. T^3; sprd; prn mildew, canker; gd res scab. C gd; prn frost damage. P l-Sept/e-Oct. S l-Oct-Jan.
Improved colour selections:
   HOLSTEIN DR SCHULTZ
   HOLSTEIN GRIEMSMANN I
   HOLSTEIN GRIEMSMANN II
   HOLSTEIN MAHLER
   HOLSTEIN PALLOKS
   HOLSTEIN STEINHAUSEN
   HOLSTEIN SPORT, NFT

HOMMET syn Hommé 5 L D cider
France; desc Boré (1997) as sweet cider apple found Normandy, Picardy, Brittany.
F 23; trip. T^2; sp; sprd. P l-Oct. S Dec-Mar.

HONEYGOLD 5 L D
USA; raised 1935 Univ Minnesota, Hort Res Centre.
Golden Delicious X Haralson. Introd 1969.
Looks, tastes like Golden Delicious.
Raised for Minnesota, South Dakota, where climate marginal for Golden Delicious.
F 13. T^2; v hrdy. C gd. P l-Oct. S Nov-Dec/Jan.

HONEY PIPPIN 7 L D
UK; raised by J. Brooke, Newmarket, Suffolk. Received 1981.
Cox-like appearance. Honeyed, sweet, juicy, crisp, almost yellow flesh.
F 10. T^2. C gd; prn bitter pit. P m-Sept. S Sept-Oct/Nov.

HOOD'S SUPREME 4 E D
UK; raised 1924 by Miss B. Y. Hood, Duriehill, Edzell, Angus, Scotland.
Large, handsome. Very sweet, white flesh; almost no acidity.
F 12. T^2. C pr. P use Aug-Sept.

HOPE COTTAGE SEEDLING 5 M D
UK; raised c1900 by Mrs Oakely, Hope Cottage, Tenbury Wells, Worcs.
Princess Pippin (prob King of the Pippins) seedling.
Shape, colouring of King of the Pippins.

Pleasant quite brisk taste; slightly chewy flesh.
F 9. T^2; sp. P e-Sept. S Sept-Oct.

HOREI 4 L D
Japan; raised 1931 at Aomori Apple Exp St. Ralls Janet X Golden Delicious. Introd 1949.
Sweet, firm flesh but often woody, tough in England.
F 16. T^1. C hvy; frt sml unless thinned. P l-Oct. S Jan-Mar/Apr.

HORMEAD PEARMAIN 1/5 L C
UK; believed raised at Hormead, Herts. Recorded 1826 by London Hort Soc. RHS AM 1900.
Sharp white flesh. Cooked keeps some shape; pleasant, sweet flavour.
Re-popularised by Bunyard's Nursery in early 1900s and planted again in gardens.
F 11. T^2. C hvy; frt sml unless thinned. P e-Oct. S Nov-Mar.

HORNEBURGER PFANNKUCHEN 3 L C
Germany; chance seedling arose c1850 in garden on the Marschdamm, Horneburg. Marschdamm leads directly to Alte Land, fruit region of north west Germany. First village is Neuenkirchen, where variety was first brought into commercial cultivation by Jacob Schliecher. Poss Boïken seedling.
'Pancake apple of Horneburg'; large, heavily ribbed. Makes good sauce; sharp, fruity. Not as juicy or acidic as Bramley.
Premier North German cooker in 1939, prized for heavy, regular crops; but declined by 1960s. Formerly imported to UK, when Bramley in short supply.
F 16; trip. T^2; sprd. C hvy. P e-Oct. S Oct-Jan.

HORSFORD PROLIFIC 5 L D
UK; found 1913 in vicarage garden, Horsford, Norwich, Norfolk by Rev. Mountford.
Large. Crisp, juicy, quite sharp, but mellowing.
F 16. T^2. P m-Oct. S Dec-Mar.

HORSKREIGER 1/3 L C
Received 1950 from Kendall, Cumberland; prob not Hoskreiger of Hogg (1851).
Large. Cooks to well-flavoured, sharp, savoury, lemon purée. Mellowing, but still strong, brisk in Feb.
F 15. T^2. P m-Oct. S Nov-Mar.

HORST NO 2 see Golden Delicious

HOSSZÚFALUSI 5 M CD
Hungary; received 1948 from Univ Agric, Budapest.
Quite richly flavoured. Cooked keeps shape, pleasant flavour.
F 9. T^2. P e-Sept. S Sept-Oct.

HOUBLON 7 L D
UK; raised by Charles Ross, HG, Welford Park, Newbury, Berks.
Peasgood Nonsuch X Cox's Orange Pippin. Introd 1901. RHS AM 1901.
Named after Ross's employers, Houblon banker family. Rich, aromatic flavour; rather coarse yellow flesh.
F 14. T^2; prt tip. **C** lght. **P** l-Sept. **S** Oct-Dec.

HOUNSLOW WONDER 3 M C
UK; introd Spooner, nursmn of Hounslow, Middx.
Believed from Dumelow's Seedling. Recorded 1910; RHS AM 1910.
Bright red flush, stripes. Cooks to creamy purée, like Dumelow's Seedling; good, sharp, savoury flavour; needs sugar.
F10. T^2. **C** gd. **P** m-Sept. **S** Sept-Dec.

HOWGATE WONDER 3 L C
UK; raised 1915–16 by G. Wratten, Howgate Lane, Bembridge, Isle of Wight. Blenheim Orange X Newton Wonder. Introd 1932 by Stuart Low Co. RHS AM 1929.
Famed for its large size. Eaten fresh, quite sweet, juicy, pleasant. Cooked, keeps shape but very light; insipid by comparison with Bramley.
Exhibition, garden variety, also grown small extent commercially.
Frt Col flushed brownish rd, rd stripes, over pale grn; turning bright orng rd over yell; becomes greasy. *Size* lrg/v lrg. *Shape* shrt-rnd-con; ribbed; 5 crowned; flt base, tapering to flt apex. *Basin* brd, dp; ribbed, puckered; sometimes ltl russet. *Eye* lrg, clsed or hlf open; sepals lrg, brd, v downy. *Cavity* brd, dp; russet lined. *Stalk* shrt, thck. *Flesh* pale crm.
F 12. T^3; sprd; res mildew. **C** hvy. **P** e/m-Oct. **S** Nov-Mar.

HUBBARD'S PEARMAIN 7 L D
UK; arose Norfolk; known before 1800. Exhib London Hort Soc 1819 by Norwich nursmn G. Lindley.
Sweet, rich, nutty; recalls Adams's Pearmain. 'Real Norfolk apple' claimed Lindley; became quite widely grown in gardens C19th.
F* snowy wht; 10. T^1. **C** gd. **P** e/m-Oct. **S** Nov-Feb/Mar.

HUBBARDSTON NONSUCH 7 L D
USA; arose Hubbardston, Mass. Recorded 1832.
Sweet, juicy, lightly flavoured.
Grown for autumn markets in Massachusetts, New York in C19th.
F 13. T^1. **C** gd. **P** m-Oct. **S** Nov-Jan.

HUME 6 M D
Canada; raised 1898, Exp Farm, Ottawa. McIntosh X. Introd 1921.

McIntosh type. Sweet, soft, juicy, white flesh; vinous flavour.
F 4. T^3; v hrdy. **P** e-Oct. **S** Oct-Nov.

HUNTER KINKEAD SPY *see* Northern Spy

HUNTER MELBA *see* Melba

HUNTER OTTAWA 244 4 L D
Canada; received 1977. Tetrap form of Ottawa. 'Spicy', good quality; long keeping.
F 7. T^2. **P** e-Oct. **S** Nov-Jan.

HUNTER SANDOW *see* Sandow

HUNTER'S MAJESTIC 3 M CD
UK; raised before 1914 by Miss E. Balding, Upwell, Wisbech, Cambs.
Worcester Pearmain X. Introd 1928 by W. Hunter, Wisbech.
Large. In Sept, cooks to brisk, bright yellow purée. Mellows to almost rich taste, eaten fresh.
F 10. T^2. **P** m-Sept. **S** Sept-Nov/Dec.

HUNTER SPARTAN *see* Spartan

HUNTINGDON CODLIN 5 M D
UK; introd by nursmn Wood & Ingram, Huntingdon; exhib by them 1883; desc Hogg.
Ribbed, angular. Sharp, soft fleshed in Sept. Cooks to brisk purée.
F 10. T^3. **C** hvy. **P** e-Sept. **S** Sept-Oct.

HUNT'S DUKE OF GLOUCESTER 8 M D
UK; raised by Dr. Fry of Gloucester from Nonpareil seed. Sent 1820 to London Hort Soc by Thomas Hunt, Stratford upon Avon; then named.
Cinnamon russet over gold. Sweet-sharp quality of Nonpareil family; rich, good; firm, rather dry flesh.
F 7. T^1. **P** m-Sept. **S** Oct-Nov

HUNT'S EARLY 4 E D
UK; poss raised by Thomas Hunt, Stratford-upon-Avon. Desc 1884 Hogg.
Very sharp, crisp; becoming less acidic, soft fleshed with hint of balsam.
Formerly grown Kent to supply early London markets.
F 7. T^3; sprd. **C** hvy. **P** use Aug.

HUSVÉTI ROZMARIN syn Entz-Rosmarin

I. B. REID 1 M C
UK. Received 1987 from Miss I. B. Reid, Welwyn Garden City, Herts.
Large. Very sharp, fresh; cooks to good, intense, sharp, fruity purée.

Resembles Bramley in quality, but the advantage of making small tree.
F 15. T^1. **P** e-Oct. **S** Oct -Dec.

IDAGOLD 5 L D
USA; raised by Leif Verner, Idaho Agric Exp St, Moscow, Idaho.
Esopus Spitzenburg X Wagener. Introd 1944.
Sweet, slightly fruity in Feb; often poor quality in England.
F 10. T^2. **P** m-Oct. **S** Jan-Mar.

IDAHO DELICIOUS *see* Delicious

IDAJON 6 L D
USA; raised by Leif Verner, Idaho Agric Exp St, Moscow, Idaho.
Wagener X Jonathan. Introd 1949.
Sweet, juicy cream flesh, but rather flat, flavourless; tough skin.
F 13. T^2. **P** l-Oct. **S** Dec/Jan-Mar.

IDARED 6 L D
USA; raised by Leif Verner, Idaho Agric Exp St, Moscow, Idaho.
Jonathan X Wagener. Introd 1942. RHS AGM 1993.
At best, crisp with sprightly taste, but often flavourless, chewy. In Michigan, however, fruit enthusiast Robert Nitschke finds 'high flavour and aromatic' content, 'excellent for sauce and pie'. Austrian chefs use it for 'Apple Strüdel'. Vit C 18–23mg/100g.
Grown US, Canada, France, Germany, Austria, Poland, Switzerland; small extent England.
Frt Col bright rd flush, indistinct rd stripes over grnish yell/yell; smooth, shiny skin. *Size* med. *Shape* flt-rnd; trace ribs, sltly crowned. *Basin* qte brd, dp; ribbed, puckered. *Eye* sml, prt clsd/open; sepals, sml. *Cavity* nrw, dp; russet lined. *Stalk* shrt, qte thin. *Flesh* wht, tinge grn.
F 7. T^2; sprd; prn canker, mildew. **C** gd. **P** e-Oct. **S** Nov-Mar; commercially-Jun.

IDUNA 5 L D
Switzerland; raised Fed. Agric. Res. St. Wädenswil.
Glockenapfel X Golden Delicious. Received 1999.
Golden; perfect even shape. Firm, juicy, cream flesh; some honeyed quality, but more piquant and savoury than Golden Delicious.
F early. T^2. **P** e-Oct. **S** Nov-Jan.

IJZERAPPEL syn Marie-Joseph d'Othée

ILDRØD PIGEON 6 L D
Denmark; prob arose Isle of Fyn. Known by 1840. Propagated by Hans Olsen, Ormslev, nr Korsør. Introd c1873 by Mathiesens Nursery.
Danish Christmas apple; its name means 'Fire-red' Pigeon. Crackling, crisp, juicy

white flesh; slightly scented, strawberry flavour; tough skin.
F 10. **T**[1]. **C** gd. **P** e/m-Oct. **S** Dec-Feb/Mar.

IMPÉRIAL 4 L D
France; received 1957 from INRA, Angers. Prob seedling of York Imperial raised at INRA.
Dark red flush; pure white flesh. Sweet, soft, juicy, but no real flavour.
F 9. **T**[2]. **P** e-Oct. **S** Nov-Jan/Mar.

IMPERIAL GALA *see* Gala

IMPROVED ASHMEAD'S KERNEL (BUNYARD) 8 L D
UK; received 1939 from E. Bunyard, Maidstone, and variety desc by Bunyard. Prob not variety exhib 1883 by nursmn Lee & Son.
Prob seedling of Ashmead's Kernel.
Larger, more russeted than Ashmead's Kernel, but same sweet-sharp, acid drop flavour.
F 12. **T**[3]. **P** m-Oct. **S** Dec/Jan-Mar.

IMPROVED COCKPIT 4 M DC
UK; recorded 1902; exhib 1934.
Larger than Cockpit; pink flushed; sweeter; ripening earlier.
F 15. **T**[2]. **P** m-Sept. **S** Sept-Dec

INDO 2 L D
Japan; recorded 1930.
Poss seedling of White Winter Pearmain.
Very strange taste; extremely sweet, no acidity; very firm, greenish yellow flesh.
Grown commercially Japan, China.
F 9. **T**[2]. **P** m/l-Oct. **S** Jan-Mar.

INGALL'S PIPPIN 7 M D
UK; raised 1915 by William Ingall, Grimoldby, Lincs. Introd 1928.
Large. Sweet, light flavour.
F 21. **T**[2]; prt tip. **P** l-Sept. **S** Sept-Nov.

INGALL'S RED 6 E D
UK; raised by William Ingall, Grimoldby, Lincs. Received 1930.
Dark maroon with bloom. Very sweet, strawberry flavour and texture of Worcester Pearmain, but almost no acidity; firm white flesh.
F 8. **T**[1]; sprd. **P** eat l-Aug-e-Sept.

INGOL 4 M D
Germany; raised 1954 at Fruit Res St, Jork, nr Hamburg.
Ingrid Marie X Golden Delicious.
Large; bright red flush over yellow. Good, rich taste; strong pineapple acidity; soft, juicy flesh.
Grown Germany, mainly for processing as juice, pie filling.
F 16. **T**[2]; weeping. **C** hvy. **P** e-Oct. **S** Oct-Nov/Dec; commercially-Apr.

INGRID MARIE 6 M D
Denmark; raised c1910 at Flemlöse, Westfyn Island.
Believed Cox's Orange Pippin seedling. Desc 1924.
Named after daughter of Mr Madsen, a teacher, who found the tree.
At best, quite rich, lightly aromatic flavour, soft juicy flesh; but in England often rather bleak.
Grown commercially and in gardens in northern Europe – southern Sweden, Denmark, Alte Land of Germany and Baltic coast; formerly to small extent England.
Frt Col dark rd flush, rd stripes; grnish yell/yell background; lenticels prom as lrg russet dots. *Size* med/lrg. *Shape* flt-rnd. *Basin* brd, shal; sltly ribbed. *Eye* hlf/fully open; sepals pointed. *Cavity* brd, dp; russet lined. *Stalk* med, qte thin. *Flesh* pale cream.
F 14. **T**[2]. **C** gd; frt cracks round eye. **P** l-Sept. **S** Oct-Dec.
> **RED INGRID MARIE**; syn Karin Schneider; more highly coloured. Discv Denmark; received 1953.

IRISH PEACH 4 E D
Ireland; poss arose Sligo. Sent 1819 to London Hort Soc by John Robertson, nursmn of Kilkenny.
Ideally eaten straight from tree. Rich balance of sweetness and acidity, slightly perfumed, juicy flesh, but on brisk side. Often difficult to capture at best.
Popular Victorian and Edwardian variety; forming 'a beautiful dish for dessert'. Also grown by London market gardeners and in Kent in C19th.
Frt brown rd flush, flecks over pale yell. *Size* med/sml. *Shape* rnd to con; rnded ribs, flt base; viewed from base appears hexagonal. *Basin* brd, med dpth; sltly ribbed. *Eye* sml, clsd; sepals, shrt, brd. *Cavity* sml, shal; sltly russeted. *Stalk* shrt, thck. *Flesh* pale crm.
F 5. **T**[3]; tip. **C** gd. **P** eat m-Aug.

IRON PIN 1 L C
UK; poss arose Dorset. Desc 1884 (Hogg).
Hard flesh. Poor culinary flavour.
F 12. **T**[3]. **P** l-Oct. **S** Nov-Apr.

ISAAC NEWTON'S TREE 3 M C
UK; propagated from tree growing in Isaac Newton's garden at Woolsthorpe Manor, nr Grantham, Lincs. Appears identical to Flower of Kent, which was listed 1629 by Parkinson, but not mentioned again until 1802 by Forsyth.
Large, heavily ribbed. Cooks to sweet, delicately flavoured purée.
Newton's early work on gravitation was done 1665–66 when he was staying at his mother's house, Woolsthorpe, where he had gone to avoid the plague. While sitting under an apple tree, the 'notion of gravitation came into his mind occasion'd by the fall of an apple'. This tree, eventually supported by props, died in 1814. Its wood was used to make a chair for the Woolsthorpe Library. Some years previously the Manor owners had repropagated the tree and planted it in Lord Brownlow's garden at nearby Belton. Material from this tree, via Kew Gardens and EMRS, Kent has provided scion wood for Isaac Newton trees at National Physics Laboratory Teddington and sites in Cambridge, USA, New Zealand.
F 20. **T**[2]; sprd; prt tip. **P** m-Oct. **S** Nov-Jan.

IVETTE 5 L D
Netherlands; raised 1950 by A. A. Schaap at IVT, Wagcningen.
Cox's Orange Pippin X Golden Delicious. Introd 1966.
Honeyed, crisp; more acidity than Golden Delicious and elements of Cox's aromatic flavour. Grown Rhineland.
F 13. **T**[2]. **P** e-Oct. **S** Nov-Jan/Feb.

IVÖ 4 L D
USA; syn Monroe Seedling. Found growing in wood nr Washington by trapper in 1920. Scions from tree sent 1930 to Ivö, Sweden. Bright red flush, which seeps in to stain flesh. Sweet, rich, delicate flowery, aromatic flavour in Nov; quite soft fleshed.
F 17. **T**[3]. **P** m-Oct. **S** Nov-Dec/Jan.

ÍZLETES ZÖLD 4/7 L D
Uncertain origin; but widely grown Transylvania C19th. Received 1948 from Hungary.
Very sweet, scented, like sugared almonds; firm, pale cream flesh.
Formerly market apple.
F 4. **T**[3]. **P** m-Oct. **S** Nov-Dec/Jan.

JACQUES LEBEL 1 M C
France; raised c1825 by nursmn Jacques Lebel at Amiens (Picardy). Introd 1849 by nursmn, Leroy.
Large. Crisp, juicy, lots of acidity. Cooks to quite sharp, yellow purée.
Grown France, Belgium; used for cooking, juice (vit C 10mg/100g) and processing for pectin. In Belgium favoured for 'siroop' (apple paste). Also grown Switzerland and formerly Germany. Appearance resembles Bramley's Seedling and imports to UK were confused with it in 1890s.
F 12; trip. **T**[3]. **C** hvy. **P** Sept. **S** Sept-Dec.

JACQUET 3 L C
France; desc 1948 (*Verger Français*), when grown Indre (Val de Loire).
Cooked, retains shape; good, sweet, rich flavour.
F 22. **T**[2]. **P** l-Oct. **S** Dec-Mar

JACQUIN 2/5 L D
France; raised by Boisbunel. Recorded
1872. RHS FCC 1893.
Fresh, fruity nr Feb; sweet, very crisp, white
flesh; tough skin.
Grown Meurthe et Moselle (Lorraine) in
1940s.
F 12. T². P l-Oct. S Jan-May; claimed keeps
2 yrs.

JAFRA 4 L D
Netherlands; found 1971 by dairy farmer,
P. van Dueren.
Quite intense, rich taste; juicy but rather
chewy flesh.
Grown commercially Netherlands; valued
for ease of cultivation, good flavour.
F 9. T². P e-Oct. S Oct-Dec.

JAMBA 69 4 M D
Germany; raised by D. E. Loewel, Jork
Fruit Res St, nr Hamburg.
Melba X James Grieve. Named 1969.
Beautiful, red flushed over cream. Sweet,
soft, juicy flesh.
Planted commercially in Alte Land, where
replaced James Grieve.
F 14. T³. C gd. P e-Sept. S Sept-Oct.

JAMES GRIEVE 4 E DC
UK; raised by James Grieve in Edinburgh.
Believed, either from Pott's Seedling or
Cox's Orange Pippin. Introd by Dickson's
Nursery, where Grieve was manager.
Recorded 1893. RHS AM 1897, FCC 1906,
AGM 1993.
Savoury, juicy, crisp yet melting flesh with
strong acidity, which can be overpowering
early in its season. Acidity mellows and flesh
becomes soft and marrowy, yet flavour does
not fade. Picked early, or grown in north,
cooks well; keeps shape; makes sweet,
delicate 'stewed apple'.
Remains valued garden variety. Formerly
market apple, but bruises easily; also
formerly pollinator for Cox orchards.
Widely distributed over Northern Europe,
grown commercially and gardens in
Netherlands, Belgium, Denmark, Sweden
and Germany, particularly in Alte Land,
mainly for juice manufacture, but now
being overtaken by Jamba.
Frt *Col* rd flush, stripes over pale
grn/yellow. *Size* med. *Shape* rnd-con; sltly
ribbed. *Basin* med wdth, qte dp; sltly
ribbed; puckered. *Eye* qte lrg, clsed; sepals,
lng, qte downy. *Cavity* brd, dp; lined scaly
russet. *Stalk* med/lng, qte thin. *Flesh* pale
crm.
F 10. T²; sprd; hrdy; prn canker; res mildew.
C hvy, frt bruises. P e/m-Sept. S Sept-Oct;
keeps-Dec.
    **ERICH NEUMANNS ROTER
JAMES GRIEVE**; darker flush, stripes;
ripening later. Arose 1953 with E.
Neumanns, Germany

LIRED (ROOSJE); flushed rather than
striped.
    REDCOAT GRIEVE (Iliffe); heavier
cropper. Discv 1916 by H. Jones at
Letchworth, Herts; introd 1921.

JAMES LAWSON 6 E D
UK; Cellini, Gravenstein cross. Exhib 1918
by nursmn, Cannel, Eynsford, Kent. RHS
AM 1918.
Curious balsam taste, like Cellini. Quite
sharp, soft, juicy, white flesh.
F 9. T²; sp. C gd. P l-Aug. S Aug-Oct.

JANSEN VON WELTEN 7 L D
Germany; raised by Jansen at Welten nr
Aachen. German pomologist Diel received
samples in 1823.
Rich, with strong raspberry flavour. Crisp
yet melting flesh, plenty of sweetness,
acidity, but can be only fair flavour.
F 16. T². P m-Oct. S Dec-Jan/Mar.

JAN STEEN 6 L D
Netherlands; raised by Piet de Sonnaville
(1917–95) in his family orchards at Puiflijk
and Altforst, nr Nijmegen.
Sternreinette (Reinette Rouge Étoilée) X
Cox's Orange Pippin. Received 1955.
Crimson flush. Lightly aromatic recalling
Cox, but quite bold acidity; finely textured
flesh.
F 18. T³. C lght; frt sml. P m-Oct. S Nov-Jan.

JÁSZ VADÓKA 2/5 L D
Hungary poss; unknown origin. First desc
1886 by Hungarian pomologist Bereczki,
who obtained it 1876/78 from friends in
Lekelhalma and Halas under this name.
Sweet, white flesh, but in England often
little flavour or acidity.
Market apple in C19th Hungary.
F 6. T¹. P m-Oct. S Nov-Feb/Apr.

JEANNE HARDY (BUNYARD) 3/6 L CD
France; received 1939 from E. Bunyard,
Maidstone, Kent; very similar to Belle de
Pontoise (Luxembourg) in NFC. Acc may
be variety raised 1879 by M. Hardy at
School of Hort, Versailles; named after his
daughter; Alexander seedling.
Large; crimson flushed. Quite sweet, slight
vinous flavour; soft, juicy, white flesh.
Cooks to cream purée, but very mild taste.
Jeanne Hardy was grown around Paris in
1940s. Formerly exhibition apple in
England.
F 12. T³. C gd. P m-Oct. S Nov-Jan/Mar.

JEAN TONDEUR 7 L D
France; desc 1948 (*Verger Français*), when
grown Marne and Argonne (Champagne).
In New Year, quite fruity, sharp; soft flesh.
F 30. T². P l-Oct. S Jan-Mar/Apr.

JEFFERIS 4 M D
USA; arose c1830 on farm of Isaac Jefferies,

Newlin Township, Chester County,
Pennsylvania. Awarded premium 1848 by
Pennsylvania Hort Soc for best seedling.
Prettily dark red flushed, striped. Sweet,
juicy, melting cream flesh; light aromatic
quality.
Remains highly regarded garden fruit
in US.
F 9. T²; hrdy; claimed res scab, mildew.
C hvy; frt sml. P m-Sept. S Sept-Dec.

JENNETING syn Joaneting

JENNIFER 4 E D
UK; raised by F. W. Wastie, Eynsham,
Oxford.
Duchess Favourite X Beauty of Bath.
Received 1944.
Sharp, firm cream flesh; low on juice,
flavour.
F 10. T³. P Aug. S Aug-Sept.

JENNIFER WASTIE 6 M D
UK; raised by F. W. Wastie, Eynsham,
Oxford.
Ribston Pippin X Barnack Beauty. Received
1945.
Sweet, little acidity; chewy, white tinged
green flesh.
F 12. T². P l-Sept. S Oct-Nov.

JERSEY BEAUTY 4 L D
UK; received 1926 from Newport, Isle of
Wight.
Very pretty, pinky red flush. Fairly acid,
firm, crisp flesh.
F 16. T². P m-Oct. S Nov-Feb/Mar.

JERSEY BLACK 6 L D
USA; pres arose New Jersey. Recorded 1817
by Coxe as 'Black Apple'.
Almost black flush. Very sweet juice but
flesh has curious resinous taste.
'Much admired table fruit' in early C19th
New Jersey; also used for cider.
F 11. T³. P l-Oct. S Nov-Jan/Mar

JERSEYMAC 6 M D
USA; raised 1956 Rutgers Univ, Coll of
Agric, New Brunswick, New Jersey.
(Melba X (Wealthy X Starr) X Red Rome X
Melba) X Julyred. Fruited 1961. Introd
1971.
Dark colour, bloom of McIntosh, but
ripening month earlier. Perfumed with
slight strawberry or vinous flavour; sweet,
melting, snow white flesh; tough skin.
Often tastes green, unattractive in England.
Grown eastern States, Italy.
F 9. T¹; prn scab; v prn canker. C gd; frt
bruises. P e-Sept. S Sept-Oct.

JESTER 7 L D
UK; raised 1966 by Dr F. Alston, EMRS,
Kent.
Worcester Pearmain X Starkspur Golden
Delicious. Introd 1981.

Attractive bright red on yellow. Light refreshing quality; lots of fruit, juicy. Can be rather 'empty'.

F 15; res frost. T^2. C hvy. P e-Oct. S Oct-Dec.

JEWETT'S FINE RED syn Nodhead 6 M D

USA; believed arose Hollis, New Hampshire. Recorded 1842.

Quite rich, good taste; firm, crisp flesh. Esteemed Maine, New Hampshire in C19th.

F 10. T^3; sprd. P l-Sept. S Oct-Nov.

JIM-BRIAN 2 L D

South Africa; raised by B. C. du Toit.

Golden Delicious X Granny Smith. Received 1995.

Crisp, green tinged cream flesh, hard and acidic; texture and sharpness of Granny Smith predominates.

F mid. T^2. P m-Oct. S Dec-Mar.

JINCOA ZAGARRA (Jinco Sagarra) 2/4 L D

France; old variety of Landes (Acquitaine), where still grown.

Firm fleshed, quite fruity in New Year. Valued for long keeping qualities.

F 19. T^2. P l-Oct. S Jan-Mar/Apr.

JOANETING syn Geneting, Joaneting, Jenetting, White Joaneting 5 E D

UK; acc is Joaneting of Hogg, which he believed to be very old. Many syns. 'Ginnitings', mentioned by Francis Bacon (1561–1626) in essay *On Gardens*. Juniting noted 1665 by J. Rea. Small, crisp, brisk, eaten straight from tree, but season over in few days.

Grown gardens and for market for centuries, and up to 1920s marked opening of apple season.

Frt *Col* grn/yell. *Size* sml. *Shape* rnd-con; trace smooth ribs. *Basin* qte brd, shal. *Eye* hlf open to clsd. *Cavity* qte brd, shal. *Stalk* lng, thin. *Flesh* pale crm.

F 1. T^2; uprt. C gd. P eat l-July.

JOHN DIVERS 3 M C

UK; raised 1905 prob by nursmn, former HG, W. H. Divers, Surbiton, Surrey.

Jenkinson's Seedling X Baumann's Reinette. Very large; round, regular shape. Cooks to sharp, juicy purée.

F 10. T^2; sp. P e-Sept. S Sept-Nov/Dec.

JOHN HUGGETT 7 E CD

UK; raised 1940 by John Huggett, Grange-over-Sands, Lancs.

Allington Pippin X.

Quite sweet and rich. Cooks to pale cream purée; rich, good with lots of flavour.

F 8. T^2. C frt often pr col. P Aug. S Aug-Sept.

JOHNSON McINTOSH *see* McIntosh

JOHN STANDISH 6 L D

UK; prob raised c1873 by nursmn John Standish, Ascot, Berks. Introd 1921 by Isaac House & Sons, nursmn, Bristol; brought prominence by exhibit at Imperial Fruit Show 1922. RHS AM 1922.

Bright scarlet flush. Intensely fruity; firm, crisp flesh.

Formerly grown commercially as late red apple.

F 9. T^3; uprt. C gd; frt sml. P m-Oct. S Dec-Feb.

JOHN WATERER 1 M C

UK; introd 1920 by John Waterer Sons & Crisp, Twyford, Berks.

Large, regularly shaped. Cooks to pretty lemon froth, sharp, needing extra sugar in Sept. Becomes less sharp, cooks more firmly.

F 11. T^1. P e-Sept. S Sept-Nov.

JOLÁNKA 4 M D

Hungary; received 1948 from Agric Univ, Budapest.

Quite intense, brisk fruity taste; firm, crisp flesh.

F 10. T^3. P m-Sept. S Oct-Dec.

JOLIBOIS syn Alfred Jolibois

JOLYNE 5 L D

France; received 1950 from Corrèze (Limousin).

Clear yellow. Quite fruity in Dec; crisp, juicy.

F 21. T^2. P m-Oct. S Nov-Jan/Mar.

JOMURED *see* Jonagold

JONADEL 6 L D

USA; raised 1923 by H. L. Lantz, Iowa Agric Exp St, Ames, Iowa.

Jonathan X Delicious. Introd 1958.

Red Jonathan colour, but larger fruit, with sweetness of Delicious. Very honeyed, but not cloying; juicy, finely textured almost yellow flesh.

Grown Germany and France to small extent.

F 16. T^2. P m-Oct. S Nov-Feb; cold store-Mar.

JONAFREE 6 L D

USA. Raised 1965 by co-operative breeding programme of Illinois, Purdue & Rutgers Univ. Complex parentage includes Crandall's Rome X *Malus floribunda*, Jonathan, Gallia Beauty, Red Spy, Golden Delicious. Introd 1979.

Firm crisp, juicy pale yellow flesh.

F mid. T^3; res scab, cedar apple rust; slight suscept mildew, fireblight. P e-Oct. S Oct-Dec.

JONAGOLD 4 L D

USA; raised 1943 at NYSAES, Geneva.

Golden Delicious X Jonathan. Introd 1968. RHS AM 1987; AGM 1993.

Attractive with rich, honeyed, almost aromatic flavour; crisp, juicy, nearly yellow flesh. Acidity of Jonathan gives less cloying, better balanced taste than Golden Delicious. Vit C 10–21mg/100g.

Grown commercially all over northern Europe, particularly Germany, also Austria, Belgium, France, Holland, Italy, Poland, Switzerland; Russia, Japan, China, northern States; small extent England.

Frt *Col* ornge rd flush, rd stripes over gold; numerous fine crm or russet dots; sltly greasy skin. *Size* med/lrg. *Shape* rnd-con; trace ribs, flt sides; sltly crowned. *Basin* lrg, brd, dp; ribbed. *Eye* sltly open; sepals, qte downy. *Cavity* brd, dp; russet lined. *Stalk* med/qte lng; thin. *Flesh* crm.

F 15; trip. T^3. C v hvy; prn canker; frt can be pr col in England. P m-Oct. S Nov-Jan/Feb; commercially-Apr, but softens, bruises.

Many highly coloured sports found include:

CROWNGOLD; discv NFC 1979; more flushed form.

DECOSTA; received 1988 from H. Decoster, Belgium. Dark red, slightly striped form.

EARLY QUEEN; received 1993 from J. Boerekamp, Netherlands. Dark red flush; matures earlier.

JONAGORED; discv at Halen, Belgium By J. & R. Morren; introd 1985. Original dark red striped sport. Several times winner of 'Tastiest Apple Competition' at National Fruit Show, Kent.

JONICA; discv Germany; received 1996. Light red flush, unstriped.

JORAYCA; received 1996 from C. Carolus, Belgium. Light red flush; ripens earlier.

JORED; discv J. Nicholaï, St Truiden, Belgium; received 1986. Light flush, unstriped.

JUMURED; received 1989 from J. Nicholaï, Belgium. Dark flush, unstriped.

KING JONAGOLD; discv 1985 by J. Nicholaï, Belgium. Even red flush.

NEW JONAGOLD; sport from Japan; discv 1980.

RED JONAPRINCE; discv 1994 by Princen Bros, Weert, Netherlands. Dark red flush, colouring early.

RUBINSTAR; discv 1980 at Gaiberg, Germany. More intense red, less conical shape.

JONAGORED *see* Jonagold

JONALICIOUS 6 L D

USA; discovered 1933 in Abilene, Texas by Anna Morris Daniels.

Believed Jonathan, Red Delicious cross. Introd 1960 by Stark Bros Nurseries, Missouri.

Bright red flush stripes over clear yellow. Slightly honeyed, but often rather bland, yellow flesh.
F 20. **T**^{2}. **P** m-Oct. **S** Nov-Jan; commercially-Mar.

JONAMAC 6 M D
USA; raised 1944 at NYSAES, Geneva.
McIntosh X Jonathan. Introd 1972.
McIntosh type, but ripening earlier. Very scented, perhaps hint of strawberry flavour; lots sugar, acid; white melting flesh.
Grown small extent West New York.
F 12. **T**^{2}; sprd. **P** m-Sept. **S** Oct-Nov.

JONARED *see* Jonathan

JONATHAN 6 L D
USA; arose on farm of Philip Rick, Woodstock, Ulster County, New York.
Believed Esopus Spitzenburg seedling. Desc 1826. Introd England prob by nursmn Thomas Rivers, who exhib RHS 1864.
In good year, crisp, juicy, sweet with plenty of refreshing acidity until Feb, but can taste woody metallic in England. In US of 'princely flavour'; also dual purpose valued for sauce, pies.
Owed early popularity to Judge Buel, Pres of Albany Hort Soc. He named it after Jonathan Hasbrouck, who had drawn his attention to tree growing amongst some rough scrub on Rick's farm in foothills Caskill Mountains. In 1829 Buel sent specimens to first meeting of Massachusetts Hort Soc, who declared it most promising new variety of year. Spread to south and west of USA; to Europe and Australia; became widely grown throughout warmer apple producing regions of world. Now surpassed but still important commercial variety in US, Canada; also grown Italy, Austria, Poland.
Frt Col bright crimson flush, some broken rd stripes; yell ground colour. *Size* med. *Shape* oblng to rnd-con; ribbed; sltly 5 crowned. *Basin* qte nrw, dp; ribbed or puckered. *Eye* qte sml, clsd; sepals, brd based, shrt; qte downy. *Cavity* nrw, dp; russet lined. *Stalk* shrt/med, thin. *Flesh* wht.
F 10; claimed self fertile. **T**^{2}; weeping; prn mildew; claimed res scab. **C** gd. **P** e-Oct. **S** Nov-Jan.
> **BLACKJON**; brighter red. Found 1929 by A. T. Gossman, Wenatchee, Washington; introd 1931.
> **JONARED**; even red flush. Discv 1930, by W. Uecher, Peshastin, Washington; introd 1934.
> **JONATHAN** (Korea)
> **KAPAI RED JONATHAN**; even red flush. Discv 1929 by A. Tomlinson, Frimley, Hastings, Hawkes Bay, New Zealand.
> **JONATHAN** (4N A); tetrap.
> **JONATHAN** (4N b); tetrap.
> **15 WELDAY JONATHAN**; tetrap.

19 WELDAY JONATHAN; tetrap.

JONGRIMES 4 L D
USA; raised by R. S. Rogers, Bloomfield, Indiana. Introd 1920s; renamed, assigned 1948 to Stark Bros, Missouri.
Parentage unknown.
Refreshing, fruity in Nov; crisp, juicy flesh.
F 6. **T**^{2}. **C** frt cracks. **P** m-Oct. **S** Nov-Jan.

JONICA *see* Jonagold

JORAYCA *see* Jonagold

JORED *see* Jonagold

JOSEPH MUSCH 4 L DC
Belgium; found by Joseph Musch; introd Galopin, Liege; recorded 1872.
Large; in style of Blenheim with crumbly texture, quite sweet, nutty. Cooked, keeps shape, sweet, delicate.
In Belgium, RGF variety, recommended for cultivation by Agric Res Centre, Gembloux because of disease resistance.
F est 6. **T**^{2}; gd res scab, mildew. **P** e-Oct **S** Nov- Mar.

JOSÉPHINE 8 L D
Received from France in 1947.
Attractive; brick red flush under netting of russet. Intense, sweet-sharp, fruit drop or lemony flavour; recalls Nonpareil family. Firm, rather chewy, deep cream flesh.
F 14 **T**^{3} **P** m-Oct. **S** Dec-Feb/Mar.

JOSÉPHINE KREUTER (?) 6 M D
Austria; acc may be variety that arose St Florian; desc 1889.
Deep red, tall, conical. Strong raspberry flavour, recalls Reinette Rouge Étoilée. Densely sweet, firm, rather dry flesh.
F 6. **T**^{2}. **P** l-Sept. **S** Oct-Nov.

JOYBELLS 4 L D
UK; raised prob by Robert Lloyd, HG at Brockwood Hospital and landscape gardener, Woking, Surrey; named Lloyd's Joybells 1900–10. Prob introd by Will Tayler, Godalming, Surrey; presented to RHS Fruit Committee 1922 by Tayler; RHS AM 1922.
Carmine striped over dusky pink. Delicate, almost ethereal taste; sweet, melting, juicy, flesh.
F 11. **T**^{2}; sprd. **C** gd. **P** l-Sept. **S** Oct-Dec.

JOYCE 4 E D
Canada; raised 1898 Exp Farm, Ottawa.
McIntosh X Livland Raspberry. Introd 1924.
Large; prominently red striped. Raspberry flavoured, sweet, juicy, melting, white flesh.
F 5. **T**^{2}; hrdy. **P** e-Sept. **S** Sept; v easily bruised.

JUBILÉ syn Delgolloune

JUBILÉ D'ARGOVIE 1 L CD
Desc Vercier (1948), when grown France, Switzerland.
Large, ribbed. Crisp, juicy, some sweetness, plenty acidity, fruit. Cooked, keeps shape; low in flavour.
F 11. **T**^{3}. **P** m-Oct. **S** Nov-Mar.

JUBILEE syn Royal Jubilee

JULGRANS 6 L D
Sweden prob; received 1921. Acc may be Farmors Juläpple; 'Granny's Christmas apple'; arose Bjärtå, Angermanland, N Sweden.
Dark maroon flushed. Intense, sweet-sharp taste, but inclined to be sour; tough skin.
Grown northern Sweden.
F 11. **T**^{2}. **P** e-Oct. **S** Dec-Feb/Mar.

JULYRED 6 E D
USA; raised 1949 by G. W. Schneider at New Jersey Agric Exp St, New Brunswick.
(Petrel X Early McIntosh) X (Williams X Starr). Introd 1962.
Very attractive, large. Strongly scented with vinous or loganberry flavour; sweet, soft, juicy, white flesh.
Raised to give early apple for New Jersey, where picked mid-July; also grown Norway.
F 9, trip. **T**^{2}. **P** eat e-Aug.

JUMBO OHRIN 5 L D
Japan; discv 1985 by Sukerkuro Tanuichi.
Prob Golden Delicious X Indo.
Resembles Mutsu (Crispin). Cream, very sweet, crisp flesh; little acidity or flavour.
F mid. **T**^{2}. **P** e-Oct. **S** Nov-Jan.

JUNE CREWDSON 6 E D
UK; arose with Bernard Crewdson, Limpsfield, Sussex. Fruited 1943.
Small; dark red. Worcester-like chewy flesh.
F 13. **T**^{2}. **C** hvy. **P** l-Aug. **S** Sept.

JUNO 4 L D
Germany. Raised by M. Schmidt and H. Muraski, Inst Agric & Hort, Münchenberg-Mark.
Ontario X London Pippin. Introd 1971.
Sharp, rather coarse flesh.
F est 12. **T**^{2}. **P** e-Oct. **S** l-Oct-Jan; commercially-May.

JUPITER 7 M D
UK; raised 1966 by Dr F. Alston, EMRS, Kent.
Cox's Orange Pippin X Starking Delicious. Introd 1981. RHS AGM 1993.
Intensely flavoured, aromatic; more robust taste than Cox. Sweet, juicy, loosely textured flesh.
F 14; trip. **T**^{3}. **C** hvy, bien; frt often misshapen. **P** e-Oct. **S** l-Oct-Jan.

JUPP'S RUSSET 8 L D
Received 1951 from New Zealand.

Quite rich and good; sweet, balanced, crisp flesh.
F 15. T^2; sprd. P l-Oct. S Jan/Feb-Apr.

JUST GYULA 6 E D
Received 1948 from Univ Agric, Budapest, Hungary.
Dark carmine; heavily ribbed, crowned. Very sweet, soft, juicy flesh, but little acidity or flavour in England.
F 12. T^2. P e-Sept. S Sept.

KAISER FRANZ JOSEPH syn François Joseph 5 L D
Germany prob; known before 1876.
Resembles smaller, brighter yellow, Calville Blanc d'Hiver. Good intense taste of fruit; sweet, modestly juicy.
F 10. T^1. C gd. P e-Oct. S Nov-Jan.

KAISER WILHELM 6 L D
Germany; found c1800 by head teacher, Hesselman, in Witzhelden, Kreis Solingen, Rhur.
Poss Harberts Reinette seedling. Introd by mid-1800s.
Resembles colourful Golden Reinette. Sweet, plenty of fruit, acidity, quite nutty. Grown Germany mainly for juice.
F 15. T^2. P e-Oct. S Nov-Feb/Mar.

KALCO syn Carola 4 M D
Germany; raised by M. Schmidt in 1930s at Inst of Agric & Hort Mücheberg-Mark.
Cox's Orange Pippin seedling. Introd 1962.
Lightly aromatic, quite rich, sweet, but coarsely textured flesh; tough skin. Some years unattractive, little flavour.
F 10. T^2; sprd. P m-Sept. S Oct-Nov.

KALTERER BÖHMER syn Rosa del Caldaro

KANDILÉ 4 L D
Bulgaria; old regional variety; widely grown before 1900.
Fruity, crisp, juicy.
Main variety 1950–60s in Kjustendil in southwest and Plovdiv in central southern region; now minor importance. Used also for drying, pectin.
F 18. T^2. C gd; bien. P m-Oct. S Dec-Apr.

KANDIL SINAP 4 L D
Russia; prob arose early 1800s in Crimea.
Remarkable shape, tall, slender, like spire with rounded base. Sweet, plenty of fresh acidity; juicy, crisp.
Grown Crimea, Transcaucasia, Turkey. Also grown France – around Paris especially Montreuil in 1920s as luxury fruit; now found at high altitudes in Franche-Comté, Puy de Dôme (Auvergne); used also juice, cooking. Widely known Europe, US as

decorative as well as fruiting tree. Popularised England 1920s, by Bunyard Nursery, Kent, but rarely seen now.
F 11. T^1; uprt. C hvy. P m-Oct. S Dec-Feb.

KANSAS QUEEN syn Nouvelle Europe 6 M D
USA; originated before 1870 in Kansas; introd France c1940s and renamed Nouvelle Europe.
Deep maroon, almost black flush. Juicy, crisp to chewy white flesh. Hint of strawberry flavour; becomes soft, winey. In southern US and France ripens August, where formerly grown for holiday sales.
F 15. T^2; tip. C hvy. P e-Sept. S Sept.

KAPAI RED JONATHAN see Jonathan

KARASTOYANKA 4 L D
Turkey prob; introd C19th to Bulgaria from Edirne, Turkey. Acc received 1957 from Bulgaria.
Quite strong taste of fruit, but inclined to be sharp. Widely grown after 1900 in Plovdiv region, central southern Bulgaria; now low commerical importance.
F 10. T^2; some res scab, mildew. C gd; bien. P m-Oct. S Nov-Jan/Mar.

KARINA 6 M D
Received 1996 from Switzerland; unknown parentage.
Most beautiful apple; pale yellow with rose pink flush. Reminiscent of Milton; crisp, juicy, savoury.
F mid. T^2. P l-Aug. S Sept.

KARINABLE 4 E DC
Scandinavia. Prob Stäringe Karin; found at Stäring, Sörnland, N Sweden; desc 1902; popularised by Gustav Lind.
Flushed in rose pink. Juicy, pleasant, some sugar but quite sharp. Cooked, very mild taste.
F 9. T^2. P use Aug-Sept.

KARIN SCHNEIDER syn Red Ingrid Marie

KARMEN 4 L D
Czech Republic; raised at Univ of Agric, Strizovice. Lord Lambourne X Linda. Select 1962. Received 1982.
Good, refreshing taste of fruit in New Year, but can be over-sharp; soft, juicy flesh.
F 10. T^2. P m-Oct. S Dec/Jan-Mar.

KARMIJN DE SONNAVILLE 7 L D
Netherlands; raised 1949 by Piet de Sonnaville in his family orchards at Puiflijk and Altforst, nr Nijmegen.
Cox's Orange Pippin X Jonathan or Belle de Boskoop. Introd 1971.
Intensely flavoured, rich, aromatic, masses of sugar and acidity; crisp, juicy flesh. More robustly flavoured than Cox, slightly

honeyed; very good. Rather large for eater. Grown commercially Netherlands; used also for juice.
F 12; trip. T^3; sprd. C gd; frt prn crack, unsightly russet. P e-Oct. S l-Oct- Dec.

KATJA syn Katy 6 E D
Sweden; raised 1947 at Balsgård Fruit Breeding Inst.
James Grieve X Worcester Pearmain. Introd 1966. Renamed Katy by Kent fruit grower, Sir James Mount.
Sweet, juicy, some of strawberry flavour and firm flesh of Worcester parent, but plenty of James Grieve acidity. Once picked soon goes soft.
Grown Sweden, commercially and in gardens. Grown England for Farm Shop trade; used also for juice.
Frt Col bright Worcester rd flush, rd stripes; grnish yell turning yell background; becomes greasy. Size med. Shape rnd-con; flt base. Basin qte nrw, shal; pinched; sometimes beaded. Eye clsd to hlf open; sepals, sml, nrw, v downy. Cavity brd, dp; russet lined. Stalk med/qte lng, qte thck. Flesh wht.
F 12. T^3. C gd. P e-Sept. S Sept-e-Oct.

KATY see Katja

KEED'S COTTAGE 4 M D
UK; received 1947 from v old tree known by this name and growing at Jessamine Cottage, Pulborough, Sussex.
Quite rich, lightly aromatic.
F 4. T^3. P l-Sept. S Oct-Dec.

KEMP 4 E C
Ireland; arose prob Ulster. Known 1837.
Syn May Bloom.
Sharp, juicy, eaten fresh, but mellowing. Cooks to quite sweet purée.
Early, dual purpose apple of County Armagh.
F* 13. T^3; uprt. P l-Aug. S Sept-Oct/Nov.

KENDALL 6 L D
USA; raised 1912 by R. Wellington at NYSAES, Geneva.
McIntosh X Zusoff. Introd 1932.
McIntosh type, but larger, keeping longer. Sweet, soft, juicy flesh.
Grown commercially US, often sold as K-McIntosh.
F 7. T^2; sp; hrdy. C hvy; prn bitter pit. P e-Oct. S Nov-Jan/Feb.

KENNETH 7 L D
UK; raised 1920 by Kenneth McCreadie at Rhyl, Wales.
Quite rich, good taste of fruit, recalls Blenheim Orange, but juicier. Drab appearance.
F 14. T^2. P m-Oct. S Dec-Feb/Apr.

KENT syn Malling Kent 7 L D

UK; raised 1949 by H. M. Tydeman at EMRS, Kent.

Cox's Orange Pippin X Jonathan. Named 1974.

At best, rich with lots of sugar and acidity, quite aromatic. Mellows to become sweeter, quite scented by New Year, but rather coarse flesh; tough skin. In sunless year, can have metallic taste, betraying its Jonathan parent and poor flavour.

Grown small extent commercially Kent; also in gardens.

Frt *Col* dark orng rd flush, rd stripes over yell; some netting, patches of russet. *Size* med. *Shape* rnd-con to con; sltly crowned. *Basin* brd, dp; ribbed; some russet. *Eye* prt open; sepals, brd based, lng, v downy. *Cavity* med wdth, dpth; often russet lined. *Stalk* shrt/med, qte thck. *Flesh* pale crm.

F* 13. T². C hvy; prn coarse russet. P m/l-Oct. S Nov-Feb; commercially-Jun.

KENTISH FILLBASKET 3 L C

UK; pres arose Kent. Known before 1820.

Very large, red striped, flushed. Plenty of acid, sugar, fruit. Cooks to sweet, light purée. Among first varieties planted Western Australia in 1860s.

F 9. T³. C gd. P m-Oct. S Nov-Jan/Mar.

KENTISH PIPPIN see Colonel Vaughan

KENTISH QUARRENDEN 6 L D

UK; received before 1931; very similar to Winter Quarrenden, but distinct.

Dark maroon flushed. Sweet, nutty, juicy cream flesh, can have woody undertones.

F 10. T². P m-Oct. S Dec-Feb.

KENTUCKY RED STREAK (?) 6 L D

Acc received 1951 from New Zealand; may be Kentucky Redstreak desc by Downing; known before 1864.

Maroon flush, stripes. Sweet, some acidity, lightly flavoured.

F 9. T². P l-Oct. S Jan-Mar/May.

KERECSI MUSKOTALY 5 L D

Hungary; raised 1882 in Kércs, village in Szabolcs, north east Hungary.

Very sweet, quite scented, firm white flesh; recalls Edelborsdorfer.

F 5. T². P m-Oct. S Dec-Mar/May.

KERRY PIPPIN 7 E D

Ireland; recorded 1802. Sent 1819/20 to London Hort Soc by nursmn John Robertson of Kilkenny.

Quite small, coloured like tortoise-shell butterfly and 'a gem' for the Victorian dessert. Rich, densely fruity taste, sweet, plenty of acidity, firm flesh.

Widely grown in gardens and market gardens in C19th; 'favourite with London retailers'.

Frt *Col* grnsh yell/gold, flushed with light orng rd, few broken rd stripes. *Size* med.

Shape rnd-con; rounded base, flat apex. *Basin* brd shal; puckered; often beaded. *Eye* clsd; sepals, sml, shrt, qte downy. *Cavity* nrw, shal; often fleshy swelling to side; russet lined. *Stalk* med, qte thin. *Flesh* yell.

F 7. T²; uprt. C hvy. P l-Aug. S Aug-l-Sept.

KESWICK CODLIN 5 E C

UK; found on rubbish heap at Gleaston Castle, nr Ulverston, Lancs. Introd by nursmn J. Sander of Keswick; known 1793. Cooking to juicy, cream froth or purée; hardly needs any sugar. 'None better for jelly' claimed Victorians; also makes refreshing, juicy eating apple.

One of most popular early cookers of C19th; grown also for market around London and Kent in 1930s, but now only found in gardens. With profuse, early blossom and neat habit, trees are highly decorative and in 1890s recommended for planting to form arbours, tunnels etc.

Frt *Col* pale grn/pale yell, darker yell flush; usually raised, russeted hairline down side; lenticels qte conspic as russet dots. *Shape* con to oblng-con; flt at base; ribbed, sltly flt sided. *Size* med. *Basin* med wdth, dpth; much puckered, ribbed; usually beaded. *Eye* clsd, pinched; sepals, brd, lng, v downy. *Cavity* brd, shal. *Stalk* shrt, thck. *Flesh* pale crm.

F* 5. T². C hvy. P m/l-Aug. S l-Aug- Sept/Oct.

KHOROSHAVKA ALAYA 6 E DC

Kazakhstan (prob); known 1892.

Pinky red flushed over primrose, with bloom, it was most esteemed of Khoroshavka apples – 'Alaya' means scarlet. 'Vinous and sweet', but brisk, juicy, and crisp in England. Cooked, very light taste.

F 9. T². C hvy. P use Aug-Sept; ripens over lng period; stored to Dec in Russia.

KIDD'S ORANGE RED 7 L D

New Zealand; raised 1924 by James Hatton Kidd, Greytown, Wairarapa.

Cox's Orange Pippin, Delicious cross. Introd UK c1932. RHS AM 1973; AGM 1993.

Rich balance of sugar, acidity and strongly aromatic; mellows to intensely flowery or rose petal quality, some claim it tastes of Parma violets. Needs plenty of autumn sunshine to build up flavours.

Kidd was fruit farmer and amateur breeder, who in this variety achieved his aim of combining the quality of best English varieties with colour of American apples. Grown commercially small extent in England and formerly also Switzerland; popular UK garden apple.

Frt *Col* dp pinky crimson flush, some darker stripes over pale yell/gold; fine russet dots, patches esp at base. *Size* med. *Shape* con; sltly ribbed; 5 crowned. *Basin* med wdth, shal; ribbed. *Eye* clsd; sepals v downy.

Cavity brd, shal; russet lined. *Stalk* med, thck. *Flesh* dp crm.

F* 12. T²; v prn canker. C gd; frt prn coarse russet. P m-Oct. S Nov-Jan.

CAPTAIN KIDD; more highly coloured.

KILE 7 L D

Australia; received 1973 from Huon Res St, Tasmania.

Elements of Cox flavour; sweet, juicy, aromatic.

F 10. T²; frt cracks. P e-Oct. S Oct-Dec.

KILKENNY PEARMAIN 4 M D

Ireland; recorded 1831; desc Hogg (1851). Brisk, some sweetness, but little flavour.

F 14. T²; sprd. P l-Sept. S Oct-Nov.

KIM 6 M DC

Sweden; raised 1946 by P. Bergendal at Agric Coll, Balsgård.

Kortland (Cortland) X Ingrid Marie. Introd late 1970s.

Takes name from parents' initials. Large, maroon flushed. Pronounced strawberry flavour. Crisp, very juicy, quite sweet but plenty of acidity. Cooks to good, savoury purée.

Grown commercially Sweden.

F 14 T²; v hrdy; more disease res than parents. C gd. P l-Sept. S Oct-Nov/Dec, cold store-Mar.

KIMBALL McINTOSH see McIntosh

KING ALBERT 3 E C

UK; acquired before 1928.

Large; prominently red striped. Quite sweet, crisp, juicy. Cooks to yellow purée; rather insipid.

F 10. T². P m-Aug. S Aug-Sept.

KING BYERD 4 L DC

UK; received 1954 from Cornwall.

Mellowing to quite rich, sweet-sharp taste. Quite sweet cooking apple.

F 15. T³. C gd. P l-Oct. S Jan-Mar.

KING CHARLES PEARMAIN 8 L D

UK; sent 1876 to Hogg by nursmn John Smith of Worcester; desc Hogg (1884).

Coarsely russeted over gold. Slightly nutty or almost tannic, smoky flavour; sweet, plenty of acidity, rather dry, firm, crumbling flesh.

F 17. T¹. C lght; prn bitter pit. P e-Oct. S Nov-Jan.

KING COFFEE 6 L D

UK; received from Worcester. Exhib at 1934 Fruit Congress

Large; dark red flushed. Hint of coffee flavour; soft, cream flesh.

F 17. T². P e-Oct. S Nov-Dec.

KING COLE 4 L D

Australia; raised by R. G. Cole, Lang Lang,

Victoria. Believed Jonathan, Dutch Mignonne cross. Exhib 1912.
Bright red Jonathan flush. Good brisk taste of fruit in Jan.
Formerly grown Australia for export.
F 9. T². P m-Oct. S Dec-Feb/Jun.

KING COX *see* Cox's Orange Pippin

KING DAVID 6 L D
USA; found 1893 in fence row on farm of Ben Frost, Durham, Washington County, Arkansas.
Believed Jonathan cross Winesap or Arkansas Black. Introd 1902 by Stark Brothers, Missouri.
Dark maroon, nearly black flush. Plenty of fruity taste, sprightly acidity; sweet, juicy, flesh.
Promoted by Stark as replacement for Jonathan. Planted southern States early 1900s, also in Tasmania and Argentina for export, but not success as only good flavour if well ripened on tree and then only stores short time.
F 11. T³; res scab, fireblight, cedar apple rust. P m-Oct. S Oct-Nov/Mar.

KING GEORGE V 4 L D
UK; raised c1898 by Lady Thorneycroft, Bembridge, Isle of Wight. Cox's Orange Pippin X. RHS AM 1928. Introd by nursmn J. Cheal, Crawley, Sussex.
Intense, rich flavour, with strong pineapple-like acidity; crisp, juicy cream flesh. Mellowing, but sharper, thinner than Cox.
F 14. T³. C lght. P m-Oct. S Nov-Mar.

KING JONAGOLD *see* Jonagold

KING OF THE PIPPINS syn Reine des Reinettes 7 M DC
UK or France; introd and renamed early C19th by nursmn Kirke of Brompton, London. Old name, Golden Winter Pearmain. Syns numerous – 101 – include Prince's Pippin, by which still known West Midlands. Prob Reine des Reinettes of France, Gold Parmäne of Germany, which continental authors believe arose 1770s in France. RHS AGM 1993.
Well ripened, quite sweet, plenty of acidity, firm, juicy flesh, but usually underlying, slightly bitter taste. Cooked, keeps shape, bright yellow, good flavour, quite sweet yet brisk taste; well suited to open tarts etc.
Widely grown in gardens and for market in C19th. Top of dessert popularity poll at 1883 Congress; still found in many old orchards, particularly West Midlands, where also used for cider. Grown France and 'Reinette' favoured for apple pâtisserie; also used Normandy for cider. Gold Parmäne grown Germany, Switzerland.
Frt *Col* orng rd flush, some rd stripes over grnish yell/gold; lenticels appear as grey russet dots; russet at base. *Size* med. *Shape*

oblng to oblng-con; sometimes lopsided; trace rnded ribs, sltly 5 crowned. *Basin* lrg, brd, dp; puckered. *Eye* lrg, open; sepals, sml. *Cavity* med dpth wdth; russet lined. *Stalk* med; qte thck. *Flesh* pale crm.
F 13. T²; uprt; gd res diseases. C gd. P e-Oct. S Oct-Dec; keeps-Feb
BAUNEN; sport of Reine des Reinettes.
KING RUSSET; russeted form. Found late 1950s, introd by K. A. Robbins, Borden, Hants. RHS AGM 1993. Boldly flavoured with sere russet taste, associated with Egremont; quite sharp, dry.
PRINCESS; russeted form of Reine des Reinettes. Not in NFC.
ROTE GOLDPARMÄNE; more highly coloured. Arose Germany; received 1967.

KING OF TOMPKINS COUNTY 6 L D
USA; believed arose nr Washington, Warren County, New Jersey. Taken to Tompkins County, New York 1804 by Jacob Wycoff who named it King. Renamed c1855. RHS AM 1900.
Often very large; red flushed, striped. Sweet, quite juicy, lightly aromatic; plenty of acidity early in season.
Widely grown home orchards and for sale US in C19th; still valued by gardeners; used also for 'sauce, pies, "apple küchen"'. Also formerly grown small extent English gardens.
F 10; trip. T³; sprd. P m-Oct. S Nov-Mar/May.

KING RUSSET *see* King of the Pippins

KING'S ACRE BOUNTIFUL 5 M C
UK; introd 1904 by King's Acre Nurseries, Hereford. RHS AM 1904.
Milky yellow; perfectly round. Cooks to creamy purée, quite brisk; needing some sugar.
F 23. T². P e-Sept. S Sept-Oct.

KING'S ACRE PIPPIN 7 L D
UK; introd 1899 by King's Acre Nurseries, Hereford.
Believed Sturmer Pippin, Ribston Pippin cross. RHS AM 1897.
Large; rather dull appearance. Richly flavoured; sweet-sharp taste of Sturmer combined with little of aromatic quality of Ribston; crisp, juicy flesh. Formerly, valued, late keeping garden apple.
Frt *Col* grnish yell/yell, flushed sltly or more in brownish rd; russet patches, netting, veins. *Size* med/lrg. *Shape* shrt-rnd-con; irregular; rnded ribs, sltly 5 crowned. *Basin* med, qte shal; ribbed, puckered; ltl russet. *Eye* clsed/ltl open; sepals brd, qte lng, qte downy. *Cavity* brd, qte dp; part lined russet. *Stalk* med lng, thck. *Flesh* pale crm.
F* 12; trip. T³; sprd. C lght; prn bitter pit. P m-Oct. S Dec/Jan-Mar.

KINKEAD RED SPY *see* Northern Spy

KINREI syn Orei 5 L D
Japan; raised 1932 at Aomori Apple Exp St. Golden Delicious X Delicious. Named 1948.
Very sweet, honeyed; crisp juicy, yellow flesh.
F 13. T¹. C hvy. P l-Oct. S Dec-Mar/May.

KIS ERNÖ TÁBORNOK 2 L D
Hungary; received 1948 from Univ Agric, Budapest. Kis Ernö was general during Revolution and Independence War 1848–49.
Firm, quite sweet in Feb; very tough, almost metallic skin.
F 11. T². P l-Oct. S Jan-Mar/May.

KITCHOVKA 4 L D
Macedonia prob; received before 1900 in Bulgaria; formerly widely grown south west, central Bulgaria, now little importance.
Bright red flush, stripes. Sweet, vinous flavour in Feb; snow white, tough flesh.
F 11. T¹; uprt. C bien. P m-Oct. S Jan-Mar.

KLARAPFEL syns Weisser Klarapfel, White Transparent

KLUNSTER 7 L D
Germany; received 1951 from Fruit Res St, Jork, nr Hamburg.
Bright red flushed. Brisk, firm flesh.
F 11. T³. P m-Oct. S Dec-Feb/Mar.

KNOBBY RUSSET 8 L D
UK; shown 1820 to London Hort Soc by Haslar Capron, Midhurst, Sussex.
Curiosity, covered in knobs or warts, heavily russeted. Strongly flavoured, firm, rather dry flesh.
F 6. T². P m-Oct. S Dec-Mar.

KNYSCHE syn Rambour Podolskii

KOLACARA 6 L D
Serbia; received 1936.
Sharp, refeshing taste of fruit, firm white flesh; tough skin.
Major winter market apple in 1930s.
F 8. T³; sp. P m-Oct. S Nov-Feb/Mar.

KÖNIGIN SOPHIENSAPFEL syn Reine Sophie 2/5 L D
UK; origins confused with Lemon Pippin, but prob arose England late 1700s; distributed as Winter Queen by nursmn Loddiges of London. Renamed 1821 by German pomologist Diel.
Similar appearance to Lemon Pippin with swelling at base of stalk, but not the same. Modest richness, quite sweet, firm, chewy, nearly yellow flesh.
F 9. T²; tip. P m-Oct. S Nov-Dec/Mar.

KÖNINGIN JULIANA 7 M D

Netherlands; raised 1935 at IVT, Wageningen.
Reinette Rouge Étoilée X Cox's Orange Pippin. Introd 1952.
Handsome. Intense, rich, aromatic flavour, recalls Holstein or Ribston Pippin. Plenty of sugar and pineapple like acidity; crisp, firm, deep cream flesh. Rather tough skin.
F 11. T^2; sp. C gd; slow to bear. P l-Sept. S Oct-Nov, cold store-Mar.

KOREI 5 L D
Japan; raised 1935 at Aomori Apple Exp St. Golden Delicious X Indo. Named 1949.
Golden Delicious type, lightly honeyed, but often rather woody taste in England; tough skin.
F 11. T^3. C hvy. P m-Oct. S Dec-Feb.

KOROBOVKA syn Cardinal 4 E D
Russia; arose Novgorod Province; desc 1855. Renamed Pierre le Grande after Russian Emperor by Leroy. UK syn Cardinal; RHS AM 1896 as Cardinal. Cardinal of NFC is not Korobovka syn Cardinal.
Beautiful; red flush, stripes over primrose. Sweet, soft flesh; balsam flavour. Name means laundry basket, which was used to carry apples to market.
F 8; res frost. T^1; v hrdy. C gd, bien. P use Aug-Sept.

KORTEGÅARD see Cox's Orange Pippin

KOSMONAUT 4 L D
Czech Republic; raised 1965 by Frantisek Vancura, Prague.
Cox's Orange Pippin X Haptmanova.
Perfectly round. Refreshing taste of fruit; soft, juicy flesh.
Trialled Czech, but not grown as fruit storage short, trees prone disease.
F 10. T^1. P l-Oct. S Nov/Dec-Jan.

KOUGETSU 6 M D
Japan; raised by S. Taniuchi, in Nanbu-machi, Aomori.
Golden Delicious X Jonathan. Introd 1981.
Crisp, juicy, sweet and good, quite rich taste. Acidity of Jonathan balances sweetness of Golden Delicious.
F mid. T^2. P l-Sept. S Oct-Nov.

KRASAVA 6 E D
Czech Republic; raised 1970 at Inst Exp Botany, Strizovice. Otcovo X Wagener.
Sweet, refreshing, savoury; crisp yet soft, melting juicy, white flesh.
Trialled in Czech; not grown commercially, but popular with amateurs.
F 4. T^2. C hvy; precocious. P l-Aug. S Aug-Sept.

KRASNYI SHTANDART 6 M
Russia; raised 1915 by I. V. Michurin at Michurinsk.
Pepin Shafrannyi X Rubynovoye

(Antonovka X central Asian species *Malus sieversii ssp niedzwetzkyana*).
Ornamental; whole tree suffused with red – leaves, bark, blossom. Small dark red apples; pink, very sharp flesh.
F*12. T^2. C hvy. P e-Oct. S Oct-Nov.

KRONPRINZ RUDOLF 6 L D
Austria; arose 1863 with I. Klöchner Steiermark.
Scented, sweet, white flesh; recalls Edelborsdorfer.
F 5. T^1. P e-Oct. S Nov-Jan/Feb.

KRÜGERS DICKSTIEL 7 L D
Germany; arose Mecklenburg; known 1852; first fruited with Revenue Officer Woltmann, in Zeven nr Bremen.
Brisk, savoury taste; rather like sharp Golden Reinette; firm flesh.
F 21. T^3. P m-Oct. S Nov-Jan.

KULDZHINKA KRUPNOPLOD AYA 6 L D
Russia; prob seedling of *Malus sieversii ssp niedzwetzkyana*, central Asian species. Received 1975 from Vavilov Inst, St Petersburg.
Bright red; tall, oblong (Krupnoplod aya means long fruited). Very sweet, white, pink stained flesh.
Grown Russia for blossom, beauty of fruit.
F* purple; 5. T^3. P m-Oct. S Nov-Dec.

KYOKKO 6 L D
Japan; raised 1931 at Aomori Apple Exp St. Ralls Janet X McIntosh. Named 1938.
McIntosh red, but more regular shape. Quite rich, sweet; soft, deep cream flesh; tough skin; can be insipid in England.
F 14. T^3. C gd. P l-Sept. S Oct-Nov/Dec.

LADY APPLE syn Api

LADY DERBY syn Thorle Pippin

LADY HENNIKER 7 L CD
UK; raised 1840–50 from cider pomace on Lord Henniker's estate, Thornham Hall, Eye, Suffolk. Introd 1873 by his HG, John Perkins. RHS FCC 1873.
Brisk, crumbling, deep cream, green tinged flesh. Cooks to pale yellow quite strongly flavoured purée; hardly needing extra sugar. Used by Perkins for dessert when 'large handsome dishes of mixed fruit were required [as] its appearance by lamplight is most telling'.
Quite popular garden variety, widely distributed East Anglia, southern England; also recommended 1890s for 'artistic' orchard.
F 13. T^3. C gd. P e-Oct. S Nov-Jan.

LADY HOLLENDALE 4 E D
UK; arose East Anglia. Recorded 1918.
Bright red flush. Sharp, rather tough flesh. Grown for market in Wisbech area 1920–30s.
F 7. T^2; sprd. P use Aug.

LADY HOPETOWN 2/5 L D
UK; received 1950 from Bexhill, Sussex.
Sharp, fruity crisp in Feb.
F 11. T^3. P m-Oct. S Jan-Mar/May.

LADY ISABEL 4 L CD
UK; found 1939 on compost heap by Mrs Reading at Guildford, Surrey.
Reminiscent of James Grieve. Sweet yet savoury, plenty of fruit, acidity; quite soft, juicy flesh. Light cooker.
F 9. T^2. P e-Oct. S Oct-Dec/Mar.

LADY LAMBOURNE see Lord Lambourne

LADY OF THE LAKE 4 L D
UK; received 1958 from Grane, nr Errol, Scotland.
Savoury, plenty of fruit; soft juicy cream flesh; recalls James Grieve.
F 12. T^2. P m-Oct. S Nov-Dec/Jan.

LADY OF THE WEMYSS 4 L C
UK; received 1949 from HG, Brotherston, Tyninghame Gardens, East Lothian, Scotland; acc may be old Scottish variety first recorded 1831.
Large; orange to inky red flush. Plenty of sugar, acidity early in season; firm flesh. Cooked keeps shape; quite rich taste.
F 9. T^3; sprd. P m-Oct. S Nov-Jan.

LADY'S DELIGHT (?) 3 M C
UK; acc may be variety desc 1851 (Hogg) when 'highly esteemed about Lancaster'.
Large; bold red stripes. Cooks to brisk, well-flavoured purée.
F 12. T^2. P m-Sept. S Sept-Oct/Nov.

LADY'S FINGER OF OFFALY 4 M D
Ireland; desc 1951 (Lamb).
Lady's Finger is name given to many tall, oval or conical shaped apples. Characteristic, long, oval shape. Slightly savoury; soft flesh.
Known Offaly and Monaghan in 1950s.
F 11. T^2; sprd. P e-Sept. S Sept-Oct/Nov.

LADY SUDELEY 4 E D
UK; raised c1849 by Mr Jacobs, prob when farm bailiff at Sharsted Farm, Chatham, Kent. Taken 1850s to Petworth, Sussex, where he set up as 'sheep doctor' and taxidermist. Originally Jacob's Strawberry. Renamed and introd 1885 by nursmn George Bunyard, Kent. RHS AM 1884.
Quite sweet, strongly flavoured, but underlying astringent taste; firm yellow flesh. Difficult to catch Edward Bunyard's

'very highly aromatic juicy' prime.
Jacobs' dish of his seedling at London show in 1884 caught nursmn George Bunyard's eye and he straight away obtained grafts and renamed it after his best customer – Sudeley estate – which had just bought over half million trees to set up Toddington Fruit Company in Gloucs. Heavily promoted in 1890s for profit; as colourful as 'Yankie' imports and also reputedly as dresses Lady Sudeley wore at court. Became garden and market fruit; on sale up to 1930s. Widely grown, even in north, Scotland; many old trees remain. Highly decorative tree in blossom and fruit, especially the red sport.
Frt *Col* bold rd stripes, pink flush over grnish yell/gold; russet at base. *Size* med/lrg. *Shape* rnd-con; ribbed; 5 crowned; flt base. *Basin* qte brd, med dp; ribbed. *Eye* hlf open to clsd. *Cavity* med wdth, dpth; ltle russet. *Stalk* shrt, qte thin. *Flesh* crm.
F* 13. T^2. **C** gd. **P** e-Sept. **S** Sept.
> **RED SUDELEY**; even red flush. Received from Scotland.

LADY WILLIAMS 4/6 L D
Australia; arose c1935 with A. R. Williams, Bononia Farm, Paynedale, Donnybrook, Western Australia. Poss Granny Smith, Jonathan or Rokewood cross.
Bright red flush. Firm crisp flesh. Usually fails to develop in England.
Original tree only just survived after being accidentally cut down by Williams' son, but it grew up again and in 1943 they began selling fruit. Apples were so bright and kept so well that by 1970s variety become quite widely planted in Western Australia. Named after Mrs Williams, who was called Lady Williams by neighbour's children who could not pronounce her name – Maud.
F 7. T^2. **C** hvy; hangs v late. **P** Nov. **S** Jan-May.

LAGRÉE 1/5 L C
Received 1948 from Loire Atlantique, France.
Some sweetness, plenty of fruit, acidity. Cooked, sweet, rather insipid.
F 19. T^2; sprd; res mildew. **P** late Oct. **S** Nov-Mar.

LAKELAND 6 M D
USA; raised c1907 Univ Minnesota Fruit Breeding Farm, Excelsior.
Malinda X. Introd 1950.
Beautiful; maroon flush, stripes. Light aromatic flavour; sweet, crisp, juicy flesh; tough skin. Dual purpose in US.
F 9. T^2; sprd; hrdy; some res scab. **P** e-Sept. **S** Sept-Oct; cold store-Dec.

LAKE'S KERNEL 4 L D
UK; brought by Mr Lake to Ashleworth, Gloucs, where known as Ashleworth. Exhib 1905.
Delicate, aromatic flavour; quite rich,

sweet, juicy, crisp flesh.
F 17. T^2. **P** e-Oct. **S** Oct-Dec.

LAMB ABBEY PEARMAIN 7 M D
UK; raised 1804 by Mrs Mary Malcolm, Lamb Abbey, Dartford, Kent, from pip of Newtown Pippin. RHS AM 1901.
Small, but intensely flavoured. Rich balance of sweetness and acidity, hint of pineapple flavour; firm flesh.
F 7. T^2; sp. **C** gd. **P** l-Sept. **S** Oct-Jan/Apr.

LAMB'S SEEDLING 3 L CD
UK; raised 1866–67 by Joseph Lamb, HG to Mr Meynell, Meynell Langley, Derby, from pip of Northern Greening.
Large; boldly red striped, flushed. Slightly sharp, but prized for its 'tender flesh in New Year'. Cooked, sweet, lightly flavoured.
F 11. T^2. **P** e-Oct. **S** Oct-Jan.

LA NATIONALE 4 L D
France; prob variety now known Bresse and Maçon region; raised by M. Roux at Saint-Romain-Mont-d'Or, Rhône. Fruited 1871; desc 1948 (*Verger Français*).
Large; dark red flush. Quite rich, sweet, lightly aromatic, with good, strong taste of fruit early in season.
F 26. T^3; prt tip. **C** gd. **P** l-Oct. **S** Dec-Feb/Apr.

LANCASHIRE PIPPIN 4 M C
UK; received 1950, from Westmorland.
Large; bright red flush, stripes. Slightly rich flavour. Cooked, bland.
F 18; trip. T^2; sprd. **P** m-Sept. **S** Oct.

LANCRAIG *see* Delicious

LANDSBERGER REINETTE 5 L D
Germany; raised c1840 by Councillor Burchardt, Landsberg, Brandenburg. RHS FCC 1882.
Slight savoury flavour, sweet, juicy, soft, white flesh; can be empty.
Formerly widely grown Northern Europe. Grown eastern Germany for juice processing, but in decline.
F 9. T^2; sprd. **C** gd. **P** l-Sept. **S** Oct-Jan.

LANE'S PRINCE ALBERT 3 L C
UK; raised before 1841 by Thomas Squire, Berkampstead, Herts.
Believed Russet Nonpareil X Dumelow's Seedling.
Introd c1850 by nursmn John Lane, Berkhampstead. Exhib British Pom Soc 1857. RHS FCC 1872; AGM 1993.
Cooks to lemon coloured purée, brisk but not as strong as Bramley. Becomes sweeter, milder, but can be rather tough cooked.
Original tree transplanted 1841 by Squire to his front garden on day that Prince Albert and Queen Victoria called at King's Arms, for change of horses. Accordingly, he named his seedling Victoria and Albert.

Tree regularly bore heavy crops, but remained small and neat, impressing John Lane, who obtained grafts and introduced it as Lane's Prince Albert. The tree remained in garden until 1958, when house was demolished to make way for new buildings. Popular garden apple by1880s, also planted for market. Fruit bruises easily which led to its demise, but still highly valued by amateurs.
Widely distributed throughout northern Europe; grown eastern Germany for juice and sharp eating apple in Mar.
Frt *Col* shiny grn, sltly flushed orng rd, red stripes; turning yell; greasy skin. *Size* med. *Shape* rnd-con. *Basin* qte brd, dp; sltly ribbed. *Eye* sml, clsd; sepals, sml, qte downy. *Cavity* brd, dp; ltl scarf skin. *Stalk* shrt, thin. *Flesh* grnish wht.
F 12. T^2; sprd; hrdy; prn mildew; res scab. **C** gd. **P** m-Oct. **S** Nov-Mar.

LANGE'S PERFECTION 6 L D
UK; received 1983 from Wigan, Lancs.
Striking, dark maroon, almost black flush; markedly crowned shape. Sweet, firm, deep cream, flesh, but tends to be chewy, low key.
F 10. T^2. **P** e-Oct. **S** Oct-Dec.

LANGLEY PIPPIN 6 E D
UK; raised by Messrs Veitch at Langley Nursery, Bucks.
Cox's Orange Pippin X Gladstone. RHS AM 1898.
Dark red flush. Slight raspberry flavour; quite sharp.
F 13. T^2; sprd. **P** eat Aug-Sept.

LAPPIO 2/5 L D
Italy; prob arose Sicily. Received 1958 from Catania.
Sharp, crisp, refreshing taste of fruit.
F 7. T^3; uprt. **P** m-Oct. **S** Nov-Mar.

LARGE TRANSPARENT (Perrine) syn Perrine Yellow Transparent

LASS O' GOWRIE 3 E C
UK; received 1949 from HG, Brotherston, Tyninghame Gardens, East Lothian; prob variety exhib 1883 Congress from Scotland.
Tall, ribbed. Sharp, crisp juicy flesh. Cooked keeps shape; sweet, delicate flavour.
F 8. T^3. **P** use Aug-Sept.

LAVINA 5 L D
Italy; arose nr Modena, Po Valley. Desc 1949 but much older.
Sweet, yet not sickly in Jan; firm, deep cream flesh.
Italian market apple.
F 18. T^2; prt tip. **P** l-Oct. **S** Jan-Mar/Jun.

LAWFAM 6 M D
Canada; raised 1898 at Exp Farm, Ottawa.

Lawyer X Fameuse. Introd 1922.
Very dark maroon flush. Sweet, light, hint of strawberry turning vinous flavour; melting flesh; tough skin.
F 10. T³; sprd; hrdy. C hvy. P l-Sept. S Oct-Nov.

LAWYER NUTMEG 5 M D
USA; arose with David Lawyer in Plains, Montana. Prob Wismer Dessert, 'Apple crab' cross. Received 1975.
'Definite nutmeg, spicy flavour' in Montana. Savoury, soft, juicy flesh in England.
F* 14. T²; prt tip. P e-Oct. S Oct-Nov.

LAXTON'S ADVANCE syn Advance

LAXTON'S EARLY CRIMSON 6 E D
UK; raised by Laxton Bros Nursery, Bedford.
Worcester Pearmain X Gladstone. Introd 1931.
Brilliant crimson flush. Very sweet, little acidity; soft, white tinged green flesh.
F 10. T³; sprd. P eat e-Aug.

LAXTON'S EPICURE syn Epicure

LAXTON'S EXQUISITE syn Exquisite

LAXTON'S FAVOURITE 7 E D
UK; raised 1925 by Laxton Bros, Bedford. Cox's Orange Pippin X Exquisite. Introd 1951.
Attractive, bright red flush over gold. Sweet, juicy.
F 10. T³; sprd; bien. P l-Aug. S Sept-Oct.

LAXTON'S FORTUNE syn Fortune

LAXTON'S HERALD 6 E D
UK; raised 1906 by Laxton Bros, Bedford. Worcester Pearmain X Beauty of Bath. Introd 1939.
Worcester flavour but ripening earlier. Sweet, rather chewy flesh.
F 8. T²; sprd. P l-Aug. S Sept.

LAXTON'S LEADER 6 E D
UK; raised 1905 by Laxton Bros, Bedford. Gladstone X Worcester Pearmain. Introd 1939.
Dusky red flush of Gladstone, but much sweeter, some strawberry flavour; soft white flesh.
F 5. T². P use e-Aug.

LAXTON'S PEARMAIN syn Bedford Pearmain 6 L D
UK; raised in 1897 by Laxton Bros, Bedford. Cox's Orange Pippin X Wyken Pippin. Introd 1922. RHS AM 1922.
Dark maroon flush. Quite rich, juicy, sweet, plenty of strong Wyken Pippin acidity; deep cream flesh.
F 16. T². P m-Oct. S Oct-Jan/Mar.

LAXTON'S PEERLESS 6 M D
UK; raised 1900 by Laxton Bros, Bedford. Introd 1922. RHS AM 1920.
Bright red flush. Quite rich, lightly aromatic flavour; sweet, juicy, deep cream flesh.
F 8. T²; sprd. C prn bitter pit. P m-Sept. S Sept-Nov.

LAXTON'S PIONEER syn Pioneer

LAXTON'S REARGUARD 7 L D
UK; raised 1907 by Laxton Bros, Bedford. Court Pendu Plat X Ribston Pippin.
Rich, intense flavour; sugary acidity; firm, crisp flesh.
F 9. T². C gd, frt sml unless thinned; pr col. P e-Oct. S Nov-Jan.

LAXTON'S REWARD 4 M D
UK; raised 1925 by Laxton Bros, Bedford. Epicure X.
Lightly aromatic, sweet, juicy.
F 14. T¹; sprd. C bien; frt sml. P e-Sept. S Sept-Oct.

LAXTON'S ROYALTY 7 L D
UK; raised 1908 by Laxton Bros, Bedford. Cox's Orange Pippin X Court Pendu Plat. Introd 1932.
Aromatic, rich, some of delicate quality of Cox, yet stronger acidity of Court Pendu Plat.
F 21. T². C lght. P l-Oct. S Jan-Mar.

LAXTON'S SUPERB 7 L D
UK; raised 1897 by Laxton Bros, Bedford. Wyken Pippin X Cox's Orange Pippin. Introd 1922. RHS AM 1919; FCC 1921.
Some of richness and complexity of Cox, but sweetness is main feature; finely textured, quite juicy flesh.
Formerly grown for market, but its biennial bearing led to demise; remains popular garden apple; will grow where Cox fails to thrive.
Frt *Col* dep rddish purple/pinky purple flush, rd stripes over grnish yell; fine russet dots, some russet netting. *Size* med. *Shape* rnd-con to con. *Basin* med brd, qte dp; sltly puckered. *Eye* hlf-open; sepals brd, lng. *Cavity* med wdth, qte shal; russet lined. *Stalk* med/lng, qte thck. *Flesh* wht.
F 13. T³; sprd; whippy new growth. C hvy, bien. P e/m-Oct. S Nov-Jan.
    LAXTON'S SUPERB NFT CLONE; more highly coloured.
    CRIMSON SUPERB; more red col. Arose with J. Anderson, Driffield, Yorks; exhib 1950.
    MAXTON; more regular cropper. Discv 1939 by R. Heseltine, Assington, Suffolk.
    RED SUPERB (Fox-Den); more highly coloured.
    RUSSET SUPERB; russeted form.

LAXTON'S TRIUMPH 7 L D
UK; raised 1902 by Laxton Bros, Bedford.

King of the Pippin X Cox's Orange Pippin. Introd 1930.
Intensely flavoured, aromatic, but rather sharp.
F* 14. T². P l-Oct. S Nov-Jan.

LAXTON'S VICTORY 7 M D
UK; raised 1926 Laxton Bros, Bedford. Cox's Orange Pippin X Exquisite. Introd 1945.
Lightly aromatic, quite rich, sweet, juicy.
F 11. T². C bien. P e-Sept. S Sept-Oct.

LEATHERCOAT RUSSET (BROTHER-STON) 8 L D
UK; received 1949 from HG, Brotherston, Tyninghame Gardens, East Lothian. Acc is not Leathercoat of Hogg and prob not Royal Russet of Hogg (syn Leathercoat Russet). Leathercoat, Leathercoat Russet, Royal Russet mentioned in C16th, but unlikely acc is ancient variety.
Heavily russeted. By Jan, intensely sweet-sharp, with leathery pale cream, green tinged flesh.
F 8. T³. P m-Oct. S Dec-Mar.

LEGANA 6 L D
Australia; raised c1940 by J. Bulman, Legana, Tasmania.
Democrat X Delicious. Introd c1950.
Bright pinky red flush. Sweet, but in England often tastes green, undeveloped.
Grown Tasmania for export to mainland.
F 13. T¹; sprd. P l-Oct. S Jan-Mar/Apr.

LEMOEN 5 L D
Netherlands; raised late C19th by notary J. H. Th. van den Hamm of Lunteren, Veluwe, north west Arnhem.
Large; lemon yellow. Strong lemony acidity; mellows to intense sweet-sharp taste.
One of many seedlings raised by van den Hamm and Lunteren villagers as part of project to improve local farmers' conditions. Van den Hamm founded Lunteren Hort Soc in 1873 to help select good varieties. Lemoen was awarded first prize 1899 at Society's national exhibition. Remains valued Netherlands, Belgium. In memory of van den Hamm's work a viewing tower which is also a museum has been built in Lunteren
F 10. T²; sprd. P m-Oct. S Nov-Jan.

LEMON PIPPIN 5 L C
UK; believed English, but history confused; poss introd from Normandy. Mentioned 1744 by agricultural writer Ellis, but prob known earlier.
Lemon coloured and shaped, especially king fruits which have fleshy swelling at side of stalk; also said to smell strongly of lemon, but this is not obvious. Quite sweet, some lemony acidity, rather dry, firm, coarse, yellow flesh.

Grown for market in C18th, C19th in Kent, around London and Normandy. Recommended by Forsyth, for drying; others found it 'good eating, though excellent for tarts'. Victorian chef, Francatelli used it for 'Apple Jelly' and sweetmeats.
F 15. T^2. P e-Oct. S Oct-Dec/Mar.

LEMON QUEEN 5 M D
UK; received 1950 from Scotland.
Lemon yellow. Quite sweet, light, pleasant taste; firm flesh.
F 14. T^1. P m-Sept. S Sept-Oct/Nov.

LEONARD LUSH 3 L DC
UK; received 1950.
Bright red flush, stripes. Brisk, fruity, soft white flesh in Dec, but very lightly flavoured when cooked.
F 9. T^2. P m-Oct. S Nov-Jan/Apr.

LEONIE DE SONNAVILLE 7 L D
Netherlands; raised after 1952 by Piet de Sonnaville in his family orchards at Puiflijk and Altforst, nr Nijmegen; named after his mother. Cox's Orange Pippin X Jonathan. Received 1974.
Delicately flavoured, aromatic, scented, sweet, juicy.
F 18. T^2; weeping; prt tip. P e/m-Oct. S Nov-Dec/Jan.

LEVERING LIMBERTWIG see Limbertwig

LEWIS'S INCOMPARABLE 3 L C
UK; prob variety desc 1831 (Ronalds) and known since late 1700s.
Large; boldly red striped. Crisp, brisk, deep cream flesh. Cooked keeps shape, but very light taste.
F 18. T^2. P m-Oct. S Dec-Feb.

LIBERTY 6 L D
US; raised by R. Lamb, NYSAES, Geneva. Macoun X Purdue 54-12 introd 1978.
First, and regarded as best, of US disease-resistant apples. Refreshing, quite sweet, soft, juicy, cream flesh.
F est 10. T^3; sprd; res scab, mildew, fireblight, cedar apple rust. C hvy; sml fruit unless thinned. P l-Sept. S Oct-Dec; commercially-Feb.

LIBOVICKÁ ORANZOVA RENATA
4 L D
Czech Republic; raised 1950s by V. Bláha, Libovice, North Bohemia.
Fruity, sweet; soft, juicy cream flesh. Formerly grown by amateurs; used also for juice.
F 8. T^2. P e-Oct. S Oct-Dec.

LIDDELL'S SEEDLING 5 E CD
UK; received 1949 from Scotland; may be variety exhib 1934 from Cumberland.
Large; heavily ribbed. Sweet, soft flesh, little

acidity.
F 7. T^2; sp; sprd. P use Aug/Sept.

LILLE 4 L D
France; desc 1948 (*Verger Français*), when grown Pyrénées-Roussillon.
Brisk, some taste of fruit in Feb; hard flesh.
F 16. T^2. P l-Oct. S Jan-Apr.

LILY BOXALL 3 L CD
UK; raised 1952 by T. H. Boxall, Bromley, Kent.
Granny Smith seedling.
Sharp, eating apple in Oct, but large and can be used earlier for cooking.
F 12; trip. T^3. P l-Sept. S Oct-Dec.

LIMBERTWIG
USA; believed arose C18th in southern US. Limbertwig varieties were grown throughout southern states, most originated in remote areas of southern Appalachians. Valued for keeping qualities, 'storing cellars, pits, caves' until April; dual purpose, used cider, apple butter. Represented in NFC by sport.

    LEVERING LIMBERTWIG 2/4 L D
    USA; arose with S. R. Levering, Ararat, Virginia. Larger form of Limbertwig.
    Crisp, hard in Feb; v tough skin.
    F 12; tetrap. T^3. P l-Oct. S Jan-Mar.

LIMELIGHT 2/5 M D
UK; raised c1985 by H. Ermen, Faversham, Kent.
Discovery X Greensleeves. Introd 2000 by F. W. Matthews Nursery, Tenbury Wells, Worcs.
Shape and taste of Discovery, and colour of Greensleeves.
F est 12. T^2; gd res scab, mildew. P m/l-Sept. S l-Sept-Dec.

LIMONCELLA 5 L D
Italy; old variety, prob arose Sicily.
Markedly columnar shaped. Quite intense sweet-sharp, lemon flavour; hard white flesh.
F* wht; 15. T^2. P l-Oct. S Nov-Jan.

LINDA 6 M D
Canada; raised Central Exp Farm, Ottawa. Langford Beauty X. Fruited 1908. Introd 1925.
Large, dark red flush. Light strawberry flavour; sweet, crisp juicy, white flesh; tough skin. Grown in Poland.
F 7. T^2. P l-Sept. S Oct-Nov.

LINDEL 6 L CD
Canada; raised 1939 Smithfield Exp Farm, Ontario.
Richard Delicious X Linda. Introd 1971.
Often large; heavily ribbed, crowned. Very sweet, scented, deep cream flesh. Cooked, keeps shape, very sweet.
F 12. T^2, tip. C gd, frt misshapen. P e-Oct.

S Oct-Dec; commerically-Mar.

LIPTON 6 E D
Canada; arose Central Exp Farm, Ottawa. Introd 1915.
Heavily ribbed. Lightly aromatic, quite rich, hint strawberry flavour; sweet, soft, juicy flesh.
F 11. T^2. P l-Aug. S Sept.

LIRED (Roosje) see James Grieve

LOBO 6 M D
Canada; raised 1898 at Central Exp Farm, Ottawa.
McIntosh X. Introd 1910. Wilder Silver Medal 1923.
McIntosh type; beautiful deep maroon flush, regular shape. Juicy flesh, like melting snow; sweet, light strawberry flavour; tough skin.
Grown commercially in Canada; also Scandinavia, Poland, small extent England.
F 9. T^2; hrdy. C gd. P m/-Sept. S Sept-Nov.

LODDINGTON, syn Stone's 3 M C
UK; believed brought c1820 from nursery in Bath to Mr Robert Stone's farm in village of Loddington, nr Maidstone, Kent, by his niece. RHS FCC 1877. Popularised by local orchardist, Lewis Killick; also known Killick's Apple.
Large. Cooks to well-flavoured, brisk, creamy purée; said to be excellent for dumplings. Mellows to pleasant eating apple.
Renowned Maidstone apple by 1870s, grown for London markets.
F* 12. T^2; sprd. C gd. P l-Sept. S Oct-Dec.

LODGEMORE NONPAREIL (?) 5 L D
UK; acc may be variety desc Hogg (1884) and raised c1808 by Mr Cook of Lodgemore, Stroud. Propagated and sold by nursmn Clissold, as Clissold's Seedling.
Russet freckles over gold. Intense, sweet-sharp, acid drops flavour of Nonpareil family; very good.
F 23. T^2. P e-Oct. S Nov-Jan.

LODI 5 E C
USA; raised 1911 by R. Wellington, NYSAES, Geneva.
Montgomery X White Transparent. Introd 1924.
Cooks to white froth, sweet, very juicy, but rather watery.
Larger, firmer, not bruising as easily as White Transparent, which it has replaced as earliest market cooker in States. Can be picked there by mid-July.
F 6. T^3; sprd; part tip. P l-Jly-m-Aug.

LOMBARTS CALVILLE 5 L D
Netherlands; raised 1906 by P. Lombarts, Zundert.
Believed Calville Blanche d'Hiver seedling.

Introd 1911 by Lombarts family.
Rich, sprightly, lots of flavour of fruit; sweet, juicy. Remains valued Netherlands.
F 12. **T²**. **P** m-Oct. **S** Nov-Jan/Mar.

LONDON PEARMAIN 7 M D
UK; known 1842; desc Hogg (1884).
Markedly pearmain-shaped. Strong, sweet-sharp flavour; lots of sugar and acidity.
F 15. **T²**. **P** l-Sept. **S** Oct-Nov.

LONDON PIPPIN 5 L C
UK; arose Essex or Norfolk according to Norwich nursmn, G. Lindley, who described it 1831. Poss dating from C16th. Hogg found 'grafts of Louden Peppen' mentioned in note book of Trevelyan family of Somerset, as received 1580 from Essex. Many syns.
Prominently ribbed, crowned; hence its syn Five Crowned Pippin. Sweet, plenty acidity, quite nutty, resembling Golden Reinette and considered 'serviceable' for Victorian dessert. Cooked keeps shape, bright lemon, quite rich, sweet taste.
Quite widely grown C19th; also grown in Australia and exported to UK 1920–30s.
F 15. **T²**. **P** m-Oct. **S** Nov-Mar.

LONG BIDER 3/5 L C
UK; exhib 1934 by EMRS, Kent; received from them 1948.
Sharp, lots of fruit; cooks to a well-flavoured purée.
F 18. **T³**. **P** e-Oct. **S** Nov-Feb.

LONGNEY RUSSET 8 L D
UK; received 1949.
Sharp, strong fresh taste in Feb; firm flesh.
F 15. **T²**. **P**-Oct. **S** Jan-Apr.

LONGSTART 4 M DC
UK; received 1951 from Westmorland and prob variety desc 1851 (Hogg).
Quite savoury, brisk, soft, juicy flesh.
Favourite cottagers' apple around Lancaster and in Westmorland in C19th.
F 12. **T¹**; sprd. **P** l-Aug. **S** Sept-Oct.

LOOP SPY *see* Northern Spy

LOOP WEALTHY *see* Wealthy

LORD BURGHLEY 7 L D
UK; arose c1834 in garden of Marquis Exeter, Burghley, Stamford, Northants. Rescued from waste ground by HG, Mr Matheson. RHS FCC 1865. Introd by nursmn House of Peterborough.
Dark red flushed; brisk, aromatic quality; plenty of sugar, juicy, firm flesh.
'Highly ornamental and unusual excellence for March and appears as if it would keep to June' pronounced RHS in 1865. Widely grown in gardens in C19th.
F 16. **T²**; sprd. **C** gd. **P** m-Oct. **S** Jan-Apr.

LORD CLYDE 5 L C
UK; raised by nursmn B. W. Witham, Reddish, Stockport, Cheshire; catalogued 1866.
Resembles small Golden Noble. Cooks to juicy purée, fruity, quite brisk.
F 8. **T⁷**; sprd. **P** e-Oct. **S** Nov-Jan.

LORD DERBY 1 M C
UK; raised by nursmn B. W. Witham, Stockport, Cheshire; recorded 1862. Believed, by some, to be Catshead seedling; can have similar shape.
Cooked early, when green, strong, sharp taste, needing some sugar; good for pies, as it keeps little of its form. Flavour and acidity fall after storing and best used early, but grown in north stays green to Dec with good deal more acidity. Widely grown by late C19th in gardens and for market. Remains popular garden variety, especially in north and Scotland; still seen on sale.
Frt *Col* grn, v slight pinky orng or brownish purple on cheek; turns yellow; much scarf skin at base; lenticels as grey dots. *Size* lrg. *Shape* rnd-con to oblng-con; ribbed, flt sided; crowned, but not always pron. *Basin* med wdth, dpth; ribbed; sometimes beaded. *Eye* clsd or prt open; sepals, broad. *Cavity* brd, shal. *Stalk* shrt, thck. *Flesh* wht.
F* 14. **T⁵**; hrdy. **C** gd. **P** l-Sept. **S** Oct-Dec. PLATE 13
LORD DERBY SPUR TYPE

LORD GROSVENOR 1 E C
UK; believed new in 1872. Brought to prominence at 1883 Congress; desc Hogg (1884).
Frothy, codlin-type cooker, juicy, brisk. Formerly widely grown in gardens and for market, but overtaken 1920s by Early Victoria.
F* 10. **T¹**. **C** hvy; sml frt unless thinned. **P** m-Aug. **S** Aug-Oct.

LORD HINDLIP 7 L D
UK; arose Worcs. Exhib RHS 1896 by Jno Watkins, Hereford. RHS FCC 1898.
Named after owner of Worcs estate. Very handsome, crimson flushed. Rich, aromatic, sweet, deep cream but rather coarsely textured flesh.
Still found in West Midlands gardens.
F 10. **T²**. **C** pr. **P** m-Oct. **S** Dec-Mar.

LORD LAMBOURNE 7 M D
UK; raised 1907 Laxton Bros, Bedford. James Grieve X Worcester Pearmain. Introd 1923. Bunyard Cup 1921. RHS AM 1923; AGM 1993.
Named after president of RHS. Shows some of strawberry flavour of Worcester with plenty of refreshing James Grieve acidity; sweet, juicy, crisp flesh.
Widely grown in gardens; small extent commercially.
Frt *Col* bright rd flush, broken rd stripes

over grnish yell/yell; lenticels prom as russet dots. *Size* med. *Shape* flt-rnd to rnd-con. *Basin* brd, qte dp; ribbed, sometimes beaded. *Eye* sml, prt open/clsd; sepals, v downy. *Cavity* med wdth, dpth; russet lined. *Stalk* shrt, qte thin. *Flesh* pale crm.
F 8. **T²**; prt tip; res mildew. **C** gd. **P** m-Sept. **S** late Sept-Nov.
    LADY LAMBOURNE more highly coloured. Found c1945 Appleby Farm, Kingston Bagpuize, Berks.
    RUSSET LAMBOURNE; russeted, but same flavour.

LORD LENNOX (?) 4/6 M D
UK; acc shows similarities with variety of Hogg and of Bunyard, but impossible to be certain as to identity. Acc received 1949.
Very handsome, dark red flushed. Intense flavour, rich blend of sweetness and acidity, some aromatic quality. Stronger, sharper than Lord Lambourne.
F 7. **T²**. **P** m-Sept. **S** l-Sept-Nov.

LORD NELSON 4 M D
Received 1960 from Australia via New Zealand.
Very large, red striped over primrose. Crisp juicy white flesh, brisk yet quite sweet and tasty. Cooks to good, savoury purée, hardly needing any sugar.
F 6. **T³**. **P** l-Sept. **S** Oct-Nov.

LORD PECKOVER 6 E D
UK; arose in Lord Peckover's garden, Wisbech, Cambs. Received 1926.
Large, dark red flush, dusky bloom. Pleasant, sweet yet quite brisk; crisp flesh.
F 1. **T²**. **P** use Aug.

LORD ROSEBERY 6 E D
UK; believed raised early 1900s by nursmn D. Storrie of Glencarse, Scotland.
Bright red flush. Slight strawberry flavour, quite brisk, yet plenty of sweetness; crisp, rather chewy flesh.
F 10. **T¹**; sprd. **P** l-Aug. **S** Sept-Nov.

LORD STRADBROKE 3 L C
UK; found or raised c1900 by HG, Mr Fenn, at Henham Hall, Suffolk, Lord Stradbroke's estate. RHS AM 1905.
Very large, prominently ribbed, crowned. Sweet, light cooker in Oct.
F 12. **T³**. **C** hvy. **P** e-Oct. **S** Oct-Dec/Feb.

LORD SUFFIELD 5 E C
UK; raised by Thomas Thorpe, hand loom weaver, who lived at Boardman Lane, Middleton, nr Manchester, where Lord Suffield had an estate. Introd c1836.
Codlin type, cooking to sharp, white froth. Popular Victorian variety, receiving more votes than any other cooker at 1883 Congress. Formerly widely grown in gardens and as early market apple. Also suggested for the Victorian shrubbery, as

makes small tree, always covered in pale pink blossom.
Frt *Col* pale grn turning pale yell; prom white lenticels; smooth skin. *Size* med/lrg. *Shape* con; irregular; tapering to apex; sltly ribbed, flt sided. *Basin* nrw, shal, pinched; ribbed; beaded. *Eye* qte sml, clsd; sepals, lng, v downy. *Cavity* brd, dp; grey russet. *Stalk* shrt, qte thck. *Flesh* wht.
F *4. T². C hvy. P m-Aug. S Aug-Sept.

LORNA DOONE 4 M D cider
UK; sent 1930s by E. H. Wells of Lorna Doone Cider, Wellington, Somerset to LARS, Bristol.
Sweet, very lightly flavoured, soft flesh with tannic undertones.
Received prize 1935 for good medium sharp cider, but never taken up; remains local variety.
F 12. T³. P l-Aug. S Sept-Nov/Dec.

LOUITON 5 L D
France; prob variety desc 1948 (*Verger Français*), when grown Meuse, Marne, Argonne (Champagne).
Sweet, firm, green tinged flesh.
Remains common in Argonne, where used mainly for cider.
F 35. T². P l-Oct. S Jan-Apr.

LOVACKA RENETA 4 L D
Received 1967 from former Yugoslavia via Scotland.
Quite rich in Nov; sweet, soft, juicy, deep cream flesh.
F 14. T²; sprd. C hvy. P e-Oct. S Oct-Dec.

LOVE BEAUTY 4 L D
Received 1967 from Scotland.
Handsome. Quite rich, sweet, juicy, but can taste green, woody.
F 15. T². P e-Oct. S Oct-Dec.

LOWLAND RASPBERRY 4 E D
Russia; recorded before 1870. Taken to North America, where acquired present name.
Tastes, perhaps, slightly of raspberries. Sweet, soft, juicy white flesh.
F 6. T²; hrdy; claimed res scab. C. P m-Aug. S Aug-Sept.

LUCOMBE'S PINE 5 L D
UK; raised c1800 in nursery of Lucombe Pince & Co, Exeter, Devon. Desc Hogg 1851.
Fine russet freckles over gold. Strong acidity suggestive of pineapple; firm flesh. Recommended for juice.
F 13. T³. C pr. P e-Oct. S Nov-Dec.

LUCOMBE'S SEEDLING (?) 5 M D
UK; may be the variety raised in nursery of Lucombe Pince & Co, Exeter, Devon. Desc Ronalds 1831.
Large, prominently ribbed. Sharp, crisp,

becoming savoury, soft.
Popular early eating apple in west Cornwall in 1920–30s; sometimes called Newquay Prizetaker.
F 8. T¹. P m-Sept. S Sept-Oct/Nov.

LUCULLUS 4/7 L D
Netherlands; raised 1935 at IVT, Wageningen.
Jonathan X Cox's Orange Pippin. Introd 1955.
Worthy of Lucullan banquet; deep rose red flush, stripes. Rich, aromatic, scented, crisp, juicy deep cream flesh. Can be undeveloped, green quality in poor year.
F 14. T²; sprd; prn mildew, canker. C lght. P e-Oct. S m-Oct-Dec.

LUNDBYTORP 6 L D
Denmark; arose from pip sown on farm in village of Lundbytorp, Lundy, Southern Sealand in C19th. Desc 1913.
Bright red. Very sweet, scented, firm, cream flesh.
Formerly grown all over Denmark, southern Sweden, but now little known.
F 9. T³; sp. C bien. P m-Oct. S Nov-Jan/Mar.

LUNTERSCHE PIPPELING (Lunteren Pippin) 5 L D
Netherlands; raised late C19th by notary J. H. Th. van den Hamm, or villagers of Lunteren in Veluwe, north west Arnhem.
Quite rich, sweet-sharp quality; recalling Belle de Boskoop.
Among seedlings raised to improve local farmers' conditions. Remains valued Netherlands.
F 8. T²; sp. P m-Oct. S Nov-Jan/Feb.

LUXEMBURGER RENETTE, syns Reinette des Vergers, Cwastresse Double 4 M DC
Luxembourg pres; known to Leroy (1873), also as Coastress; acc is variety now known as Cwastresse Double, long grown in standard orchards of Belgium. Name derives from Walloon word 'cwasterê' meaning sides or ribs and equivalent to Calville shape.
Crisp, sharp juicy white flesh; fully ripe full of fruit and flavour, lightly perfumed. Cooked, keeps shape.
In Belgium, as Cwastresse Double, RGF variety, recommended for cultivation by Agric Res Centre, Gembloux because of disease resistance.
F 16. T³; tip; part res scab, mildew. P m-Sept. S Oct-Nov/Dec.

LUZHANKA 5 M DC
Russia; received 1975 from Vavilov Inst, St Petersburg.
Sharp, soft flesh. Cooked keeps shape, yellow, good brisk, rich flavour.
F 5. T²; frt bruises. P m-Sept. S Sept-Oct.

LYNN'S PIPPIN 7 M D
UK; raised 1942 by William Lynn, Wisbech, Cambs.
Cox's Orange Pippin X Ellison's Orange.
At best, intensely rich, aromatic; sweet, soft, juicy, deep cream flesh. Often less interesting.
F 13. T². P m-Sept. S Sept-Oct.

LYS GOLDEN *see* Golden Delicious

MABBOTT'S PEARMAIN 7 M D
UK; introd by Lewis Killick, fruit grower, at Langley, nr Maidstone, Kent. Exhibit 1883 Congress, but grown locally for some yrs.
Solid, little apple, thickly speckled with russet over orange red flush, stripes. Brisk, good, plain taste of fruit.
Grown around Maidstone for London markets in C19th.
F 12. T²; prt tip. C hvy. P m/l-Sept. S Oct-Dec.

MACLEAN'S FAVOURITE (?) 4 L D
UK; acc may be variety raised c1820 by Dr Allan Maclean of Colchester, Essex; desc 1884 (Hogg). Maclean who was best known as raiser of peas.
Rich, sweet-sharp, almost pineapple flavour; Hogg claimed resembled Newtown Pippin.
F 16. T¹; tip. P e-Oct. S Oct-Dec/Jan.

MACOUN 6 M D
USA; raised by R. Wellington, NYSAES, Geneva.
McIntosh X Jersey Black. Introd 1923.
Named after Canadian fruit breeder, W. T. Macoun. Almost purple flush, with bloom. Very sweet, scented, hint of strawberry flavour; snow white, juicy flesh; tough skin. McIntosh type, but ripening later.
Esteemed by American connoisseurs; now renewed interest commercially in north east States.
F 11. T³; hrdy; prn canker. C hvy, bien. P m-Sept. S Oct-Dec.

MACROSS 4 E D
Canada; raised Central Exp Farm Ottawa.
McIntosh X. Introd 1925.
Crisp, juicy, melting flesh, but can be thin, sharp.
F 7. T². P eat l-Aug-Sept.

MACWOOD 6 L D
Canada; raised 1936 Central Exp Farm, Ottawa.
McIntosh X Forest. Introd 1932.
McIntosh type. Very sweet, soft, deep cream flesh; can be woody, green in England.
F 7. T². P m-Oct. S Dec-Feb/Apr.

MACY 4 L M
USA; raised by H. L. Lantz, Iowa Agric Exp

St, Ames, Iowa.
Seedling of Northwestern Greening X Wealthy. Introd 1921.
Sweet, juicy, soft cream flesh; hint of strawberry flavour.
F 13. T^2. C frt drops. P e-Oct. S Oct-Nov/Dec.

MADOUE ROUGE 6 L D
France; acc prob old variety, arose Nieulin, Haute-Vienne (Limousin); desc 1948 (*Verger Français*).
Bright red. Refreshing, brisk, fruity until Feb; tough flesh.
Formerly much grown Limousin, still found Haute-Vienne. Valued for late flowering, long keeping.
F 30. T^2. P l-Oct. S Dec-Mar/Apr.

MADRESFIELD COURT 7/8 M D
UK; prob raised by William Crump, HG, Madresfield Court, Worcs.
Introd by J. Carless, Worcester. RHS AM 1915.
Carmine flushed; ribbed, angular. Quite rich, slightly aromatic, sweet deep cream flesh, but rather tough.
F 11. T^2. tip; prn canker. C lght. P l-Sept. S Oct-Dec.

MAGDALENE 7 L D syn Saint Magdalen
UK; found c1890 by H. Bridge at St Mary Magdalene, Norfolk. Introd 1924 by H. Goude; exhib from Norfolk 1934.
Very sweet, quite rich, aromatic, but little acidity; becomes almost like sweetmeat; deep cream flesh.
F 9. T^2. P e-Oct. S Oct-Dec.

MAGGIE GRIEVE 6 M D
UK; raised at LARS, Bristol. Received 1928.
Bright red. Sweet, rather chewy, deep cream, flesh.
F 16. T^2. P l-Sept. S Oct-Nov

MAGGIE SINCLAIR 1/5 L CD
UK; received 1949 from Scotland, where prob arose.
Large. Quite sweet, crisp, fruity. Cooks to juicy, gold purée.
F 11. T^3; sprd; prt tip. P m-Oct. S Nov-Jan.

MAGLEMER 4 M D
Denmark; origin unknown; named after place nr Maribo, Lolland. Known late 1700s as Alfred Hayes Apple. Introd c1870 present name.
Very juicy, savoury, quite brisk; soft, cream flesh.
Widely planted Denmark, Sweden in late C19th; still valued as dual purpose apple; formerly used Denmark for cider.
F 7. T^2. C hvy. P e-Sept. S Sept-Oct.

MAGNOLIA GOLD 5 L D
USA; raised or found 1960 by W. J. Wilson,

Fort Valley, Georgia. Introd 1970.
Golden Delicious seedling.
Crisp, juicy, sweet like Golden Delicious, but more acidity. Less prone russeting than parent.
F 18. T^2. C gd. P l-Oct. S Nov-Jan.

MAHE 4 L D
Japan; raised 1956 Aomori Apple Exp St. (Indo X Golden Delicious) X McIntosh.
Deep maroon; prominently crowned. Sweet, scented, firm flesh; tough skin.
F* purple; 9. T^2. C hvy; frt sml unless thinned. P m-Oct. S Nov-Jan.

MAIDEN'S BLUSH 4 M D
Acc received 1945 from Norwich. Not Maiden's Blush USA, or Ireland.
Attractive carmine flush and stripes. Crisp, brisk, juicy becoming light and sweet.
F 23. T^3; tip. P m/l-Sept. S Oct-Nov.

MAID OF KENT 3 L C
UK; obtained 1979 from Marden, Kent from 60 yr old tree. Acc may be variety of this name exhib 1942.
Large; bright red flushed. Sharp, crisp, juicy. Cooks to bright lemon purée; savoury, sharp.
F 12. T^2; sprd. P m-Oct. S Nov-Jan.

MAIDSTONE FAVOURITE 4 E D
UK; raised by nursmn G. Bunyard, Maidstone, Kent.
Alexander X poss Beauty of Bath. RHS AM 1913.
Prettily striped, like brightly coloured Beauty of Bath, but ripening later. Sweet, but rather flavourless.
F 7. T^2; sprd. C v hvy. P use m-Aug-Sept.

MAIGOLD 6 L D
Switzerland; raised 1944 by F. Kobel, Fed Res St, Wadenswil.
Franc Roseau X Golden Delicious. Introd 1964. Syn Gold-N-Rose.
Rich, sweet-sharp taste, slightly honeyed; almost citrus flavour in Dec; very crisp, pale yellow flesh.
Grown commercially Switzerland, north Italy.
F 10. T^3; weeping. C hvy; frt pr col Eng. P m-Oct. S Nov-Jan; cold store-May.

MALDON WONDER 6 M D
UK; raised c1900 by A. Mynard, Heybridge, Essex. Introd 1933 by H. Brewer, Maldon.
Handsome. Lightly aromatic, sweet, juicy; like dilute Cox.
F 6. T^2. P m-Sept. S Oct-Nov.

MALLING KENT syn Kent

MALSTER 4 M DC
UK; prob arose Notts; known 1830; desc Hogg (1884).
Crisp, juicy, but low on flavour. Cooked,

sweet, light; but prob more acidity grown in north. In C19th 'much grown' around Nottingham.
F 5. T^2; sprd. C hvy. P e-Sept. S Sept-Oct/Nov.

MANDAN 4 M CD
USA; raised by W. A. Oitto, Northern Great Plains Field St, Mandan, North Dakota.
Duchess X Starking Delicious. Introd 1965.
Sweet, juicy, firm flesh, little astringency; tough skin. Dual purpose US.
F 11. T^2; v hardy. P e-Sept. S Sept-Oct.

MANKS CODLIN 5 E C
UK; raised by Mr Kewley, Ballanard, Isle of Man. Fruited 1815. Distributed throughout Europe. Syns many; include Eve Apple of Scotland, Irish Codlin, Eva's Calvill.
Sweetest of codlins and most colourful with slight red flush. Makes delicately flavoured baked apple.
Widely grown C19th, especially in north, Scotland.
F 5. T^2; sprd, hrdy. C hvy, bien. P m-Aug. S Aug-Oct.

MANNINGTON'S PEARMAIN 7 L D
UK; arose c1770 from cider pomace thrown under hedge in garden of blacksmith, Mr Turley, Uckfield, Sussex. Sent 1847 to London Hort Soc by his grandson, John Mannington, local butcher and keen fruit man.
Rich, aromatic, not intensely so, but well balanced with definite nutty quality.
Esteemed by Victorians, but considered only fair by Bunyard.
F 13. T^2. C gd. P m-Oct. S Nov-Feb.

MANTET 5/4 E D
Canada; raised Dom Exp St, Morden, Manitoba.
Tetofsky X. Introd 1929.
Sweet, juicy, light, perhaps almond flavour; soft, melting flesh.
F 11. T^2; v hrdy. P l-Aug. S Aug-Sept.

MAREDA 4 L D
UK; raised 1926 by A. C. Nash, Scutes Farm nr Hastings, Sussex, from Winesap pip.
Handsomely flushed, striped in bright red. Sweet, winey or like sweet melon; distinctive taste; firm flesh.
F 12. T^2. C bien. P l-Oct. S Dec-Feb/Mar.

MAREN NIS 4 E D
Denmark; received 1950, but prob very old variety from North Schleswig.
Tall, ribbed, box-shaped; red flushed over pale yellow. Sweet, crisp, juicy, white flesh; melon or strawberry flavour; tough skin.
F 14. T^2. C hvy, bien; frt sml. P Sept. S Sept-Nov.

MARGARET syn Red Joaneting 4 E D
UK; arose England; believed very old. Name recorded 1665 by J. Rea. Distributed throughout Europe; introd North America. Syns numerous.
Ripening around St Margaret's or St Magdalen's Day, July 22nd. Small, like highly coloured Joaneting. Quite sharp, refreshing, soft cream, green tinged flesh.
F 16. T^2; uprt. P eat l-July.

MARGARET TAYLOR 7 L D
UK; found by F. E. Taylor growing in hedge at Witley, Surrey. Exhib 1944.
Crisp, refreshing, quite rich in Dec; recalls Golden Reinette. Often dull appearance.
F 15. T^2; prt tip. C frt drops. P e-Oct. S Dec/Jan-Mar.

MARGIL 7 L D
Europe; reputedly planted by Sir William Temple in garden at Sheen, Surrey in late C17th. Poss introd from France at this time by George London of Brompton Park Nursery, which by 1750 had large stocks. Syns numerous; incl Reinette Musquée, Never Fail, Small Ribston.
Intensely flavoured, rich, aromatic, deep cream flesh. More delicate than Ribston, very like Cox, only more scented, exotic. Bunyard placed it in epicure's top ten, but can taste metallic if deprived of sun.
Perfect for Victorian dessert; widely grown C19th in gardens and by London market gardeners; often found in old gardens.
Frt *Col* orng rd flush, qte brd rd stripes over gold; russet patches. *Size* med/sml. *Shape* rnd-con; flt sided, sltly angular; sometimes 5 crowned. *Basin* nrw, qte dp; ribbed; sometimes beaded. *Eye* sml, clsd or prt open; sepals, sml. *Cavity* nrw, qte dp; russet lined. *Stalk* shrt, thin. *Flesh* crm.
F 10. T^2. C hvy; prn canker. P e-Oct. S Oct-Dec.

MARGOL 6 L D
Germany; received 1987 from Fruit Res St, Jork, nr Hamburg.
Ingrid Marie X Golden Delicious.
Bright scarlet flushed. Soft, juicy, pale cream flesh, savoury and brisk; plenty of fruit, but can be sour.
F est 12. T^2. P e-Oct. S Oct-Dec/Jan.

MARIBORKA 6 L D
Slovenia; raised 1949 by Prof J. Priol at Inst of Cultivation, Moribor.
Golden Pearmain X Jonathan. Named 1961. Striking, bright red flush. Strong, robust flavour; rich, honeyed, quite aromatic, crisp, juicy, deep cream flesh.
F 7. T^2; uprt. P e-Oct. S Nov-Jan.

MARIE DOUDOU 4 L D
France; recorded 1909 as growing region Englefontain, Nord; desc 1948 (*Verger Français*).

Rich, brisk fruity taste; firm flesh. Found as orchard standards in Avesnois-Thiérache, Nord.
F 23. T^3; prt tip. P l-Oct. S Dec-Feb.

MARIE-JOSEPH D'OTHÉE syn IJzerappel 7 L DC
Belgium; arose c1870 at Othée, Liège province; prob discv on property of Drisket by gardener Monville of Othée; dedicated to his wife.
Mellows to intense, rich, sweet-sharp flavour by Jan. Firm, rather spongey texture; like Nonpareil.
Grown orchards of East Belgium; used for cooking.
F 23. T^3; sprd. C gd, regular. P l-Oct. S Jan-Mar/Apr.

MARIE-LOUISE DUCOTÉ 6 L D
France; obtained 1895 as chance seedling by M. Ducôté of St-Cyr-au-Mont-d'Or, Rhone.
Poss seedling of Reinette Cusset. Mother tree 26 yrs in 1917; brought to notice 1927. Quite sweet, firm, white flesh, but tannic undertones.
F 20. T^2; sp. P l-Oct. S Jan-Mar.

MARIE-MADELEINE 2/5 L D
France; prob Meaux, Seine-et-Marne. Desc 1948 (*Verger Français*).
Brisk, fruity, crisp in Feb, Mar.
Remains much appreciated around Meaux; used also for cider.
F 22. T^1; sprd. P l-Oct. S Dec-Mar.

MARIENWERDER GULDERLING 5 L D
Received from Germany. Recorded France 1876 by nursmn Simon-Louis Frères.
Quite fruity, slightly perfumed; sweet, soft, cream flesh.
F 15. T^1. P l-Oct. S Dec-Feb/Apr.

MARINA 6 L D
Switzerland; raised Fed. Agric. Res. St Wädenswil.
Idared X Kidd's Orange Red. Received 1999.
Handsome, large, rose pink flush and stripes. Deep cream flesh, crisp, juicy, sweet, lightly perfumed, reminiscent Kidd's; very tough skin.
F mid. T^2. P e-Oct. S Nov-Jan/Mar.

MÁROSSZÉKI PIROS PÁRIS 7 L D
Transylvania (Romania); arose district of Maros-Torda along rivers Maros and Nyárád; believed centuries old. Syn, Grosse Pomme Pâris, recorded 1598 by J. Bauhin in Württemberg and named after fabled apple given by Paris to Helen of Troy.
Sprightly, fruity; crisp, juicy in Feb; v tough skin.
Widely grown throughout Transylvania in C19th.

F^* 6. T^3. C hvy. P m-Oct. S Nov-Mar.

MARRIAGE-MAKER 6 M D
UK; desc Hogg (1884) but believed older; sent to Hogg by nursmn Harrison of Leicester.
Showy crimson flush. Sweet, clean taste of fruit; quite juicy, firm flesh.
F 15. T^2. C bien. P l-Sept. S Oct-Nov/Jan.

MARROI ROUGE 7 L D
France; received 1948.
Large; deep red flush, stripes. Sweet crisp, juicy, quite fruity.
F 21; trip. T^3; sprd. C gd. P l-Oct. S Nov/Dec-Mar.

MARSTAR *see* Melrose

MARSTON SCARLET WONDER *see* Newton Wonder

MARTIN BECKER 8 M D
Received 1947 from France; closely resembles, not identical, Wyken Pippin. Freckled, veined in russet over gold. Brisk, fruity taste.
F 14. T^1; sprd. P l-Sept. S Oct-Nov.

MARTINI 7 L D
Germany; found 1875 nr Kollmar, Holstein.
Believed seedling of Cox's Orange Pippin, or Prinzen Apfel, or Edelborsdorfer; mother tree existed 1960s.
Named after the time of harvest, St Martin's Day, Nov 11. Rich, sweet-sharp intense flavour; crisp, cream flesh.
F 14. T^1; sprd. C gd. P e-Oct. S l-Oct-Dec; cold store-Mar.
  RED MARTINI; arose with Mr Peters, Wenzendorf, nr Hamburg; received 1938.

MARTIN'S KERNEL 6 L D
UK; raised by M. Martin, Priding, Arlingham, nr Gloucester as a cider apple; received 1953. Bright red. Crisp, juicy, but low in flavour.
F 14. T^2. P l-Oct. S Nov-Jan/Mar.

MARUBA
Malus prunifolia, native to northern China and eastern Siberia. Attractive, delicate pale pink blossom. Cultivated in parts of China, Korea, Japan, eastern Siberia. Bears small astringent fruit; widely used in China for jellies, preserves and canning in syrup. Very hardy; resistant drought; used in breeding programmes and as rootstock.

MARY BARNETT 3 L CD
UK; raised c1920 by Mrs Mary Jane Barnett, Steeple Ashton, Wilts on her wedding day with pip from Lane's Prince Albert tree growing next to Lady Sudeley.
Resembles Lane's Prince Albert in

appearance. Savoury, brisk, fruity flavour; crisp, juicy. Cooked, sweet but tough.
F 13. **T**^{2}. **P** m-Oct. **S** Dec-Feb/Apr.

MARY GREEN 3 L C
UK; raised or discv by D. E. Green, Send, Surrey. Received 1948.
Cooks to quite sharp purée.
F 11. **T**^{2}; sprd. **C** hvy; frt sml unless thinned. **P** l-Oct. **S** Dec-Mar.

MÁTÉ DÉNES DR. 2 L D
Received 1948 from Univ of Agric, Budapest.
Quite scented, firm flesh, but low on flavour, sugar, acid in England.
F 7. **T**^{3}. **P** e Oct. **S** Dec-Jan/Mar.

MAUSS REINETTE syn Reinette de Mauss 6 L D
Germany; raised by Mauss of Herrenhausen, Hannover. Recorded 1874.
Intense, rich flavour, good sprightly taste of fruit; can be rather sharp.
F 8. **T**^{2}. **P** m-Oct. **S** Oct-Dec.

MAXTON see Laxton's Superb

MAY BEAUTY 7 L D
UK; received 1945 from D. & J. Walker, Ham, Surrey.
Newtown Pippin X Golden Pippin.
Dark red flush; undistinguished flavour.
F 14. **T**^{3}. **P** l-Oct. **S** Dec/Jan-Mar.

MAYPOLE 6 M
Raised 1976, EMRS, Kent.
Wijick Mcintosh, Baskatong cross. Introd 1989.
Ornamental. Large, deep pink blossom; purple young foliage. Small dark crimson fruit suitable for apple jelly. See Ballerina.
F* 3. **T**^{2}.

MAY QUEEN 7 L D
UK; raised by Haywood of Worcester. Introd by nursmn Penwill. Recorded 1888. RHS AM 1892.
Fruity, brisk, some richness in Feb.
F 12. **T**^{1}. **C** hvy. **P** m-Oct. **S** Nov-Feb/May.

McINTOSH 6 M D
Canada; found c1811 by John McIntosh at Dundela, Dundas County, Ontario. Also claimed found 1796, but this prob when he arrived in Canada. Propagation began c1835; more widely distributed by son Allan by c1870. Believed Fameuse, or poss Saint Lawrence seedling.
In prime soon after picking with strawberry or even elderflower flavour and sweet, glistening, melting, juicy, white flesh. But in England can easily collapse into metallic, woody taste; often described as vinous and soon becomes winey with keeping; tough skin. Vit C 3–9mg/100gm.
John McIntosh was son of Scottish Highland family who had emigrated to New York State. After family disagreement, John moved to Canada, settling in 1811 in Dundela. The McIntosh apple tree, the only one of several seedlings on his land to survive, eventually led to Dundela becoming known as McIntosh Corners and monument now marks site of original tree, which survived until 1908.
Grown commercially by early 1900s, and exported to UK. Now widely planted throughout colder parts of apple growing regions – Canada, northern USA, northern Europe, Poland, Austria – but in England prone to canker; needs clear days, cool nights of Canadian-type autumn to colour well.
Frt Col deep purplish rd flush, indistinct stripes; pale grnish yell/pale yell ground col; bloom; lenticels conspic as white dots. *Size* med. *Shape* rnd to flt-rnd; sometimes lop-sided; trace rnded ribs; sometimes crowned. *Basin* med wdth, dpth; sltly ribbed; beaded. *Eye* sml, clsd or sltly open; sepals, sml, v downy. *Cavity* med wdth, dpth; ltl russet. *Stalk* shrt, qte thin. *Flesh* wht.
F 7. **T**^{2}; prn scab, canker. **C** hvy. **P** m/l-Sept. **S** Oct- Dec, commercially-Jun.
PLATE 14
  ALEXIS; more highly coloured; received 1967 from former Yugoslavia via Scotland.
  BLACKMACK; colours earlier. Found 1928 by H. Simpson, in Oliver, British Columbia; introd 1930.
  BLACK MICKEY; solid flush. Found 1929 by I. C. Rogers in Dansville, New York; introd c1930.
  DERMEN McINTOSH; tetrap.
  JOHNSON McINTOSH; tetrap.
  KIMBALL McINTOSH; tetrap.
  ROGERS McINTOSH; even, dark red flush. Found by I. C. Rogers in Dansville, New York; introd 1932.
  STARKSPUR McINTOSH; spur type; introd by Stark Bros, Missouri.
  WIJICK McINTOSH; see Ballerina.

McLELLAN (?) 6 L D
USA; acc may be variety raised c1780 at Woodstock, Connecticut; desc Beach (1905).
Good sprightly taste of fruit in Nov; crisp, firm flesh. Esteemed C19th in eastern states as 'very choice dessert apple, handsome, fragrant and tender'; now being revived by amateurs.
F 10. **T**^{2}. **C** frt pr col Eng. **P** m-Oct. **S** Nov-Jan/Apr.

McLIVER'S WINESAP 6 L D
Received 1950 from New Zealand. May be sport or seedling of Winesap.
Carmine flushed. Fruity, firm flesh, brisk but rather metallic taste.
F 13. **T**^{2}; sprd. **P** m-Oct. **S** Jan-Mar.

McSWEET 6 M D
Canada; received 1927 from Central Exp Farm, Ottawa. McIntosh X. Introd 1920.
McIntosh type. Sweet, juicy, white flesh; can be empty, metallic in England.
F 8. **T**^{1}; sprd. **P** m-Sept. **S** Sept-Oct.

MEAD'S BROADING 3 M C
UK; sent to Hogg by nursmn J. R. Pearson, Chilwell, Nottingham. Desc 1884.
Large; colourful. Cooks to light purée, sweet yet brisk.
F 8. **T**^{2}. **P** e-Sept. **S** Sept-Oct/Nov.

MEASDAY'S FAVOURITE 1 M C
UK; received 1939 from Bunyard's Nursery, Kent.
Quite sharp, soft flesh. Cooks to mildly tasting purée.
F* 12. **T**^{2}; sp. **P** e-Sept. **S** Sept-Oct.

MEDINA 4 L D
USA; raised 1911 by R. Wellington at NYSAES, Geneva.
Deacon Jones X Delicious. Introd 1922.
Crowned, Delicious shape; dark red flush. Very sweet, scented, some richness and aromatic quality; deep cream flesh; tough skin.
F 12. **T**^{3}. **P** m-Oct. **S** Nov-Jan.

MEGUMI 6 L D
Japan; raised 1931 at Aomori Apple Exp St. Ralls Janet X Jonathan. Named 1948.
Lightly honeyed, crisp, juicy in Feb.
F 14. **T**^{1}; sprd. **C** hvy. **P** l-Oct. **S** Jan-Mar/Apr.

MEKU 4/6 E D
Japan; raised Aomori Apple Exp St. Tsugara X Iwai (American Summer Pearmain). Received 1977.
Very pretty; tall, conical. Juicy, sweet, deep cream flesh.
F 14. **T**^{2}. **P** e-Sept. **S** Sept-Oct.

MELA CARLA 6 L D
Italy; origins unclear, poss arose Finale Ligure (Piedmont), or Trentino (Italian Tyrol), or Calabria (southern Italy). Known 1700s, but poss older. Desc 1817. Syns many. Widely distributed Europe; introd USA.
Very attractive, small, flushed in bright red. In good year, slightly scented, juicy, sweet, soft, white flesh. Remains bright and attractive until Feb, although flavour fades. 'Pomi Carli', grown around Turin, were admired by Tobias Smollett, the Scottish novelist and journalist in 1765 and were exported from Genoa to Nice, Marseilles, Barcelona and Cadiz. Such was its beauty and reputation that Victorian gardeners grew trees in pots under glass, as it often does not ripen properly in England. Now almost unknown in Italy.
F 7. **T**^{2}. **C** gd. **P** e-Oct. **S** Nov-Feb.

MELBA 4 E D
Canada; raised 1898 by W. T. Macoun, Central Exp Farm, Ottawa.
McIntosh X. Introd c1924. Wilder Silver Medal 1927.
Flushed, striped in pink and red over pale yellow. Very sweet, juicy, melting white flesh with strawberry to vinous flavour; can be lovely. Early summer market apple of New Brunswick, Quebec; also grown northern Europe.
F 8. T^2; prn canker. C gd, bien; bruises v easily. P m/l-Aug. S Aug-e-Sept;
HUNTER MELBA; tetrap.
RED MELBA; deeper colour, also firmer, bruises less easily. Introd 1940s.

MELKAPPEL 5 L D
Belgium; local to East Flanders. Received 1948.
Large; pale yellow, slight flush. Firm crisp flesh, quite rich, lemony.
F 10. T^3. P l-Sept. S Oct-Nov/Dec.

MELMOTH 6 L D
UK; raised c1890s by Florence Melmoth, nr Yeovil, Somerset. Received 1948, when variety of local repute.
Attractive. Very sweet juice, little richness; quite firm, white flesh.
F 15. T^2. P m-Oct. S Nov-Dec/Jan.

MELON 6 L D
USA; arose c1800 in seedling orchard of Heman Chapin, East Bloomfield, Ontario County, New York. Introd 1845.
Large; very beautiful, flushed in strawberry red over pale yellow. Crisp yet melting, juicy flesh; refreshing flavour suggesting perhaps melons or strawberries.
Grown for 'fancy trade' in Western New York and Oregon 1890s/early 1900s; exported to London. Often not at best in English climate, but Victorians grew it under glass in orchard houses and it became favoured exhibition variety.
F 14. T^1. P m-Oct. S Dec-Jan.

MELROSE syn White Melrose 5 M C
UK; acc prob variety desc Hogg (1884), but not entirely agree. Melrose prob introd by monks of Melrose Abbey, Roxburgh, Scotland. Recorded 1831; prob older.
Large, ribbed; pale milky yellow. Cooked, keeps some shape; good, sweet-sharp flavour. Most popular apple of Tweedside orchards in C19th.
F 10. T^2. C hvy. P m-Sept. S Sept-Oct.

MELROSE (AMERICA) 6 L D
USA; raised by Freeman S. Howlett, Ohio Agric Exp St, Wooster, Ohio.
Jonathan X Delicious. Introd 1944.
Shiny red like Jonathan.
Brisk, refreshing, plenty of sweetness; crisp, juicy flesh.
Grown commercially US – extensively

Ohio; France, Poland; in England can be sour, poorly coloured.
F 10. T^2. C hvy. P e-Oct. S Nov-Feb/Apr.
More highly coloured sports:
DUGAMEL; MARSTAR

MERAN 6 L D
Italy; raised 1966.
Golden Delicious X Morgenduft. Received 1983.
Sweet, quite honeyed, but little depth or length of flavour; very crisp, juicy.
F* 5. T^2. P e-Oct. S Oct-Dec.

MERCER 6 L D
Received 1948 from EMRS Kent; acc is poss French variety Mercier, known Val de Loire.
Dark reddish brown flush, stripes. Brisk, fruity, but light on flavour.
F 21. T^3. P m-Oct. S Dec-Mar/Apr.

MERCHANT APPLE 6 M D
UK; arose pres Ilminster Somerset as syn is Merchant Apple of Ilminster. Desc by nursmn Scott 1872.
Dark blood red flush. Sweet juice but tough flesh.
F 14. T^2. P l-Sept. S Oct-Nov.

MERE DE BAIA 5 M DC
Romania; received 1948 from Agric Acad School, Bucharest.
Large. Crisp, juicy, quite sharp. Cooked keeps shape, lemon colour; good, rich taste.
F 14. T^2. P m-Sept. S Sept-Oct.

MÈRE DE MÉNAGE 6 L C
Europe; known late 1700, but confused history. Syns many, incl Flander's Pippin, Combermere Apple, Queen Emma, Husmoder, but prob not Swedish, Danish Husmøder.
Boldly striped, flushed in carmine; often very large. Early in season, cooks to purée, sweet, some acidity; not very juicy and recommended for 'Apple Charlotte'.
Widely grown England in C19th, especially in East Anglia; on sale up to 1930s.
F 10. T^3. P e-Oct. S Nov-Feb.

MERI CREŢEŞTI 3 M C
Received 1951 from Fruit Res St, Jork, nr Hamburg; prob Romanian variety.
Sharp. Cooks to bright lemon purée, brisk but little flavour.
F 9. T^2. P m-Sept. S Sept-Nov/Dec.

MERI DE SĂMÎNŢĂ 3 L C
Romania; received 1948 from Agric Acad Sc, Bucharest
Crisp, firm flesh. Cooks to gold slices, sweet, pleasant.
F 18. T^2. P m-Oct. S Nov-Jan.

MERIDIAN 7 M D
UK. Raised by Dr Frank Alston, EMRS, Kent.

Falstaff X Cox's Orange Pippin. Introd 1999.
Cream flesh, sweet, balanced, aromatic Cox-like flavour, but in delicate, not intense spectrum.
F est 12. T^2. C gd. P m-Sept. S Oct-Dec.

MERLYN 6 M D
Belgium; raised by nursmn J. Nicolaï, Sint-Truiden.
Jored X Liberty. Received 1996.
Scarlet flush. Sweet, quite rich and aromatic cream flesh.
F est 12. T^2. P m-Sept. S Oct-Dec.

MERRIGOLD 5 M D
France; received 1983.
Prob Golden Delicious seedling.
Looks and tastes like Golden Delicious, but ripens earlier.
F 12. T^2. P e/m-Sept. S Sept-Oct.

MERS BRYAN 6 L CD
Received 1947 from France.
Large, handsome; crisp, juicy, quite sweet. Cooked, keeps shape; sweet some richness.
F 14. T^2. P l-Oct. S Dec-Mar.

MERTON BEAUTY 7 M D
UK; raised 1932 by M. B. Crane, JI, Merton, London.
Ellison's Orange X Cox's Orange Pippin. Released 1962.
Truly delicious, if idiosyncratic. Tastes, smells strongly of aniseed; also very rich; sweet, crisp yet melting flesh. Short season; becoming strongly medicinal.
F 14. T^2. C gd. P m-Sept. S Sept-Oct.

MERTON CHARM 7 E D
UK; raised 1933 by M. B. Crane, JI, Merton, London.
McIntosh X Cox's Orange Pippin. Released 1962. RHS AM 1960; AGM 1993.
Sweet, light aromatic flavour; crisp, juicy. Eaten early when hard and sharper, flavour is much stronger.
F 6. T^2; semi-weeping. C hvy; frt sml, pr col. P m-Sept. S Sept-Oct.

MERTON DELIGHT 4 M D
UK; raised by M. B. Crane, JI, Merton, London. Received 1951. Discarded UK, but grown USA.
Prettily flushed, striped. Very juicy, fruity, soft flesh, like sweet James Grieve.
F 11. T^2. C gd; frt bruises easily. P l-Sept. S Oct-Nov.

MERTON JOY 7 M D
UK; raised 1946 by M. B. Crane at JI, Merton, London.
(Cox's Orange Pippin X Sturmer Pippin) X Cox's Orange Pippin. Named 1965.
Boldly striped. Rich, aromatic flavour, not as intense as Cox, but good. Sweet, crisp, firm, cream flesh. Flavour quickly fades; soon becomes dull, lifeless after picking.

F 12. T^2; sprd. **P** m Sept. **S** Sept-Oct.

MERTON KNAVE 6 E D
UK; raised 1948 by M. B. Crane at JI, Merton, London.
Laxton's Early Crimson X Epicure. Named 1970.
Dark red flushed. Light strawberry flavour; sweet, soft, juicy flesh; best eaten from tree.
F 13. T^2; sprd. **C** v hvy; frt sml unless thinned. **P** use Sept.

MERTON PEARMAIN 7 L D
UK; raised 1934 by M. B. Crane at JI, Merton, London.
Laxton's Superb X Cox's Orange Pippin.
Lightly aromatic, sweet, but low in balancing acidity; slightly crumbly texture.
F 12. T^3. **C** bien. **P** m-Oct. **S** Dec-Feb/Mar.

MERTON PIPPIN 7 L D
UK; raised 1914 by M. B. Crane at JI, Merton, London.
Cox's Orange Pippin X Sturmer Pippin. Introd UK 1948; USA 1953.
Sweet, fresh taste, lots of steely Sturmer acidity; crisp, juicy.
F 12. T^2. **P** m-Oct. **S** Nov-Mar.

MERTON PROLIFIC 7 L D
UK; raised 1914 by M. B. Crane at JI, Merton, London.
Northern Greening X Cox's Orange Pippin. Named 1947. RHS AM 1950.
Good plain brisk taste of fruit; crisp, juicy.
F 11. T^2. **C** hvy; frt sml. **P** m-Oct. **S** Dec-Feb/Apr.

MERTON REINETTE 6/7 L D
UK; raised 1933 by M. B. Crane at JI, Merton, London.
Cox's Orange Pippin X Herring's Pippin.
Sweet, rich, aromatic, quite like Cox; deep cream flesh. Can have woody acidity.
F 16. T^2; prt tip. **P** m-Oct. **S** Nov/Dec-Jan.

MERTON RUSSET 8 L D
UK; raised 1921 by M. B. Crane, JI, Merton, London.
Sturmer Pippin X Cox's Orange Pippin. Named 1943.
Fine cinnamon russet over gold. Sweet-sharp, acid drop flavour of Sturmer; firm, almost yellow flesh. Initially very sharp but mellowing and ready by New Year.
F* 9. T^3. **C** hvy. **P** m-Oct. **S** Dec-Mar.

MERTON WORCESTER 6 M D
UK; raised 1914 by M. B. Crane at JI, Merton London
Cox's Orange Pippin X Worcester Pearmain. Named 1947. RHS AM 1950.
Worcester colour and light strawberry flavour, but crisper flesh and trace of aromatic quality.
Grown for sale small extent.
Frt *Col* bright rd flush, faint rd stripes;

grnish yell/gold background; trace russet. *Size* med. *Shape* rnd-con; sometimes sltly flt sided. *Basin* brd, shal; sltly wrinkled, 5 beads. *Eye* sml, clsd; sepals, shrt, qte downy. *Cavity* med wdth, dpth; some russet. *Stalk* med lng, qte thck. *Flesh* pale crm.
F 10. T^1. **C** hvy; prn bitter pit. **P** e-Sept. **S** Sept-Oct.

MICHAELMAS RED syn Tydeman's Michaelmas Red

MICHINOKU 6 M D
Japan; raised 1981 by K. Maeda, Hirosaki, Aomori.
Kitakami X Tsugaria.
Deep maroon, large. Soft, juicy flesh with sugary sweetness, yet balanced.
F mid. T^2. **P** l-Sept. **S** Oct-Dec.

MICHOTTE (?) 2/5 L D
France; acc may be variety desc 1948 (*Verger Français*), when grown Eure-et-Loir (Val de Loire).
Strong acidity, little flavour.
F 30. T^2; sprd. **P** l-Oct. **S** Jan-Mar.

MILFOR 4 M D
Canada; raised by W. T. Macoun Central, Exp Farm, Ottawa.
Milwaukee X Forest. Introd 1925.
Large, prominently ribbed, bright red flush.
Very juicy, sweet, soft cream flesh, but often green unattractive taste in England.
F 12. T^3. **P** l-Sept. **S** Oct-Nov/Dec.

MILLER'S SEEDLING 4 E D
UK; raised 1848 by James Miller, Speen Nursery, Newbury, Berks. RHS AM 1906.
Savoury, crisp yet melting, very juicy flesh. Plenty of sweetness and refreshing acidity, but must be ripe – pink flushed over cream. Popular in early 1900s; grown Kent especially, for late Aug markets but easily bruised and strongly biennial which contributed to its demise as commercial variety. Still seen on sale in Kent, Sussex and found in old orchards.
Frt flushed sltly or prom in light pink, bright rd stripes, over pale grnish yell turning pale yell. *Size* med. *Shape* rnd-con; sltly ribbed, flt sided. *Basin* med wdth, dpth; ribbed, puckered. *Eye* clsd; sepals sml, downy. *Cavity* brd, dp, wide; ltl russet. *Stalk* lng, thin. *Flesh* wht.
F 9. T^2. **C** hvy, bien; frt sml unless thinned.
P l-Aug. **S** Aug-Sept.
   RED MILLER'S SEEDLING; even red flush.

MILLET 5 E D
UK; arose Cornwall; received 1921.
Small. Sharp, savoury taste, soft flesh.
F 13. T^2. **P** l-Aug. **S** Aug-Sept.

MILLICENT BARNES 6 M D
UK; raised c1903 by N. F. Barnes, HG to

Duke Westminster, Eaton Hall, Chester. Gascoyne's Scarlet X Cox's Orange Pippin. Introd by Clibrans, Chester.
Scarlet flushed. Brisk, refreshing, quite sharp; crisp, juicy flesh.
Popular 1920–30s; found in old gardens.
F 10. T^2; uprt. **C** gd. **P** l-Sept. **S** Oct-Nov.
   MILLICENT BARNES SPORT; more highly coloured.

MILTON 6 M D
USA; raised 1909 by R. Wellington NYSAES, Geneva.
White Transparent X McIntosh. Introd 1923.
Quite large; very beautiful; flushed in bright cherry red over pale cream, with heavy bloom. Flesh is like melting snow, juicy, sweet with hint of raspberry flavour. In US dual purpose, prized for sauce.
F 6. T^3; sprd; hrdy. **C** gd; frt bruises easily. **P** e-Sept. **S** Sept- e-Oct.

MIMI 7 L D
Netherlands; raised 1935 at IVT, Wageningen.
Jonathan X Cox's Orange Pippin.
Lightly aromatic, quite rich; not as intense or complex as Cox; sweet, crisp, juicy flesh.
F 15. T^2; sprd. **P** m-Oct. **S** Nov-Jan/Mar.

MINIER'S DUMPLING (TANN) 3 L C
UK; obtained c1986 by J. Tann, fruit grower, nr Colchester, from tree growing nr Yarm, Essex. May be variety desc 1807, Hogg (1884), but difficult to be certain.
Very large. Very sharp, fresh; cooks to sharp intensely flavoured purée. Said to live up to name and make excellent dumplings.
F est 12. T^2. **P** e-Oct. **S** Oct-Jan.

MINISTER VON HAMMERSTEIN 5 L D
Germany; raised 1882 by R. Goethe in Geisenheim, Rhine.
Landsberger Reinette seedling. Named after Minister of Agriculture.
Sweet, juicy, quite fruity, but often rather flavourless. Only large fruits are said to be worth eating and develop 'flavour of cumarin'. Vit C 5mg/100gm.
Remains popular German garden apple. Formerly also grown Valais, Switzerland.
F 9. T^1. **C** gd. **P** e-Oct. **S** Nov-Jan/Mar.

MINJON 6 M D
USA; raised c1912 at Univ Minnesota Fruit Breeding Farm, Excelsior.
Prob Wealthy X Jonathan. Introd 1942.
Deep red flush, which can stain flesh. Rather bleak flavour in England; tough skin. Dual purpose in US.
F 13. T^2; v hrdy. **C** hvy, bien. **P** l- Sept. **S** Oct-Nov.

MINNEHAHA 6 M D
USA; raised at Univ Minnesota Fruit Breeding Farm, Excelsior from seed

collected from orchard of Seth Kenny, Morriston, Minnesota.
Malinda X. Introd 1920.
Brightly flushed, striped. Some richness, suggestion aromatic flavour; soft, juicy, cream flesh.
F 9. T^2; sprd; hrdy. C gd. P l-Sept. S Oct-Nov.

MINNESOTA RUSSET 8 L D
USA; arose Minnesota.
Parentage unknown. Received 1975.
Very attractive; cinnamon russet, orange red flush over gold. Sweet but rather flavourless in England.
F 18. T^2. P m-Oct. S Nov-Dec.

MINSHULL CRAB 3 L C
UK: arose in Minshull village, Cheshire.
Original tree growing in 1777; desc Hogg (1851). Syns include Lancashire Crab.
Large. Very astringent, hard. Cooked keeps shape or makes stiff purée, but very sharp. Grown in Lancs for markets of Manchester and other cotton towns in late C18th, C19th.
F 9. T^2; sprd. C gd. P e-Oct. S Nov-Mar.

MIO 6 E D
Sweden; raised 1932 by Emil Johansson at Hort Res St, Alnarp.
Worcester Pearmain X Oranie.
Bright red flush. Good sweet taste, slight strawberry flavour; soft, juicy white flesh. Grown gardens, orchards in Sweden.
F 13. T^3; sprd. C hvy; easily bruised. P use Aug.

MISEN JAROMERSKA 4 L D
Czech Republic; arose early 1800s as chance seedling in town of Jaromer, East Bohemia.
Small, markedly flat-round shape. Savoury, refreshing taste; juicy, fruity.
Formerly widely grown for home supplies, now only found old farm gardens; used also juice.
F 11. T^2. P m-Oct. S Nov-Dec.

MISKOLCI KORMOS 5/8 L D
Hungary; received 1948 from Univ Agric, Budapest.
Sweet, quite rich, firm, flesh.
F 6. T^2; uprt. P e-Oct. S Nov-Jan.

MISSING LINK 4 L CD
USA; introd c1903 by Shank & Clayton, Illinois. Recorded 1897.
Carmine flushed, striped. Sweet, firm, but little flavour.
F 12. T^2; sprd. P e-Oct. S Oct-Jan/Apr.

MISSOURI 5 L D
USA; raised Missouri State Fruit Exp St; received 1971.
Claimed Ingram X Lily of Kent, but resembles Golden Delicious; often bright orange flush and firmer, sharper flesh.

F 16. T^2. P l-Oct. S Nov-Jan.

MITCHELSON'S SEEDLING 5 L C
UK, raised by market gardener Mitchelson at Kingston-on-Thames, London. Desc Hogg (1851).
Large. Sharp, crisp, juicy. Cooks to lightly flavoured purée.
F 13. T^2. P e-Oct. S Oct-Dec.

MOBBS ROYAL 5 L C
Australia or New Zealand; known 1868 in Australia.
Large. Sharp, soft flesh. Cooks to smooth, creamy purée, brisk, light; needs no sugar.
F 7. T^2. P m-Sept. S Oct-Jan/Feb.

MØLLESKOV 5 L D
Denmark; found c1860 by A. Brun at Mølleskoven, nr Ringsted, Sjaelland.
Tall, conical; slight pink flush. Savoury, quite sweet, melting, juicy flesh.
F 8. T^2. P m-Oct. S Nov-Dec.

MOLLIE'S DELICIOUS 4 E D
USA; raised 1948 by G. W. Schneider, Rutgers Univ, Agric Exp St, New Jersey.
(Golden Delicious X Edgewood) X (Red Gravenstein X Close). Introd 1966.
Named after admirer, Mrs Mollie Whatley, wife of former Rutgers student. Large, ribbed, angular. Almost rich taste, but also recalls Golden Delicious; crisp, sweet, juicy, cream flesh.
Grown US for local sales; also Italy.
F 16. T^2. C gd. P l-Aug. S Aug-Sept, cold store-Nov.

MOLLYANNE 7 M D
UK; raised by E. Kennedy, Bournemouth, Dorset. Received 1976.
Light, quite rich, sweet, crisp, juicy; often scented.
F* purple; 14. T^2. P l-Sept. S Oct-Nov.

MONARCH 3 L C
UK; raised 1888 and introd 1918 by nursmn W. P. Seabrook, Boreham, Essex.
Peasgood Nonsuch X Dumelow's Seedling.
Cooks to juicy, brisk purée, with creamy texture of Dumelow but not as sharp, or as acidic as Bramley.
Grown Essex for London markets by 1927. Popular during WWII as needed less sugar than Bramley. Often found in old gardens.
Fruit Col pinky rd flush, darker rd stripes over pale yell. Size med/lrg. Shape rnd-con; flt base, flt apex; trace ribs sometimes; often hair line near apex, like Dumelow. Basin qte brd, qte dp. Eye qte lrg, open; sepals brd, long. Cavity nrw, dp; some russet. Stalk med, thck. Flesh wht.
F 11. T^3. C hvy, bien; frt easily bruised. P e-Oct. S Nov-Jan.

MONDIAL GALA see Imperial Gala

MONK'S GOLDEN PIPPIN 5 L D
UK; received 1948.
Elements of Golden Pippin type flavour. Rich, densely fruity, lots of sugar, acidity; crisp, white flesh.
F 11. T^1. P m-Oct. S Nov/Dec-Jan.

MONROE 6 L D
USA; raised 1910 by R. Wellington at NYSAES, Geneva.
Jonathan X Rome Beauty. Introd 1949.
Named after Monroe County, fruit growing area of West New York. Firm, sweet, juicy, flesh. In US also considered 'excellent for pie, sauce, baking'.
Grown US for fresh fruit and processing. Often lacks flavour in England.
F 10. T^2; prn mildew. C hvy. P e-Oct. S Nov-Jan.

MONTFORT 4 L D
UK; raised c1928 by A. E. Sadler at Woodford Green, Essex.
Mellowing to refreshing flavour; crisp.
F 11. T^2. C lght; v prn mildew. P mid-Oct. S Jan-Mar.

MONTMEDY 6 L DC
Switzerland; old variety; received 1982.
Crisp, brisk white flesh; sharp even in Jan. Cooked keeps shape; sharp.
F 22. T^2; claimed gd res disease. P m-Oct. S Nov-Mar/later.

MOORE'S SEEDLING (?) 6 M L
UK; received from Devon 1947. Acc may be variety desc 1851 (Hogg).
Dark red flush, stripes. Rich, sweet-sharp taste; nearly yellow flesh.
F 9. T^2. C bien. P l-Sept. S Oct-Dec.

MORGENROT 6 M D
Germany; raised Max-Planck Inst, Cologne.
Cox's Orange Pippin X. Named 1973.
Large. Very sweet, juicy, lovely flavour; rich quite aromatic, but rather coarse texture.
F 12. T^2. C lght; frt cracks, prn hvy russeting. P l-Sept. S Oct-Nov.

MORKROD see Gravenstein

MORLEY'S SEEDLING 1 L C
UK; raised by C. Morley, Ely, Cambs.
Alfriston X Lane's Prince Albert. Received 1928.
Large. Cooked, retains shape, sharp, fruity; but flavour faded by New Year.
F* lrg; 15. T^3. P m-Oct. S Nov-Jan/Mar.

MORRIS'S RUSSET 8 L D
UK; raised by market gardener Morris at Brentford, Middx. Desc 1851.
Crimson flush under russet. Rich, intense sweet-sharp flavour. Recalling Nonpareil, similar chewy texture, but sweeter, not as strong.
F* 10. T^2. P m-Oct. S Nov-Jan/Feb.

MORS DE VEAU 6 L D
Received 1948 from Switzerland.
Brisk, fruity, yet trace of nutty flavour; crisp to soft, juicy flesh.
F 16. T³. **P** m-Oct. **S** Dec-Feb/Mar.

MOSS'S SEEDLING 4 L D
UK; raised c1955 at Chetwynd End Nurseries, Newport, Shrops.
Sweet, rich, honeyed, lightly aromatic; elements of Cox flavour. Heralded in 1972 as improvement on Cox, with heavier crops and easier to grow, but proved poorly coloured, not as well flavoured.
F 11. T². **C** gd. **P** e-Oct. **S** l-Oct-Dec.

MOTHER syn American Mother 6 M D
USA; arose Bolton, Worcester County, Massachusetts. Recorded 1844; usually known American Mother; prefix given by Hogg. Introd England prob by nursmn Thomas Rivers. RHS AGM 1993.
At best, flavour is distinct, sweet, perfumed, aromatic with exotic quality; quite soft, juicy, almost yellow flesh. Hogg found 'balsamatic aroma'; Bunyard 'flavour of pear drops'; possibly hint of vanilla gives the spicy quality but it can collapse into flat, musty, sweetness in poor year.
Popular 1920–30s; in Bunyard's 'top ten'. Found in many old orchards. In US, only recently been taken up by amateurs. Identified as La Paix of Belgium and RGF variety, recommended for cultivation by Agric Res Centre, Gembloux because of disease resistance.
Frt *Col* deep rd flush, indistinct broken rd stripes; grnish yell/yell background; lenticels conspic as sml russet dots. *Size* med. *Shape* lng-con; sometimes lopsided; ribbed; sometimes flt-sided, rnded at base. *Basin* med, shal; sltly ribbed; trace russet. *Eye* sml, clsd or hlf open; sepals, sml. *Cavity* qte brd; shal to qte dp; ltl russet. *Stalk* slnt, qte thin. *Flesh* dp crm.
F 15. T²; uprt; gd res scab, mildew. **C** erratic. **P** l-Sept. **S** Oct-Dec.
PLATE 11

MOȚI (?) 4 L D
Transylvania (Romania); acc received 1958; may be variety arose Munții Apuseni, north west Transylvania.
Sweet, quite scented in Jan; firm rather dry flesh.
F 22. T². **P** l-Oct. **S** Jan-Mar/Apr.

MOUNT RAINIER 6 L D
Netherlands prob; known 1929.
Tall, conical shape. Sprightly, sugary taste of fruit, quite rich flavour.
F10. T². **P** m-Oct. **S** Nov-Jan/Mar.

MR GLADSTONE syn Gladstone

MRS BARRON 5 M CD

UK; exhib 1883 Congress; desc Hogg (1884).Named after wife of RHS Fruit Officer, organiser of 1883 Congress. Large. Lightly flavoured, rather bland cooked.
F 12. T³; sprd. **P** l-Sept. **S** Oct-Nov.

MRS LAKEMAN'S SEEDLING 5 M CD
UK; raised c1900 by Mrs Lakeman, Northumberland; introd nursmn W. Fell, Hexham.
Large. Savoury pleasant taste, recalling James Grieve. Cooked, keeps shape, or forms lemon purée; sweet, lightly flavoured.
F 11. T². **P** e-Sept. **S** Sept-Oct.

MRS PHILLIMORE 6 L D
UK; raised by Charles Ross, HG, Welford Park, Newbury, Berks.
Cox's Pomona X Mother (American). Recorded 1896. RHS AM 1899.
Boldy red striped, flushed, like Cox's Pomona with sweet taste of American Mother. Often definite hint of vanilla, but can be flat, insipid.
F 14. T²; sp. **P** e-Oct. **S** Dec-Feb.

MUNSTER TULIP 4 M DC
Ireland; received 1950, when grown in County Offaly.
Large, tall; carmine striped over yellow. Very sharp. Cooks to sharp lemon purée, little richness.
F 12. T¹. **P** m-Sept. **S** Oct-Nov.

MUNTENEȘTI 4 L D
Romania (poss); poss arose Fagaruș, Brașov region, where widely grown 1960s.
Crisp, brisk, juicy in Feb.
F 19. T². **P** l-Oct. **S** Jan-Mar/April.

MURASAKI 6 L D
Japan; raised 1935 at Aomori Apple Exp St. Jonathan X Delicious. Named 1948.
Sweet, scented, crisp.
F 11. T³. **C** hvy; frt sml. **P** l-Oct. **S** Dec-Mar.

MURFITT'S SEEDLING 1/5 L C
UK; prob introd by nursmn Wood & Ingram, Huntingdon; desc Hogg (1884).
Large. Quite rich blend of sugar, acid. Cooked keeps shape, pleasant taste. Favourite apple of Cambs when exhib 1883.
F 14. T¹; sprd. **P** m-Oct. **S** Oct-Dec/Jan.

MUSKAT RENETTE syn Margil

MUTSU syn Crispin 5 L CD
Japan; raised 1930 at Aomori Apple Exp St. Golden Delicious X Indo. Named 1948. Renamed Crispin 1968 England, but elsewhere known as Mutsu.
In good condition, sweet, honeyed, crisp, juicy flesh, similar to Golden Delicious but coarser texture. Can become quite scented, but can be much less interesting. Good used fresh in vegetable salads. Cooked, tends keep shape; sweet, light flavour.

Major Japanese variety and grown to great size. Grown US, Canada, much used in pie trade, also Italy. As Crispin, in England formerly grown small extent commercially; popular garden apple.
Frt *Col* grnish yell/yell, sometimes slight brownish flush; lenticels as grey russet dots. *Size* lrg. *Shape* oblng; tapers to apex; ribbed, sltly flt sided; 5 crowned. *Basin* brd, dp; ribbed. *Eye* lrg, clsd or sltly open; sepals, lng, downy. *Cavity* brd, dp, russet lined. *Stalk* lng, thin. *Flesh* pale yell.
F 10; trip. T³; sprd; prn canker. **C** hvy. **P** m-Oct. **S** Dec-Feb.
　　　　MUTSU SPUR TYPE

NANCY JACKSON 3 L C
UK; prob arose Yorks. Described 1875 (Hogg).
Large; colourful. Slightly astringent flesh. Cooked keeps shape, pretty lemon colour; rich, good flavour. Still good in Mar.
'Much cultivated' North Riding of Yorkshire in C19th.
F12. T¹; sprd. **P** m-Oct. **S** Nov-Mar/May.

NANNY 7 M D
UK; arose either West Sussex or Surrey. Recorded 1842; desc Hogg.
Large. Sweet, quite rich, but soon goes mealy.
F 5. T³. **P** m-Sept. **S** Sept-Oct.

NASONA 4 L D
Italy; arose prob nr Modena, Po Valley. Desc 1949, but older.
Markedly oblong. Scented, very sweet; rather dry.
F 13. T²; tip. **P** m-Oct. **S** Nov-Jan/Mar.

NEBUTA 6 M D
Japan; raised 1981 by K. Maeda in Horosaki, Aomori. Kitakami X Tsugaru.
Deep cream, red stained flesh; juicy quite rich blend of sugar and acid; slightly coarse texture.
F est 10. T². **P** e-Sept. **S** Sept.

NEILD'S DROOPER 4 M CD
UK; received from N. H. Neild, LARS, Bristol.
Propagated at Woburn, Beds 1915–16 from scion of tree labelled Calville Blanc d'Été, but does not resemble it.
Rich blend of sugar and acid; crisp, juicy.
F 19. T¹; weeping. **P** m-Sept. **S** Oct-Nov/Dec.

NELSON'S FAVOURITE 3 M C
UK; received 1958 from Kendal, Westmorland.
Cooked, keeps shape; sharp, fruity, good taste.
F 9. T¹. **P** m-Sept. **S** Sept-Oct/Nov.

245

NEMES SÓVÁRI ALMA syn Sóvári Nobil
3 L C
Hungary; prob from Sórái region. Distributed abroad 1880, when very old variety.
Ribbed, crowned, oblong. Sweet, very lightly flavoured in England.
As Sóvári, grown all over Hungary and Transylvania in C19th; still widely grown 1960s.
F 15. T^2; sprd. P l-Oct. S Nov-Mar/Apr.

NEMES SZERCSIKA ALMA 1/3 L C
Hungary; prob variety desc 1883, which arose southern Alföl, nr Drava River, on border between Hungary and former Yugoslavia.
Often very large, its name means 'Noble Apple'. Sharp; cooks to quite fruity, firm purée.
F 8. T^3. P l-Oct. S Dec-Mar.

NEMŢESC CU MIEZUL ROŞU 6 L D
Romania; old variety; still popular market and amateur apple 1960s.
Very attractive, bright red flushed. Some sweetness; crisp to quite soft, juicy.
F 10; frost res. T^2; sp; claimed gd res disease. C hvy, reg; prn bitter pit. P e-Oct. S Nov-Jan/Mar.

NEUE GOLDPARMÄNE 8/7 L D
Germany; received 1951 from Fruit Res St, Jork, nr Hamburg.
Golden Winter Pearmain, Parker's Grey Pippin cross.
Handsome. Rich, nutty flavour; can be very good.
F 9. T^3. C gd. P e-Oct. S Oct-Jan/Mar.

NEW BESS POOL 6 L D
UK; raised before 1850 by J. Stevens, Stanton-by-Dale, Yorks.
Believed Bess Pool seedling.
More highly coloured than Bess Pool. Quite sweet, light flavour.
F 19. T^3. P l-Oct. S Dec-Feb.

NEW FUJI see Fuji

NEW GERMAN 6 M DC
UK; arose Herefords. Desc 1884 (Hogg).
Large, ribbed; dark mahogany red flush, stripes. Very sharp early in season. Cooks to deep cream, purée or froth; sharp, fruity.
F 12. T^3; sprd; prt tip. C prn bitter pit. P e-Sept. S Sept-Nov/Dec.

NEW HAWTHORNDEN (?) 1/5 M C
UK; acc may be variety introd 1847 by nursmn Thomas Rivers; desc Hogg (1884); very similar to Hawthornden.
Cooks to sharp, quite creamy purée.
F 11. T^2. P e-Sept. S Sept-Oct.

NEW JONAGOLD see Jonagold

NEWPORT CROSS 7 E D
UK; raised 1920 by G. T. Spinks, LARS, Bristol.
Devonshire Quarrenden X Cox's Orange Pippin.
Similar appearance to Devonshire Quarrenden, but later season. Very sweet, crisp, juicy.
F 7. T^2. C bien. P eat Sept.

NEWQUAY PRIZETAKER see Lucombe's Seedling

NEW ROCK PIPPIN 8 L D
UK; raised by William Pleasance at Barnwell, Cambridge. Exhib 1821.
Intensely rich, sweet-sharp flavour; firm flesh. Hogg found it 'perfumed with flavour of anise' and Bunyard 'excellently flavoured'.
F11. T^2. P m-Oct. S Jan-Mar/May.

NEWTON WONDER 3 L C
UK; found c1870 by innkeeper Mr Taylor growing in thatch of Hardinge Arms, King's Newton, Melbourne, Derbyshire. He transplanted it into garden, where tree survived until 1940s. Introd by J. R. Pearson Nursery of Nottingham 1887. FCC RHS 1887; AGM 1993.
No record of parents but believed by Pearson to be Dumelow's Seedling, Blenheim Orange cross.
Cooks to cream, juicy purée, brisk and good, but mild in comparison with Bramley. Baked, creamy, quite frothy like Dumelow, but not as sharp. Later in season makes brisk, fruity eating apple; good for savoury salads.
Widely planted, formerly grown for market. Still found on sale; growing in very many gardens.
Frt *Col* brownish turning orng rd flush, rd stripes over grnish yell/yell; lenticels as conspic russet dots; becomes greasy. *Size* med to v lrg. *Shape* flt-rnd to almost rnd-con. *Basin* brd, dp; sltly puckered. *Eye* lrg, open; sepals, qte downy. *Cavity* nrw, shal; ltl russet. *Stalk* shrt, thck; king fruits have fleshy swelling at side cavity.
F 14. T^3; sprd. C hvy; bien; prn bitter pit. P m-Oct. S Nov-Mar.
PLATE 23
More highly coloured sports:
   CRIMSON NEWTON; arose 1921 with C. H. Dicker, Holywell Bury Fruit Farm, Herts.
   MARSTON SCARLET; discv 1909 by Mr Hughes, Marston, Herefords
   RED NEWTON WONDER; arose with A. Haggard, Hill Top Farm, Ledbury, Herefords; received 1958.

NEWTOSH 6 L D
Canada; raised Central Exp Farm, Ottawa. McIntosh X Yellow Newtown Pippin. Introd 1923.

Slight strawberry-like McIntosh taste with firm, crisp, juicy flesh of Newtown Pippin; inclined to be metallic in England; tough skin.
F 8. T^2. P l-Sept. S Oct-Jan/Mar.

NEWTOWN PIPPIN syn Albermarle Pippin 2 L D
USA; arose on estate of Gershom Moore, Newtown, Long Island; known by 1759. Green and Yellow Newtown Pippin differentiated by Coxe 1817. Latter believed sport. Introd England after 1759, year when parcel of fruit sent from US to Benjamin Franklyn in London. Grown 1768 by Brompton Park Nursery, London. Syns numerous. Widely distributed throughout North America, Europe.
Esteemed by all C19th-pomologists and connoisseurs, but needing warmer climate than England to thrive. In good year, developing brisk, aromatic quality, which recalls 'rich pineapple flavour' that charmed Victorians. In States also used for pies, sauce, jelly and sweetmeats.
Newtown Pippin launched American fruit export industry; grown commercially in New York, Virginia by 1830s, and Oregon, Washington by 1890; continued to be exported to UK until 1930s. Overtaken by other varieties, but still grown Oregon, although mostly for processing.
Yellow Newtown Pippin:
Frt *Col* grn/grnish yell, light orng brown flush; lenticels conspic as russet or grey dots; scarf skin. *Size* med. *Shape* flt-rnd to oblng; trace rnded ribs; sltly flt sided; crowned. *Basin* brd, dp; sltly ribbed; ltl russet. *Eye* lrg, open; sepals, separate at base, shrt. *Cavity* brd, qte dp; russet lined. *Stalk* med, qte thck. *Flesh* pale crm.
F 10. T^2; prt tip. C gd, bien. P m/l-Oct. S Dec/Jan-Mar.
PLATE 31

NEWTOWN SPITZENBURG 6 L D
USA; arose Newtown, Long Island. Recorded 1817. Introd c1830 to England by William Cobbett, as Matchless.
Colourful as Esopus Spitzenburg. Lightly aromatic, sweet, firm flesh, almost nutty quality.
F 11. T^3. P l-Oct. S Nov-Jan.

NIAGARA 6 M D
USA; raised at NYSAES, Geneva.
Carlton X McIntosh. Introd 1962.
Dark red, deep bloom; McIntosh type but ripening earlier. Very sweet juicy flesh; tough skin.
F 6. T^3. C gd. P e-Sept. S Sept-Oct.

NICO 5 E DC
Belgium; raised by nursmn Jo Nicolaï, Sint-Truiden. Received 1981.
Appearance and taste resemble White Transparent. Sharp, juicy cream flesh.

Cooks to good brisk, juicy purée.
F A30. T^2; sprd. P use Aug.

NIĞDE 6/7 L D
Turkey; received 1947.
Niğde in Cappodocia, east central Turkey, is famous for its red apples. Very sweet, almost no acidity; more like sweetmeat than apple; tough skin.
F 5. T^3. P m-Oct. S Nov-Jan.

NOBIL DE GEOAGIU 6 L D
Romania; received 1958 from Agric Acad Sc, Bucharest.
Sweet, quite rich; soft flesh.
F 4. T^3; uprt. P m-Oct. S Nov-Dec.

NODHEAD syn Jewett's Fine Red

NOLAN PIPPIN 8 L D
UK; received 1920 from Mrs Woodward, Colchester, Essex.
Cinnamon russet over gold. Intense, sweet-sharp flavour of acid drops; firm, rather spongy flesh, like Nonpareil. Still strongly flavoured in Mar.
F 13. T^3. P l-Oct. S Nov-Mar.

NONPAREIL 8 L D
France (prob); very old variety; acc is of Hogg (1884). Tradition claimed it was introd from France by Jesuit in C16th in reign of Queen Mary or Queen Elizabeth, but name not recorded until 1696 in list of Brompton Park Nursery, London.
Develops intense sweet-sharp flavour of acid or fruit drops; firm, almost spongy, cream tinged green flesh. Can be over sharp, needs plenty of autumn sunshine to build up flavour. Prized for C18th dessert; also decorative tree, grown in pots to set about fruit garden, parterre. Apples hang very late and claimed could be in fruit and flower at same time, like orange trees. Remained popular throughout C19th; grown by London market gardeners and in Kent; still 'worthy of all commendation' in 1920.
F 11. T^2; uprt. C gd. P m/l-Oct. S Dec/Jan-Mar.

NONSUCH PARK 5 L D
UK; desc 1831 by Lindley; Hogg (1884).
Resembles Golden Pippin; small, russet freckled over gold. Dense fruity taste, plenty of sugar, acidity; crisp juicy.
F 14. T^2. P m-Oct. S Nov-Dec.

NO PIP 3 L C
UK: received 1913 at RHS Wisley, from Earl Morley, Whiteway, Chudleigh, Devon via Veitch of Exeter.
Rarely sets seed and often no pips. Low in flavour, sugar, acidity; insipid cooked.
F 14. T^3. P e-Oct. S Oct-Nov.

NORFOLK BEAUTY 5 M C
UK; raised by Mr Allan, HG, Gunton Park, Norwich, Norfolk.
Warner's King X Waltham Abbey. Recorded 1901. RHS AM 1901; FCC 1902.
Large. Cooks to creamy purée, well-flavoured; needing little sugar.
F 8. T^3. C bien. P e-Sept. S Sept-Nov/Dec.

NORFOLK BEEFING 6 L CD
UK; arose Norfolk. Known late C18th, but prob older. 'Beefing' was cited in fruit tree list in notebook of 1698 belonging to Walpole family, Wolverton & Mannington estates, Norwich, Norfolk.
Used early in season for cooking, but by spring sweet enough to eat. Cooked, keeps shape, quite rich taste. Baked as for Biffins, very slowly at lowest oven setting for 24 hrs, flesh becomes thick and tastes almost of raisins and cinnamon.
Apple has tough, rather dry flesh and tough skin, which allows fruit to be baked without bursting. They were put in the bread ovens, after the bread had been removed and an iron plate was placed on top to press the air out. From early C19th, Biffins packed in boxes were dispatched by Norwich bakers to London fruiterers and also posted as presents. Commercial operations waned in 1914, but Biffins were on sale up to 1950s.
In C19th grown in gardens all over country, also for market; recommended for drying. By early 1900s, 'not much grown', but many trees still found in Norfolk.
Frt *Col* dark purplish rd flush, shrt rd stripes over grn; lenticels conspic as white dots. Ripens to crimson over gold. *Size* med/lrg. *Shape* flt-rnd or rnd; trace rounded ribs; scarf skin at base; sltly crowned. *Basin* brd, shal; sltly ribbed; ltl russet. *Eye* lrg, open; sepals, brd base, separate, lng, downy. *Cavity* med wdth, dpth; russet lined. *Stalk* shrt, v thck. *Flesh* wht, tinge grn.
F* 12. T^3. C gd. P m-Oct. S Dec-Apr.
PLATE 30

NORFOLK ROYAL 6 M D
UK; found c1908 on Wright's Nursery, North Walsham, Norfolk. Introd 1928.
Striking appearance. Sweet, crisp, juicy.
Formerly grown commercially to small extent, but skin becomes very greasy, which led to its decline in popularity. Revived as 'Pick-Your-Own' apple, to eat sharp and crisp from tree.
Frt *Col* bright rd flush; few rd stripes; pale yell background; lenticels as tiny cream dots; v greasy skin. *Size* med. *Shape* con; tends to be crowned. *Basin* med brd, dp; ribbed or puckered. *Eye* clsd; sepals qte sml; downy. *Cavity* nrw, dp; some russet. *Stalk* med, qte thck. *Flesh* wht.
F* 14. T^2. C gd. P e-Sept. S Sept-Dec.
　　NORFOLK ROYAL RUSSET; russeted sport. Found by Rev C. E. Wright, retired WWII, RAF chaplain, in his garden,

Burnham Overy Staithe, North Norfolk. Introd 1983 by Highfields Nursery, Gloucester. RHS AGM 1993. More attractive than Norfolk Royal with bright red cheek peeping through russet. Flavour also better; quite intense, good rich taste, crisp, sweet.

NORIS 6 L D
Lithuania; received 1976 from Pavlovsk Exp St of Vavilov Inst, St Petersburg.
Osennee Polosatoe X McIntosh.
Striped, flushed in pink, red. Sweet, slightly scented, crisp, juicy flesh; tough skin. Widely grown Baltic States and north west Russia.
F 7. T^2. C hvy. P m-Oct. S Nov-Dec.

NORMANDIE (SWITZERLAND) 7 L D
Received 1948 from Switzerland. Acc not Normandie of Moselle, or of Indre, desc 1948 (*Verger Français*).
Soft, sweet flesh, slight strawberry flavour, but often unattractive tannic taste.
F 27. T^2; sprd. P m-Oct. S Nov-Dec.

NORMAN'S PIPPIN 8 L D
Believed introd East Anglia from Flanders. Recorded 1900. RHS AM 1900.
Netted in russet. Soft, sweet, juicy flesh, light taste of fruit, but Bunyard found it 'rich'.
F 13. T^1; sprd. P l-Oct. S Nov-Jan/Mar.

NORTHERN GREENING 3 L C
UK; pres English. Cited 1802 by Forsyth, but prob older; desc Ronalds, Hogg.
'Valuable sauce apple'; cooked keeps little of form; juicy, sharp.
Grown in gardens all over country in C19th; also for market, but rather small for cooker.
F 15. T^2. C gd. P m-Oct. S Dec-Apr.

NORTHERN SPY 6 L DC
USA; arose c1800 in seedling orchard of Heman Chapin, East Bloomfield, New York, from seed brought from Salisbury, Connecticut. Introd c1840. Listed 1852 as new variety of promise by American Pom Soc.
At best rich, intense flavour, sweet, plenty of fruit, acidity; crisp, cream flesh. Dual purpose in N America.
Formerly major commercial and home orchard variety in north eastern states and Canada. Exported to UK in C19th; overtaken by Jonathan and Delicious, but still grown commercially and popular amateur all purpose fruit – for cooking, drying, juice.
Its resistance to root woolly aphid led to use in rootstock breeding programmes.
Frt *Col* dark rd flush, grn ground col; some rd stripes; becomes bright red over pale yell; fine russet dots; scarf skin at base. *Size* med/lrg. *Shape* rnd-con to oblng; rnded

ribs; crowned. *Basin* brd, dp; often ribbed. *Eye* sml, clsd or hlf open; sepals, shrt. *Cavity* brd, dp; russet lined. *Stalk* qte lng, qte thin. **F** 17. **T**^{2}; prt tip. **C** gd; often fails ripen Eng. **P** m-Oct. **S** Nov-Mar.

> **CRIMSON SPY**; even crimson flush. Found by S. B. Hathaway in Coburg, Ontario; introd 1931.
> **DOUBLE RED NORTHERN SPY**; even red flush. Discv 1895 by W. S. Green in Victor, New York.
> **HUNTER KINKEAD SPY**; tetrap.
> **KINKEAD RED SPY**; more highly coloured. Received from Exp Farm Ottawa, Canada.
> **LOOP SPY**; discv 1925 in orchard of A. I. Loop; North East, Pennsylvania; tetrap.

NORTHLAND SEEDLING 5 M D
UK; raised by A. P. Kitcat, Northland Cottage, Tetbury, Gloucs. Received 1930.
Crisp, brisk, savoury. Dull colour.
F 8. **T**^{2}. **P** m-Sept. **S** Oct-Nov.

NOTARISAPPEL (Notary's Apple) 4 L D
Netherlands; raised late C19th by notary J. H. Th. van den Hamm of Lunteren, Veluwe, north west Arnhem; poss seedling of Alant. Introd 1899 at National Exhibition of seedling fruits staged by Lunteren Hort Soc.
Large, 'tasty' apple, still sought by Dutch enthusiasts. In England sweet, crisp, pleasantly savoury.
Raised to provide fruit trees for local farmers to improve their livelihoods; named in honour of van den Hamm's work.
F 8. **T**^{3}. **P** l-Sept. **S** Oct-Jan/Mar.

NOTTINGHAM PIPPIN 5 L D
UK; arose pres Nottingham. Existed 1815; desc Hogg.
Strong flavour of fruit, sweet, plenty of acidity; crisp, juicy.
F 12. **T**^{1}; sprd. **P** e-Oct. **S** Oct-Dec/Feb.

NOUVELLE EUROPE syn Kansas Queen

NOVA EASYGRO 6 L D
Canada; raised 1956 by L. E. Aalders, Agric Res St, Kentville, Nova Scotia.
Spartan X scab resistant seedling. Introd 1971.
Snow white flesh, sweet, crisp, but little flavour in England. Recommended for cooking, salads and eating fresh in Canada.
F 12. **T**^{2} hrdy; res scab, mod res mildew. **C** v hvy. **P** m-Oct. **S** Nov-Dec.

NUGGET 5 L D
USA; discv 1954 in Princess Anne, Maryland by H. S. Kemp, Bountiful Ridge Nursery.
Golden Delicious seedling. Introd 1966.
Resembles Golden Delicious in appearance and taste.
F 16. **T**^{2}. **P** l-Oct. **S** Nov-Mar.

NURED ROME syn Seeand O Red Rome

NUTMEG PIPPIN 8 L D
UK; desc 1920 by Bunyard, but prob older. Distinct from Cockle Pippin.
Small, russeted over gold. Rich, almost aromatic flavour.
F 9. **T**^{2}; uprt. **C** gd. **P** e-Oct. **S** Nov-Jan/Feb.

OAKEN PIN (TAYLOR) 4 M D
UK; acc is variety of Taylor, which in 1920s was widely grown on Exmoor, Devon; may be Oaken Pin of Hogg, but prob not of earlier writers.
Egg shaped or like 'pin' used to fasten the door. Prettily flushed, quite tough apple. Rich, intense flavour, almost aromatic; sweet, firm, crisp flesh, but can be much sharper and closed up, as if not quite developed.
F 12. **T**^{2}; res red spider. **C** gd. **P** l-Sept. **S** Oct-Nov.

OBELISK syns Flamenco Crimsonspire (US) 6 L D
UK; raised EMRS Kent.
(Cox's Orange Pippin X Court Pendu Plat) X Wijcik McIntosh. Introd 1991. *See also* Ballerina.
Rather sharp, but mellowing by New Year.
F 16. **T**^{2}. **P** e-Oct. **S** Dec-Feb; commercially-Apr.

OBERLAUSITZER MUSKATRENETTE 7 L D
Germany; received 1951 from Fruit Res St, Jork, nr Hamburg.
Handsome. Lightly aromatic, quite rich, sweet; tough skin.
F 12. **T**^{2}. **P** e-Oct. **S** Oct-Dec.

ODIN 6 M D
Netherlands; raised 1953 by A. A. Schaap, IVT, Wageningen.
Golden Delicious X Ingrid Marie. Introd 1966.
Rich, intense, some aromatic quality; sweet, juicy, but rather coarsely textured flesh.
F 15. **T**^{3}; sprd. frt cracks. **P** m-Sept. **S** Oct-Nov, cold store-Dec.

OHINEMURI syn Dunn's Seedling

OHIO NONPAREIL 4 M DC
USA; arose in orchard of Mr Bowman, Massillon, Ohio. Desc 1848.
Similar to Gravenstein, but firmer fleshed. Soft, juicy, savoury, quite sweet. Dual purpose in US. Popular in mid-west in C19th; remains favourite with amateurs.
F 6. **T**^{2}. **P** e-Sept. **S** Sept-Oct/Nov.

OHM PAUL 1/5 E C
Germany; arose c1865–70, Eilenburg, Saxony; discv, named c1900, by nursmn K. Grothe from Lemgo, Lippe district, Westphalia.
Named after Uncle Paul (1825–1904), president of the Farmers' Assocation. Ripens to pale milky yellow like White Transparent. Sharp but pleasant to eat fresh. Cooks to very lightly flavoured, cream purée.
F 7. **T**^{3}. **C** hvy; frt sml. **P** use July-Aug.

OLD ENGLISH ROUND 4 M C
UK; received 1981 from Somerset.
Cooks to a sweet, quite rich purée.
F mid. **T**^{2}. **P** e-Sept. **S** Sept-Oct.

OLD FRED 4 L D
UK; raised by F. W. Wastie, Eynsham, Oxford.
Allington Pippin X Court Pendu Plat. Exhib 1944.
Sweet, moderately juicy, crisp, but lacks interest.
F 12. **T**^{2}; sprd. **P** l-Oct. **S** Nov-Jan/Mar.

OLD PEARMAIN (KELSEY) 4 L D
UK; acc received 1924 from Mr Kelsey, Surrey under this name, but prob not variety of Hogg. Old Pearmain is syn of Pearmain, mentioned France and UK c1200, but most unlikley that acc is ancient Pearmain.
Pear shaped, well coloured; quite richly flavoured.
F 11. **T**^{2}. **P** e-Oct. **S** Nov-Dec/Jan.

OLD ROCK PIPPIN 5 L D
UK; received 1981 from Glastonbury, Somerset; tree said to date from early 1900s.
F mid. **T**^{2}. **P** m/l-Oct. **S** Nov-Jan; said to keep v late.

OLD SOMERSET RUSSET 8 L D
UK; grown by nursmn Pope Brothers, Wokingham, Berks. Received from them 1949, but older.
Brisk, good taste of fruit with pineapple acidity; quite crumbling cream flesh.
F 17. **T**^{1}. **P** m-Oct. **S** Nov/Dec-Jan.

OLDENBURG syn Geheimrat Doktor Oldenburg

ONTARIO 4 L CD
Canada; raised c1820 by Charles Arnold, Paris, Ontario.
Wagener, Northern Spy cross. Introd Britain, Europe. RHS AM 1898.
Fairly sharp eaten fresh. Cooked keeps shape or makes brisk, well-flavoured, juicy lemon purée.
Formerly quite widely grown in Canada as dual purpose apple. Grown Saxony, Germany, where valued for juice, spring eating apple and source of vit C

(21mg/100gm); also grown northern France.
F* ivory/pink 12. T²; res rosy apple aphid.
C gd; frt bruises easily. P l-Oct. S Nov-Apr.

GENEVA ONTARIO; syn Ontario sport (Geneva); tetrap. Arose NYSAES, Geneva.

ONTARIO; (4X 2-4-4); tetrap.

OPAL 6 M D
UK; raised by W. P. Seabrook, Boreham, Essex.
Rival X Worcester Pearmain. Received 1936.
One of series, named after semi-precious stones. Crimson flushed, striped, with bloom. Quite rich, hint of strawberry flavour; crisp, juicy; can be only moderately flavoured.
F 9. T². P e-Sept. S Sept-Oct.

OPALESCENT 6 M D
USA; found by George Hudson, Barry County, Michigan, originally Hudson's Pride of Michigan; introd, renamed 1880 by Dayton Star Nurseries.
Large. Sweet, pleasant, but coarsely textured; often poor flavour in England.
Formerly widely grown New England.
F 11. T². P m-Sept. S Oct.

OPETIAN 1/5 L CD
Received 1953 from JI, Merton, London.
Large. Crisp, juicy, quite sweet. Cooked, keeps shape, sweet, lightly flavoured.
F 9, T², P e-Oct. S Oct-Nov.

ORANGE GOFF 4 M C
UK; prob arose Kent. Recorded 1842; desc Hogg (1884).
Cooked, keeps shape, very sharp with definite taste, almost marmalade colour.
Popular C19th cooker, also known as Pork Apple. 'Goffs' were grown around Maidstone for market; also bought up by the 'smashers' for mixed fruit jams, consignments were even sent off to Dundee to 'help' marmalade producers until practice was stopped by Adulteration Act.
F 6. T³. P m-Sept. S Oct-Nov/Dec.

ORANGENBURG 4 L D
Germany; received 1976 from Max Planck Inst, Cologne.
Cox's Orange Pippin X Geheimrat Doktor Oldenburg.
Some of Cox's flavour; scented flowery quality; quite rich, aromatic. Crisp, but coarse texture.
Grown in Rhineland.
F 14. T². C pr. P m-Oct. S Nov-Jan.

ORANJE DE SONNAVILLE 7 M D
Netherlands; raised after 1952, by Piet de Sonnaville in his family orchards at Puiflijk and Altforst, nr Nijmegen. Cox's Orange Pippin X Jonathan. Introd 1971.
Sweet, rich, some aromatic quality. Reminiscent of Cox, but rather coarsely

textured.
F 18; trip. T²; prn canker. C lght. P m-Sept. S Oct.

ORBAI ALMA 4 M CD
Hungary; recorded 1882. Named after first propagator – Orbai.
Crisp, juicy, sharp, but quite lot of sugar.
Formerly popular Hungarian garden apple.
F 6. T². P m-Sept. S Oct-Dec.

OREI syn Kinrei

OREGON SPUR *see* Delicious

ORENCO 6 M D
USA; believed arose nr old pioneer cemetery, Oregon City, Oregon. Poss 'Oregon City' tree, fruit of which was being sold 1840–1850s. Suggested first propagated by Oregon Nursery of Orenco and Salem. Trade of this nursery helped found Orenco town.
Beautiful, bright red flush. Very sweet, scented; juicy, melting cream flesh.
Recently rediscovered by American fruit enthusiasts.
F 14. T²; claimed res scab. C lght. P m-Sept. S Oct-Nov.

ORIENT 6 L D
Japan; raised by Junichiro Nishiya, Aomori Prefecture.
Red Delicious X. Introd 1935.
Delicious shape, boldly striped. Sweet, perfumed, firm flesh. Often fails ripen in England.
F 12. T³. C hvy. P l-Oct. S Dec-Feb/Apr.

ORIN 5 L D
Japan; raised before 1942 by T. Otsuki, apple grower in Fukushima Prefecture.
Golden Delicious X Indo.
Sweet, very honeyed, pale yellow flesh, but hardly any taste of fruit or acidity.
F 12; trip. T². P l-Oct. S Nov-Jan.

ORIOLE 4 E D
USA; raised 1914 Univ Minnesota Fruit Breeding Farm, Excelsior.
Parentage unknown. Introd 1949.
Large. Sweet, soft, juicy, very lightly flavoured. Dual purpose US.
F 9. T² hrdy. C hvy. P eat Aug.

ORLEANS 7 M D
USA; raised 1911 by R. Wellington, NYSAES, Geneva.
Deacon Jones X Delicious. Introd 1924.
Delicious type, but larger. Very sweet, quite rich, intense but not complex flavour; tough skin.
F 9. T¹; sprd. P l-Sept. S Oct-Nov/Dec.

ORLEANS REINETTE 7 L D
France prob; described 1776 by Dutch botanist Knoop. History confused especially

with Golden Reinette. Widely distributed Europe; 121 syns. RHS AM 1914 as Winter Ribston; 1921 as Orleans Reinette.
At best, aromatic, nutty; sweet, firm, rather dry flesh. Like richer form of Golden Reinette or an aromatic Blenheim Orange; in appearance very similar and often confused with these varieties. Windfalls, early fruit can be cooked – slices keep shape; makes sweet, baked apple.
Long known in Europe, but not popular in England until early 1900s. Edward or George Bunyard found it among box of Blenheims sent in to their nursery for identification and introduced it as Winter Ribston, but in 1916 under correct name. Among Edward Bunyard's top six apples; became widely grown garden variety.
Frt slight orng rd flush, shrt rd stripes, over gold; many russet dots, netting, patches.
Size med. *Shape* flt rnd to oblng; sltly 5 crowned. *Basin* brd, med to shal; sltly ribbed; russet rings. *Eye* lrg, open. *Cavity* med wdth, dpth; russet. *Stalk* shrt, thck.
Flesh pale crm.
F* 16. T²; some res scab. C gd; needs warm spot for gd flavour. P m-Oct. S Nov-Jan.

ORTLEY syn Cleopatra 5 L D
USA; arose in orchard of Michael Ortley, New Jersey. Desc 1817 by Coxe as Woolman's Long Pippin; Woolman was famous Quaker preacher. Sent 1825 to London Hort Soc as Ortley Pippin. Renamed Cleopatra 1872; known by this name in Australia, New Zealand.
Savoury, crisp, juicy, cream flesh; tough skin.
Grown 1920–30s in all Australian apple states, but mainly Tasmania, also New Zealand; exported to UK. Still grown Australia.
F 12. T¹; sprd. C hvy; prn bitter pit. P l-Oct. S Nov-Mar.

OSENNEE DESERTNOE 4 E D
Estonia; raised Estonian Agric Plant Breeding Res Inst.
Antonovka X Uelsi (Wealthy). Received 1975.
Attractive pink flush, stripes. Juicy, refreshing.
F 6. T³. C hvy; res scab. P l-Aug. S Aug-Sept/Oct

OSENNEE POLOSATOE 4 M D
Germany poss; old variety introd Russia via Baltic States. Received 1975 from Pavlovsk Exp St, St Petersburg.
Refreshing, light taste; crisp flesh.
F 6. T³. P e-Sept. S Sept-Oct

OSLIN 4 E D syn Arbroath Pippin
UK; believed arose Arbroath, Scotland or introd from France by monks of abbey, which used to exist there. Known 1815, but prob much older.

Primrose yellow, dotted with pink, russet. Scented skin. Brisk and sparkling; distinctive taste, perhaps hint of aniseed. Claimed, roots from cuttings which earned it name Original Apple.

F 5. T². P l-Aug. S Aug-Sept.

OSNABRÜCKER REINETTE
syn Reinette d'Osnabrück 7 L D
Germany; arose Osnabrück, Hannover. Well known 1802.
Scented, flowery quality; sweet, some balancing acidity; quite juicy.
F 11. T¹; sprd. P m-Oct. S Nov-Jan/Mar.

OWEN THOMAS 4 E D
UK; raised 1897 by Laxton Brothers Nursery, Bedford. Introd 1920.
Named after Queen Victoria's last head gardener, a renowned fruit man. Intense, rich sweet-sharp flavour; soft, creamy, juicy flesh. At best, very good, but rapidly going over once picked.
F 10. T²; sprd. P eat Aug.

OXFORD BEAUTY 6 M D
UK; raised by F. W. Wastie, Eynsham, Oxford.
Gascoyne's Scarlet X Scarlet Nonpareil. Recorded 1944.
Beautiful, clear red flush. Sweet, lightly flavoured white flesh.
F 10. T²; sprd. P m-Sept. S Oct-Nov.

OXFORD CONQUEST 7 L D
UK; raised 1927 by F. W. Wastie, Eynsham, Oxford.
Blenheim Orange X Court Pendu Plat. Large. Sharp, chewy flesh, mellowing by Jan to rich, sweet-sharp taste.
F 10. T². C pr. P m-Oct. S Dec-Feb/Mar.

OXFORD HOARD 5 L D
UK; raised by F. W. Wastie, Eynsham, Oxford.
Sturmer Pippin X Golden Russet. Exhib 1943.
Develops quite intense rich, sweet-sharp taste, lots of fruit; like Sturmer Pippin.
F 9. T³. P m-Oct. S Dec-Feb.

OXFORD SUNRISE 5 L D
UK; raised by F. W. Wastie, Eynsham, Oxford.
Cox's Orange Pippin X Lane's Prince Albert. Exhib 1942.
Plenty of savoury acidity, yet some delicacy of flavour; crisp, juicy.
F 5. T¹. P e-Oct. S Nov-Dec.

OXFORD YEOMAN 3 L C
UK; raised 1922 by F. W. Wastie, Eynsham, Oxford.
Blenheim Orange X Lane's Prince Albert. Exhibit 1942.
Refreshing, crisp juicy. Cooks to slightly sharp, but bland purée.

F* 8. T². P m-Oct. S Nov-Jan.

OZARK GOLD 5 L D
USA; raised by P. Shepherd, Mountain Grove Fruit Exp St, Missouri.
Golden Delicious X (Red Delicious X Conrad). Introd 1970.
Named after fruit growing region of Midwest. Golden Delicious type, but ripening month earlier, firmer, less prone to russeting. Quite honeyed, sweet, very juicy, but often little acidity or taste of fruit; open textured flesh.
Planted throughout States, especially Washington, southern Illinois; also grown Italy.
F 20. T²; hrdy; reported gd res disease. C gd. P l-Sept/e-Oct. S Oct-Dec.

PAGSUP SPUR TYPE *see* Delicious

PALMER'S ROSEY 6 E D
UK; arose garden of H. I. Palmer, Whyteleafe, Surrey. Received 1965.
Large, handsome. Sweet, balanced; flavour of aniseed; soft, juicy flesh.
F 12. T². P use l-Aug-Sept/Oct.

PAPIROVKA syn White Transparent
Considered identical, except in Russia where regarded as distinct.

PARADISE APPLE
Name recorded 1398 in list of trees for Hôtel St Pol, Paris. By C15th quite widely cultivated in Normandy; on sale Rouen market. Planted in C16th Italian gardens, served at banquets and then believed to be of Roman origin. Known 1629 in England. Widely recommended as dwarfing rootstock by late C17th. A number of Paradise type rootstocks were in use by C19th on the Continent and in England. These were rationalised in early 1900s into the East Malling series, numbered from 1–9. French Paradise, which is M8, may be the Paradise apple of the literature, but it is impossible to be sure. Paradise rootstock M8 produces suckers and develops burrs at the base branches, which will root. It blossoms very early with small pink flowers and produces yellow fruit, ripe in August. This rootstock is no longer used; it is similar in vigour to the most popular Paradise dwarfing root, M9, which is Paradis Jean de Metz.

PARADIS DE LIMOUSIN (?) 5 L D
France; acc may be Paradis of Limousin, syn Vigneronne. Old variety now found throughout Limousin.
Yellow with pinky, orange brown flush. Sweet, nutty; rather dry, slightly crumbly texture. Paradis of Limousin used for baking, juice (vit C 20mg/100); principal

cider variety of Limousin.
F 31. T². C hvy. P l-Oct. S Dec-Mar.

PARKER'S PIPPIN syn Pépin Gris de Parker 8 L D
UK; recorded early 1800s by Diel in Germany as 'received from England', but better known on continent under syn.
Orange flush under cinnamon russet. Intensely flavoured, sweet-sharp acid or fruit drop quality of Nonpareil family.
F 10. T¹; sp. P e-Oct. S Nov-Feb.

PAROQUET 6 L D
UK; raised by Charles Ross, HG, Welford Park, Newbury, Berks.
Believed Peasgood's Nonsuch X Cox's Orange Pippin, but appears doubtful. Recorded 1897. RHS AM 1899.
Strong, raspberry flavour; sweet, juicy, quite firm flesh.
F 13. T¹; sp; sprd. P l-Sept. S Oct-Jan/Mar.

PATRICIA 6 M D
Canada; raised 1898 at Central Exp Farm, Ottawa.
McIntosh X. Introd 1921.
McIntosh red with bloom. Refreshing, distinct flavour, hint of melons or strawberries; crisp, juicy, white flesh.
F 7. T²; hrdy. C hvy. P m-Sept. S Oct-Nov.

PATRICK 4 L D
UK; received 1945; acc not Patrick's exhib 1883.
Recalls Blenheim Orange in taste. Sweet, plain, firm flesh; can be empty.
F 13. T². P e-Oct. S Nov-Dec.

PATTE DE LOUP 8 L D
France; acc prob variety of Leroy (1873) which prob arose late 1700s around Beaupréau, Maine-et-Loire (Western Loire).
Sharp, fruity, but mellowing. Many varieties in France bear this name, but legend maintains only true one carries 'scratch of the wolf'.
F 16. T². C hvy. P l-Oct. S Jan-Apr.

PAULARED 6 M D
USA; found 1960 by Lewis Arends, Ravine-Sparta township, Kent County, Michigan. Tree then c14 yrs old. Introd 1968.
McIntosh type, but ripens earlier. Strong vinous or strawberry flavour; sweet yet sprightly, quite firm white flesh.
Grown Michigan, West New York, Italy. Dual purpose in US; popular amateur's fruit.
F 9. T². C hvy. P l-Sept. S Oct-Nov.

PAYETTE 3 L C
USA; raised by L. Verner at Agric Exp Stat, Moscow, Idaho.
Ben Davies X Wagener. Introd 1944.
Large. Brisk, juicy, soft melting flesh. Cooks

to deep cream purée, well-flavoured, juicy.
F 9; trip. T^3. P mid-Oct. S Nov-Mar/Apr.

PEACEMAKER 4 M D
UK; raised by Charles Ross, HG, Welford
Park, Newbury, Berks.
Houblon X Rival. Recorded 1913.
Light aromatic flavour; sweet, soft, juicy
flesh.
F 10. T^2. C bien. P e-Sept. S Sept-Oct/Nov.

PEARL 7 M D
UK; raised by W. P. Seabrook, Boreham,
Essex.
Worcester Pearmain X Rival. Introd 1938.
One of a series named after semi-precious
stones. Handsome, lightly aromatic; sweet,
crisp; can be chewy.
Planted commercially 1950s; still grown
East Anglia.
F 9. T^1; prt tip; hrdy. C gd. P l-Sept/e-Oct.
S Oct-Nov.

PEARMAIN
*Name given, mainly in the UK, to varieties
that display marked pearmain or pear shape,
that is broad at the base and narrow at the
apex. Pearmain was mentioned c1200 in
both UK and France, but very doubtful that
the ancient Pearmain still exists. Hogg
claimed in 1870s that trees of 'true Old
Pearmain' were believed to be growing in
Herefords. No accs matching Hogg's
description have been received at NFC up to
now.*

PEASGOOD'S NONSUCH 4 M CD
UK: raised 1853 or 1858 by Mrs Peasgood,
when child living in Grantham, Lincs.
Later taken by her to Stamford. Exhib 1872
at Stamford and RHS London in Stamford
Hort Soc exhibit organised by nursmn
Thomas Laxton. Introd by Laxton. RHS
FCC 1872; AGM 1993.
'One of the most handsome apples
in cultivation' proclaimed RHS Fruit
Committee. Cooks to sweet, delicately
flavoured purée; makes generous baked
apple. Brisk, juicy eaten fresh; good in
vegetable salads.
Popular exhibition and garden variety of
C19th; also recommended as decorative
tree in 1890s. Remains popular UK garden
variety. As Sans Pareille de Peasgood grown
gardens northern France, Belgium,
Netherlands.
Frt *Col* orng rd flush, brd broken rd stripes
over pale grn becoming yell; some russet
patchs; lenticels conspic as russet dots; some
bloom. *Size* v lrg. *Shape* rnd to rnd-con;
regular; flat base, apex; trace rnd ribs, flt
sides. *Basin* brd, dp; sltly ribbed or
puckered. *Eye* lrg, open. *Cavity* brd, dp;
russet lined. *Stalk* shrt, qte thck; often with
fleshy knob. *Flesh* wht.
F11. T^2; sprd; res mildew, red spider; mod

res scab; prn canker. C gd. P m-Sept.
S Sept-Dec.
   CRIMSON PEASGOOD; discv before
1931.

PÊCHE MELBA 7 E D
Ireland; brought to England from County
Kilkenny 1930; then named.
Extraordinary colour; pink, orange skin
with peachy coloured flesh. Sharp, not at all
like peach.
F 13. T^2. P l-Aug. S Aug-Sept.

PECK'S PLEASANT 2/4 L D
USA; prob arose Rhode Island. Recorded
1832.
Rich savoury taste in New Year. Favourite
for New England home orchards in C19th.
F 11. T^2. P m-Oct. S Jan-Mar.

PEDERSTRUP 7 M D
Denmark or Germany; grown on Fynn
Island, Denmark by 1858.
Named after village on Fynn Island. Quite
large. Crisp, sharp.
F 10. T^1; sprd. P l-Sept. S Oct-Nov.

PEDRO 4 E D
Canada; raised 1898 at Exp Farm, Ottawa.
McIntosh X. Introd 1913.
Bright McIntosh red, with bloom. Sweet,
light; soft, white juicy flesh; often sour,
metallic undertones in England.
F 12. T^2. P use Sept.

PEGGY'S PRIDE 4 M D
UK; raised by F. W. Wastie, Eynsham,
Oxford.
Allington Pippin X Golden Spire. Received
1944.
Refreshing, crisp, juicy, but little of flavour
of parents.
F 8. T^2. P m-Sept. S Oct-Nov.

PENCO *see* Golden Delicious

PÉPIN OR PIPPIN
*Pépins were recorded as on sale in Rouen
markets in 1360 and it is presumably grafts
of these or related varieties Henry VIII's
fruiterer introduced to England in 1533.*

PÉPIN DE BOURGUEIL 2 L DC
France; desc 1948 (*Verger Français*); prob
originated Indre-et-Loire (Val de Loire).
Brisk, sweet-sharp, quite rich taste in New
Year.
Grown Indre-et-Loire, Vienne, primarily as
pollinator for Reinette du Mans (syn de
Jaune); used also apple pastries, juice.
F 20. T^2. C hvy; reported gd res disease.
P l-Oct. S Jan-Mar/May.

PÉPIN DE BOVELINGEN 5 L D
Received 1947 from France. Acc may be
Pépin d'Or de Bovelingen of nursmn
Bruant; introd 1915. Bovelingen is village

in Sint Truiden, centre of Belgian fruit
growing.
Sweet, crisp, juicy. Becomes much sweeter;
almost honeyed sweetmeat by Dec.
F 11. T^2. P e-Oct. S Nov-Dec.

PÉPIN GRIS DE PARKER syn Parker's
Pippin

PEPINO JAUNE 5 M D
Received 1948 from Côtes du Nord,
Brittany.
Sharp, astringent; poss cider variety.
F 15. T^3. P l-Sept. S Oct-Nov.

PÉPIN SHAFRANNYI 4 M D
Russia; raised 1907 by I. V. Michurin at
Michurinsk.
Orleans Reinette X (Glogerovka X Kitaika/
Malus prunifolia). Fruited 1915.
'Saffron Pippin'; pretty, strawberry red flush.
Refreshing, firm crisp, juicy flesh; high vit
C content.
F 6. T^2; prt tip; res scab. C gd. P l-Aug.
S Sept.

PERO DOURADO 5 L D
Portugal; received 1952.
'Typical Portuguese apple'. Sweet, winey,
firm flesh.
F 10. T^2. P e-Oct. S Oct-Dec.

PERO MINGAN 2/5 L D
Spain; received 1947.
Modestly fruity; quite sweet, soft flesh, but
lacking flavour in England.
F 12. T^2. P m-Oct. S Nov-Jan.

PERRINE YELLOW TRANSPARENT
see White Transparent

PERRINE YORK *see* York Imperial

PETER LOCK 4 L CD
UK; found early 1800s by villager, Peter
Lock in Dean Woods, Buckfastleigh,
Devon. RHS AM 1922.
Large. Sweet, slightly scented, deep cream
flesh; good fresh in Dec. Cooks to very
sweet, bright gold, smooth purée.
F 6. T^3. P m-Oct. S Nov-Feb/Apr.

PETIT PIPPIN 5 L D
Received 1948 from EMRS, Kent. Acc very
similar to Rosemary Russet.
F 12. T^2. P e-Oct. S Oct-Jan/Mar.

PEUPION syn Saint-Baussan

PEWAUKEE 4 L D
USA; raised by George P. Peffer, Pewaukee,
Wisconsin.
Duchess of Oldenburg X Northern Spy.
Recorded 1870.
Sweet, light, fruity; soft, juicy cream flesh.
F 9. T^2. P e-Oct. S Oct-Nov/Dec.

PFIRSICHROTER SOMMERAPFEL
6 E D
Germany; prob arose Thuringia. Desc early 1800s. Syns many; widely distributed in Europe.
'Peach Red Summer Apple'; bright pink flush over pale cream; prominently ribbed, crowned. Slightly scented; quite sharp, soft flesh.
Much grown in Denmark, Switzerland in 1940s.
F 4. T^2. P eat Aug.

PHILADELPHIA 5 M DC
UK; received 1983 from old tree in orchard, first planted 1840, belonging to Quantrell family, Alford, Lincs.
Rosy flush over cream. Sharp, juicy, crisp. Cooks to smooth purée; good brisk taste.
F* 14. T^2. P m-Sept. S Oct-Nov.

PICKERING'S SEEDLING 4 M D
UK; prob arose Notts. Brought to Hogg's notice in 1869 by Mr W. H. Caparn of Newark.
Quite rich, scented; sweet, plenty of acidity; soft, finely textured, cream flesh. Flavour can be only 'fair'.
F 11. T^2. P l-Sept. S Oct-Nov.

PIERRE syn Belle des Buits

PIGEON
Name given in France to group of apples characterised by tall, markedly conical shape; tradition claimed name derived from their bluish bloom similar to pigeon's plumage.

PIGEON DE JÉRUSALEM (?) 4 L D
France; acc may be Pigeon de Jérusalem of Leroy (1873) and prob is Pigeon/Pigeonnet Commun now known in Normandy, which may be variety recorded late 1600s. But confused literature and many syns.
La Quintinye grew 'Jerusalem Apple' at Versailles, which was 'almost red all over [with] firm pulp, but of little taste'. Leroy found it 'sweet, with exquisite aroma'. Collection fruit is crisp, savoury.
Pigeon Commune now grown Normandy is used for cooking and juice (vit C 20mg/100g).
Connection with Jerusalem was due to fact that core contained only four cells and when cut across took form of cross, but this is not constant.
F 12. T^2. P l-Oct. S Nov- Feb.

PIG'S NOSE PIPPIN 7 L D
UK; prob arose Hereford. Desc 1884 (Hogg). Attractive; small, with wide, shallow basin making top of fruit flat, like pig's snout. Lightly aromatic; sweet, crisp flesh.
F 14. T^2. P m-Oct. S Nov-Dec/Feb.

PIKKOLO 6 L D
Germany; raised Inst Fruit Res, Dresden-Pillnitz.
Clivia, Auralia cross. Introd 1993.
Pinky red flush and stripes over gold. Sweet, modestly rich; crisp to quite soft deep cream flesh.
F est 10. T^2; low susceptibility disease.
C hvy. P e-Oct. S Nov-Jan; commercially-Apr.

PILOT 6 L D
Germany; raised Inst Fruit Res, Dresden-Pillnitz.
Clivia X Undine. Introd 1988.
Quite sweet and fruity, deep cream firm flesh.
F est 12; frost res. T^2; res fireblight; low susceptibility disease. P m-Oct. S Jan/Feb-Mar; commercially-July.

PILTSAMASSKOE ZIMNEE 5 L D
Estonia; old variety arose Pilts.
'Winter Pilts'. Sweet, quite savoury taste; firm flesh.
F 1. T^2. P e-Oct. S Nov-Jan/Mar.

PINE APPLE RUSSET 8 M D
UK; desc 1831 by nursmn G. Lindley of Norwich; prob arose Norfolk; known since 1730.
Definite pineapple flavour, sweet, yet plenty of acidity, firm, pale yellow flesh. Can be dry, flavourless; very tough, coarse skin.
F 13. T^2. C lght. P e-Sept. S Oct-Nov.

PINE GOLDEN PIPPIN 8 L D
UK; recorded 1861; desc Hogg.
Hint of pineapple, but 'fine sprightly distinct pineapple flavour' according to Hogg. Sweet, plenty of acidity, juicy, crisp; but often underlying resinous taste.
F 18. T^1. C gd; frt sml. P e-Oct. S Oct-Nov/Dec.

PINK LADY 4/6 L D NOT IN NFC
Australia; raised Western Australia Dept Agric.
Golden Delicious X Lady Williams.
Select 1979 by J. Cripps. Introd 1989; syn Cripps Pink.
Prettily flushed. Firm solid apple; crackling cream flesh; sweet, quite honeyed.
Grown Australia, New Zealand; increasingly planted warmer apple regions: South Africa, US, France, Italy; imported UK. English climate unsuitable to grow this variety. Cultivation and marketing carefully controlled; only best fruit sold as Pink Lady, less perfect apples marketed as Cripps Pink.

PINNER SEEDLING 8 L D
UK; acc may be variety raised by nursmn James Carrel, Pinner, Middx; first fruited 1818; exhib 1820. Acc very similar to Wheeler's Russet (Potter).
Handsome, cinnamon russet, slight flush. Crisp, juicy full of fruit.
F 12. T^2. P l-Oct. S Nov-Feb; said keep to May.

PINOVA 6 L D
Germany; raised Inst Fruit Res, Dresden-Pillnitz.
Clivia X Golden Delicious. Introd 1986.
Red flushed over gold. Juicy, crisp, firm, slightly coarse flesh; fruity, tasty.
The most successful of new Pillnitz varieties, with commercial plantings in England, Europe, US, South Africa.
F est 12. T^2; med, spreading; low susceptibility disease. C hvy. P e/m-Oct. S Nov-Jan; cold store-Jun.

PIONEER 6 M D
UK; raised by Laxton Brothers Nursery, Bedford.
Cox's Orange Pippin X Worcester Pearmain. Recorded 1934. Syn Laxton's Pioneer.
Large; bright Worcester colour. Quite rich with zing of strawberry flavour; juicy, crisp, but becoming soft after storing.
F 6. T^2. P l-Sept. S Oct-Nov.

PIRJA 6 E D
Finland; raised 1961 at Inst Hort, Agric Res Centre, Piikki.
Huvitus X Melba.
Crisp, pleasant; recommended for amateurs in Finland.
F mid. T^1; hrdy. P eat Aug.

PIROS 6 E D
Germany; raised Inst Fruit Res, Dresden-Pillnitz. Helios X Apollo (Cox's Orange Pippin X Oldenburg). Introd 1985.
Attractive pinky red flush. Crisp, juicy, plenty of sugar and acid and distinctive almost cidery taste.
F est 10. T^2; low susceptibility disease. P eat Aug.

PITCHERS
Varieties which root from cuttings, that is can be propagated by sticking branches in the ground. See Ben's Red, Burr Knot.

PITMASTON NONPAREIL syn Pitmaston Russet Nonpareil

PITMASTON PINE APPLE 8 L D
UK; raised c1785 by Mr White, steward to Lord Foley of Stoke Edith, Herefords.
Golden Pippin seedling. Exhib 1845 at London Hort Soc by J. Williams, Pitmaston, Worcester.
In good year, intensely flavoured, sweet yet sharp, rich, nutty, honeyed. Hogg found 'distinct pineapple flavour' and Bunyard 'remarkable blend of honey and musk', but some years not nearly so good. Re-popularised by Bunyard Nursery in 1920s.
Frt *Col* golden, fine netting, dots of russet. *Size* sml. *Shape* oblng-con. *Basin* brd, shal; russet lines; sltly ribbed or beaded. *Eye clsd*

or hlf open; sepals downy. *Cavity* qte nrw, shal; russet lined. *Stalk* med lng, thck. *Flesh* yell.
F* 10. T^2; uprt. C hvy, bien. P e-Oct. S l-Oct-Dec.
PLATE 22

PITMASTON RUSSET NONPAREIL
8 L D
UK; raised by John Williams, Pitmaston, nr Worcester.
Exhib 1818 at London Hort Soc. Syn Pitmaston Nonpareil.
Slight reddish flush under russet, like Nonpareil, with similar sweet-sharp, fruit drop flavour, but sweeter, ripening earlier. Firm, rather dry, deep cream flesh, which can become little nutty with some of sere 'russet' flavour found in Egremont Russet.
F 11. T^2. P m-Oct. S Nov-Jan.

PIXIE 7 L D
UK; raised 1947 at NFT, RHS Wisley.
Prob Cox's Orange Pippin or Sunset seedling. RHS FCC 1972; AGM 1993.
Intensely aromatic, rich with plenty of sugar, acidity. Harder, sharper than Cox. Small fruit weighed against its market potential, but popular garden variety.
Frt *Col* ornge rd flush, rd stripes over grnish yell/yell; russet patches, dots. *Size* med. *Shape* flt-rnd. *Basin* med wdth, dpth; sltly ribbed; ltl russct. *Eye* clsd or prt open; sepals, shrt, downy. *Cavity* med wdth, dpth; russet present. *Stalk* med lng, qte thin. *Flesh* dp crm.
F 16. T^2 sprd. C gd; frt sml. P m-Oct. S Dec-Mar.
PIXIE RED SPORT

P. J. BERGIUS *see* Savstaholm

PLADEI 3 L C
Belgium; unknown origin; arose before 1940.
Large; ribbed, prominently crowned; Calville shape. Rich, fairly sharp. Cooked keeps shape; quite sweet.
Common in orchards of East Flanders up to 1960s.
F 9. T^3. P m-Oct. S Nov-Jan/Mar.

PLYMOUTH CROSS 7 E D
UK; raised by 1916 by G. T. Spinks, LARS, Bristol.
Allington Pippin X Star of Devon.
Interesting taste, quite rich, pineapple flavour; firm flesh. Allington Pippin acidity, but more sweetness.
F 4. T^2. C lght; frt sml. P l-Aug. S Sept-Oct.

POBEDA CHERNENKO 5 M D
Russia; raised by Chernenko, Vavilov Inst, St Petersburg.
Antonovka X Pepin Londondski. Known 1953.

Chernenko's Victory; savoury, refreshing taste, rather sharp. Juicy, crisp, firm flesh. Valued Russian garden apple.
F 13. T^2. P l-Sept. S Oct-Dec.

POELMAN'S RODE JAMES GRIEVE
see James Grieve

POHORKA 4 L D
Serbia; unknown parentage. Received 1978 from Inst Fruit Research, Čačak.
Red streaked, flushed. Rich, full of fresh flavour in Dec; sweet, often quite scented; very juicy, crisp flesh.
F19. T^2. P l-Oct. S Nov-Jan.

POLKA syn Trajan

POLLY 5 E DC
UK; arose Cornwall. Received 1954.
Sweet, light taste; soft, juicy flesh. Early used as cooker.
F 12. T^3; sprd. P e-Sept. S Sept-Oct.

POLLY PROSSER 7 L D
UK; raised 1946 by J. H. Cooper, Maidstone, Kent.
Cox's Orange Pippin X Duke of Devonshire.
Even more deeply flavoured than Ribston and stronger than Cox's Orange Pippin. Intense aromatic flavour; rich blend of sugar and acidity; crisp, juicy, nearly yellow flesh.
F 11. T^2. C lght; frt often coarsely russeted. P e-Oct. S l-Oct-Dec.

POMEROY OF SOMERSET 7 L D
UK; desc 1851 (Hogg). Pomeroy apples were famed for their fine flavour and three distinct Pomeroys existed in C19th – Somerset; Herefordshire and Worcester-shire; Lancashire.
Handsome, scarlet flushed. Rich, aromatic, sweet with strong pineapple-like acidity; crisp, quite juicy, cream flesh.
F 13. T^2. P m-Oct. S Nov-Jan/Mar.

POMME À CÔTES 1/5 L CD
France; desc 1948 (*Verger Français*), when widely grown Haute Savoie (Rhône Alps).
Good fresh in Dec; fruity, almost nutty. Cooks to pale lemon slices, sweet, pleasant.
F 8. T^2. P l-Oct. S Nov-Mar/May.

POMME CROTTE 6 L D
France; received 1947.
Wine red flush. Quite sweet, juicy in Jan, but often rather metallic taste.
F 15. T^2. P m-Oct. S Dec-Mar/May.

POMME D'AMOUR 7 L D
France; received 1947.
Attractive. Lightly flavoured, crisp juicy deep cream flesh.

F 10. T^2. P e-Oct. S Oct-Dec/Mar.

POMME DE CHOUX À NEZ CREUX
1 L C
France; received 1948 from Cher (Val de Loire).
Prominently ribbed, crowned. Soft, juicy, sharp. Cooked, keeps shape; light flavour.
F* 23; trip. T^3; sprd. C hvy; frt sml. P l-Oct. S Dec-Mar.

POMME DE FER 7 L D
France; acc prob Fer of Cher (Val de Loire); desc 1948 (*Verger Français*). Several regions of France have Fer apple.
Very sharp in Dec.
F 10. T^2. P m-Oct. S Dec/Jan-Mar/Apr.

POMME DE FEU (INDRE) 1/5 L C
France; received 1948 from Indre (Val de Loire).
Firm, sharp flesh. Cooked keeps shape, but little flavour.
F 21. T^1. C hvy. P l-Oct. S Dec-Mar/Apr.

POMME DE GLACE (?) 1/5 L C
France; received 1948. Uncertain identity; Pomme de Glace is syn Blandurette of Corrèze, which exists in several forms in south west France. Acc may be Coquet de Meilhards, syn Blandurette (Gros), desc 1948 (*Verger Français*), when grown Corrèze (Limousin), particularly in commune of Meilhards.
Sharp, fruity. Cooks to gold purée or keeps shape; good, intense, brisk taste.
F 22; trip. T^2. P l-Oct. S Dec-Feb/Mar.

POMME D'ENFER 6 L D
France; acc may be Pomme d'Enfer, syn Bordes, old traditional variety of south west. Flame red flush. Very sharp, crisp in Feb. Very long keeping, giving out strong perfume and formerly used to scent linen cupboards.
F 14; trip. T^2. P l-Oct. S Jan-Mar.

POMME DU VIEZ 4 M D
Received 1947 from Switzerland.
Slight honeyed taste, but fairly plain; sweet, juicy, firm flesh.
F 16. T^3; prt tip. P l-Sept. S Oct-Nov/Dec.

POMME NOIRE 6 L D
France; received 1973.
Extraordinary colour, virtually black. Sweet but no flavour; pale green flesh.
F 5. T^2. C gd. P l-Oct. S Dec-Mar.

POMME ROYALE syn Dyer 5 M D
USA prob; although held to be French by some writers. Recorded 1835; renamed Dyer 1850 by Massachusetts Hort Soc.
Large, heavily ribbed. At best, quite rich; sweet, juicy; rather coarse flesh.
Well known in Rhode Island C19th when highly regarded; recently revived by

enthusiasts.
F 9. T^2; tip. **P** e-Sept. **S** Sept-Oct.

POMMERSCHER KRUMMSTIEL
4 L D
Germany; known 1826; unknown origin; local to coastal area of north Germany. Quite rich, lightly aromatic.
F 11. T^2. **P** e-Oct. **S** Nov-Dec.

POMME VIOLETTE syn Violette

PONSFORD 1 L C
UK; received 1950 from Crediton, Devon. Cider variety of this name, also used for cooking, known 1888, whether this is same impossible to know as no descs published. Large, grass green. Sharp, soft juicy flesh. Cooked, keeps shape; sharp, quite good flavour.
F 12. T^3. **P** m-Oct. **S** Nov-Feb/Mar.

PÓNYIK ALMA 5 L C
Czech Republic or Transylvania (Romania). Commonly believed ancient Transylvanian variety, which arose Alsó-Fehér county. Name first mentioned 1530 in book written in town of Érsekújvár (now Nové Zamky, Czech Republic) and also known Balkan Peninsular.
'Queen of Transylvanian apples' in early 1900s. Large, golden; quite sweet; little honeyed, firm, pale yellow flesh.
F 14. T^3. **P** m-Oct. **S** Nov-Jan/Feb.

POOR MAN'S PROFIT 3 L C
UK; acc is variety now known in Somerset; desc 2001 (Copas); may be variety recorded 1824 by William Forsyth, but impossible to know as desc given too brief.
Large, handsome. Crisp, juicy, quite sharp. Cooked, keeps shape; sweet, little rich.
F 14. T^2. **C** prn bitter pit. **P** e-Oct. **S** Oct-Dec.

POPE'S SCARLET COSTARD 6 L DC
UK; prob raised early C20th by nursmn William Pope, Newbury, Berks.
Large, costard shape, but almost entirely red flushed. Pale yellow, crisp, plenty of sugar and acid and some richness. Cooked slices become yellow, sweet and rich.
F est 8. T^3. **P** e-Oct **S** Oct-Jan.

PORT ALLEN RUSSET 7 M D
UK; arose Port Allen, Errol, Scotland. Received 1958.
Strong, brisk flavour of fruit, some richness; crisp, juicy.
F 5. T^3. **P** m-Sept. **S** Oct-Nov.

PORTER 5 E D
USA; raised c1800 by Rev Samuel Porter, Sherburne, Mass. Syn Yellow Summer Pearmain.
Prominently ribbed, flat sided. Quite juicy, soft, sweet flesh, slightly perfumed, but low

in acidity.
Main September apple in Boston markets of 1850s. Later valued for canning as kept its shape when cooked, but variable size and irregular ripening led to commercial demise. Recently revived by amateurs.
F 19. T^2. **P** Sept. **S** Sept-Oct.

POTTS' SEEDLING 1/5 E C
UK; raised c1849 by Mr Samuel Potts of Ashton-under Lyme, Lancs, from seed of poss American apple. Popularised by nursmn John Nelson of Rotherham, who noticed it in gooseberry grower's garden at Oldham in 1850s.
Cooks to juicy froth, or later to cream purée; brisk, good flavour, needs hardly any sugar. Widely grown in gardens and by market gardeners in C19th, especially recommended for smoky districts.
F* 12. T^2; sprd. **C** gd; frt bruises. **P** e-Sept. **S** Sept-Oct.

PRAIRIE SPY 4 L D
USA; raised 1914, Univ Minnesota Fruit Breeding Farm, Excelsior. Introd 1940.
Very sweet, honeyed, almost sickly taste; little acidity; firm, cream flesh.
F 11. T^2; hrdy. **P** m-Oct. **S** Nov/Dec-Mar/Apr.

PRESENT VAN ENGELAND 5 L C
Netherlands; recorded 1864.
Very large, tall, ribbed. Cooks to rather insipid, cream, purée. Remains valued Netherlands.
F 13. T^2; tip. **P** e-Oct. **S** Oct-Dec.

PRESENT VAN HOLLAND 4 L D
Netherlands; raised 1914 by Ludwig & Co Nursery, Hillegom.
Present van Engeland X Brabant Bellefleur. Introd 1934 by van Rossem Nursery.
Markedly oval. Quite sweet, pleasant taste of fruit; crisp, juicy. Remains valued Netherlands.
F 7. T^1; sprd. **P** e-Oct. **S** Nov-Jan/Mar.

PRESIDENT BOUDEWIJN 7 L D
Netherlands; raised 1935 at IVT, Wageningen. Jonathan X Cox's Orange Pippin. Introd 1952.
Rich, aromatic, intense flavour; almost yellow flesh. More robust flavour than Cox, but rather large unattractive appearance.
F 14. T^1; sprd. **P** e-Oct. **S** Nov-Jan.

PRIMA 6 M D
USA; raised 1957 by co-operative breeding programme of Purdue, Rutgers and Illinois Universities. Complex parentage involving *Malus floribunda* (carries V_f gene for scab resistance). Introd 1970.
Striking red flush. Good taste of fruit; sweet juicy, rather coarse, deep cream flesh. Regarded as 'excellent' amateur fruit in US; planted commercially Italy.

F 9. T^3; sprd; res scab, mod res mildew, fireblight. **C** gd. **P** l-Sept. **S** Oct-Nov.

PRIMATE 5 M D
USA; believed arose c1940 with Calvin D. Bingham, Camillus, Onondaga County, New York. Listed 1854 by American Pom Soc.
Sweet, quite sprightly acidity, almost winey flavour; soft, juicy, white flesh.
Popular C19th home orchard fruit in New York State; remains valued by lovers of old American apples.
F 7. T^3; sprd. **P** e-Sept. **S** Sept-Oct.

PRIME GOLD 5 L D
USA; discovered by B. Hoekman in Zillah, Washington.
Golden Delicious seedling. Introd 1965. Resembles Golden Delicious, but free of russet. Slightly honeyed; crisp, juicy, pale yellow flesh.
F 20. T^2. **P** l-Oct. **S** Nov-Jan/Feb.

PRIMROUGE see Akane

PRIMUS 6 L D
Netherlands; raised 1935 at IVT, Wageningen. Reinette Rouge Étoilée X Cox's Orange Pippin.
Colour of Reinette Rouge Étoilée, but none of raspberry flavour. Very sweet, little juice, almost no acidity.
F 12. T^2. **P** e-Oct. **S** Oct-Dec.

PRINCE ALFRED 4 L D
UK; recorded 1933.
Large; handsome. Slight strawberry flavour; quite sweet, fruity; crisp, juicy flesh.
F 8. T^2. **P** e-Oct. **S** Dec-Feb.

PRINCE CHARLES 4 M D
UK; raised 1940–45 by Herbert Robinson, Victoria Nurseries, Burbage, Leicester. Lord Lambourne X Cox's Orange Pippin. Beautiful, broad red stripes, pinky red flush over gold. Rich blend of sugars and acid, crisp, quite juicy. Sweeter than Lord Lambourne, but not really aromatic or complex like Cox.
F 5. T^3; sprd; prn midlew. **C** lght. **P** e-Sept. **S** Sept-Oct.

PRINCE EDWARD 4 L D
UK; raised at Sawbridgeworth Nurseries, Herts by Messrs Rivers.
Believed Cox's Pomona X Cellini. Introd 1897. RHS AM 1897.
Bright red flush, stripes over lemon. Light, refreshing taste, juicy; tough skin.
F 10. T^2. **P** e-Oct. **S** Oct-Dec/Jan.

PRINCE GALA see Gala

PRINCE GEORGE 1 L C
UK; prob raised by former HG, nursmn W. H. Divers, Surbiton, Surrey.

Lane's Prince Albert X Peasgood Nonsuch. Received 1935.
Large. In early Dec cooks to sharp purée but keeps some form.
F 14. T^2; sprd. P m-Oct. S Nov-Dec.

PRINCE NICOLAS 7 L D
Received 1949 from Switzerland.
Quite rich, pleasant taste of fruit; juicy, firm flesh.
F 9. T^3. P e-Oct. S Nov/Dec-Mar.

PRINCESS *see* King of Pippins

PRINS BERNHARD 6 L D
Netherlands; raised 1935 IVT, Wageningen. Jonathan X Cox's Orange Pippin.
Bright Jonathan colour with much Cox flavour. Aromatic, rich, sweet; not as complex as Cox, but its juicy, finely textured flesh, and scented quality like Lucullus.
F 14. T^2; sprd; prn mildew. C gd. P e-Oct. S Nov-Dec.

PRINSES BEATRIX 4 L D
Netherlands; raised 1935 at IVT, Wageningen.
Cox's Orange Pippin X Jonathan.
Quite rich, intensely flavoured, but not aromatic quality of Prins Bernhard. Sweet, juicy, crisp flesh.
F 10. T^2; sprd. P e-Oct. S Nov-Dec/Jan.

PRINSES IRENE 1 L D
Netherlands; raised 1935 at IVT, Wageningen.
Jonathan X Cox's Orange Pippin. Introd 1955.
Sweet with delicate, flowery aromatic quality in Nov. Elements of Cox flavour, but rather coarse texture; tough skin.
F 7. T^1; sprd. C frt sml. P e-Oct. S Nov-Dec/Jan.

PRINSES MARGRIET 7 L D
Netherlands; raised 1935 at IVT, Wageningen.
Jonathan X Cox's Orange Pippin. Introd 1955.
Quite rich, some aromatic quality, but more in style of Jonathan than Cox. Crisp, fairly sharp in Dec.
F 13. T^2. P e-Oct. S Dec-Mar.

PRINSES MARIJKE 6 L D
Netherlands; raised 1935 at IVT, Wageningen.
Jonathan X Cox's Orange Pippin. Introd 1952.
Beautiful deep rose flush. Rich aromatic, scented flavour; resembling Lucullus, but smaller.
F 11. T^1; sprd. C frt sml. P e-Oct. S Nov-Dec.

PRINZ ALBRECHT VON PREUSSEN 4/6 L D
Germany; raised 1865 by Braun, at Kamenz, east of Dresden, Silesia.
Alexander seedling.
Sweet, juicy, slightly perfumed quality, like Alexander, but later season. Refreshing in Jan. Grown as standard tree in Central Germany.
F 13. T^1; sprd. C hvy. P l-Sept. S Oct-Jan/Feb.

PRINZEN APFEL 4 M D
Germany; known C18th. Syns many.
Prized in Germany for its scent of pineapples. Tall, ribbed, flat sided. Quite rich, sweet, fruity taste, but not noticeably scented in England.
Grown by few devotées in Germany; formerly also grown Norway, Sweden.
F 12. T^3. C bien. P l-Sept. S Oct-Dec.

PRIOLOV DELISES 5 L D
Slovenia; raised 1947 by Prof J. Priol at Inst Hort, Maribor.
Golden Delicious X Yellow April. Introd 1967.
Sweet, slightly honeyed; soft, juicy, nearly yellow flesh. Style of Golden Delicious, but better, more acidity.
F 13. T^1; claimed gd res disease. P e-Oct. S Nov-Dec.

PRISCILLA 6 L D
USA; raised 1961 by co-operative breeding programme of Purdue, Rutgers and Illinois Univ. Complex parentage involving *Malus floribunda* (carries V$_f$ gene for scab resistance). Introd 1972.
Named after wife of F. D. Hovde, president of Purdue University. Very sweet, scented; crisp, juicy.
F 17. T^2; res scab; mod res mildew, fireblight; res rosy apple aphid. C hvy. P m-Oct. S Nov-Dec.

PROCTOR'S SEEDLING 7 L D
UK; prob arose Lancs; exhib 1934; desc Taylor.
Quite rich, aromatic; reminiscent Ribston Pippin, but sharper, less interesting. Formerly widely grown Lancs; well known Liverpool markets.
F 14. T^2. P e-Oct. S Nov-Jan.

PUCKRUPT PIPPIN 8 L D
UK; received 1942 from R. Fairman, Crawley, Sussex. Acc not Puckrupp Pippin of Scott (1872); exhib 1883.
Developing rich, sweet-sharp flavour of acid drops. Firm, deep cream flesh.
F 11. T^2. P e-Oct. S Dec/Jan-Mar.

PUFFIN 3 M D
UK; exhib 1883 as Puffin Sweet by nursmn R. H. Poynter, Taunton, Somerset; desc 1884 (Hogg).
Large. Soft, white flesh, but little flavour. Cooked, very light taste.

F 8. T^3. P e-Sept. S Sept-Oct.

PUMPKIN SWEET 2 L DC
USA; arose in orchard of S. Lyman, Manchester, Connecticut. Recorded 1834.
Heavy, large, often sold as Pound Sweet. Very sweet, little acidity, low in flavour. Esteemed as sweet, baking apple, for 'canning or stewing with quinces' in C19th US, but 'too coarse, strangely flavoured for dessert'.
F* lrg; 13. T^3. C gd; v prn water core. P e-Oct. S Oct-Dec.

PURPURROTER COUSINOT 6 L D
Netherlands or Germany; desc 1766. Syns numerous, include Cousinotte Rouge d'Hiver. Many 'Cousinotte' varieties were cited by old authors. Acc is prob Cousinotte Rouge d'Hiver now found Franche-Comté, France.
Sweet, tough flesh, lacking flavour, but in France 'agreeable acidity, sweet and perfumed'.
F 11. T^2. P m-Oct. S Dec-Mar.

PUSZTAI SARGA 5 L D
Hungary; known in C19th.
Sweet, fruity, firm; rather dry flesh.
Grown for market in C19th, but now largely forgotten.
F 10. T^2. P e-Oct. S Nov-Jan/Feb.

QUEEN 3 M C
UK; raised 1858 by W. Bull, farmer at Billericay, Essex. Exhib 1880 by nursmn William Saltmarsh, Chelmsford. RHS FCC 1880.
Cooks to bright yellow purée, sharp, well-flavoured, juicy, translucent; sharper than Golden Noble. Makes good baked apple. Formerly widely planted in gardens, especially in Essex, where remains popular.
Frt *Col* rd flushed, many broken rd stripes over pale grnish yell/pale yell. *Size* lrg. *Shape* flt to flt-rnd; flt base; trace rnded ribs, sltly flt sided; sltly crowned. *Basin* brd, qte dp; sltly ribbed. *Eye* lrg, hlf to fully open; sepals, sml, downy. *Cavity* v brd, dp; russet lined. *Stalk* med lng, med thck. *Flesh* wht.
F 9. T^2. C gd. P m-Sept. S m-Sept-Dec.

QUEEN ALEXANDRA 3 L C
UK; received 1919 from HG, William Crump, Madresfield Court, Malvern; exhib 1902.
Handsome. Cooks to brisk yet rich purée; makes very good sauce.
F 17. T^2; sprd. P e-Oct. S Nov-Jan.

QUEENBY'S GLORY 6 M D
UK; received 1949 from Wrest Park, Bedford, but prob old variety.
Sweet, lightly aromatic, but not intense

flavour.
F 8. T^2. P e-Sept. S Sept-Oct.

QUEEN CAROLINE 5 M C
UK; raised by nursmn T. Brown, Measham, Ashby-de-la-Zouch, Leics. Named after George IV's wife; tree first fruited 1820, year Queen's case was in law courts.
Large. Cooks to creamy purée, brisk, well-flavoured. Similar to, but not as sharp as, Queen.
F 11. T^2. P m-Sept. S Sept-Nov/Dec.

QUEEN COX see Cox's Orange Pippin

QUEENIE syn Devon Crimson Queen

QUEENING or **QUOINING**
Group of apples which are prominently ribbed. Name derives from coin or quoin, which signified corner or angle. Old writers appear to have equated French Calvilles with Queenings.

QUINDELL 4 L D
USA; discv 1934 by R. Banta, Green Forest, Arkansas. Unknown parentage. Introd 1965.
Sweet, honeyed, Delicious type; deep cream, rather coarsely textured flesh.
F 11. T^1. P l-Oct. S Nov-Jan.

QUINTE 6 E D
Canada; raised at Exp Farm, Ontario.
Crimson Beauty (Early Red Bird) X Red Melba. Introd 1964.
Named after Bay of Quinte, fruit growing area on Lake Ontario between Toronto and Kingston. Beautiful appearance, flushed in pinky red over cream with dusky bloom. Sweet, scented, snow-like flesh. Very good of type, but difficult to catch at best.
Developed for eastern Canada; ripe before Melba. Major commercial apple in Norway.
F 6. T^3; hrdy. C frt drops. P eat e-Aug.

RABALEYZE 3 L C
Received 1950 from Lozère, France.
Markedly ribbed, flat sided. Sharp, but mild cooked flavour.
F 16. T^2. P m-Oct. S Dec-prob Mar.

RACINE 1/5 L DC
France; acc prob variety desc 1948 (*Verger Français*), when grown Allier (Auvergne).
Light taste of fruit, but bland cooked.
F 27. T^2. P l-Oct. S Dec-Mar.

RACINE BLANCHE 1/5 L C
France; acc is prob variety desc 1948 (*Verger Français*) when grown Vendée (Western Loire). Racine, Racine Blanche are syns, but accs are distinct.
Sharp, firm flesh. Cooked, keeps shape;

bright yellow, some brisk flavour.
F 20. T^2. P l-Oct. S Dec-Mar/May.

RAFZUBIN syn Rubinette 7 M D
Switzerland; raised 1966 by Walter Hauenstein, Rafz, nr Schaffhausen, on German border.
Golden Delicious X prob Cox's Orange Pippin.
Handsome. Rich blend of sugar, acid, honeyed intense flavour, nearly yellow flesh.
As Rubinette, planted commercially in Switzerland; small extent England.
F 11. T^2. C hvy; frt sml, but improves as tree matures; mod res scab. P l-Sept. S Oct-Nov.

RALLS JANET 4 L D
USA; arose on farm, or nursery, of Mr Caleb Ralls, Amherst County, Virginia. Known c1800. Believed, by some, brought from France for Pres Jefferson by M. Genet, a French minister. Syns many; include Geniton, Jefferson Pippin, Never Fail because of late flowering.
Carmine and pink flush, stripes. Sweet, crisp.
Formerly grown Southern States, but displaced by Ben Davis by 1900 and later by Delicious. Taken to Australia, Japan, where formerly widely grown and used in breeding programmes.
F 18. T^2. C hvy; frt sml unless thinned. P l-Oct. S Jan-May.

RAMBOUR
Name given in France to group of varieties characterised by large size and red colour. Rambour apples were widely grown throughout continent and still found in French gardens, but never highly valued in England. Rambour Franc is best-known and oldest of these, but not in Collection. Believed to have arisen village of Rambure, nr Abbeville, Picardy. 'De Rambure' was recorded 1535 by botanist Jean de la Ruelle and this may be Rambour Franc syn Rabour d'Eté, which is still grown in France.

RAMBOUR D' AUTOMNE (BEL-GIUM) 4 M D
Received from Belgium 1948.
Large; ribbed, box-like shape. Sweet, plenty of acidity, faint hint of strawberry flavour; firm, rather dry flesh.
F 12. T^3. P e-Sept. S Sept-Oct/Nov.

RAMBOUR PAPELEU 4 L CD
Russia; raised by N. A Hartwiss, Director of Botanic Gardens, Nikita, Crimea. Introd to Belgium c1853 by nursmn A. Papeleu of Wetteren.
Large. Crisp, juicy, quite sharp. Cooked, slices keep shape; sweet, slightly rich.
Grown Belgium, France – Normandy and north east; considered 'excellent for juice, cider, tarts'.

F 11. T^2. C hvy. P m-Oct. S Nov-Jan/Mar.

RAMBOUR PODOLSKII syn Knysche 4 L D
Ukraine; known by 1899.
Large, ribbed. Sweet, light pleasant taste of fruit; crisp, juicy.
Formerly much grown in Winnitker district of Podoloka.
F 9. T^3. P m-Oct. S Nov-Jan.

RAMPALE 4 L D
France; old variety, arose Corrèze, Limousin; desc 1948 (*Verger Français*).
Intense sweet-sharp flavour in early Nov; firm flesh.
Widely grown throughout Limousin, also Charente (Poitou-Charentes), Dordogne (Aquitaine).
F 22. T^3. P e-Oct. S Nov-Jan/Mar.

RANGER 4 E D
Canada; raised at Exp Farm, Ottawa.
Crimson Beauty (Early Red Bird) X Melba. Introd 1964.
Carmine stripes, flush over creamy yellow, with bloom. Very scented, sweet, soft white flesh, little acidity. Ripe before Melba; developed for eastern Canada.
F 5. T^2; hrdy. C gd, bien. P eat l-July-Aug.

RANK THORN 4 M D
UK; old Westmorland variety; received 1951.
Bold red stripes. Soft white fesh; sharp, juicy.
F 16. T^3. P l-Sept. S Oct-Nov.

RARITAN 4 E D
USA; raised 1949 by G. W. Schneider, Rutgers Univ, Agric Exp Stat, New Jersey.
(Melba X Sonora) X (Melba X (Williams X Starr)). Named 1966.
Named after Raritan River, in year of Rutgers bicentenary. Boldly striped in carmine, which also stains cream flesh. At best, scented, with flavour of raspberries; sweet, plenty of balancing acidity; crisp, juicy. Flavour fades once picked.
F 14. T^2; reported res scab. P m-Aug. S Aug-Sept; cold store-Nov.

RATHE RIPE 5 E D
UK; acc received 1947 when listed by Barnham Nursery; impossible to know if variety recorded 1831.
Savoury, brisk taste; full of good, fruity flavour by mid-Aug.
F 10. T^2. P m-Aug. S Aug-Sept.

RÉALE D'ENTRAYGUES 4 L D
France; considered local variety of Aveyron (Midi-Pyrénées). Also known as Reinette de Pons, after village in Aveyron, where commercially important in 1920–30s.
Refreshing, crisp, quite sweet, fruity in Feb.
F 15. T^2. P l-Oct. S Jan-Mar/May.

RÉAUX 6/7 L D
France; desc 1948 (*Verger Français*) when grown Meuse (Lorraine), Marne, Argonne (Champagne).
Intense, brisk taste of fruit; quite rich.
F 27. T^2. C hvy; frt sml. P l-Oct. S Dec-Mar/Apr.

RED ARMY 6 M D
UK; raised by F. W. Wastie, Eynsham, Oxford.
Chatley's Kernel X Worcester Pearmain. Received 1945.
Sweet, but moderate flavour.
F 10. T^2. P m-Sept. S Oct-Nov.

RED ASTRACHAN 6 E DC
Russia; prob arose Astrachan on Volga. Mentioned 1780 by Swedish botanist P. J. Bergius, but prob grown much earlier in Sweden. By 1816, grown England, by Mr William Atkinson, Grove End, Paddington. By 1835 grown by Massachusetts Hort Soc. Widely distributed in Europe, North America. Syns numerous.
Deep rose crimson with bloom. Crisp, juicy, white flesh with strong acidity, yet plenty of sweetness. Good flavour, but well ripened specimen is often difficult to find; quickly goes over once picked. Cooks to juicy, light purée.
One of first Russian varieties tested in North America, where became valued home orchard fruit used for pies, sauce, jelly, canning. Still grown by amateurs in States, Northern Europe, especially Sweden, where can be planted as far north as Västerbotten. In England, grown by London market gardeners in C19th, but never highly esteemed.
F 1. T^3; v hrdy. C gd, bien. P e/m-Aug.

RED BEAUTY OF BATH see Beauty of Bath

RED BELLE DE BOSKOOP see Belle de Boskoop

RED BENONI see Benoni

RED BLENHEIM (Passey) see Blenheim Orange

RED BLENHEIM (F. W. Wastie) see Blenheim Orange

RED CHARLES ROSS see Charles Ross

REDCOAT GRIEVE (Iliffe) see James Grieve

RED DELICIOUS see Delicious

RED DEVIL 6 D L
UK; raised 1975 by H. F. Ermen, Faversham, Kent.

Discovery X Kent. Introd c1990 by F. W. T. Matthews Nursery, Worcs.
Striking scarlet flush. Strong, fruity taste, some strawberry flavour; crisp, juicy, pink stained flesh. Produces pink juice. Highly decorative.
Popular garden apple; also introd France, Holland, Germany, Belgium.
F 8. T^2; gd res scab, mildew. C hvy; P l-Sept. S Oct-Dec.

RED DOUGHERTY 6 L D
New Zealand; highly coloured sport of Australian variety Dougherty; found by C. F. Bixley at Twyford, Hawkes Bay. Introd 1930.
Very sweet, little acidity in Feb
Formerly Dougherty and its sport were exported to UK.
F 11. T^1; sprd; tip. C hvy; frt hangs v late. P l-Oct. S Jan-Apr.

RED ELLISON see Ellison's Orange

RED ELSTAR see Elstar

RED FALSTAFF see Falstaff

RED FAMEUSE 6 L D
Received 1967 from former Yugoslavia via Scotland.
Very similar to Fameuse in appearance and taste, but more highly coloured and prob seedling rather than sport of Fameuse.
F 10. T^2. P l-Sept/e-Oct. S Oct-Dec.

RED FORTUNE see Fortune

REDFREE 6 M D
USA; raised 1966 by co-operative disease resistant programme of Illinois, Indiana, New Jersey Exp Sts.
Raritan X scab resistant selection (Melba X *Malus floribunda*). Introd 1981.
Deep maroon. Crisp to soft, juicy, light good flavour with hint of strawberry; tough skin.
F early. T res scab, cedar apple rust; mod res mildew, fireblight; C uneven ripening. P e-Sept. S Sept; short season.

RED GEEVESTON FANNY see Geeveston Fanny

RED GEORGE CAVE see George Cave

REDGOLD 6 L D
USA; raised by F. A. Schell, Cashmere, Washington.
Golden Delicious X Richard Delicious. Introd 1946 by Stark Bros, Missouri.
Amazingly sweet, honeyed, rich, scented; hardly any acidity, yet not sickly. Juicy, soft, nearly yellow flesh, like Red Delicious; tough skin.
Formerly grown commercially in Italy.
F10. T^3. C gd. P e-Oct. S Nov-Dec.

RED GRANNY SMITH 4 L D
Australia; arose with Herbert Batt, an engine driver at timber mill in Banksiadale, Murray-Wellington, Western Australia; select 1935 by H. Birmingham, fruit grower at nearby Dwellingup, south of Perth. Introd 1945. Known as Red Gem in Australia.
Poss Granny Smith, Jonathan cross.
Quite sweet, plenty of acidity, crisp, juicy flesh; can have some flavour, but usually not ripening fully in England. Grown small extent Australia.
F 14. T^2. P l-Oct. S Dec-Mar/Apr.

RED HAWTHORNDEN syn Hawthornden

REDHOOK 6 E D
USA; raised 1923 by R. Wellington NYSAES, Geneva.
McIntosh X Carlton. Introd 1938.
McIntosh type; heavy bloom. Sweet, melting, juicy flesh.
F 8. T^3. P l-Aug. S Sept-Oct/Nov.

RED INGRID MARIE syn Karin Schneider

RED JAMES GRIEVE see James Grieve

RED JOANETING syn Margaret

RED JONATHAN see Jonathan

RED JONAPRINCE see Jonagold

RED MARTINI see Martini

RED MELBA see Melba

RED MILLER'S SEEDLING see Miller's Seedling

RED MUSK 6 M C
UK; local variety of Limpsfield, Surrey, when received 1951. Not Herefords cider variety of this name.
Cooked keeps shape; quite sharp, fruity.
F 14. T^3. P m-Sept. S Oct-Nov.

RED NEWTON WONDER see Newton Wonder

RED PIPPIN syn Fiesta

RED PIXIE see Pixie

RED ROME (Australia) see Rome Beauty

RED SAELET 4 L D
Belgium; arose with Edwin Saels, Herk-de-Stad.
Prob Golden Delicious seedling. Introd 1976.
Resembles Golden Delicious, but flushed,

firmer, ripening later. Recommended for processing.
F 13. T². P l-Oct. S Nov-Jan.

RED SAUCE 6 M D
USA; raised 1910 by R. Wellington, NYSAES, Geneva.
Deacon Jones X Wealthy. Introd 1926.
Pink stained flesh when really ripe. Cooks to sharp, well-flavoured pink purée. Makes attractive pink jelly.
F 16. T². C gd. P m-Sept. S Sept-Nov.

REDSLEEVES 6 E C
UK; raised by Dr F. Alston, EMRS, Kent.
Exeter Cross X (scab resistant seedling). Introd 1986.
Sweet, lightly aromatic, crisp, juicy flesh; can be weakly flavoured.
F 11; frost res. T²; some res scab, mildew. C hvy. P l-Aug. S Sept.

REDSTART 6 E D
UK; chance seedling, arose c1950 in Oxford garden.
Sweet, juicy, plenty of fruit, acidity; soon going soft.
F 11. T². C hvy. P use e-Aug.

RED STATESMAN *see* Statesman

RED SUDELEY *see* Lady Sudeley

RED SUPERB (Fox-Den) *see* Laxton's Superb

RED TÖNNES syn Rød Tønnes

RED TRANSPARENT 6 E D
Russia prob; recorded 1872.
Brisk, crisp, white flesh, with strong acidity; quickly going soft.
F 3. T². P use e-Aug.

RED VICTORIA 6 E C
UK; found c1884 nr Wisbech, Cambs.
Introd by Messrs Miller, Wisbech. RHS AM 1908; FCC 1910.
Large; dark red flushed. Cooks to sharp froth.
F 9. T². P use Aug.

REDWELL 4 M D
USA; raised c1911 at Univ Minnesota Fruit Breeding Farm, Excelsior, Minnesota.
Scott's Winter X. Introd 1946.
Sprightly, crisp, juicy. Dual purpose US.
F 15. T²; hrdy. P l-Sept. S Oct-Nov.

RED WINDSOR *see* Red Alkmene

REDWING 6 M D
UK; prob raised by Charles Ross, HG at Welford Park, Newbury, Berks. Recorded 1908.
Bright cherry red flush. Crisp, juicy, but little flavour; low in sugar, acidity; tough

skin.
F 8. T². P l-Sept. S Oct-Nov.

RÉGALI *see* Delkistar

REGENT 6 L CD
USA; raised 1924 University Minnesota, Fruit Breeding Farm, Excelsior.
Daniels Red Duchess X Delicious. Introd 1963.
Honeyed, plenty of acidity; crisp, crackling, juicy flesh. Cooked, keeps shape; sweet, fruity; delicately flavoured. Developed for northern regions.
F 11. T²; hrdy; res cedar apple rust. P m-Oct. S Nov-Jan.

REID'S SEEDLING 6/7 M D
Ireland; raised c1880–90 by Mr Reid of Drumart Jones, Richill, County Armagh.
Boldly striped in carmine. Sweet, light flavour, hint of strawberries; pink stained flesh.
F 16. T²; reported res scab. P l-Sept. S Oct-Nov.

REINE DES REINETTES syn King of the Pippins

REINE SOPHIE syn Königin Sophie-nsapfel

REINETTE
Name given to group of varieties, characterised by good flavour, late keeping and russeting on surface. Most originated in France but name used by pomologists to include other varieties with these properties. Name prob first recorded 1540 by Charles Estienne.

REINETTE À LA REINE 3 L DC
Belgium; local Flanders variety, syn Wijning; known 1940s.
Large. Quite sweet, juicy, plenty of fruit, crisp to crumbling texture; little culinary flavour. Vit C 17mgs/100g.
F 11. T². P m-Oct. S Nov-Jan.

REINETTE À LONGUE QUEU (?) 2/5 L D
Received from France; may be variety recorded 1831, but insufficient details available to confirm.
Long stalk; markedly conical. Brisk, crisp in Feb, some sweetness and fruit.
F 14. T²; sprd. P l-Oct. S Jan-Mar.

REINETTE ANANAS syn Ananas Reinette

REINETTE CLOCHARD syn Clochard 8 L DC
France; poss known mid-1800s; desc 1948 (*Verger Français*).
Intense, rich, quite aromatic flavour; firm, cream flesh.

Formerly one of best-known market apples, widely grown – traditionally in north west, but also south west; still found Vendée (Western Loire), Charente, Deux-Sèvres (Poitou-Charentes). Valued also for cooking; recommended for 'Crêpes à la Normande', 'Tarte aux Pommes'.
F 11; trip. T²; sp; gd res scab. P m-Oct. S Nov/Dec-Mar.

REINETTE COULON 7 L D
Belgium; raised c1850 by nursmn Coulon, Liège. Fruited 1856.
Large. Rich with almost pineapple-like acidity; quite intense flavour. Crisp, cream flesh. Dual purpose in Belgium; grown as orchard standards in Liège province; also known Normandy.
F 6. T². P e-Oct. C prn bitter pit. S Nov-Dec/Jan.

REINETTE COURTHAY 6 L D
Received 1948 from Switzerland.
Refreshing, quite sweet, plenty flavour of fruit; crisp, juicy.
F 12. T³. P l-Oct. S Nov-Mar/Apr.

REINETTE D'AMÉRIQUE 1/3 L C
Received 1947 from France.
Brisk, crisp, juicy, but often tannin undertones.
F 12. T². P m-Oct. S Dec-Mar.

REINETTE D' ANJOU 7 L D
France or Germany; mentioned 1817 Germany, 1848 France.
Resembles Golden Reinette. Quite rich, lightly aromatic.
F 11. T². P e-Oct. S Nov-Mar.

REINETTE D'ANTHÉZIEUX (?) syn Reinette de Demptézieu 2/5 L D
France; acc may be variety desc 1948 (*Verger Français*); long grown Demptézieu (Isère); distributed by nursmn Babout of Thoissey, Ain (Rhône).
Brisk, crisp, juicy, plenty of fruity taste in Dec.
F 18. T³. P e-Oct. S Dec-Mar.

REINETTE D'ARMORIQUE 8 L D
France; acc variety now known Brittany, where prob arose; desc 1948 (*Verger Français*).
Sharp, firm, but mellowing; said to become slightly aromatic. Also known small extent throughout area north of Loire, Massif Central, Massif Armoricain. Formerly sold Paris markets as Canada de Bretagne, to distinguish it from Reinette du Canada.
F 18. T³. P l-Oct. S Dec/Jan-Mar.

REINETTE D'AUTOMNE DE WILK-ENBURG syn Wilkenburger Herbstreinette

REINETTE DE BAILLEUL syn Gros-Hôpital

REINETTE DE BRUCBRUCKS 7 L D
Received from France c1947.
Large, handsome. Good rich, intense flavour, aromatic with pineapple acidity; coarse texture. Bears burrs, prob roots from cuttings.
F 10. T^3. P e-Oct. S Nov-Feb/Mar.

REINETTE DE BURCHARDT syn Burchardt's Reinette

REINETTE DE CAUX syn Dutch Mignonne

REINETTE DE CHAMPAGNE 1/5 L C
France; prob arose c1770 in Champagne, but widely distributed over Europe. Syns many.
Markedly flat-round. Quite sharp, soft, juicy flesh. Cooked, keeps shape; juicy, sweet, but not bland.
Formerly widely grown northern Europe, also Italy. Remains valued France, Belgium, Germany where grown mainly for juice.
F15. T^2; sp. C gd; bien. P l-Oct. S Dec-Mar.

REINETTE DE DAMASON syn Reinette de Mâcon

REINETTE DE DEMPTÉZIEU syn Reinette d'Anthézieux

REINETTE DE FRANCE 7 L D
Belgium (prob); Syn Court-Pendu de Tournay, Reinette d'Orléans. Desc 1853 by A. Bivort, when known around Tournai, Belgium.
Resembling Court Pendu Plat in appearance and flavour. Intensely, sweet-sharp with pineapple-like acidity. Mellowing, but still richly flavoured in Feb; firm, deep cream flesh. Grown Belgium, northern France.
F 31; trip. T^2. P l-Oct. S Nov-Mar.

REINETTE DE GEER 5 L D
Belgium; raised by fruit breeder Jean-Baptiste Van Mons; fruited 1815. Named after Baron de Geer.
Rich, intensely flavoured; reminiscent of Golden Reinette.
F 11. T^2. P e-Oct. S Nov-Feb/Mar.

REINETTE DE GRANVILLE (?) 5 L D
France; acc prob variety desc Mas (1865–74), which arose Granville, Manche, Normandy; first recorded 1842.
Sharp, fruity, firm; mellowing with keeping.
F 16. T^3. P l-Oct. S Dec-Mar.

REINETTE DE LUCAS 6 L D
Belgium prob; believed raised by fruit breeder de Jonghe; recorded 1872. Named after German pomologist Eduard Lucas.
Deep pinky maroon flush. Slightly vinous

flavour.
F 11. T^2. P e-Oct. S Nov-Jan/Mar.

REINETTE DE MÂCON syn Reinette de Damason, Gris Braibant 8 L D
France; prob arose Mâcon, Saône-et-Loire (Burgundy). Leroy believed that the Reinette de Damason he grew in 1868 was 'Double Reinette de Mascon' recorded 1628 by Le Lectier.
Darkly russeted over slight red flush. Quite richly flavoured, fairly sharp 'crumbly', rather dry flesh.
Widely distributed over northern France and Belgium, principally Liège, where known now as Gris Braibant, RGF variety selected for cultivation by Agric Res Centre Gembloux due to resistance to disease.
F 9; trip. T^2; gd res scab. P e-Oct. S Dec-Mar.

REINETTE DE MAURS 2 L D
France; desc 1948 (*Verger Français*) when grown Maurs, Cantal (Auvergne).
Sweet, rich, but modest flavour; firm flesh.
F 16. T^2. P l-Oct. S Dec-Feb/Apr.

REINETTE DE MAUSS syn Mauss Reinette

REINETTE DE METZ 7 L D
France (prob); acc prob variety desc 1948 (Vercier).
Like sweetmeat; sweet, quite rich, firm, cream flesh.
F 7. T^2; sp; sprd. P e-Oct. S Nov-Dec/Jan.

REINETTE DE PLOUERC 6 L D
Received from France 1948.
Purplish red flush. Brisk, flavour of fruit, some sweetness; firm, quite juicy flesh.
F 16. T^2; sp. P l-Oct. S Nov-Jan.

REINETTE DE PLUVIGNER
France; received 1948 from Morbihan and prob arose Pluvigner, Brittany.
Hard, quite sharp. Cooks to purée; light flavour.
F 14. T^2. P m-Oct. S Nov-Jan/Mar.

REINETTE DE PONS see Reale d'Entraygues

REINETTE DE RAFFRAY 5 L D
France; received 1948 from Côte du Nord, Brittany.
Sweet, quite savoury; firm flesh.
F 19. T^1. C hvy. P l-Oct. S Jan-Mar.

REINETTE DESCARDRE 7 L D
Belgium; raised c1820 by nursmn Benoît Descardre of Chênée. Introd c1834.
Acc is v similar to Blenheim Orange.
F 12, trip. T^3. P l-Sept/e-Oct. S Oct-Dec.

REINETTE DES VERGERS syn Luxemburger Reinette

REINETTE D'ETLIN syn Etlins Reinette

REINETTE DE VERSAILLES 8 L D
France; syn Reinette Grise de Champagne, which arose c1730, prob in Champagne and well known around Paris by 1750.
Sweet, quite scented; firm to crumbly rather dry flesh.
F 8. T^3. P e-Oct. S Nov-Feb.

REINETTE DE ZUCCAMAGLIO syn Von Zuccamaglios Renette

REINETTE DORÉE DE BOEDIKER syn Bödikers Gold-Reinette

REINETTE D'OSNABRÜCK syn Osnabrücker Reinette

REINETTE DUBUISSON 5 L D
Received from France 1950.
Sweet, plenty of acidity, quite juicy.
F 16; trip. T^3. P m-Oct. S Nov-Jan/Mar.

REINETTE DU CANADA 8 L DC
France (prob); poss arose Normandy. Origins obscure but mentioned under this name 1771. Widely distributed throughout Europe; introd North America by early 1800s. Syns, numerous – 152. RHS AM 1901.
Cooked, keeps shape or makes stiff purée; sweet, quite rich. Best used early for cooking, when acidity is high. By Dec, sweeter with crumbling flesh, rather like Blenheim Orange.
Favoured exhibition variety in England; often found in old gardens. Still widely grown in Europe and esteemed in France for making tarts and very late eating apple. Vit C 17mg/100g.
Frt Col slight orng flush over grnish yell/gold; much russet patches, netting; lenticels conspic as star shaped dots. *Size* med/lrg. *Shape* rnd-con to oblng; rnded ribs, flt sides; 5 crowned. *Basin* brd, dp; russet lined, ribbed. *Eye* lrg, prt to fully open; sepals qte downy. *Cavity* brd, dp; russet lined. *Stalk* shrt, thck. *Flesh* pale crm.
F* 14; trip. T^3. C gd. P m-Oct. S Nov-Mar.
  REINETTE GRISE DU CANADA; more uniformly russeted.

REINETTE DU MANS see de Jaune

REINETTE GRISE DE BILLON 8 L D
Received 1948 from Puy de Dôme (Auvergne), France.
Small. Quite brisk, some flavour; cream tinged green flesh.
F 11. T^2. P m-Oct. S Nov-Jan.

REINETTE GRISE DE CHAMPAGNE syn Reinette de Versailles

REINETTE GRISE DE LA CREUSE
5/8 L D

France; received 1950; pres arose Creuse (Limousin).

Dense taste of fruit, plenty of sugar, acid; deep cream flesh. Mellows to become sweeter, almost aromatic.

F 25. **T**^{2}. **P** l-Oct. **S** Nov-Feb/May.

REINETTE GRISE DE PORTUGAL (?)
8 L D

France (prob); first recorded France by Duhamel in 1768, but much confused with German Lederapfel. Acc may be Reinette Grise de Portugal of C19th French pomologists.

Sharp, fruity in Jan; firm, leathery, cream flesh.

F 6. **T**^{2}; sprd. **P** m-Oct. **S** Jan-Mar.

REINETTE GRISE DE SAINTONGE
8 L D

France; desc *Pomologie de la France* (1863-73). Old variety, long known around Bordeaux; takes its name from old province of Saintonge, along the Gironde (now Charente-Maritime and Charente) where widely grown C19th and earlier. French pomologists now conclude that Leroy incorrectly made it synonymous with Haute Bonté, which is a different apple.

Quite rich flavour; firm, rather leathery flesh. Remains well known France.

F 13. **T**^{3}. **P** m-Oct. **S** Dec-Mar.

REINETTE HARBERT syn Harberts Reinette

REINETTE MARBRÉE 8 L D
Netherlands; as syn 'à Caractères' described 1760 by Dutch botanist, Knoop.

Cobweb of russet accounts for name 'marbled' or Character Apple. Light flavour, quite sweet, but low in fruit, acidity; fairly soft, juicy.

Grown Auvergne, Creuse (Limousin), Indre (Val de Loire); used also juice, cooking.

F13. **T**^{2}. **P** m-Oct. **S** Nov-Jan/Feb.

REINETTE ROUGE ÉTOILÉE 6 L D
Belgium or Netherlands; known since 1830s, when noticed in Maastricht (Netherlands), Liège and Sint Truiden (Belgium). Syns include Calville Étoilé; Early Red Calville (Hogg), Sterappel.

Vivid flush and star shaped russet freckles accounts for name. Intense, quite sharp flavour of raspberries; juicy firm flesh, often stained pink under skin and around core. With keeping, becomes drier, sweeter and flavour seems almost distilled into raspberry essence. Bunyard, however, found only 'slight strawberry flavour'.

Formerly widely grown in Europe; small extent England; grown all over Belgium up to 1950s, now produced for luxury Xmas trade; remains valued also Netherlands; north France, especially Aisne (Picardy).

Frt Col dark rd flush, pinky rd at edges; gold background; rd stripes visible edge of flush; many star-shaped russet dots. *Size* med. *Shape* rnd-con to rnd; regular. *Basin* qte brd, med dp. *Eye* lrg, open; sepals nrw, lng, sep at base. *Cavity* nrw, shal. *Stalk* shrt/med, thck. *Flesh* wht, rd tinge.

F* 14. **T**^{2}; uprt; part tip; res scab, mildew. **C** hvy, bien. **P** l-Sept/e-Oct. **S** Oct-Nov. PLATE 18

REINETTE SANGUINE DU RHIN
7 L D

Pres arose on Rhine. Desc c1840.

Deep carminc to purple flush. Quite sweet; plain fruity taste.

F 15. **T**^{2}. **P** l-Oct. **S** Dec-Mar.

REINETTE SIMIRENKO 2/5 L D
Ukraine; arose on farm of horticulturalist Platon F. Simirenko. Desc 1895; also claimed to be American variety, Wood's Greening.

In England, savoury taste, quite soft flesh, but in California 'fragrant, very juicy, pleasantly vinous'.

Widely grown in Ukraine and popular Russian variety; now taken up by amateurs in California.

F 10. **T**^{1}. **P** m-Oct. **S** Dec-Mar/May.

REINETTE THOUIN 2/8 L D
France; arose in garden of M. Gillet de Laumont, nr Montmorency, north of Paris. Fruited 1822.

Named after botanist André Thouin, Director of Jardin des Plantes, Paris.

Sharp, fruity in Jan.

F 15. **T**^{2}. **P** l-Oct. **S** Jan-Mar/Jun.

REINETTE VAN BERK'S 6 L D
Received from France 1947.

Quite intense raspberry flavour develops; firm, rather coarsely textured flesh.

F 10. **T**^{2}. **P** l-Oct. **S** Dec-Feb/May.

REINETTE VAN EKENSTEIN 5 M D
Netherlands; raised c1830 by Jhr Albarda van Ekenstein, Appingedam.

Sweet, slightly honeyed; firm, cream flesh. Remains valued Netherlands.

F 9. **T**^{2}. **P** m-Sept. **S** Oct-Nov.

REINETTE VERTE (?) 2 L D
France; acc may be variety desc 1948 (*Verger Français*), when grown Haute-Vienne (Limousin) and Pyrénées-Atlantiques (Aquitaine).

Brisk, savoury in Jan, quite firm flesh.

F 13. **T**^{2}. **C** hvy; frt sml. **P** l-Oct. **S** Jan-May.

REINETTE WEIDNER syn Weidners Goldreinette

REINETTE VON ZUCCALMAGLIO
syn Von Zuccalmaglio's Renette

RENETTA GRIGIA DI TORRIANA
8 L D

Italy; arose Torriana di Barge, Cuneo, West Piedmont. Distributed c1905.

Cinnamon russet over gold. Scented, sweet, hardly any acidity, but 'aromatic' grown in Italy. Torriana is long famed for its fruit, where variety was popular in 1920s, but no longer grown.

F 15. **T**^{2}; gd res scab. **P** l-Oct. **S** Nov-Feb/Apr.

RENORA 6 L D
Germany; raised by C. Fischer, H. Murawski, Inst Fruit Res, Dresden-Pillnitz. Clivia X seedling of *Malus floribunda* (carrying scab resistant V_f gene). Introd 1996.

Deep red flush over gold. Crisp solid deep cream flesh; plenty sugar, acid and fruit, but lacks distinction, can be sour.

F est 10. **T**^{2}; res scab. **C** hvy. **P** mid-Oct. **S** Nov-Jan; commercially-May.

RENOWN 6 M D
UK; raised by Charles Ross, HG, Welford Park, Newbury, Berks.

Peasgood's Nonsuch X Cox's Orange Pippin. Desc 1908. RHS AM 1908.

Lightly aromatic, sweet, quite juicy; like dilute Cox.

F 14. **T**^{2}. **P** l-Sept. **S** Oct-Nov.

RESISTA syn Golden Resista

RETINA 6 M D
Germany; raised by C. Fischer, M. Fischer, H. Murawski, Inst for Fruit Res, Dresden-Pillnitz.

Apollo (Cox's Orange Pippin X Oldenburg) X seedling of *Malus floribunda* (carrying scab resistant V_f gene) Introd 1991.

Crisp, juicy, sweet, pleasant.

F est 10. **T**^{2}; res scab. **C** hvy. **P** l-Aug. **S** Aug-Sept.

REVEREND W. WILKS 5 E C
UK; raised by J. Allgrove, manager of Langley Nurseries, Slough of Messrs James Veitch.

Believed Peasgood's Nonsuch X Ribston Pippin. RHS AM 1904; FCC 1910. Introd 1908.

Named after distinguished RHS secretary and vicar of Shirley parish, South London, where he raised first Shirley poppies. Cooks to pale lemon purée, light, quite sweet. Baked, juicy, soft and translucent; hardly needing sugar.

Valued exhibition apple; remains favourite garden variety. Also valued France – grown Vienne, Deux-Sévres (Poitou-Charentes), Vendée, Maine-et-Loire (Western Loire) Recently introd to Japan, to Obuse, Nagano prefecture.

260

Frt *Col* pale crm mottled in light orng, splashed rd stripes. *Size* lrg. *Shape* con; rndded; regular. *Eye* lrg, clsd. *Basin* brd, dp; sltly ribbed. *Cavity* brd, dp; ltl russet. *Stalk* shrt, thin. *Flesh* wht.
F* 7. T². C gd, bien. P l-Aug/e-Sept. S Aug-Nov.

REWENA 4 L D
Germany; raised by H. Murawski, C. Fischer, M. Fischer at Inst for Fruit Res, Dresden-Pillnitz.
(Cox's Orange Pippin X Oldenburg) X seedling of *Malus floribunda* (carrying scab resistant V_f gene). Introd 1991.
Pleasant taste of fruit but can be rather sour.
F est /10. T²; res scab, mildew, fireblight. C hvy. P e-Oct. S Oct-Dec; commercially-Feb.

RHEINISCHER KRUMMSTIEL 3 L C
Germany; desc 1821.
Cooked, keeps shape; fairly brisk taste.
'Much grown between Cologne and Bonn' in C19th.
F 15. T². C hvy. P l-Oct. S Dec-Mar.

RHODE ISLAND GREENING 1 L CD
USA; prob arose nr Green's End, Newport, Rhode Island where inn was kept by Mr Green. Well known by early 1700s. Sent to London Hort Soc and introd Europe early 1800s.
Cooked, keeps shape; sweet, plenty of briskness and flavour. Mellowing to quite rich, but rather large, eating apple.
Widely grown in C19th North America, highly esteemed for pies and sauce and as dessert fruit for 'sprightliness of its abundant juice, and delicacy of fine flavour'.
Major market apple of eastern states up to 1930s; still grown and valued. Never achieved great popularity outside US.
Frt *Col* bright grn/yell, slight brownish rd flush; lenticels conspic as lrg russet dots. *Size* lrg/v lrg. *Shape* oblong-con; sltly ribbed, crowned. *Basin* med brd, dp; ribbed, puckered; ltl russet. *Eye* clsd or prt open; sepals lng, downy. *Cavity* brd, dp; some russet; some scarf skin. *Stalk* shrt qte thck. *Flesh* pale crm.
F 9; trip. T³; sprd. C gd. P e-/m-Oct. S Dec-Apr.
RHODE ISLAND GREENING tetrap.

RIBONDE 5 M D
Received from France 1947.
Brisk, good plain taste of fruit; rather dry flesh.
F 24. T². P e-Oct. S Oct-Nov/Dec.

RIBSTON PIPPIN 7 L D
UK; believed raised c1707 from pip brought from Rouen by Sir Henry Goodricke to Ribston Hall, nr Knaresborough, Yorks. Listed 1769 by nursmn William Perfect of

Pontefract, Yorks; well known by early 1800s. Introd early 1800s to Europe, N America, later to Antipodes. Syns many. RHS AM 1962; AGM 1993. Original tree died 1835, but shoot grew up from roots to give another tree which lived until 1928 when blown over and died c1932.
Intense, rich, aromatic flavour; juicy, firm deep cream flesh. More acidity than Cox, not as delicate, but stronger with great depth and length. Most highly esteemed Victorian dessert apple.
Grown gardens all over country, also for market around London, Kent and elsewhere in C19th. By 1890s declining in popularity, minor market variety by 1930s, but remains valued garden variety. Exported to UK from Canada, Australia and New Zealand up to 1930s. Remains valued Swedish apple.
Frt *Col* brownish orng flush, rd stripes over yell/grn; ripening to brighter rd over gold; some russet patches. *Size* med. *Shape* rnd-con to oblng-con; qte ribbed, flt sided; sltly 5 crowned; can be irregular, almost misshapen. *Eye* lrg, open or hlf open; sepals qte lng, brd base; v downy. *Basin* brd, dp; ribbed, puckered; ltl russet. *Cavity* nrw, dp; russet lined. *Stalk* shrt/med, qte thin. *Flesh* pale yell. Characteristic, fat, downy fruit buds.
F* 8; trip. T²; uprt; res scab, prn mildew, canker. C gd. P l-Sept/e-Oct. S Oct-Jan.
PLATE 20

RICHARDSON (IRELAND) 5 M D
Ireland; brought from orchard of Richardson, Bessborough, Kilkenny to England c1930 by T. E. Tomalin.
Brisk savoury taste, but astringent quality.
F 12. T¹; sprd. P l-Sept. S Oct-Nov/Dec.

RICHARDSON (USA) 6 M D
USA; discv 1956 by L. Richardson at Hammond, New York.
Poss Northern Spy or St Lawrence X. Introd 1959 by nursmn F. L. Ashworth, Heuvelton, NY.
Quite sweet, firm flesh, but 'rich, mellow flavour' in New York.
F* 9. T³. P e-Sept. S Sept-Oct.

RICHARED DELICIOUS *see* Delicious

RINGER 5 M C
UK; recorded 1864; desc Hogg.
Cooks to brisk, juicy purée; lightly flavoured.
F 9. T². P e-Sept. S Sept-Oct.

RINGSTAD 6 E D
Sweden; arose at Ringstad mansion, nr Norkoping, Ostergotland, mid-Sweden. Believed raised c1818 by owner, Notary Frans Rundberg; originally called Notary's Apple.
Dark red flushed over pale yellow, with

bloom. Soft, juicy, lightly flavoured; can be rather metallic in England.
Long esteemed in Sweden; still grown in gardens in mid-Sweden; also Finland.
F 10. T². C gd. P eat Aug.

RIVAL 4 M D
UK; raised by Charles Ross, HG, Welford Park Gardens, Newbury, Berks.
Peasgood's Nonsuch X Cox's Orange Pippin. RHS AM 1900.
Introd by Clibrans, Altrincham, Ches.
Crisp, firm, juicy, but rather bleak and empty compared with Cox. Also recommended for cooking; keeps shape, but very lightly flavoured.
Popular in 1920–30s when also grown for market.
F 11. T²; res red spider. C gd, bien. P l-Sept. S Oct-Dec.

RIVERS' EARLY PEACH 5 E D
UK; raised by Thomas Rivers & Sons, Sawbridgeworth, Herts. Introd 1893.
Similar appearance to Irish Peach, but firmer fleshed. Eaten early, strong peach-like taste; soon becoming very sweet, little acidity, rather bland.
F 4. T³. P eat e-Aug.

RIVERS' NONSUCH 6 M D
UK; selected mid-1800s by nursmn Thomas Rivers as rootstock; desc Hogg; widely used C19th.
Fruit quite pleasant to eat.
F 10. T². P m-Sept. S Sept-Oct.

ROANOKE 6 L D
USA; raised 1949 by G. D. Oberle, Virginia Polytechnic Inst, Blacksburg.
Red Rome X Schoharie (Northern Spy X). Introd 1967.
Named after Virginian town. Crisp, juicy nearly yellow flesh, often rather weak watery flavour in England.
F 20. T². C hvy. P e-Oct. S Oct-Dec.

ROBA *see* Erwin Bauer

ROBERT BLATCHFORD 5 L C
UK; raised by F. Chilvers, nursmn of Hunstanton, Norfolk.
Believed Blenheim Orange X Rev. W. Wilks. Introd 1914.
Pale milky yellow, like Reverend W. Wilks. Cooked, keeps shape; brisk, light flavour.
F 16. T²; sprd. P e-Oct. S Oct-Dec.

ROBIN PIPPIN 6 E D
UK; arose 1951 with R. & B. Wickham, Gate House Farm, Brenchley, Kent.
Bright red Worcester colour, some of its strawberry flavour, but ripening earlier.
F 11. T²; prt tip. C gd. P l-Aug. S Aug-Sept.

ROCK 5 M C
UK; received 1947 from Scotland; may be variety exhib 1883.
Small. Quite sharp, crisp, white flesh.
F 8. T^2; sprd. P e-Sept. S Sept-Nov.

RODE WAGENAAR 6 L D
Netherlands; raised at IVT, Wageningen. Received 1963.
Large, dark, red flush. Crisp, refreshing; little flavour.
F* 6. T^2. P m-Oct. S Nov-Jan.

RØD TØNNES syn Red Tönnes 6 L D
Denmark; arose with Rasmus Roed c1942 at Kolbjerk Mark. Sport of Tönnes.
Quite rich, plenty of sweetness, acidity and fruit; crisp, juicy.
F 9. T^2. P m-Oct. S Nov-Feb/Mar.

ROGERS McINTOSH *see* McIntosh

ROKEWOOD 6 L D
Australia; arose 1870s with John Bullock at Dereel, Rokewood district, nr Ballarat, gold rush city of Victoria. Syn Bullock's Seedling.
Usually not ripening fully in England; sharp, little flavour.
Grown most Australian states in 1950s; exported to UK, but now overtaken.
F 6. T^1. C hvy. P l-Oct. S Jan-Mar/May.

ROME BEAUTY 4 L D
USA; arose as shoot from seedling rootstock of grafted apple on farm of Zebulon and Joel Gillett, Rome Township, Lawrence County, Ohio.
Tree had been bought 1817 from Israel Putnam, nursmn, Marietta. Introd 1848 at Ohio Fruit Convention. Introd early 1900s to Europe, Australasia.
Famed for its colour and storage qualities, but only mildly flavoured and in US considered 'passable' for cooking. Usually does not colour or ripen well in England. Formerly widely grown in southern States; recently regained prominence mainly for processing.
Grown Italy, Australia, New Zealand, Brazil; needs no winter chilling, and succeeds in tropics – grown Indonesia.
Frt *Col* deep rd flush, some rd stripes, grnish yell background; brightens and lightens as matures; lenticels qte conspic as wht dots. *Size* med/qte larg. *Shape* rnd to rnd-con; flt base; trace ribs at apex. *Basin* brd, qte shal; sltly puckered. *Eye* lrg, hlf /fully open; sepals, lng, grn. *Cavity* brd, qte dp; grey/grn russet lined. *Stalk* lng, med thck. *Flesh* pale crm.
F 19. T^2; res red spider. C hvy. P m-Oct. S Dec-Apr.

> **BARKLEY RED ROME;** more highly coloured. Discv 1944 by G. L. Barkley, in Manson, Washington. Introd 1953.
> **DOUBLE RED ROME BEAUTY;**

even flush, stripes. Discv c1920 by E. A. Cowin, Wapato, Washington; introd 1927.
> **GLENGYLE RED;** more highly coloured. Arose 1914 at Balhannah, South Australia.
> **NURED ROME;** syn SeeandO Red Rome; solid red flush. Discv 1941 by J. A. Snyder in Wenatchee, Washington; introd 1943.
> **RED ROME (AUSTRALIA);** redder flush. Discv Barnley; received 1950.
> **RUBY ROME BEAUTY;** syn Black Rome; solid red flush. Discv c1912 at Milton-Freewater, Oregon; introd 1928.

ROSA DEL CALDARO syn Kalterer Böhmer 6 L D
Italy; arose district Caldaro, Trentino, South Tyrol. Desc 1889.
Dark red flushed. Quite rich, sweet, juicy; tough skin.
Speciality of Bolzano, Trentino and still grown.
F 7. T^2; sp. P e-Oct. S Nov-Jan.

ROSA DU PERCHE (?) 6 L D
France; acc may be variety desc 1948 (*Verger Français*), when grown Indre (Val de Loire).
Bright red flush. Brisk refreshing taste of fruit; melting, white flesh.
F 12. T^2; sp. P e-Oct. S Nov-Feb/Mar.

ROSANNE 6 M D
Australia; arose with Mr Waller of Hobart, Tasmania.
Devonshire Quarrenden X Democrat.
Deep maroon flush. Sweet, scented; soft, juicy, pink stained flesh; often green, woody taste in England.
F 8. T^2. C lght. P m-Sept. S Oct-Nov.

ROSE DE BÉNAUGE 4 L D
France; prob arose C19th; traditional variety of Gironde, Lot-et-Garonne (Aquitaine), where still grown.
Crisp, juicy, pleasantly fruity.
F 22. T^2. P l-Oct. S Jan-Mar/May.

ROSE DE BERNE syn Berner Rosen

ROSE DE BOUCHETIÈRE 8 L D
Received 1948 from Isère (Rhône), France.
Scented, quite sharp, but plenty sugar in Feb; firm, white, tinged green flesh.
F 9; trip. T^3. P l-Oct. S Jan-Mar/May.

ROSE DE SAINT-FLORIAN syn Florianer Rosenapfel

ROSE DE STAEFA syn Stäfner Rosen

ROSE DOUBLE 4 L D
Received from France 1948.
Small; red flushed over lemon. Brisk refeshing taste of fruit; crisp.

F 26. T^2. P m-Oct. S Nov-Feb.

ROSEMARY RUSSET 7 L D
UK; arose England. Desc 1831 by nursmn Ronalds of Brentford, Middx. RHS AGM 1993. Intense, sweet-sharp taste of acid drops; not as sugary as Ashmead's Kernel, but Nonpareil family. 'One of best late sorts' claimed Bunyard. Remains popular garden apple.
Frt *Col* orng or rdish brown flush over grnish yell/gold; russet as patches, dots, veins, usually at apex. *Size* med. *Shape* con; flt sided, ribbed; sltly 5 crowned; flt base; often irregular. *Basin* med wdth, dpth; ribbed, sltly puckered; russet present. *Eye* clsed or sltly open; sepals, brd based. *Cavity* qte brd, qte dp; russet lined. *Stalk* lng, thck. *Flesh* pale crm.
F* 10. T^2; uprt. C gd. P e/m-Oct. S Nov/Dec-Mar.
PLATE 26

ROSE ROUGE 6 L D
Received 1950 from France.
Sprightly savoury taste of fruit; firm, crisp, white flesh.
F 12. T^2. P l-Oct. S Dec-Feb/Mar.

ROSIOARE CĂLUGĂREŞTI 6 L D
Romania; local to montainous zone of Moldavia, where widely grown 1960s in gardens, roadsides, field boundaries.
Strong, brisk taste of fruit, almost rich.
F 6. T^2. P l-Oct. S Dec-Feb/Mar.

ROSMARINA BIANCA syn Weisser Rosmarin 5 L CD
Italy; arose Bolzano, South Tyrol. Known early 1800s. Widely distributed over Europe. Syns many.
Named for its aroma of rosemary, but this is not obvious in England. Tall, conical, pale gold. Firm, sweet flesh. Cooked, keeps shape; mildly flavoured.
F 14. T^2; sp. P m-Oct. S Nov-Jan.

ROSSIE PIPPIN 1/3 M C
UK; received from G. Bunyard Nursery, Kent. Catalogued 1890.
Large. Quite sharp. Cooks to purée, but rather flavourless.
F 11. T^3. P e-Sept. S Sept-Oct.

ROSS NONPAREIL 8 L D
Ireland; known 1802 in County Meath. Sent 1819 by nursmn John Robertson of Kilkenny to London Hort Soc.
Ruddy cheeked, russeted. Intense, sweet-sharp taste, reminiscent of Ashmead's Kernel and Nonpareil family. Good Oct to early Nov; becoming drier, almost scented.
Re-popularised early 1900s by Bunyard's Nursery, Kent, as 'perfect shape, pretty size and appearance for dessert, much relished by those who cannot digest a crisp, hard, apple'.

F 8. T^2. C gd. P l-Sept. S Oct-Dec.

ROSSO DEL POVERO 4 L D
Italy; received 1958 from Turin Univ.
Sweet, juicy, slightly scented, white tinged green flesh; usually undeveloped in England.
F 13. T^1. P m-Oct. S Nov-Jan/Feb.

ROSY BLENHEIM 7 L D
UK; raised by F. W. Thorrington, retired London Customs Officer, Hornchurch, Essex.
Cox's Orange Pippin X. Received 1925.
Large. Resembles Blenheim Orange; plain taste, deep cream, crumbling flesh.
F 11. T^2. P e-Oct. S Nov-Jan.

ROTE GOLDPARMÄNE see King of the Pippins

ROTER ANANAS syn Ananas Rouge 6 M D
Germany; claimed found at Sieglitzerberg on Elbe by Richter, gardener to Court at Dessau, Saxony. Desc 1854 by Lucas.
Beautiful, bright scarlet flush. Strong pineapple acidity eaten fresh from the tree, but mellows after storing to definite strawberry flavour; pink stained flesh under skin.
F 14. T^2. P l-Sept. S Oct-Nov/Dec.

ROTER EISERAPFEL 6 L D
Germany; arose early 1700s. Widely distributed over Europe by C19th. Syns numerous, incl Eiser Rouge.
Very attractive, ripens to pinky red flush over pale yellow with deep bloom. Only modestly flavoured; sweet, slightly scented, firm flesh.
F 15; trip. T^3. P m-Oct. S Nov-Feb/Apr.

ROTER MÜNSTERLÄNDER BORS-DÖRFER 6 L D
Germany; received 1951 from Fruit Res St, Jork, nr Hamburg.
Pres arose Münsterländer, Westphalia.
Deep red flush. Sprightly, refreshing taste of fruit; crisp, juicy.
F14. T^2. P l-Oct. S Jan-Mar.

ROTER SAUERGRAUECH syn Sauer-grauech Rouge 6 L D
Switzerland; received 1947. Red sport of Sauergrauech – recorded 1842, prob arose Bern.
Piquant taste of fruit, almost redcurrant or raspberry flavour; plenty sugar, acid; crisp, juicy. Formerly also used for cider in Bern.
F 9. T^2. P e-Oct. S Nov-Dec/Jan.

ROTER STETTINER syn Török Bálint 6 L D
Germany; believed ancient variety; acquired

present name 1776. As syn 'Vineuse Rouge d'Hiver', listed 1598 by Jean Bauhin. As 'De Signeur d'Hiver' catalogued 1628 by Le Lectier. Widely distributed over Europe. Syns numerous – 170.
Dark red flush, ribbed, irregularly shaped. Vinous quality, like a cross between melon and strawberry; in Feb, sweet, quite intense taste, crisp.
F 10. T^2; uprt. P l-Oct. S Jan-Mar/Apr.

ROTE STERNRENETTE syn Reinette Rouge Étoilée

ROUGET 6 M Cider
France; arose Côtes-du-Nord, Brittany. Received 1948.
Brittany cider apple; astringent, firm, white flesh.
F 13. T^2. P l-Sept. S Oct.

ROUGET DE BORNE 6 L D
France; old Cévennes variety; arose Villefort (foot of Mt Lozère), at Pied-de-Borne where Borne joins river Chassezac.
Sweet, scented, nutty, rather like sugared almonds, but not intense flavour; firm, cream flesh.
Grown Aveyron, Lozère; suited to high altitudes, up to 1000m. Traditionally picked on All Saints Day, Nov 1; used for cooking, preserves.
F 17. T^3; sp; claimed gd res pest, disease.
C hvy. P m-Oct. S Nov-Feb/Mar.

ROUGH PIPPIN (CORNWALL) 4 L D
UK; received 1984 from Cornwall, where found by James Evans near Stoke Climsland.
Acc not variety desc Hogg (1884).
Dotted with russet over bright red flush, stripes. Sweet, light, pleasant taste.
Found west country; propagated by cuttings pitchers.
F8. T^2. P l-Sept. S Nov-Dec.

ROUNDWAY MAGNUM BONUM 7 L D
UK; raised at Roundway Park, Devizes, Wilts, by HG, Mr Joy. Recorded 1864. FCC RHS 1864.
Large, prominently ribbed. Definite pear-like flavour. Re-popularised early 1900s by Bunyard's nursery who recommended small fruits for dessert and larger apples for kitchen. Loses flavour when cooked, more attractive fresh.
F 11. T^3; sprd. C lght. P e-Oct. S l-Oct-Jan.

ROUGEMONT (?) 6 E D
UK; acc may be variety raised by nursmn J. Cheal, Crawley, Sussex; recorded 1888.
Scarlet flushed, tall and ribbed. Decorative tree in fruit; apples ripen over long period, but rather sour with soft, green tinged cream flesh.
F mid. T^2. P l-Aug-Sept. S l-Aug-Oct.

ROXBURY RUSSET syn Boston Russet 8 L D
USA; believed arose early 1600s in Roxbury, Massachusetts. Spread to Connecticut; 1796 taken to Putnam Nursery, nr Marietta, Ohio. Syns include Putnam's Russet.
Sweet, quite aromatic; cream, crumbly flesh, slight green tinge.
Most popular russeted apple of New York and New England, especially around Boston and also grown Ohio until modern storage lessened demand for late keepers. Also esteemed in C19th England.
F 12. T^2. C gd. P m-Oct. S Jan-Mar/Apr.

ROYAL GALA see Gala

ROYAL GEORGE 3 L C
UK; raised prob 1840s by George Clark, master bricklayer of East Bridgeford, Notts. Originally Clark's Seedling, renamed late C19th. Acc is variety of Taylor (1946) and of trees still known in Notts.
Large. Brisk, pleasant. Cooks to well-flavoured, sharp lemon purée; creamy, translucent like Dumelow's Seedling, but not as sharp.
F* 7. T^3. P e-Oct. S Oct-Jan.

ROYAL JUBILEE syns Jubilee, Graham 5 M C
UK; raised by John Graham, Hounslow, Middx. Recorded 1888. Introd 1893 by nursmn George Bunyard in Queen Victoria's Jubilee year.
Large. Cooked, keeps shape, quite sharp, slight pear like quality.
Formerly recommended for frost pockets, but now largely forgotten in England. As Graham, valued in Germany and grown gardens and farms in cold areas, foothills of Alps. Used for topping 'Apple Cake', juice, storing as purée and processing. Seed used for rootstocks.
F 22. T^2; spd. C gd. P l-Sept. S Oct-Dec.

ROYAL SNOW 6 L D
UK; received 1933 from Hants.
Red flush over bright yellow, but disappointing flavour. Sweet, light, low in acidity, often sickly green taste.
F 16. T^2. P e-Oct. S Oct-Dec.

ROZOVOE IZ TARTU 6 E D
Estonia; old variety; arose Tartu. Received 1976 from Vavilov Inst, St Petersburg.
'Pink apple from Tartu', blood red flush. Rather insipid in England.
F 1. T^2. P Aug. S Aug-Sept.

RUBENS 7 L D
Netherlands; raised IVT, Wageningen.
Reinette Rouge Étoilée X Cox's Orange Pippin. Recorded 1954.
Rubens red flush. Raspberry flavour of Reinette Rouge Étoilée, but not nearly as intense; sweet, plenty of acidity; firm, rather

dry flesh.
F 11. **T**^{3}; uprt; some res mildew. **C** gd, frt sml unless thinned. **P** e-Oct. **S** Nov-Jan.

RUBIN 6 L D
Czech Republic; raised Univ Agric, Prague. Lord Lambourne X Golden Delicious. Select 1960.
Honeyed, intense sweet, sharp taste; crisp, juicy, but coarse texture.
F 10. **T**^{2}. **C** hvy. **P** e-Oct. **S** Oct-Dec.
   BOHEMIA darker red, less striped sport.

RUBINETTE syn Rafzubin

RUBINSTAR *see* Jonagold

RUBY (SEABROOK) 6 M D
UK; raised by nursmn W. P. Seabrook, Boreham, Essex.
Prob Worcester Pearmain, Rival cross. Received 1936.
Bright red Worcester colour and strawberry flavour; sweet, juicy cream flesh.
F 10. **T**^{2}. **P** e-Sept. **S** Sept-Oct.

RUBY (THORRINGTON) 6 M D
UK; raised by F. W. Thorrington, retired London Customs officer, Hornchurch, Essex.
Cox's Orange Pippin X. Received 1925.
Some strawberry flavour, but inclined to be merely sweet, rather empty, little acidity; firm flesh.
F 11. **T**^{3}. **P** e-Sept. **S** Sept-Oct.

RUBY ROME BEAUTY syn Black Rome

RUNSÈ 4 L D
Italy; received 1958 from Turin Univ.
Intense, rich flavour, plenty of sugar, some acidity, but often rather cloying; firm, quite juicy flesh.
F 13. **T**^{2}; sprd. **P** l-Oct. **S** Dec-Mar/Apr.

RUSSET LAMBOURNE *see* Lord Lambourne

RUSSET SUPERB *see* Laxton's Superb

RUZENA BLAHOVA 4 L D
Czech Republic; raised 1945 by V. Blaha at Libovice, North Bohemia.
Mother (American) X James Grieve.
Sweet, plenty of fruit, slightly scented; soft, juicy, flesh.
Trialled Czech Republic 1960–70s but not introd as short season, often poor colour.
F 8. **T**^{2}. **C** gd; precocious cropping; bien. **P** m-Oct. **S** Nov-Jan.

SABAROS 3 L C
France; arose Ile de Ré, Charent-Maritime (Poitou). Recorded by nursmn Simon-

Louis Frères 1895.
Cooked keeps shape, good fruity taste. Mellows to become sharp eater by Jan. Famed for withstanding salt winds of Il de Ré, island opposite La Rochelle, where it was propagated by suckers.
F 12. **T**^{2}; sp. **C** hvy, frt sml. **P** l-Oct. **S** Nov-Mar/Apr.

SACRAMENTSAPPEL 4 L D
Netherlands; raised at IVT, Wageningen. Received 1955.
Ribbed, misshapen. Intense, rich, aromatic; plenty of sugar, almost pineapple-like acidity. Reminiscent of Holstein, but sharper. Mellows and becomes scented.
F 27. **T**^{2}. **P** m-Oct. **S** Dec-Feb.

SAINT AILRED 4 M D
UK; raised 1942 at Mount St Bernard Abbey, Coalville, Leics.
James Grieve X Ellison's Orange.
Resembles Ellison's Orange in appearance; tastes like ripe, savoury James Grieve. Quite sweet, juicy, soft, cream flesh, but can have underlying metallic taste; rather tough skin.
F 12. **T**^{2}; sprd. **C** frt cracks. **P** e-Sept. **S** Sept.

SAINT ALBANS PIPPIN 4 M D
UK; exhib 1883 by J. Veitch & Sons, Chelsea, London.
Brisk, but plenty of sweetness; crisp, juicy. 'Grown about Brenchley, Kent' in C19th; brought to Hogg's notice by Harrison Weir, the artist.
F 8. **T**^{2}; sprd. **P** m-Sept. **S** Sept-Oct.

SAINT-BAUSSAN 6 L D
France; recorded 1872.
Striking, blood red flush, even darker stripes. Very sweet in New Year, but often metallic undertones; quite coarse texture. Formerly popular north east France.
F 15. **T**^{2}. **P** l-Oct. **S** Nov-Feb/Apr.

SAINT CECILIA 7 L D
UK; raised 1900 by John Basham & Sons, Bassaleg, Mons.
Cox's Orange Pippin seedling. FCC RHS 1919.
Intense, rich, aromatic. Sweet, juicy, crisp, but can be less exciting; rather tough skin. Popular garden variety in west of England.
Frt *Col* dp crimson flush, rd stripes over grnish yell/yell; lenticels as fine grey dots; shiny, smooth skin. *Size* med. *Shape* rnd-con; sometimes ltl lop-sided, sltly ribbed. *Basin* med wdth, qte dp; ribbed; ltl russet. *Eye* clsd, often pinched; sepals lng, brd based. *Cavity* nrw, qte dp; russet lined. *Stalk* shrt, thck. *Flesh* wht, grn tinge.
F 7. **T**^{3}. **C** hvy. **P** e/m-Oct. **S** Dec-Mar.

SAINT EDMUND'S PIPPIN syn Saint Edmund's Russet 8 M D
UK; raised by Mr R. Harvey at Bury St Edmunds, Suffolk. RHS FCC 1875; AGM

1993.
Sweet, juicy, rich; densely textured, pale cream flesh. Picked too early, when green, hard and disappointing, but really ripe can be ambrosial, like pear flavoured vanilla ice cream. Popularised by Bunyard's Nursery, Kent, early 1900s. Grown commercially small extent in East Anglia; remains valued garden variety.
Frt *Col* light russet with silvery sheen over grnish yell/gold ground colour, burnished cheek. *Size* med. *Shape* flt-rnd to oblng-con. *Basin* med wdth, dpth; sltly puckered; russet lined. *Eye* qte sml, clsd; sepals, shrt; v downy. *Cavity* nrw, dp. *Stalk* qte lng, thin.
F 8. **T**^{2}. **C** gd; frt bruises. **P** m-Sept. **S** l-Sept-Oct.
PLATE on front cover

SAINT EVERARD 7 E D
UK; raised by C. Terry, HG, at Papworth Everade, nr Cambridge.
Cox's Orange Pippin X Margil. Introd 1910 by J. Veitch & Sons, Chelsea. RHS AM 1900; FCC 1909.
Handsomely red striped; resembles Margil. Good, rich, aromatic flavour; plenty of sugar, acidity; juicy.
F 9. **T**^{2}; sprd. **P** e-Sept. **S** Sept.

SAINT LAWRENCE 4 M D
Canada; prob arose around Montreal. Known 1835.
Carmine striped, deep bloom. Sweet, crisp, juicy, melting flesh in late Sept; very like McIntosh.
Grown C19th in Saint Lawrence Valley for market.
F 13. **T**^{3}; sprd. **C** frt pr col Eng. **P** m-Sept. **S** Sept-Oct.

SAINT MAGDALEN syn Magdalene

SAINT MARTIN'S 7 L D
UK; raised by nursmn Thomas Rivers or his son at Sawbridgeworth, Herts. RHS AM 1896.
Dusky maroon. Very sweet, tastes of sugar and lemon in late Oct; soft, melting flesh. Becomes even sweeter, drier; more like sugar soaked cotton wool than an apple.
F 12. **T**^{3}. **P** e-Oct. **S** Oct-Dec.

SALOME 4 L D
USA; arose c1853 in nursery in Ottawa, Illinois. Introd 1884 by Arthur Bryant, Princeton, Illinois.
Sweet with good taste of fruit; crisp, quite juicy, deep cream flesh.
F 7. **T**^{2}. **P** e-Oct. **S** Nov-Jan.

SALTCOTE PIPPIN 7 L D
UK; raised by James Hoad at Rye, Sussex. Either Ribston Pippin or Radford Beauty seedling. Recorded 1918. RHS AM 1928.
Large; handsome. Rich, brisk, aromatic; sharper, not as intense as Ribston, but

similar; rather open texture. Becomes sweeter, milder.
F 11. **T**^{3}; uprt. **C** gd. **P** e-Oct. **S** Nov-Jan.

ŠAMPION 6 L D
Czech Republic; raised 1960 by Otto von Louda in Penecin, Bohemia.
Golden Delicious X Lord Lambourne. Introd 1976.
Pleasant, light taste of fruit; sweet, juicy (vit C 8mg/100mg). Grown Czech republic, Germany, Netherlands.
F 8. **T**^{1}. **P** m-Oct. **S** Nov-Jan.

SAM YOUNG 8 L D
Ireland. Exhib 1818 by Earl Mount Norris at London Hort Soc; sent 1819 to Soc by nursmn John Robertson of Kilkenny.
Heavily russeted, often slight orange flush. Intense sweet-sharp, fruit drop flavour of Nonpareil family; firm texture. Becomes sweeter, drier, quite nutty; still good in Feb. Esteemed in C19th Ireland, but condemned as too small by 1890s in England.
F 9. **T**^{2}; sprd. **C** gd; frt sml, cracks. **P** e-/m-Oct. **S** Dec-Feb.

SANDEW 3 M CD
UK; received 1950 from Axbridge, Somerset, from very old tree.
Large. Cooks to pale cream froth or purée; sweet, delicate taste.
F 11. **T**^{3}. **P** m-Sept. **S** Sept-Oct.

SANDLIN DUCHESS 4 L D
UK; raised c1880 by H. Gabb at Sandlin, Malvern. Introd by William Crump, HG at Madresfield Court, Worcs. RHS AM 1914.
'Improved Newton Wonder' in Crump's opinion; more colourful, sweeter. Brisk, plain taste of fruit; juicy, crisp, rather open texture.
F 10. **T**^{3}. **P** e-Oct. **S** Nov-Jan/Mar.

SANDOW 6 L D
Canada; raised at Ottawa, Central Exp Farm.
Northern Spy X. Select 1912. Introd 1935.
Bright scarlet stripes, flush. Sweet, juicy, melting flesh; definite flavour of raspberries; can be very good of its type.
Hardier than Northern Spy, favoured by New Brunswick, Ontario fruit growers, but now little planted.
F 12. **T**^{3}. **C** lght. **P** m-Oct. **S** Nov-Jan/Mar.
   HUNTER SANDOW; tetrap.

SANDRINGHAM 3 L D
UK; raised by Mr Perry, HG, at Sandringham, Norfolk, Prince of Wales's estate. Exhib 1883 Congress. Believed Winter Pearmain seedling. Introd by Messrs Veitch. RHS FCC 1883.
Cooks to sweet, quite richly flavoured cream purée. Becomes lighter, more delicate as cooker, but makes pleasant, if large, eating apple. Popular early 1900s.

F* maroon/dp pink; 19. **T**^{2}; sprd. **P** m-Oct. **S** Nov-Feb/Mar.
PLATE 1 (blossom)

SAN PEINTE 6 M D
No record of acc.
Sweet, strawberry flavour like Melba; soft, juicy, white flesh.
F 10. **T**^{3}. **P** e-Sept. **S** Sept.

SANSPAREIL 4 L DC
Known England late 1800s. FCC RHS 1899.
Sweet, slightly honeyed, plenty of taste of fruit, balancing acidity; crisp, juicy. Retains fresh quality, until Feb and later. Cooked, keeps shape; sweet, delicate flavour.
F* 9. **T**^{2}. **C** hvy. **P** m-Oct. **S** Nov-Apr.

SATURN 6 L D
UK. Raised by Dr F. Alston, EMRS, Kent. Falstaff X Cox's Orange Pippin. Introd 1997.
Attractive deep rose pink flush. Cream, crunchy flesh; sweet, honeyed, quite rich, but can be merely sweet and low in acidity.
F est 11. **T**^{2}; res scab; some res mildew. **C** hvy. **P** Sept. **S** Oct-Dec; commercially-Jan.

SAUERGRAUECH ROUGE syn Roter Sauergrauech

SÄVSTAHOLM 4 E D
Sweden; believed arose 1830s at Sävstaholm, mansion in Södermanland, south of Stockholm. Rescued from shrubbery between kitchen and flower garden by gardener Wrongstein. Desc 1859. Original tree still survives, preserved in park area.
Pink striped over cream. Sweet, juicy, soft, melting flesh, but quickly going over once picked.
Distributed all over Scandinavia, northern Germany in C19th. Remains highly valued in Sweden; grown northern Sweden, Finland.
F 4. **T**^{3}; v hrdy. **C** gd. **P** use Aug.
   P. J. BERGIUS; more even, highly coloured flush. Found 1912 Bergianska, Stockholm's Botanic Gardens; named after pupil of Linneas.

SCARLET CROFTON 7 M D
UK or Ireland; well known early 1800s, when sent to London Hort Soc by nursmn John Robertson of Kilkenny. Tradition claims introd Ireland late 1500s by founder of Crofton family, Longford House, County Sligo.
Striking dark red flush, under network of russet. Sweet, almost scented taste.
Still found County Sligo in 1950s.
F 12. **T**^{2}; rep gd res scab. **P** m-Sept. **S** Sept-Oct/Nov.

SCARLET NONPAREIL 7 L D
UK; raised c1773 in garden of inn at Esher,

Surrey. RHS AM 1901.
More colourful than Nonpareil, but similar sweet-sharp flavour, firm flesh.
F 10. **T**^{2}. **C** gd. **P** m-Oct. **S** Jan-Mar.

SCARLET PEARMAIN 6 M D
UK; introd c1800 by Bell, land steward to Duke of Northumberland, Sion House, Middx. Widely distributed in Europe in C19th. Syns many; incl Bells' Scarlet Pearmain.
Striking scarlet flush. Distinct flavour, poss of raspberry, but Edward Bunyard found it closer to taste of crab apple. Crisp, juicy, quite sweet, nearly yellow flesh.
F 9. **T**^{2}. **P** l-Sept. **S** Sept-Oct.

SCARLET PIMPERNEL syn Stark's Earliest

SCARLET STAYMARED see Stayman Winesap

SCHMIDTBERGERS ROTE REIN-ETTE 7 L D
Austria; raised c1832 by Liegel at Braunau; named after Josef Schmidtberger, pomologist of St Florian monastery. Widely distributed over Europe. Syns many.
Rich with sugary acidity, but can be thin, sharp. French pomologist Leroy found it 'sweet, delicious acidity and perfume'.
F 9. **T**^{3}. **P** e-Oct. **S** Nov-Jan/Mar.

SCHÖNER AUS EXTERNAL 4 L D
Germany; received 1951 from Fruit Res St, York, nr Hamburg.
Large. Sharp, fruity in late Jan, but can be fairly bleak in England.
F 9. **T**^{3}. **P** m-Oct. **S** Jan-Mar/May.

SCHÖNER AUS HERRNHUT syn Herrnhut 4 M D
Germany; discv by A Heintze, in Herrnhut, nr Zittau, Saxony. Introd c1900.
Sweet, vinous; soft, juicy. Grown eastern Germany mainly for juice manufacture.
F 6. **T**^{2}; hrdy. **C** hvy. **P** m-Sept. **S** Sept-Oct/Nov.

SCHÖNER VON NORDHAUSEN syn Belle de Nordhausen 5 L D
Germany; raised before 1850 in garden of Kaiser at Nordhausen. Introd 1892.
Refreshing, sharp fruity taste; soft juicy flesh. Grown eastern Germany mainly for juice manufacture.
F 8; frost res. **T**^{2}. **P** e-Oct. **S** Oct-Jan.

SCHÖNER VON BOSKOOP see Belle de Boskoop

SCHOOLMASTER 1/5 L C
UK; believed raised c1855 either from seed of imported Canadian apple in Old Stamford Grammar School garden, Stamford, Lincs or arose in Herefords. Sent

to RHS 1880 from Stamford by nursmn Thomas Laxton. RHS FCC 1880.
Cooks to sharp, white froth or purée; good, fruity, brisk taste.
Grown Herefords 1920–30s for canning industry as easily cooked, never discoloured.
F 13. **T**^{2}; hrdy. **C** hvy. **P** m-Oct. **S** Nov-Jan.

SCHURAPFEL 5/7 L D
Germany; received 1951 from Fruit Res St, Jork, nr Hamburg.
Good strong flavour of fruit; sweet, balanced; crisp. Greasy skin.
F9. **T**^{1}. **C** gd, bien. **P** e-Oct. **S** Nov-Mar/May.

SCHWEIZER ORANGE syn Swiss Orange 7 L D
Switzerland; raised 1935 at Wädenswil Res St.
Ontario X Cox's Orange Pippin. Released 1955.
Crisp, juicy in Jan, but low in flavour in England; pale yellow, coarsely textured flesh.
F 15. **T**^{1}; sprd. **P** m-Oct. **S** Nov-Feb.

SCILLY PEARL 1/5 E DC
UK; received 1924 from Isles of Scilly, where prob arose.
Markedly angular, flat sided. Brisk eater cooks to sweet, lightly flavoured purée.
F 10. **T**^{3}. **P** use Aug.

SCOTCH BRIDGET 4 L C
UK; prob arose Scotland. Desc 1851 (Hogg).
Oblong-conical, prominently ribbed, crowned. Quite rich, cream, crisp, flesh, but rather insipid cooked.
Much grown around Lancaster in C19th; still found Scottish, northern, West Midland gardens.
F 11. **T**^{2}; hrdy. **P** e-Oct. **S** Oct-Dec.

SCOTCH DUMPLING 1/3 M C
UK; received from Scotland 1949.
Large. Cooks to brisk, well-flavoured froth.
F* 5. **T**^{2}; sp. **P** l-Aug. **S** Sept-Nov.

SCOTIA 6 M D
Canada; raised at Kentville Res St, Nova Scotia.
McIntosh X. Introd 1961.
McIntosh red with dusky bloom. Sweet, strawberry flavour; soft, melting white flesh.
F 9. **T**^{2}. **P** l-Aug. **S** Sept.

SCRUMPTIOUS 4/6 E D
UK; raised c1985 by H. Ermen, Faversham, Kent.
Starkspur Golden Delicious X Discovery. Introd 2000 by F. W. Matthews Nursery, Tenbury Wells, Worcs.
Rose pink flush. Crisp, juicy, plenty of fruit, with good sweet yet balanced taste. Advantage over most earlies in that it can be

picked from tree late Aug to mid-Sept, does not drop.
F est 10. **T**^{2}; gd res scab, mildew. **P** m-Aug-m-Sept. **S** l-Aug-Nov; commercially-Dec.

SEABROOK'S RED 6 M D
UK; raised by nursmn W. P. Seabrook, Boreham, Essex. Received 1925.
Crisp, juicy white flesh; faint strawberry flavour.
F 14. **T**^{2}. **P** m-Sept. **S** Sept-Oct.

SEATON HOUSE 3 L C
UK; raised at Seaton House, Arbroath, Scotland. Introd 1860.
Large. Cooked keeps shape; sharp.
F 14. **T**^{1}; sprd. **P** e-Oct. **S** Oct-Dec/Jan.

SEEANDO RED ROME syn Nured Rome

SEIDENHEMDCHEN syn Weisses Seidenhemchen

SENSYU (Senshu) 6 M D
Japan. Raised at Fruit Tree Exp St, Akito Prefecture.
Toko X Fuji. Introd 1980.
Firm, crisp.
F 11. **T**^{2}; spding; res mildew. **P** l-Sept. **S** Oct-Dec/Jan.

SEPTEMBER BEAUTY 4 M D
UK; raised by nursmn Thomas Laxton, Bedford. RHS FCC 1885.
Delicate aromatic quality, like Cox, but not as intense; sweet, juicy, crisp, finely textured flesh.
F 11. **T**^{2}. **P** m-Sept. **S** Sept-Nov/Dec.

SEPTER syn Septerappel 4 M D
Netherlands; raised after 1952 by Piet de Sonnaville in his family orchards at Puiflijk and Altforst, nr Nijmegen.
Jonathan X Golden Delicious. Received 1977.
Jonathan colour, Golden Delicious taste. Sweet, honeyed, weak flavour; tough skin. Grown Netherlands.
F 18. **T**^{1}; sprd. **P** l-Sept. **S** Oct-Nov.

SERGEANT PEGGY 5 L C
UK; raised 1922 by F. W. Wastie, Eynsham, Oxford.
Blenheim Orange X Gloria Mundi.
Large. Firm, sweet, but little acidity or flavour. Insipid cooked.
F 8. **T**^{2}. **P** e-Oct. **S** Oct-Dec.

SERVEAU 1/5 L C
France; acc prob variety desc 1948 (*Verger Français*), when grown Hautes-Alpes, Alpes de Hte-Provence.
Firm, sharp, astringent. Cooked keeps shape; sweet-sharp taste.
F 13. **T**^{2}. **C** hvy; hangs late. **P** l-Oct. **S** Nov-Jan.

SEVERN BANK (?) 5 E C
UK; acc may be variety desc 1884 (Hogg).
Cooks to sharp purée. Firm, travelling well; grown C19th in Severn Valley, Gloucs for Midlands markets.
F 8. **T**^{3}. **P** l-Aug. **S** Sept-Oct.

SHAMROCK 2/4 L D
Canada. Raised 1970 by K. Lapins, CDARS, Summerland, British Columbia.
McIntosh X Starkspur Golden Delicious. Introd 1986.
Shamrock green yet sweet and honeyed; soft cream flesh. Combines texture of McIntosh with sweetness of Golden Delicious.
F est 7. **T**^{2}; bears early. **P** l-Sept. **S** Oct-Dec; commercially-spring.

SHARLESTON PIPPIN (?) 5 M D
UK; acc may be variety recorded 1888 Conference from Doncaster, which prob arose Sharlston Village, nr Wakefield, Yorks. Impossible to know if acc is true as published desc too brief.
Light, savoury flavour, quite sweet, soft, juicy flesh.
F 16. **T**^{2}. **P** m-Sept. **S** Sept-Oct.

SHARON 4 L D
USA; raised 1906 by S. A. Beach, Iowa State Agric Exp St, Ames.
McIntosh X Longfield. Introd 1922.
Sweet, soft, juicy, McIntosh type; can lack acidity, taste woody, green.
F 7. **T**^{2}; hrdy. **C** bien. **P** e-Oct. **S** Oct-Dec.

SHEEP'S NOSE 3 M C
UK; received 1951 from Surrey. Sheep's Nose (3) of *Apple Register*; not Sheep's Nose of Hogg.
Large, prominently ribbed, crowned, like sheep's nose. Cooks to pleasant, brisk, purée.
F 12. **T**^{2}; sprd. **P** m-Sept. **S** Sept-Oct.

SHENANDOAH 6 L CD
USA; raised 1942 by R. C. Moore, Virginia, Polytechnic Inst, Blacksburg.
Winesap X Opalescent. Introd 1967.
Named after West Virginian fruit growing area. Rich, slightly scented and aromatic; crisp, quite juicy. Also recommended for processing as pie filling, as keeps shape when cooked.
F 8. **T**^{2}. **C** gd. **P** e-Oct. **S** Oct-Dec/Jan.

SHINFIELD SEEDLING 4 M D
UK; received 1944 from Reading Univ, Shinfield, Berks.
Quite rich, lightly aromatic; like sharp, low key Cox, but can be merely sharp, hard.
F 13. **T**^{2}. **P** l-Sept. **S** Oct-Dec.

SHIN INDO 4 L D
Japan; raised 1930 at Aomori Apple Exp St.
Indo X Golden Delicious. Named 1948.

Extraordinary flavour, like sugary sweet-meat, perfumed, very sweet, no acidity; rather dry, firm yellow flesh. Resembles Indo but larger, more colour.
F 11. T³. P l-Oct. S Nov-Feb.

SHINKO 6 M D
Japan; raised 1931 at Aomori Apple Exp St. Ralls Janet X Jonathan.
Quite rich, slightly honeyed, crisp, juicy; often poor quality in England.
F 12. T³. C hvy, frt sml. P l-Sept. S Oct-Nov/Dec.

SHINSEI 5 M D
Japan; raised 1930 at Aomori Apple Exp St. Golden Delicious X Early McIntosh. Named 1948.
Resembles Golden Delicious, but harder. Very sweet, scented, rather sickly flavour; crisp, juicy.
F 12. T³. C hvy, bien, frt sml. P l-Sept. S Oct-Nov.

SHOESMITH 3 L C
UK; raised by G. Carpenter, Byfleet, Surrey. Lane's Prince Albert X Golden Noble. Exhib 1930.
Large, handsome, exhibition apple. Cooks to firm purée; good, sweet, balanced flavour.
F 14. T². C bien. P e-Oct. S Oct-Dec.

SHOREDITCH WHITE 5 M D
UK; arose Somerset. Desc 1884 (Hogg).
Sweet, little acidity.
F 10. T². P m-Sept. S Sept-Oct.

SIDDINGTON RUSSET *see* Galloway Pippin

SIDNEY STRAKE *see* Tom Putt

SIGNE TILLISCH 4 M CD
Denmark; prob raised c1866 by Councillor Tillisch of Bjerre, nr Horsens; named after his daughter.
Believed Calville Blanc d'Hiver seedling. Desc 1889.
Pale milky yellow, slightly flushed; ribbed. Sweet, light taste; soft, juicy cream flesh; can lack acidity in England. Cooked, very light flavour.
Formerly esteemed in Denmark. Also grown Sweden; Germany, where remains valued garden variety, and considered equal of Gravenstein and James Grieve; France; Belgium.
F 10. T². C gd. P e-Sept. S Sept-Oct.

SIKULAI-ALMA 6 L D
Hungary; first noted 1875. Sikulai Alma was found in village of Sikula, south eastern Alföld, reputedly in C16th or C17th.
Large. Robust, rich, taste; can be quite honeyed, almost aromatic.
F 10. T². P e-Oct. S Nov-Jan/Mar.

SILVA 4 E D
Sweden; raised 1945 at Alnarp Agric Coll. Melba X Stenbock. Introd 1970.
Dark pink flush over pale yellow. Sweet, juicy, crisp, hint strawberry flavour. Grown northern Sweden.
F 10. T²; v hrdy. P eat Aug.

SIMONFFY PIROS 6 L D
Hungary; origin uncertain, takes its name from prominent family who lived in town of Debrecen. Recorded 1876.
Brisk but mellowing to quite sweet with raspberry-like acidity, firm flesh. Valued C19th Hungarian variety.
F 9. T². P m-Oct. S Nov-Jan/Mar.

SINTA 5 L D
Canada; raised 1955 by Dr. K. Lapins, CDARS, Summerland, British Columbia. Golden Delicious X Grimes Golden. Introd 1970.
Sweet, honeyed, juicy, deep cream flesh. In US 'superior to both parents'.
F 10. T²; sprd. P m-Oct. S Oct-Jan.

SIR JOHN THORNYCROFT 6 L D
UK; raised by Collister, HG to Sir John Thorneycroft, Bembridge, Isle of Wight. Recorded 1911. RHS AM 1911. Introd 1913 by G. Bunyard & Sons, Maidstone, Kent.
Fragrant, sweet, light flavour; 'pleasantly aromatic' in Edward Bunyard's opinion. Crisp, juicy, but rather coarse texture; tough skin.
F 13. T². P e-Oct. S Oct-Dec/Feb.

SIR PRIZE 5 L D
USA; raised by co-operative breeding programme of Purdue, Rutgers and Illinois Universities. Complex parentage includes *Malus floribunda*, (carries V_f gene for resistance to scab). Fruited 1961. Released 1975.
Resembles large Golden Delicious. Very sweet, scented taste; crisp, juicy.
F 17; trip. T³; sprd; res scab, mod res mildew. C hvy; frt bruises. P m-Oct. S Nov-Jan.

SISSON'S WORKSOP NEWTOWN 5 L D
UK; raised c1910 by Mr Sissons, Worksop, Notts, from pip of Canadian Newtown Pippin.
Resembles Newtown Pippin but ripens earlier. Savoury taste; sweet, juicy.
F 11. T². P e-Oct. S l-Nov-Dec/Feb.

SKINLITE 5 L D
Italy; raised 1947 by Prof R. Carlone, Univ Turin.
Golden Delicious X Commercio (Collins). Introd 1973.
Russet free Golden Delicious type. Sweet, honeyed, but little acidity; coarse yellow

flesh.
F 13. T². C hvy. P l-Oct. S Nov-Jan.

SKOVFOGED 6 E C
Denmark; mentioned by priest Wöldike, who saw original tree growing in forest at Lov in 1830. Named after Christer Skovfoged, who propagated it from root cutting before mother tree died.
Sweet, light strawberry flavour; soft, juicy, white flesh.
Formerly major Danish variety.
F 5. T²; sprd. P eat Aug; v shrt season.

SLAVYANKA 5 L D
Russia; raised 1889 by I. V. Michurin at Michurinsk.
Antonovka X Ananas Reinette. Fruited 1896.
Sweet, soft, juicy, lightly flavoured.
F 12. T²; part tip. C bien. P e-Oct. S Oct-Nov.

SLEEPING BEAUTY 3 L C
UK; arose Lincs; desc Hogg (1851).
Sharp, astringent. Cooked, good flavour, making firm purée or keeping shape. Grown for Boston market, Lincs in 1850s.
F 9. T²; sprd. P m-Oct. S Nov-Mar.

SLOTØAEBLE syn Annie Elizabeth

SMALL'S ADMIRABLE (?) 5 L C
UK; acc received 1949 from Tyninghame Gardens, East Lothian; may be variety raised c1850 by F. Small, nursm of Colnbrook, Slough, Berks; desc Hogg (1884).
Large. Cooks to juicy purée, sharp, fruity.
F 10. T³; sprd. C hvy. P m-Oct. S Nov-Dec.

SMART'S PRINCE ARTHUR 3 L C
UK; raised by Mr Smart, nr Sittingbourne, Kent. Exhib 1883 by nursmn G. Bunyard, Kent; desc 1884.
Very tall, conical, often waisted; dark maroon flush over yellow. Brisk, deep cream flesh; mellowing by Jan and quite rich. Little cooked flavour.
Firm, tough apple, grown around Maidstone for London markets in C19th.
F 14. T². C gd. P e-Oct. S Nov-Apr.

SMILER 6 M D
UK; raised by A. Simmonds, Clandon, Surrey. Recorded 1934.
Good tangy sweet-sharp taste; can be over sharp.
F 10. T¹; sprd. P m-Sept. S Oct-Nov/Dec.

SMOOTHEE *see* Golden Delicious

SNÖVIT 6 E D
Sweden; raised 1936 by Emil Johansson at Agric Coll, Alnarp.
Stenbock X Persikerött Sommaräpple (Peach Red Summer Apple).

Beautiful, pink flush, darker stripes over pale cream, with bloom. Sweet, scented, soft white flesh, which accounts for name – Snövit means 'snow white'.
Valued Swedish garden variety.
F 6. **T**[3]; hrdy. **P** use Aug-Sept.

SOUTH PARK 7 L D
UK; raised c1940 by gardener W. Barkway at South Park, Penshurst, Kent.
Cox's Orange Pippin X Winter Queening.
Delicate aromatic flavour, very like Cox, but sharper, not as juicy; can be over-acidic.
F 11. **T**[2]. **C** lght. **P** e-Oct. **S** Nov-Jan/Mar.

SÓVÁRI COMUN 3 L C
Transylvania (Romania); believed ancient variety. Tradition claims arose village of Sóvár which was destroyed in Turkish conquest of C16th. Widely grown in region for centuries. Recorded 1876.
Purplish red stripes over green. Sharp, soft, pale green flesh; cooked, insipid, woody.
F 18. **T**[2]. **P** m-Oct. **S** Nov-Mar/May.

SÓVÁRI NOBIL syn Nemes Sóvári Alma

SOWMAN'S SEEDLING 1 E C
UK; raised 1914 by Mr Sowman at County Agric St, Hutton, Lancs.
Grenadier X Ecklinville. Acquired by Messrs Rivers, Sawbridgeworth, Herts. Introd 1927.
Large, even in early Aug. Cooks to sharp froth or purée; needs extra sugar.
F 12. **T**[2]; sprd. **P** use Aug-Sept/Oct.

SPAANSE KEIING 6 L D
Belgium; origin unknown, but formerly grown all over northern Belgium, especially East Flanders.
Firm fleshed, very tough skinned; its name means Spanish pebble. Quite fruity, flavoursome in Jan.
F 16. **T**[3]. **P** l-Oct. **S** Dec-Mar/Apr.

SPARTAN 6 L D
Canada; raised 1926 by R. C. Palmer, CDARS, Summerland, British Columbia.
McIntosh X Yellow Newtown Pippin. Introd 1936.
At best, perfumed with flavour like a cross between strawberry and melon; usually described as vinous. Sweet, some acidity, juicy, crisp, white flesh; can lapse into metallic blandness; tough skin. Needs to be well coloured before picked for good flavour.
Major Canadian market apple, also grown Switzerland, Poland and small extent England.
Frt Col dp maroon flush with bloom; grnish yell ground colour; becoming lighter and brighter. *Size* med. *Shape* rnd-con; ribbed; sltly 5 crowned. *Eye* sml, clsd or part open; sepals shrt, v downy. *Basin* med wdth, dpth;

ribbed. *Cavity* nrw, dp; ltl russet. *Stalk* med lng, qte thin. *Flesh* wht.
F 12. **T**[2]; v prn canker. **C** gd. **P** e-Oct. **S** Nov- Jan; commerically-Feb.
PLATE 21
Sports include:
   HUNTER SPARTAN; tetrap.
   SPARTAN COMPACT; spur type.
   SPARTAN 4N SCOTLAND; tetrap.
   SPARTAN 4N SWEDEN; tetrap.
   SPARTAN 4N No 3; tetrap.

SPÄTBLÜHENDER TAFFETAPFEL 5 L D
Germany; raised in nursery of Hohenheim Agric Inst. Recorded 1872. Introd by pomologist, Eduard Lucas.
Resembles Weisser Winter Taffetapfel. Sweet juice but astringent flesh.
F 42. **T**[3]. **P** l-Sept. **S** Oct-Dec.

SPENCER 6 M D
Canada; raised 1926 by R. C. Palmer, CDARS, Summerland, British Columbia.
McIntosh X Golden Delicious. Introd 1959.
Crimson flushed. Honeyed, very sweet in early Oct; crisp, juicy.
F 12. **T**[2]; v. hrdy. **C** gd. **P** e-Oct. **S** Oct-Dec; cold store-Mar.

SPENCER SEEDLESS 2 L D
USA; known 1929.
Curiosity, very large, open eye; rarely sets seed. Poor flavour.
F 21. **T**[1]. **P** m-Oct. **S** Nov-Jan.

SPIGOLD 4 L D
USA; raised by R. D. Way, NYSAES, New York.
Red Spy X Golden Delicious. Introd 1962.
Very large. Sweet, honeyed, more acidity than Golden Delicious, but similar; very juicy, crisp, yellow flesh.
Highly esteemed by American enthusiasts.
F 17; triploid. **T**[3]; sprd. **C** hvy; bien; prn bitter pit. **P** m-Oct. **S** Nov-Jan.

SPIJON 6 L D
USA; raised 1944 by R. D. Way, NYSAES, Geneva.
Red Spy X Monroe. Introd 1968.
Bright red Northern Spy colour, with 'sprightly juiciness of Spy, yet sweeter'. In England, sweet, pleasant, juicy.
Highly regarded by American enthusiasts; also valued Netherlands.
F 11; trip. **T**[2]. **C** gd. **P** e-Oct. **S** l-Oct-Dec/Jan.

SPITZLEDERER 8 L D
Austria (prob); acc is prob Tiroler Spitzlederer, which arose South Tyrol (now Italy); first desc 1855.
Cinnamon russeted over gold. Firm, sweet-sharp, Nonpareil type flavour; rather leathery flesh; hence the name.
Grown southern Tyrol for market and

export 1920–30s and planted throughout Austria; valued for its good storage, attractive even shape.
F 15. **T**[3]. **P** l-Oct. **S** Jan-Mar.

SPIZA 3 L CD
Canada; raised 1916 Ottawa Exp Farm.
Northern Spy X. Introd 1922.
Quite intense, almost rich flavour; plenty of acid, sugar. Cooks to well-flavoured sweet, purée.
F 14. **T**[2]. **P** e-Oct. **S** Nov-Dec/Jan.

SPLENDOUR 7/6 L D
New Zealand; discovered 1948 in garden at Napier by Charles L. Roberts.
Introd 1967 by Stark Bros, Missouri.
Attractive, bright red flush. Crisp, sweet, good taste of fruit. In Feb tastes like cross between Idared and Delicious.
Grown around Napier.
F 15. **T**[2]. **P** m-Oct. **S** Jan-Mar.

STÄFNER ROSEN syn Rose de Staefa 6 L D
Switzerland or USA; tree found by Pfenniger of Stäfa in 1840, but in 1912 believed identical to American variety. Acc is variety known by this name in Switzerland; desc 1924.
Refreshing, savoury, crisp, firm flesh.
F 12. **T**[3]. **P** m-Oct. **S** Nov-Dec.

STANWAY SEEDLING 5 L C
UK; arose Essex; recorded 1899. RHS AM 1899.
Poss named after place nr Colchester.
Large. Quite sharp. Cooked, sweet, pleasant.
F 16. **T**[2]; sprd. **P** e-Oct. **S** Oct-Jan/Mar.

STARK BLUSHING GOLDEN 5 L D
USA; discovered by R. Griffith of Cobden, Illinois.
Believed Jonathan, Golden Delicious cross. Introd 1968 by Stark Bros Nursery, Missouri.
Slight pink blush. Honeyed, sweet, juicy, firm yellow flesh; can be metallic, green in England; tough skin. Dual purpose in US.
F 17. **T**[2]. **P** l-Oct. **S** Nov-Jan.

STARK EARLIBLAZE 4 M D
USA; discv 1949 by W. B. Mooney, New Lennox, Illinois. Introd 1957 by Stark Bros, Missouri.
Dark red Wealthy colour over gold, with bloom. Sharp, soft, juicy, but often green, empty taste in England. In States dual purpose; 'makes excellent sauce'.
F 8. **T**[3]. **C** gd. **P** e-Sept. **S** Sept-Oct.

STARKING see Delicious

STARKRIMSON see Delicious

STARK'S EARLIEST syn Scarlet Pimpernel 6 E D
USA; discv 1938 by Douglas Bonner at Orofino, Idaho. Introd 1944 by Stark Bros, Missouri.
Sharp yet plenty of sweetness; juicy, soft, melting flesh. Best eaten from tree.
Formerly grown for early English markets, also on continent, but small, easily bruised fruit and short season led to commercial demise.
Frt Col bright scarlet flush over pale crmy wht; lenticels conspic as brown spots. *Size* med/sml. *Shape* flt-rnd to rnd-con. *Basin* qte brd, med dpth; sometimes sltly puckered. *Eye* clsd or prt open; sepals brd. *Cavity* med wdth, dpth; russet streaks. *Stalk* med lng, qte thin. *Flesh* wht.
F 3. T². sprd; prn mildew. C hvy. P eat e-m-Aug.

STARK'S LATE DELICIOUS *see* Delicious

STARKSPUR GOLDEN DELICIOUS *see* Golden Delicious

STARKSPUR McINTOSH *see* McIntosh

STAR OF DEVON 6 L D
UK; Devon, raised by J. Garland, at Broadclyst, Devon. Introd by nursmn George Pyne, Topsham. RHS AM 1905.
Pretty, little red apple. Refreshing, quite sweet, crisp, juicy, white flesh.
F 9. T². P m-Oct. S Nov-Feb/Apr.

STARR 2/4 M D
USA; found in grounds of Judge J. M. White, Woodbury, New Jersey. Property later belonged to Mrs Starr. Propagated 1865 by William Parry.
Large. Quite sharp, juicy, soft, melting flesh in late Sept. Valued by American enthusiasts for 'very juicy, sprightly and aromatic' taste.
F 8. T². P m-Sept. S Sept-Oct/Nov.

STATEFAIR 6 E/M D
USA. Raised 1949 Minnesota Univ. Mantet X Oriole. Select 1959; introd 1979. Brilliantly striped red/orange over yellow. Juicy aromatic sweet and firmer than most earlies.
F 11. T²; v hrdy. P eat late Aug/Sept.

STATESMAN
Australia; raised late 1800s by William Chandler, nursmn of Malvern, Melbourne, Victoria; listed 1884; syn Chandler's Statesman. Represented in NFC by sport.
   RED STATESMAN 6 L D; more highly coloured; discv c1940 in Lenswood, South Australia; introd 1950.
   Brisk, crisp, full of fruit in March. Statesman and sport, formerly,

extensively grown in Victoria, South Australia, Tasmania and New Zealand for export.
F 13. T². C gd. P l-Oct. S Jan-Apr.

STAYMAN'S WINESAP 6 L D
USA; raised 1866 by Dr. J. Stayman, physician and plant breeder at Leavenworth, Kansas. Winesap seeding. Introd 1895 by Stark Brothers Nursery, Missouri. Represented in NFC by sports.
Deep red, like Winesap. Sprightly, light aromatic quality; juicy, pale yellow flesh.
In US dual purpose. Grown in milder climates of Shenandoah Valley of West Virginia, North Carolina, Pennsylvania; also Australia, Turkey, Italy, France.
   BLAXTAYMAN; more even flush. Discv 1926 by J. H. Dickey, Columbia & Okanogan Nursery, Wenatchee, Washington. Introd 1930.
   DARK RED STAYMARED; darker red. Arose with B. C. Moomaw Jr., Barber County, Virginia; introd 1929.
   SCARLET STAYMARED; colours earlier. Arose 1930 with J. H. Dickey, Wenatchee, Washington; introd 1936.
   F 12; trip. T³. C hvy. P m-Oct. S Jan-Mar; commercially-Jun.

STEARNS 4 L D
USA, raised 1880–90s by C. L. Stearns, North Syracuse, New York from pip of Esopus Spitzenberg. Distributed by Van Dusen, Geneva. Recorded 1900.
Very large. Good savoury taste; plenty of sweetness, acidity, juice. Repopularised 1984 by Michigan nursmn, Robert Nitschke, who also recommended it for cooking.
F 14. T³. P m-Oct. S Nov-Dec/Jan.

STEIRISCHER ROTER MARSCH-ANSKER 5 L D
Austria; appears to be form of Steirischer Marschansker, which prob arose Steirmark. Received 1951 from Fruit Res St, Jork, nr Hamburg, Germany.
Pleasant sweet, slightly honeyed taste.
F 10. T²; sp; sprd. P e-Oct. S Nov-Feb/Mar.

STERAPPEL syn Reinette Rouge Étoilée

STERN APFEL syn Reinette Rouge Étoilée

STEVENSON WEALTHY *see* Wealthy

STEYNE SEEDLING 4 L D
UK; raised c1893 at Steyne, Isle of Wight, home of Sir John Thorneycroft. RHS AM 1912.
Delicate aromatic, quite intense flavour, reminiscent of Cox; sweet, well balanced. Soft, juicy cream flesh.
F 9. T². P e-Oct. S Nov-Jan/Apr.

STINA LOHMANN 7 L D
Germany; found 1841 on property in Kellinghusen, Holstein by Stina Lohmann; tree then 40/50 yrs old.
Quite rich; strong blend of sugars and acid; pale yellow flesh.
F 14. T². P e-Oct. S Dec-Feb/Mar.

STIRLING CASTLE 1 M C
UK; raised 1820s by John Christie, who kept toy shop and grew apples in his back garden at Raplock, or Causey Head, nr Stirling, Scotland. Introd by Drummonds Nursery, Stirling.
Cooks to sharp, white purée; plenty of fruity flavour.
Widely planted C19th in gardens; among top ten cookers in 1883. Also grown by market gardeners and orchardists in Kent, where often planted between Bramley trees and removed when these reached full size. Damaged by lime suphur sprays used 1920–30s which led to its demise, but remains popular with gardeners, especially in north, Scotland.
Frt Col grn, slt orng flush; becoming yell. *Size* lrg. *Shape* rnd to flt-rnd; regular. *Basin* lrg, v brd, dp; sltly puckered. *Eye* open; sepals, shrt, brd base. *Cavity* brd, dp; russet lined. *Stalk* med lng, qte thck. *Flesh* wht.
F 11. T¹. C hvy. P m-Sept. S Sept-Dec.

STOBO CASTLE 3 E C
UK; prob arose Scotland. Introd c1900 by nursmn D. Storrie of Glencarse.
Cooks to astringent, sharp, cream froth.
F 10. T³; sprd. P use Aug.

STOKE ALLOW 4 M CD
UK; received 1950 from Somerset.
Soft, juicy, slightly sharp. Cooks to sweet, smooth purée.
F 11. T³; sprd. P e-Sept. S Sept-Oct.

STOKE EDITH PIPPIN 5/8 L D
UK; prob arose at Stoke Edith, Herefords, which in C19th was famed for its fruit collection. Recorded 1872 (Scott); desc Hogg.
Russet freckled over gold. Sweet, light, slightly scented in Nov; quite soft, juicy, pale yellow flesh.
F 11. T². P e-Oct. S Nov-Jan/Mar.

STONECROP 4 L D
Canada; received 1927 from Ottawa Exp Farm.
Stone X Introd 1925.
Sweet, rich, delicately aromatic, like light Cox. Crisp to soft, juicy, flesh.
F 11. T². P e-Oct. S Nov-Dec/Mar.

STONEHENGE 4 L D
Canada; received 1927 from Ottawa Exp Farm.
Stone X. Introd 1925.
Monolithic, very large, ribbed, crowned.

Refreshing, light taste; quite sweet, crisp, juicy flesh.
F 5. T^2. P e-Oct. S Nov-Dec/Mar.

STONE'S syn Loddington

STONE'S MOSAIC 5 L C
UK; received 1990 from LARS, Bristol.
Large, green turning yellow. White soft sharp flesh. Cooks to sharp juicy purée; with added sugar good intense flavour.
F md. T^{2}. P e-Oct S Oct-Nov/Dec.

STONETOSH 6 L D
Canada; raised 1909 at Ottawa Exp Farm.
Stone X McIntosh. Introd 1922.
Sweet, scented, McIntosh type. Soft, juicy, white flesh; tough skin.
Formerly grown commercially in Canada.
F 11. T^3. P e-Oct. S Oct-Jan/Mar.

STOREY'S SEEDLING 2 L D
UK; raised 1927 by R. O. C. Storey, Northholt Park, London, from Newtown Pippin pip.
Flavoursome, crisp, juicy. Dull appearance.
F 13. T^3. P l-Oct. S Dec-Mar/Apr.

STOWELL 7 L D
Australia; raised at CSIRO, Stowell Avenue, Hobart, Tasmania.
Cox's Orange Pippin X Democrat. Received 1974.
Handsome, quite large. Honeyed, rich; firm, deep cream flesh.
F 17. T^3. P m-Oct. S Nov-Jan.

STRAUWALDTS GOLDPARMÄNE syn
Neue Goldparmäne

STRAWBERRY PIPPIN (?) 4 M D
UK; received 1982 from Stoke on Trent.
Acc may be Strawberry Pippin of C19th, but descs too brief to be certain.
Pretty red flush, stripes. Crisp, quite sweet, juicy; tough skin.
F 14. T^2; sp. P m-Sept. S Sept-Oct.

STRIPED BEEFING 3 L C
UK; found 1794 in garden of William Crowe, Lakenham, Norwich by nursmn George Lindley. Distributed 1847 by Hogg, when partner in Brompton Park Nursery, London.
Large, handsome, striped in dark red. Cooking to cream purée, quite rich, brisk. Makes well-flavoured baked apple. Sweetens, pleasant to eat fresh by Jan.
F 9. T^2; sprd. P e/m-Oct. S Dec-Apr/May.

STRIPPY 4 E C D
UK; arose County Armagh, N Ireland. Received 1949.
Boldly striped, small. Quite sweet, soft, juicy. Cooking to sweet, pleasantly flavoured froth.
F 14. T^3. P use Aug.

STURMER PIPPIN 2 L D
UK; arose garden of Mr Dillistone, Rectory House, Sturmer, nr Haverhill, Suffolk. Believed Ribston Pippin, Nonpareil cross; pip planted c1800. Introd by Dillistone's sons, of Sturmer Nurseries; recorded 1831. Strong characteristic taste, like cold steel; sweet, crisp, juicy, but needs plenty of autumn sunshine to build up sugar and flavour.
Popular Victorian garden variety, keeping longer even than Nonpareil and obvious choice for planting in Australia and New Zealand as export variety. Laid foundation of Antipodean fruit trade. Grown Tasmania, New Zealand for export, also used for processing.
Frt slt orng brown flush over grnish yell, turning by Feb to pinky brown over gold. *Size* med. *Shape* oblng-con; qte ribbed, 5 crowned. *Basin* brd, qte dp; ribbed. ltl russet. *Eye* clsd or sltly open; sepals, brd base, v downy. *Cavity* brd, dp; russet lined. *Stalk* med/lng, qte thck. *Flesh* wht, tinge grn/pale crm.
F^* 10. T^2; some res scab. C gd; needs warm spot. P l-Oct. S Jan/Feb-Apr/Jun.
PLATE 32

S. T. WRIGHT 3 M C
UK; raised by J. Allgrove, manager of Veitch Nurseries, Langley, Berks.
Peasgood's Nonsuch X Bismark. RHS AM 1913.
Large, highly coloured like Bismark. Named in honour of Fruit Officer, RHS, Wisley. Cooks to sweet, lightly flavoured yellow purée.
F 12. T^2. P m-Sept. S Sept-Nov.

SUMMER APPLE 4 E D
UK; received 1931 from Cobham, Surrey.
Like tall Beauty of Bath, with similar, sharp taste, soft flesh.
F 12. T^2. P eat Aug.

SUMMER BLENHEIM 7 M D
UK; received 1949 from R. Tincknell, Wisley village, Surrey.
Large, resembling Blenheim Orange, with similar crumbly flesh, but ripens earlier. Quite rich, but not nutty.
F 14. T^3; sprd. P m-Sept. S Sept-Oct.

SUMMER-BROADEN 5 M C
UK; first desc 1831 by Norwich nursmn G. Lindley.
Very sharp in Sept; firm, fruity.
Grown for Norwich markets in late C18th.
F 9. T^3. P m-Sept. S Sept-Oct.

SUMMERGLO 6 M D
USA; discv 1935 in Calhoun County, Michigan by F. H. McDermid. Introd 1967 by Stark Bros, Missouri.
McIntosh type, but ripening earlier.
F 13. T^3; sprd; hrdy. P e-Sept. S Sept-Oct.

SUMMER GOLDEN PIPPIN (?) 5 E D
UK; acc may be variety of Ronalds and Hogg, which prob arose England, known before 1800. Acc is golden, slightly bronzed cheek. Intense fruity tang of Golden Pippin family; rich, concentrated blend of sweetness and acidity; quite juicy, firm, nearly yellow flesh.
Summer Golden Pippin was highly esteemed in C19th and widely grown in gardens. Often confused with Yellow Ingestrie, which were sold as 'Summers', but richer flavour.
F^* 7. T^3; uprt. C gd. P l-Aug. S l-Aug-e-Sept.

SUMMER JOHN 5 M D
UK; arose County Fermanagh, N Ireland. Received 1948, but believed old variety.
Sweet, juicy, soft white flesh. Poss cider variety.
F 8. T^2. C rep res scab. P m-Sept. S Sept-Nov.

SUMMERLAND 6 L D
Canada; raised 1926 at CDARS, Summerland, British Columbia.
McIntosh X Golden Delicious. Introd 1969. McIntosh type. Brisk raspberrry flavour; very juicy, crisp yet melting flesh; can taste metallic, woody in England.
F 11. T^2. C hvy, bien. P l-Sept/e-Oct. S Oct-Dec.

SUMMERLAND McINTOSH see
McIntosh

SUMMER PEARMAIN syn American
Summer Pearmain

SUMMERRED 6 E D
Canada; raised by K. O. Lapins, at CDARS, Summerland, British Columbia.
Summerland X. Introd 1964.
Dark red McIntosh type, but ripening earlier. Sorbet-like white flesh; sweet, soft, juicy, hint of strawberry or vinous flavour. Grown Canada, Italy, Switzerland, Netherlands.
F 5. T^2. C hvy; frt col dulls after picking. P l-Aug. S Sept-Oct.

SUMMER ROSE 4 E D
USA; arose New Jersey, mentioned 1806. Syns incl Harvest Apple, Woolman's Early. Pinky red flush over cream. Lightly flavoured; quite sweet, melting, juicy flesh. In New Jersey, early 1800s, sold as Harvest Apple in July for stewing.
F 10. T^2. C frt bruises. P use Aug.

SUMMER STIBBERT (?) 5 E D
UK; acc may be variety of Hogg (1884) which arose West of England; recorded 1831. Good, brisk, savoury taste; soft, juicy white flesh. Popular in Devon, Somerset, Cornwall in C19th.
F 8. T^3. P eat Aug.

SUNBURN 7 L D
UK; raised by F. W. Thorrington, retired London Customs Officer, Hornchurch, Essex.
Cox's Orange Pippin X. Received 1925.
Delicate aromatic quality; sweet, quite richly flavoured, but thin by comparison with Cox; juicy, cream flesh.
F 11. T[1]; sp; part res mildew. C lght; frt sml. P e-Oct. S Nov-Jan/Mar.

SUNGOLD 5 L D
USA; arose with A. & A. Caggiano, Bridgeton, New Jersey.
Believed Golden, Red Delicious cross. Introd 1963.
Resembles Golden Delicious, but less russet prone.
Extensively planted New Jersey; grown Italy small extent.
F 13. T[1]. C hvy. P m-Oct. S Nov-Jan.

SUNRISE 6 M D
UK; poss raised by Charles Ross, HG, at Welford Park Gardens, Newbury, Berks.
Northern Spy seedling. Recorded 1897.
Soft, white, juicy flesh, but little flavour; often metallic taste; tough skin.
F 9. T[2]. P m-Sept. S Oct-Nov.

SUNRISE (CANADA) 6 E D
Canada. Raised by D. Lane, R. MacDonald, CDARS, British Columbia.
(McIntosh X Golden Delicious) X chance seedling. Introd 1991.
Attractive deep red flush over pale yellow. Very crisp texture; plenty sugar and acid; hints of strawberry flavour.
F early. T[3]; spd; some res mildew. C hvy. P and eat l-Aug-m-Sept.

SUNSET 7 M D
UK; raised c1918 by G. C. Addy, Ightham, Kent.
Cox's Orange Pippin seedling. Named 1933. RHS AM 1960; AGM 1993.
Like small, early Cox; aromatic, intensely flavoured; sharper, not as complex as Cox, but robust and good.
Too small to be commercial but easier to grow than Cox; widely grown in gardens.
Frt *Col* diffuse orng rd flush, many rd stripes over gold; ltl russet patches; lenticels as fine russet dots. *Size* sml. *Shape* rnd or flt-rnd. *Basin* med wdth, dpth; sltly ribbed. *Eye* hlf-open; sepals, qte sml. *Cavity* brd, qte dp; ltl russet. *Stalk* lng, qte thck. *Flesh* pale crm.
F* 10. T[2]; prn mildew, canker; res scab. C hvy. P l-Sept. S Oct-Dec.
> **SUNSET SPORT**; more highly coloured. Arose at Loughgall Hort Centre, Armagh; N Ireland; received 1963.

SUNTAN 7 L D
UK; raised 1956 by H. M. Tydeman, EMRS, Kent.

Cox's Orange Pippin X Court Pendu Plat. RHS AM 1980; AGM 1993.
Handsome, robustly aromatic, rich, sweet, masses of pineapple-like acidity. Can be too sharp in Nov; best kept to Dec and later.
Frt *Col* bright orng rd, rd stripes over gold; russet patches, lenticels conspic as russet dots. *Size* med. *Shape* flt-rnd. *Basin* brd, shal; sltly ridged; ltl russet. *Eye* lrg, hlf to fully open; sepals, lrg, brd, downy. *Cavity* med wdth, dpth; russet lined. *Stalk* shrt, thck. *Flesh* dp crm.
F 20; trip. T[3]; sprd; prn canker. C gd; prn bitter pit. P e/m-Oct. S Nov/Dec-Feb.

SUPERB syn Laxton's Superb

SURE CROP 5 L CD
UK; received 1905 from nursmn Clibrans, Cheshire. Exhib 1919.
Cooked, keeps shape or makes stiff purée; bright yellow; sweet, rich flavour. Culinary flavour fades by New Year but then serves as pleasant eating apple.
F 7. T[2]; sp. C hvy. P m-Oct. S m-Oct-Feb/Mar.

SURPRISE 5 M CD
UK; received 1905; desc 1920 by Bunyard.
Cooked, keeps shape; sweet, delicate taste of pears. Mellows to quite sweet eating apple with definite pear-like flavour.
F 15. T[3]. C lght; frt drops, bruises. P m-Sept. S Sept-Oct.

SUSSEX FORGE syn Forge

SUSSEX MOTHER 2/4 M D
UK; grown around Heathfield, Sussex in 1800s. Desc 1884 (Hogg).
Strongly ribbed, crowned; brownish orange flush over yellow. Distinctive flavour, sweet, quite spicy hint of aniseed. Still found in Sussex gardens.
F 12. T[2]; sprd. P e-Sept. S Sept.

SWAAR 2/8 L D
USA; believed raised by Dutch settlers nr Esopus on Hudson River, New York. Recorded 1804.
Name means heavy; solid apple. Rich, slightly aromatic, almost nutty flavour; firm, cream flesh.
In US, 'celebrated winter table fruit' by 1817; grown New Jersey, New York home orchards in C19th. Recently regained popularity with amateurs.
F 10. T[3]; sprd. P e-Oct. S Oct-Dec.

SWEET CAROLINE 6 L D
Netherlands; raised IVT, Wageningen.
Golden Delicious X McIntosh. Received 1968.
Poor flavour in England, sweet, bland. Grown Netherlands.
F 14. T[2]; tip. P e-Oct. S Oct-Dec; cold store-Feb.

SWEET CORNELLY 6 L D
Netherlands; raised IVT, Wageningen.
Golden Delicious X McIntosh. Received 1968.
Deep carmine flush, tall. Extraordinary flavour; sweet, can be intensely scented, almost of elderflowers, but little acidity or taste of fruit.
F 11. T[2]. P e-Oct. S Nov-Dec.

SWEET DELICIOUS 7 L D
USA; raised 1911 by R. Wellington at NYSAES, Geneva.
Deacon Jones X Delicious. Introd 1922.
Very sweet, scented like sugary sweetmeat in early Dec; rather lovely in its way; yellow, melting flesh.
F 15. T[2]. P e-Oct. S Nov-Jan.

SWEET ERMGAARD *see* Zoete Ermgaard

SWEET MERLIN 6 M D
UK; received from Cornwall 1954.
Very sweet, almost no acidity, slightly scented. Prob cider apple.
F 12. T[2]. P m-Sept. S Sept-Oct/Nov.

SWEET-TART 6 L D
USA; raised by R. B. Gorden, California. Received 1982.
Delicious type. Very sweet, soft, juicy flesh.
F 10. T[2]. P e-Oct. S l-Oct-Dec.

SWISS ORANGE syn Schweizer Orange

SYLVIA 4 E D
Sweden; raised at Ramlösa Nursery on behalf of Balsgård Fruit Breeding Inst.
Gyllenkroks Astrakan X Worcester Pearmain. Introd 1973.
Ripens on August 8th, which is 'Sylvia's Day' in Swedish calendar. Pink flushed over cream; tall, ribbed. Sweet yet sprightly, intense taste of fruit; crisp, juicy.
F 11. T[2]. P use e-Aug.

SYMOND'S WINTER 6 L D
Received 1953 from New Zealand.
Bright red flush. Honeyed, sweet in Dec; slight piquancy.
F 9. T[3]. P m-Oct. S Dec-Mar.

SZABADKAI NAGI SZERCSIKA 2 L D
Hungary; well known C19th.
Some richness in Feb; juicy, crisp, firm flesh.
Formerly widely planted southern Hungary on plain between Duna and Tisza rivers around towns of Szeged and Szabadka (now Subotica, Serbia). Highly valued C19th, when sold by single fruits.
F 6. T[3]. P l-Oct. S Jan-Mar/Apr.

SZACSVAY TAFOTA 5 E D
Transylvania (Romania); named after Szacsvay Zsigmond, who propagated it early

C19th.
Bright yellow, slight orange flush. Savoury, brisk, quite juicy, white flesh.
F 11. T^1; sprd. P eat Aug.

SZÁSZPAP ALMA 7 L D
Transylvania (Romania); believed ancient variety, widespread in Transylvania in C19th. Its name means Saxon Priest's apple.
Sweet, quite rich, refreshing, plenty of fruit.
F 17. T^2. P l-Oct. S Dec-Feb.

SZÉCHENYI RENET 4 M D
Hungary; recorded 1876 by French nursmn Simon-Louis Frères.
Savoury, crisp, juicy; becoming sweeter, almost scented.
F 5. T^2. P l-Sept. S Sept-Nov/Dec.

TAM MONTGOMERY syn Early Julyan

TAMPLIN syn Cissy

TARE DE GHINDA 2/5 L D
Romania; arose Ghinda, Bistriţa, Moldavia, where widely grown in 1960s. Believed very old variety.
Very crisp in Jan, but only light flavour; pale green flesh.
F 19. T^3. C hvy. P l-Oct. S Jan-Mar/May.

TASMA see Democrat

TASMAN PRIDE 6 L D
Australia; arose late 1800s on farm of Thomas Young, Margate, Channel district, south Tasmania.
Shiny red. Very sweet, honeyed; firm, crisp, deep cream flesh.
Extensively planted in Tasmania in 1920s; still grown Channel district 1970s; no longer commercially important.
F 9. T^2; sp. C gd. P m-Oct. S Nov-Jan/Mar.

TAUNTON CROSS 4 M D
UK; raised 1919 by G. T. Spinks, LARS, Bristol.
Wealthy X.
Deep maroon flush. Vinous flavour; sweet yet sharp.
F 14. T^3. P m-Sept. S Oct-Nov/Dec.

TEINT FRAIS 3 L C
France; known late 1700s around Quimperlé, Finistère, Brittany under local name, Kerlivio, after its propagator.
Large, beautiful, salmon pink flush over yellow. Sharp, white flesh. Cooks to soft juicy purée, well-flavoured, brisk.
Formerly grown for market Finistère; remains valued Brittany, Normandy.
F 13; trip. T^3; sp. C gd. P m-Oct. S Nov-Mar/Jun.

TELAMON syns Waltz, Ultraspire (US) 6 M D
UK; raised 1976, EMRS, Kent.
Wijcik McIntosh, Golden Delicious cross. Introd 1989. *See also* Ballerina.
Dark red McIntosh type, but firmer fleshed; sweet, crisp, juicy.
F 10. T^2. P Sept. S Sept-Oct.

TELLINA 4 L D
Italy; received 1948 from Univ Turin.
Small, bright red flushed. Light taste; can be green, woody, undeveloped in England.
F 14. T^2; sp. P l-Oct. S Dec-Feb/Apr.

TELLISAARE 5 L D
Estonia; raised before 1947 at Estonia Plant Breeding Inst Exp St from cross involving Kitaika/*Malus prunifolia*. Received 1976.
Pink blush over pale yellow. Brisk, savoury, soft white flesh. Standard commercial variety in Estonia.
F 14. T^2; res scab. C hvy. P e/m-Oct. S Oct-Dec.

TELSTAR 4/7 L D
New Zealand; raised 1934 by J. H. Kidd at Grey Town, Wairarapa Valley.
Golden Delicious X Kidd's Orange Red. Kidd's seedlings, after his death, passed to DSIR, Havelock North. Select by Dr D. W. McKenzie. Named 1965.
Very sweet, honeyed, like very good Golden Delicious, yet also scented, reminiscent of Kidd's Orange Red. Juicy, cream, firm flesh.
F* 13. T^2; sprd. C hvy; frt sml, unless thinned. P e-Oct. S Nov-Jan.

TEN COMMANDMENTS 4 M D
UK; arose Herefords. Desc Hogg (1884).
Red stained core with ten points gives name. Small, dark red flush. Weak flavour. Still found Herefords; used for cider.
F* 13. T^2; res mildew. P l-Sept. S Oct-Nov.

TENROY see Gala

TENTATION syn Delblush

TESTERSPUR GOLDEN DELICIOUS
see Golden Delicious

TÉTON DE DEMOISELLE 6 L D
France prob; recorded 1895 by nursmn Simon-Louis Frères.
Breast shaped as name suggests; pinky red flush, stripes. Quite rich, sweet, scented; nearly yellow flesh.
F 11. T^2; sprd. P l-Sept. S Oct-Nov.

TEWKESBURY BARON 6 E D
UK; arose pres Tewkesbury, Gloucs. Exhib 1883 by nursmn Wheeler, Gloucester.
Dark crimson flush, resembles Devonshire Quarrenden, but ripens later. Little flavour.
F 7. T^2. P l-Aug. S Aug-Oct.

TEXOLA 4 L D
USA; raised in Utah by Mr Broome.
Ben Davies seedling. Received 1930.
Good taste of fruit, plenty of sweetness, acidity; soft, juicy flesh.
F 8. T^1. P e-Oct. S Nov-Jan/Mar.

THODAY'S QUARRENDEN 6 L D
UK; found 1949 by R. E. Thoday at Reed-ground Farm, Willingham, Cambridge.
Prob Devonshire Quarrenden seedling.
Sweet, refreshing; juicy, quite soft flesh.
F 8. T^1. P e-Oct. S Nov-Feb.

THOMAS JEFFREY 6 E D
UK; raised Edinburgh by nursmn D. W. Thomson. Received Royal Caledonian Society award some years before 1923, when sent to NFC.
Named after nursery's foreman. Small, resembling Devonshire Quarrenden. Fairly sharp, firm, chewy white flesh.
F 7. T^2. C hvy. P Sept. S Sept-Oct.

THOMAS RIVERS 3 M C
UK; raised at Rivers Nursery, Sawbridgeworth, Herts.
American Mother seedling. RHS FCC 1892 as River's Codlin. Introd 1894; renamed 1897.
Named after Victorian nurseryman and fruit breeder. Cooks to juicy purée with brisk, intense taste of pears; retaining aromatic qualities. Makes splendid sauce or pie; good stir-fried with cabbage.
Drab colour, and irregular ribbed shape, like American Mother, probably accounts for its lack of popularity.
F 14. T^2; reported res scab. C gd. P e/m-Sept. S Sept-Oct/Nov.

THOMPSON'S APPLE 4 E D
Ireland; received 1950 when known County Monaghan and north west Tyrone as The Smeller.
Attractive peachy coloured flush over pale yellow, but little flavour.
F 17. T^2. P use Sept.

THORLE PIPPIN syns Lady Derby, Whorle Pippin 4 E D
UK; arose Scotland; desc 1831 (Ronalds).
Hogg claimed Thorle was corruption of Whorle, the correct name, which derived from its round, flat shape which resembled whorle or fly-wheel of spindle used for spinning wool. Small, bright red flushed, striped. Good, strong, savoury tang; quite sharp, juicy, white flesh.
F 6. T^3; sprd. P l-Aug. S Aug-Sept.

THORPE'S PEACH 6 M D
UK; raised 1899 by Miss Goodwin at E. J. Thorpe's Nursery, Brackley, Northants. Introd 1927. Sweet, gentle flavour; crisp yet melting flesh. Can be watery, weak.
F 10. T^2. P m-Sept. S Sept-Oct.

THURGAUER WEINAPFEL 4 L C
Switzerland; known in Thurgau since 1860s.
Quite rich, fresh fruity taste in Dec; firm flesh; tough skin.
F 10. T^3. **P** l-Oct. **S** Dec-Apr.

THURSO 6 M D
Canada; raised 1898 Central Exp Farm, Ottawa.
Northern Spy X. Introd 1908.
Bright red, like Northern Spy, but earlier season. Pleasant, slightly perfumed; sweet, juicy.
F 19. T^2. **P** l-Sept. **S** Oct-Nov/Dec.

TIFFEN 3 M C
UK; prob arose north west England. Exhib 1883 from Westmorland.
Red flush over green. Cooks to sharp, juicy purée.
F 13. T^2. **C** bien. **P** e-Sept. **S** Sept-Oct.

TILLINGTON COURT 6 L DC
UK; acc prob variety exhib 1934; received 1988 from old trees in Burghill village Herefords, where Tillington Court is situated.
Handsome, colourful. Juicy, quite sharp yellow flesh.
F mid. T^2 **P** l-Sept. **S** Oct-Dec.

TINSLEY QUINCE 2/5 L D
UK; received 1942 from R. Fairman, Crawley, Sussex.
Claimed quince-like in taste but not noticeably so, although turning yellow by Nov. Sweet, firm flesh.
F 8. T^2; sprd. **P** e-Oct. **S** Nov-Dec.

TIROLER SPITZLEDERER syn Spitzlederer

TOBIÄSLER 4 L C
Switzerland; arose with Tobias Schmidheini in Canton of Baselland. Known 1805.
Deep maroon flush over dark green. Sharp, firm. Cooked, keeps shape; quite sweet. Formerly used also for cider.
F 15. T^2. **P** m-Oct. **S** Nov-Mar/May.

TOMMY KNIGHT 4 L D
UK; acc is variety now known St Agnes, Cornwall; desc 1946 (Taylor).
Brisk taste of fruit; firm, cream flesh. Probably also used for cider.
Formerly common in Cornwall.
F 10. T^2. **P** l-Oct. **S** Nov-Mar/Jun.

TOM PUTT 3 M C Cider
UK; raised late 1700s, either by Rev. Thomas Putt, rector of Trent, Dorset, (formerly Somerset) or by his barrister uncle, Tom Putt, at family estate, Gittisham, nr Honiton, Devon, and brought to vicarage.
Bright red flush, stripes. Quite sharp, but sweet, light, cooked.
Widely grown in west country and west Midlands for cider up to early 1900s; also sold as 'pot' fruit.
F 10. T^3; sprd; some res scab. **C** gd. **P** e-Sept. **S** Sept-Nov.

> **SIDNEY STRAKE**; striped sport, also paler red. Received from Cornwall 1954.

TOPSY 7 L D
UK; received 1948 from EMRS, Kent.
Sweet, pleasant; can be woody.
F 16. T^2. **P** e-Oct. **S** Oct-Dec/Jan.

TORDAI ALMA 3 E C
Transylvania (Romania); arose c1771 nr Torda; recorded 1881.
Large; handsome, slight flush over primrose. Cooks to sharp froth or purée, but soon fades once picked.
F 4. T^3. **P** use m-Aug-Sept.

TORDAI PIROS KÁLVI 4 E C
Hungary; received 1948 from Univ Agric, Budapest, but prob arose nr Torda, Transylvania (Romania).
Pretty, pink flush, stripes over primrose. Soft, astringent. Cooks, juicy, pleasant purée.
F 4. T^2. **C** hvy, frt sml. **P** use Aug/Sept.

TOWER OF GLAMIS (SCOTT) 1/5 L C
UK; acc received 1976 from J. Scott Nursery, Somerset; may be old Scottish variety known before 1800, desc Hogg (1884).
Large, heavy. Cooks to pale lemon, soft, sweet purée.
Tower of Glamis was local to orchards of Clydeside and the Carse of Gowrie in C19th.
F 5. T^3; sprd. **C** gd. **P** e-Oct. **S** Oct-Dec/Mar.

TRAJAN syns Polka, Scarletspire (US) 6 L D
UK; raised 1976, EMRS, Kent.
Wijcik McIntosh, Golden Delicious cross. Introd 1989.
Dark red McIntosh type, but firmer flesh; crisp, juicy, sweet. *See also* Ballerina.
F 5. T^2. **P** Sept. **S** Sept-Oct.

TRANSPARENTE DE BOIS-GUILLAUME 4 E DC
France; obtained 1895–1923 by M. Vilaire, Professor of Arboriculture, Jardins des Plantes, Rouen, Normandy; also known as Pomme Vilaire. Name suggests arose Bois-Guillaume, nr Rouen.
Transparente de Croncels seedling.
Large, ribbed. Refreshing with plenty of acidity; firm flesh. Cooked slices keep shape; quite rich, sweet taste.
Local to area around Rouen, where valued for early ripening – late July – and freedom from problems.
F 10. T^3; rep gd res disease, aphids. **P** l-Aug. **S** Aug-Sept.

TRANSPARENTE DE CRONCELS 5 M C
France; raised 1869 by nursmn Ernst Baltet of Croncels, Troyes, Champagne. Exhib 1881; recommended by French Pomological Congress.
Calville type; pale yellow with slight brownish red. Cooked, slices keep shape; bright yellow with intense, rich, sweet-sharp taste. Ideal for September 'Tarte aux Pommes'; makes sweet, rich baked apple, but delicate taste, easily overpowered. Later serves as eating apple.
Commercial variety in France, also Holland in 1920–1930s; still grown northern Europe and valued by French cooks.
F 7. T^3; sprd; hrdy. **C** gd; frt bruises. **P** e-Sept. **S** Sept-Oct.

TRÉLAGE 5 L C
France; desc 1948 (*Verger Français*), when grown Creuse, Corrèze (Limousin).
Cooked, keeps shape; brisk taste in Dec.
F 20. T^2; uprt. **C** hangs late. **P** l-Oct. **S** Dec-Mar/Apr.

TREZEKE MEYERS 4 L DC
Belgium, origin uncertain, local to East Flanders; received 1981 from Univ Antwerp.
Large; prominently ribbed. Can have intense vinous taste; juicy, quite sweet, firm flesh. Also 'first quality for kitchen' in Belgium.
F 14. T^2. **P** m-Oct. **S** Nov-Jan/Feb.

TRIERER WEINAPFEL syn Gelber Trierer Weinapfel

TROPICAL BEAUTY 4 L D
South Africa; raised c1930 by M. B. Strapp, Maidstone, nr Durban, Natal. Distributed 1953 by F. B. Harrington. Introd Australia 1958.
Slightly sweet, hard, crisp.
Planted subtropics in 1960s, in South Africa, Queensland, New South Wales, New Guinea, South America; now discarded as too poorly flavoured, too small.
F 11. T^2. **P** e-Oct. **S** Nov-Jan.

TROTUŞE (?) 8 L D
Romania; acc may be old variety that arose Trotuşul Valley, Moldavia.
Sweet, rather dry.
Trees long lived, 150–300 yrs old recorded. Used for roadside planting, high ground and now as rootstock.
F 13. T^2; v hrdy; rep gd res pests, disease. **C** hvy. **P** l-Oct. **S** Dec-Mar/Apr.

TSUGARA 5 L D
Japan; raised Aomori Apple Exp St.

Golden Delicious X. Introd 1975.
Resembles Golden Delicious, but firmer fleshed; weakly flavoured.
Main Japanese commercial variety.
F mid. T² . C hvy; frt sml unless thinnd.
P l-Oct. S Dec-Mar.

TUKKER 6 M D
Netherlands; raised by N. Hubbeling, Wageningen.
Lunteraan X Lonneker Apple. Received 1994.
Attractive, red/pink flushed and striped. Crisp, firm flesh.
F est 10. T² ; spd; res scab, mildew. P m-Sept. S Sept-Nov/Dec.

TÜKÖR ALMA 1/3 L C
Hungary; received 1948 from Univ Agric, Budapest.
Juicy, quite tannic. Cooked keeps shape, sweet, soft.
F 14. T² ; sprd. C hvy; frt var size. P m-Oct, S Nov-Jan/May.

TUMANGA syn Auralia

TUPSTONES 7 L D
UK; arose Worcs. Received 1945.
Sweet, firm flesh, but said to be nutty.
F 12. T² . P e-Oct. S Nov-Jan/Mar.

TUSCAN syns Bolero, Emeraldspire (US) 2 M D
UK; raised 1976, EMRS, Kent.
Wijcik McIntosh, Greensleeves cross. Introd 1989. *See also* Ballerina.
Crisp, juicy early September.
F 1. T² . P e-Sept. S Sept.

TWENTY OUNCE 4 L CD
USA; believed arose New York or Connecticut. Brought notice c1844. Syns many.
Very large. Quite sweet, savoury. Cooked, keep shape; lemon colour; sweet, but very light taste.
Grown for processing in West New York fruit belt and home orchards in early 1900s; still planted.
F 11. T² . P e-Oct. S Oct-Dec/Jan.

TYDEMAN'S EARLY syn Tydeman's Early Worcester

TYDEMAN'S EARLY WORCESTER, syn Tydeman's Early 6 E D
UK; raised 1929 by H. M. Tydeman, EMRS, Kent. McIntosh X Worcester Pearmain. Introd 1945.
Really ripe, shows strawberry flavour of Worcester and very juicy quality of McIntosh. Early commercial variety in British Columbia and France; small extent in England; also valued Netherlands.
Frt *Col* bright crimson, some darker rd stripes; pale yell ground; lenticels conspic as

white dots. *Size* med. *Shape* rnd-con; slt suggestion flt sides. *Basin* nrw, med dp; sltly puckered; trace beading. *Eye* sml, tightly clsd; sepals shrt, qte downy. *Cavity* nrw, dp; russet lined. *Stalk* lng, qte thin. *Flesh* white.
F 12. T² ; sprd; prt tip. C gd. P l-Aug/e-Sept. S Sept.

TYDEMAN'S EARLY SPUR TYPE

TYDEMAN'S LATE ORANGE 7 L D
UK; raised 1930 by H. M. Tydeman at EMRS, Kent.
Laxton's Superb X Cox's Orange Pippin. Introd 1949. RHS AM 1965.
Intensely rich and aromatic in Dec; sharper, stronger than Cox. Sweetens and by Mar only lightly aromatic, but still good.
Frt *Col* Dark purplish rd flush, like L Superb, darker rd stripes, over grnish yell/yell; some russet dots, patches, netting. *Size* med/sml. *Shape* rnd-con to con. *Basin* med wdth, dpth; sltly puckered; some russet. *Eye* lrg, open; sepals, lng, sep at base. *Cavity* med wdth, qte shal; russet lined. *Stalk* shrt/med; qte thck. *Flesh* yell.
F* 13. T³ ; lng whippy new growth like Laxton's Superb. C hvy; frt sml, unless thinned. P m-Oct. S Dec-Apr.

TYDEMAN'S MICHAELMAS RED syn Michaelmas Red 6 M D
UK; raised 1929 by H. M. Tydeman, EMRS, Kent.
McIntosh X Worcester Pearmain.
Deep red flush with bloom. Juicy, sweet, slight raspberry flavour; becomes strongly winey and scented, but often underlying metallic taste; tough skin.
F 12. T¹ . P e-Sept. S Sept-Oct.

TYLER'S KERNEL 3 L C
UK; exhib 1883 Congress by Mr Tyler of Hereford. RHS FCC 1883; desc Hogg.
Large; handsome. Quite rich flavour in Nov; deep cream flesh. Cooked keeps shape, plenty of flavour.
F 10. T² ; sp. P m-Oct. S Nov-Jan.

UDARRIA ZAGARRA syn Anisa

ULAND 1 M C
UK; raised 1922 by nursmn William Ingall, Grimoldby, Lincs.
Large. Very sharp, pale green flesh. Cooks to bright lemon purée, juicy, sharp; needs sugar.
F 7. T³ . P m-Sept. S Sept-Nov.

ULMER REINETTE 5 L D
Germany; received 1951. Pres arose in Ulm on Danube, nr Swiss border.
Perfectly conical, slight orange flush. Brisk with intense taste of fruit, but often sour.
F 9. T² . P e-Oct. S Nov-Jan/Feb.

UNDERLEAF 5 M D
UK; acc may be variety of Taylor (1946). Blenheim type flavour; sweet, plain taste of fruit, slightly crumbly texture. Underleaf is remembered as good Gloucs variety.
F 17. T² ; tip. P l-Sept. S Oct-Nov.

UNDINE 7 L DC
Germany; raised 1930s by M. Schmidt, Inst Agric & Hort, Müncheberg-Mark.
Jonathan X. Received 1977.
Refreshing, very juicy in Dec, plenty of sweetness, acidity; soft flesh; very tough skin. Grown eastern Germany, mainly as culinary apple.
F 8. T² . P m-Oct. S Nov-Jan/Mar.

UPTON PYNE (?) 4 L DC
UK; acc may be variety raised by nursmn Pyne of Topsham, Devon. Introd 1910.
Prominently pink striped over yellow. Cooks to purée, brisk, slightly pineapple-like flavour.
F* cerise; 11. T² . P e/m-Oct. S Nov-Feb.

VAJKI ALMA 5 M D
Hungary; found by G. Bencsik in garden of Count Boronkay in Vajk; sent 1871 to pomologist Bereczki.
Refreshing, good taste of fruit; sweet, crisp.
F 12. T³ . P m-Sept. S Oct-Dec.

VANDA 4 M D
Czech Rep; raised Exp Botany St, Strizovice.
Jolana (scab res selection) X Lord Lambourne; fruited 1983.
Intense savoury taste, brisk, juicy, soft pale cream flesh; reminiscent James Grieve or Gravenstein.
F mid. T² ; res scab, mildew. C hvy. P l-Sept. S Oct-Nov.

VENUS PIPPIN 1 E C
UK; believed raised c1800 either in Devon, or Tamar Valley, Cornwall.
Cooks to sharp purée. By Sept mellowing to sharp, eating apple.
F 14. T³ . P e-Aug. S Aug-Sept.

VERALLOT 3 L C
France; acc may be variety now known Yonne (Burgundy); desc 1948 (*Verger Français*) when also grown Aube (Campagne-Ardenne).
Sharp. Cooked keeps shape, but little flavour.
F 34. T¹ ; sprd. P l-Oct. S Dec-Mar.

VERDE DONCELLA 2/5 L D
Spain; received 1957 from France.
Light taste of fruit, hard white flesh.
Much grown in Midi-Pyrénées in 1950s
F 14. T² ; sprd. P m-Oct. S Dec-Mar.

VERDESE 2/5 L C
Italy; received 1958 from Univ Turin.
Cooks to gold purée, brisk, fruity.
F 14. T³. P l-Oct. S Jan-Mar.

VERDONA 3 L CD
Italy; received 1958; believed to be very old variety, known C17th.
Markedly conical. Sweet, juicy, fruity in March. Very light cooked flavour.
F 12. T². P l-Oct. S Jan-May.

VÉRITÉ 7 L D
France; acc is variety now grown Seine-et-Marne; prob not Vérité recorded 1876, but difficult to be certain.
Small, dark red flushed. Intense, brisk taste of fruit in Jan, but sharp; white, crisp, flesh. Used also for juice.
F 26. T²; sprd. C hvy. P l-Oct. S Jan-May.

VERNADE 8/7 L D
France; desc 1948 (*Verger Français*), when grown Cher and Indre (Val de Loire).
Intense sweet-sharp taste in Feb; firm, deep cream flesh.
F 20. T². P l-Oct. S Feb-May.

VERNAJOUX 2/5 L DC
France; desc 1948 (*Verger Français*), when grown Haute Vienne (Limousin).
Crisp juicy flesh, quite sweet in Jan.
Remains well known Haute Vienne. Used also for cooking, conserves, juice (vit C 15mg/100g).
F 22. T². C hvy. P l-Oct. S Dec-Mar/Apr.

VICAR OF BEIGHTON 6 L D
UK; recorded 1894; believed raised Beighton, Norwich.
Large; handsome. Sweet, soft, crumbling flesh.
F 14. T². P e-Oct. S Oct-Dec/Jan.

VICKING 6 E D
USA; raised by nursmn G. W. Gurney, Yankton, S. Dakota.
Parents include Jonathan, Delicious, Williams Early Red, Early McIntosh, Starr; introd 1925.
Brisk, soft, juicy flesh.
F est mid. T²; v hrdy. C hvy. P l-Aug. S Aug-Sept.

VICTORY (GEORGE CARPENTER)
3 M C
UK; raised by George Carpenter, HG, West Hall, Byfleet, Surrey.
Bismark X Blenheim Orange. Received 1923.
Large; colourful, like Bismark. Cooks to bright yellow purée, sharp; needs sugar.
F 13. T³. P m-Sept. S Oct-Nov.

VICTORY (USA) 6 L D
USA; raised by C. Haralson, Minnesota

Agric Exp St, Excelsior.
McIntosh X. Fruited 1918. Introd 1943.
McIntosh type. Strawberry flavour; sweet, plenty of acidity; crisp, juicy, glistening snow-like flesh; tough skin.
F 9. T². P e-Oct. S l-Oct-Dec/Jan.

VINCENT 7 L D
France; desc 1948 (*Verger Français*) when grown Seine et Marne.
Intense taste of fruit, quite rich, but sharp.
F 23. T². P m-Oct. S Jan-Mar.

VINNOE 5 L D
Russia; old variety. Received 1976.
'Wine Apple', but does not seem vinous in England. Strong, savoury taste; crisp becoming soft, juicy.
Commercial variety of St Petersburg area.
F 30; res frost. T²; res scab. C hvy. P m-Oct. S Nov-Jan.

VIOLETTA 6 L D
Italy; received 1958 from Univ of Bologna.
Deep scarlet flush with bloom. Crisp, fruity, refreshing in New Year.
F 16. T². P m-Oct. S Jan-Mar/Apr.

VIOLETTE 6 M D
France; acc is variety now known France; believed to be very old. Name mentioned 1628 by Le Lectier; reached England by 1670s. Syns many. Identical to Black Apple, RHS Wisley.
Beautiful blue black bloom like a violet; also claimed flesh was perfumed with violets, but this is not obvious. Brisk, crisp, white flesh.
Grown northern Franche Comté; used also cooking and source of pink juice.
F 4. T³. C hvy; frt drops. P m-Sept. S Sept-Oct.

VISTA BELLA 6 E D
USA; raised 1956 at Rutgers Univ, Agric Exp St, New Jersey.
Complex parentage involving July Red, Williams Early Red and Starr. Introd 1974.
Named because so successful in Guatemalan highlands. Dark red with bloom. Scented, flavoured with raspberries or loganberries; sweet, juicy, melting white flesh.
Grown eastern States, Canada, Guatemala, Italy, Norway, small extent England.
F 6. T³; prn canker. C gd, bien. P eat e-Aug.

VIRGINISCHER SOMMER ROSEN-APFEL see Vitgylling

VITGYLLING (?) syn Virginischer-Sommer Rosenapfel 1/5 E CD
North Europe; acc may be old variety, first desc 1816; believed introd Sweden from Netherlands. Syns many. Acc very similar to White Transparent.
Milky white, slight flush; its name means

white in Swedish. Brisk, plenty fruity flavour; soft flesh. Cooked, slices just keep shape; yellow, pleasant flavour.
Formerly widely grown in Sweden, where many old trees remain.
F 8. T². P l-Aug. S Aug-Sept.

VLAANDERNENS ROEM 6 L D
Belgium; received 1948.
Flame of Flanders. Rich, quite fragrant, fruity; sweet, juicy, soft flesh.
F 10. T². P e-Oct. S Oct-Dec/Jan.

VON ZUCCALMAGLIO'S RENETTE
syns Reinette von Zuccalmaglio, Zuccalmaglio 5 L D
Germany; raised 1878 by Diedrich Uhlhorn jn, engineer, fruit breeder, at Grevenbroich, Rhineland.
Ananas Reinette X Purpurroter Agatapfel.
Named after noted folk story writer and researcher into origins of fables. Russet freckled over gold, like Ananas Reinette with same intense taste of pineapple acidity, but not as strong. Crisp, juicy, deep cream flesh.
In Germany grown for juice manufacture, home use; also Netherlands.
F 9. T³. C bien. P m-Oct. S Nov-Mar.

VOYAGER 4 L D
UK; raised 1952 by A. R. King, Barnet, Herts.
Laxton's Superb X poss Monarch.
Quite intense vinous flavour; sweet, soft flesh.
F 10. T². P m-Oct. S Nov-Dec.

WADEY'S SEEDLING 6 L D
UK; raised 1919 by W. J. Wadey at Caterham, Surrey.
Handsome, wine red flush. Sweet, juicy, hard flesh; undistinguished flavour.
F* 19. T². P m-Oct. S Dec-Feb/Mar.

WADHURST PIPPIN (?) 4 L D
UK; acc may be variety of Hogg, Bunyard, which arose early 1800s, Wadhurst, Sussex.
Savoury, quite sharp, hard flesh.
F 10. T³. P e-Oct. S Nov/Dec-Feb.

WAGENER 4 L DC
USA; raised in 1791 by George Wheeler at Penn Yan, New York. His farm nursery was bought 1796 by Abraham Wagener, who brought variety to notice of New York Agric Soc in 1847. Exhib England by Messrs Thomas Rivers at 1883 Congress. RHS AM 1910.
Quite sweet, crisp. Early in season used for cooking. Famed for its very late keeping qualities. Standard market apple in USA by late C19th; exported to UK. No longer commercially important. In England, quite

popular garden apple 1920–30s.
Frt *Col* dull rddish brown flush, indistinct rd stripes over grnsh yell/gold; ltl russet. *Size* med. *Shape* flt-rnd, traces rnded ribs; hammered surface; qte irregular. *Basin* med wdth, dpth; ribbed; sometimes ltl russet. *Eye* sml, clsd; sepals shrt; qte downy. *Cavity* med wdth, dpth. *Stalk* med lng, qte thin. *Flesh* pale crm.
F 6. **T**^{1}; res rosy apple aphid. **C** gd, bien. **P** e/m-Oct. **S** Dec-Apr.

WALTZ syn Telamon

WANG YOUNG 6 L D
South Korea; received 1967.
Dark maroon with bloom. Sweet quite honeyed, like Delicious; soft juicy, deep cream, tinged green flesh.
F 14. **T**^{2}. **P** e-Oct. **S** Nov-Jan.

WARDEN 6 L D
UK; received 1967 from former Yugoslavia via Scotland.
Quite sweet, soft, juicy white flesh, but often green, woody.
F 13. **T**^{2}. **P** e-Oct. **S** Nov-Dec.

WARDINGTON SEEDLING 2 E D
UK; raised by D. Burchnall, HG to Lady Wardington at Banbury, Oxon.
Cox's Orange Pippin X. Fruited c1938.
Lightly aromatic, but dull appearance.
F 8. **T**^{1}; sprd. **P** m-Aug. **S** Aug-Sept.

WARNER'S KING 1 L C
UK; known late 1700s. Originally King Apple, but sent by Mr Warner to nursmn Thomas Rivers, who gave it prefix. Also known around Maidstone as Killick's Apple after fruit grower, and further north as D. T. Fish, after HG. RHS AGM 1993.
Cooks to sharp, strongly flavoured purée, but never as sharp as Bramley; mellows with keeping.
One of most popular Victorian cookers, grown in gardens all over Britain and for market; also recommended in 1890s for 'artistic' orchard.
Declining in commercial importance by early 1900s, as fruit easily bruised, but still found in many old gardens.
Frt pale grn turning pale yell, slight brownish pink flush; becomes greasy when stored. *Size* v lrg. *Shape* flt-rnd or con; rnded ribs, sltly flt sided; often lopsided; irreg. *Basin* qte brd, dp; sltly puckered. *Eye* lrg, clsd usually; sepals lrg, brd based, downy. *Cavity* brd, dp; russet lined. *Stalk* med lng, qte thck. *Flesh* wht, tinge grn.
F* 7; trip. **T**^{3}. **C** hvy; prn bitter pit. **P** l-Sept. **S** l-Sept-Dec.

WARREN'S SEEDLING (?) 1/5 L C
UK; acc may be variety exhib 1934 from Cambridge; desc Taylor (1946).
Large. Cooks to quite sharp, fruity purée.

F 23. **T**^{2}. **P** e-Oct. **S** Oct-Dec/Mar.

WASHINGTON STRAWBERRY 4 M D
USA; arose on Job Whipple's farm, Union Spring, Washington County, New York. Exhib 1849 at State Agric Soc Fair, Syracuse. Introd UK by nursmn Thomas Rivers.
Sweet, sprightly taste; perhaps hint strawberry flavour; soft, juicy, cream flesh. Home orchard fruit of C19th America. Also popular Victorian exhibition variety, grown under glass to achieve a beautiful red flush and 'rich flavour with fine perfume'.
F 7. **T**^{1}; sprd. **P** m-Sept. **S** Oct-Nov/Dec.

WAYNE 6 M D
USA; raised at NYSAES, Geneva.
Northwestern Greening X Red Spy. Selected 1951. Introd 1962.
Named after Wayne County in West New York fruit belt. Large; bright red Northern Spy colour. Intense taste of sweet strawberries; some acidity; soft cream flesh. Grown northern States; also used for processing.
F 16. **T**^{3}. **P** l-Sept. **S** Oct-Nov.

WAYSIDE 7 M D
UK; raised 1930 by Miss Cunningham, Wayside, Huntingdon Rd, Cambridge. Claimed Charles Ross seedling.
Rich, sweet-sharp taste, with definite raspberry flavour; soon becomes merely sweet.
F 11. **T**^{2}. **P** l-Sept. **S** Oct-Nov.

WEALTHY 4 M D
USA; raised c1861 by Peter Gideon on his farm in Excelsior Minnesota, from seed of Cherry Crab sent from Bangor, Maine. Named c1868. Exhib UK at 1883 Congress by nursmn George Bunyard. RHS AM 1893.
At best, sweet with strawberry flavour, soft, juicy, cream flesh, but can be much less exciting; tough skin.
Raised for hard Minnesota winters and named after Gideon's wife. Formerly important commercial variety; still grown north central America and New England; favourite with amateurs, used early for pies, sauce, preserves. Popular English garden variety 1920–30s, 'valuable for decoration alone'.
Frt *Col* blood rd flush, brd rd stripes; pale yell ground col. *Size* med. *Shape* flt-rnd; regular; trace rnded ribs. *Basin* brd, dp; ribbed; puckered. *Eye* sml, clsd; sepals, sml, qte downy. *Cavity* nrw, dp; russet lined. *Stalk* med, qte thin. *Flesh* wht.
F 12; gd pollinator. **T**^{1}; sprd; v hrdy; res red spider; res scab. **C** gd. **P** m/l-Sept. **S** l-Sept-Dec.

> **CASE WEALTHY**; syn Double Red Wealthy; deeper red; more open growth habit allowing fruit to colour. Discv 1933 by J. G. Case, in Sodus, New York; introd

1940.
> **LOOP WEALTHY**; discv by H. S. Loop in orchard of G. C. Smith & Son, North East Philadelphia; tetrap.
> **STEVENSON WEALTHY**; tetrap; discv c1934 by J. R. Stevenson in Cayuga, New York.

WEBSTER 6 L D
USA; raised 1912 by R. Wellington NYSAES, Geneva.
(Ben Davies X Jonathan) X (Ben Davies X Jonathan). Introd 1938.
Large; bright red. Fruity, sharp cream flesh in Nov. Dual purpose in US.
F 7; trip. **T**^{1}. **P** e-Oct. **S** l-Oct-Dec.

WEIDNERS GOLDREINETTE syn Reinette Weidner 7 L D
Germany; raised 1844 by Weidner, miller and keen fruit man at Grasmühle, nr Nürnberg. Orleans Reinette seedling.
Savoury, yet also 'nutty', reminiscent of Orleans Reinette; quite soft, juicy, cream flesh.
F 14. **T**^{1}. **P** m-Oct. **S** Dec-Mar/Apr.

WEIGELTS ZINSZAHLER 4 M D
Germany; received 1951 from Fruit Res St, Jork, nr Hamburg.
Rich, sweet-sharp taste; mellowing to very sweet; curious crumbling texture.
F 10. **T**^{2}. **P** m-Sept. **S** Sept-Nov.

WEIGHT 3 L C
Received 1967 from former Yugloslavia via Scotland.
Sharp, becoming sweeter; plenty of fruit. Cooked, sweet, lightly flavoured.
F 13. **T**^{2}. **P** e-Oct. **S** Oct-Dec.

WEISSER KLARAPFEL syn White Transparent

WEISSER ROSMARIN syn Rosmarina Bianca

WEISSER WINTERGLOCKENAPFEL syn Glockenapfel

WEISSER WINTER TAFFETAPFEL 5 L DC
Germany; recorded 1800. Syns many.
Strong savoury taste of fruit; quite sharp early Nov; firm white flesh.
Grown particularly in Hannover in C19th; popular and dual purpose throughout Germany.
F 11. **T**^{3}. **P** l-Sept. **S** Oct-Jan/Feb.

WEISSES SEIDENHEMDCHEN syn Seidenhemdchen 4/6 M D
Europe; Chemisette Blanche of Leroy (1873), which he believed to be very old. As syn 'de Demoiselle' recorded by Bauhin (1598) and known in Würtemberg and Switzerland. Chemisette de Soie of C18th

German and Dutch pomologists.
Its pale cream and pink skin is perhaps evocative of lady's silk chemise. Sweet, refreshing taste; soft juicy, white flesh.
F 10. T³. P l-Sept. S Oct-Dec/Jan.

WELCOME 4 E D
Received from New Zealand 1953.
Small. Some aromatic quality; sweet, soft juicy flesh.
F 7. T². P l-Aug. S Aug-Sept/Oct.

15 WELDAY JONATHAN *see* Jonathan

19 WELDAY JONATHAN *see* Jonathan

WELLINGTON syn Dumelow's Seedling

WELLINGTON BLOOMLESS 4 L D
USA; known 1929.
Oblong, waisted shape. Quite sweet, pleasant flavour; deep cream flesh.
Flowers have no petals.
F 24. T². P e-Oct. S Oct-Dec.

WELLSPUR DELICIOUS *see* Red Delicious

WELSCHISNER 1/3 L C
Germany; recorded 1889.
Large. Sharp, cream tinged green flesh.
Cooks to brisk, fruity, yellow purée.
F 6. T². P l-Oct. S Dec-Mar.

WESTFIELD SEEK-NO-FURTHER 7 M D
USA; arose Westfield, nr Springfield, Mass in 1700s. Spread to Connecticut, taken to Ohio 1796.
Sweet and nutty, reminiscent of old English varieties, such as Adams' Pearmain. Favourite eating apple of C19th New England; now revived by amateurs.
F 12. T²; uprt. P l-Sept. S Oct-Dec.

WESTON'S SEEDLING 3 L C
UK; received from EMRS; exhib by them 1934.
Large. Deep cream, soft, sharp flesh.
Cooked, very light, little flavour.
F 12. T³. P e-Oct. S Oct-Nov.

WEST VIEW SEEDLING 7 L D
UK; found 1932 by F. W. Rainbird in Billericay, Essex.
Some sweet- sharp quality in Dec; very low key.
F 15. T². C lght. P l-Oct. S Dec-Mar.

WHEELER'S RUSSET (POTTER) 8 L D
UK; acc received 1944 when identified by J. Potter as Wheeler's Russet, but small differences with variety of Hogg, Bunyard. Wheeler's Russet, according to Hogg, was presumed raised by nursmn James Wheeler of Gloucester, but listed 1717 by Brompton Park Nursery, London, which was before Wheeler's time.
Cinnamon russeted. Brisk in Jan, some fruit and hint of perfume, which becomes stronger.
F* 12. T². P l-Oct. S Jan-Mar/Apr.

WHITE ASTRACHAN 5 E DC. Not in NFC.
Russia or Sweden; recorded 1748 or earlier.
Widely grown over Northern Europe and Russia where still valued.
Pale yellow, ribbed. Ripening in August, with juicy, sharp white flesh.

WHITE JOANETING syn Joaneting

WHITE MELROSE syn Melrose

WHITE QUARRENDEN (?) 5 E D
UK; acc may be variety recorded 1831; exhib 1888 by Veitch, Exeter. Acc received 1979 from collection of O. D. Knight, Norfolk; originally from Hanniford Nursery, Paignton, Devon in c1918.
Large; pale yellow. Soft, quite sweet but definite acidity.
F 11. T². P use Aug.

WHITE TRANSPARENT syn Weisser Klarapfel 5 E CD
Russia; arose Russia or Baltic States. Introd Europe early 1800s; North America 1870. Syns numerous; include Yellow Transparent, USA; Klarapfel, Germany; Transparente blanche, Sweden.
Refreshing, well-flavoured, soft pale cream flesh; plenty of acidity, but can be too sharp for pleasure. Cooks to cream purée, sweet, balanced, flavoursome.
Ripening by 5th of July, in States where, as Yellow Transparent, became early market dual purpose apple, now replaced by firmer fleshed Lodi. Widely grown Europe in C19th; still grown Russia, Sweden, Austria, Germany, but replaced in Alte Land by Astramel; never very popular in England.
F 6. T²; v hrdy. C gd. P use l-July-eAug.
  PERRINE YELLOW TRANS-PARENT; tetrap. Discv c1930 by D. B. Perrine in Centralia, Illinois. Introd 1961. Ripens week earlier.

WHITE WINTER CALVILLE syn Calville Blanc d'Hiver

WHITE WINTER PEARMAIN (SOUTH AFRICA) 2/5 L D
Acc is not White Winter Permain of US, described by Beach (1905), catalogued 1858, but variety grown in South Africa under this name in 1962, when sent to NFC by D. Graaf.
Sweet, honeyed, plenty of balancing acidity, fruit; deep cream, firm flesh. Formerly South African export variety, but overtaken by Granny Smith.
F 10. T². P m-Oct. S Dec-Mar.

WHORLE PIPPIN syn Thorle Pippin

WICKHAM GREEN 1 L C
UK; grafts taken 1913 by nursmn William Pope of Wokingham from tree in Wickham, Berks, which was prob growing c1860.
Cooks to cream purée, full of fruit, sharp, strong in Dec.
F 18. T². C lght. P m-Oct. S Dec-Mar/Apr.

WIDDUP 7 L D
Received 1961 from New Zealand.
Delicately flavoured, slightly scented by Dec.
F 8. T²; sprd. P e-Oct. S Nov-Jan/Mar.

WILHELM LEY *see* Belle de Boskoop

WIJICK *see* Ballerina, McIntosh

WILKENBURGER HERBSTREINETTE syn Reinette d'Automne de Wilkenburg 7 E D
Recorded 1861; listed 1895 by nursmn Simon-Louis Fréres, France.
Small. Intense, rich, quite aromatic flavour.
F 3. T². P l-Aug. S Sept-Oct.

WILLIAM CRUMP 6/7 L D
UK; claimed raised by Carless, foreman at Rowe's Nurseries, Worcester, who introduced it, but when shown to RHS in 1910 recorded as raised on Madresfield Court estate where the distinguished fruit man, William Crump was HG
Cox's Orange Pippin X Worcester Pearmain. RHS AM 1908; FCC 1910.
Intense, aromatic flavour; rich, sweet with masses of pineapple acidity in Nov. Mellows, becoming more like Cox.
Frt *Col* bright orng rd flush, rd stripes, over grnish yell/yell; much russet as dots, netting, patches. *Size* med/lrg. *Shape* rnd-con to shrt-rnd-con; sltly ribbed. *Basin* med wdth, dpth; sltly ribbed; ltl russet. *Eye* hlf open or clsd; sepals qte lng, v downy. *Cavity* brd, dp; russet lined. *Stalk* shrt/med, thck. *Flesh* nearly yell.
F 11. T³; uprt. C gd. P m-Oct. S Dec-Feb.

WILLIAM PETERS 4 M D
UK; raised by Peters of Leatherhead, Surrey. RHS AM 1917.
Like early, brisk Cox; rich, quite intensely flavoured.
F 13. T²; sprd. P e-Sept. S Sept-Oct.

WILLIAMS FAVOURITE 4 E D
USA; arose c1750 in Roxbury, Mass. Brought notice of Massachusetts Hort Soc 1850, then named Williams, but previously known as Queen, Lady's Apple. Introd England c1828. Also known on continent. Syns many. RHS AM 1895.
Boldly red striped, flushed. Sweet, light raspberry flavour, which rapidly becomes winey; soft, juicy, cream tinged green flesh.

Early market apple of C19th New England, especially around Boston; long superseded. F 13. T². C hvy. P eat Aug.

WILLY SHARP 1 E C
New Zealand (prob); desc 1917.
Quite sharp. Cooked keeps shape; little flavour. Formerly grown New Zealand, New South Wales.
F 9. T². C hvy; frt sml. P Aug. S Aug-Sept.

WILSTEDTER 7 L D
Germany; received 1951 from Fruit Res St, Jork, nr Hamburg.
Sprightly, fruity, refreshing; soft flesh.
F 17. T². P l-Oct. S Dec-Feb/Apr.

WINESAP 6 L D
USA; arose New Jersey before 1800; first recorded by Dr James Mease in 1804 as cultivated by Samuel Coles of Moore's Town, NJ. In 1817 Coxe described it as 'favorite cider fruit of West Jersey'.
Dark red flush. Sweet, juicy, fruity until spring in good year, but in English climate usually does not fully ripen or colour well. In US 'named for its sprightly, winey flavor'. Widely planted southern States in C19th, especially Virginia; dual purpose, also used for cider. Valued for its good crops and keeping qualities and major variety up to 1950s. Overtaken by Stayman's Winesap, Red Delicious, but still grown; also in Turkey, Italy, France.
F 10. T². P l-Oct. S Dec-Mar/Apr.
   DERMEN WINESAP; tetrap.

WINSTON 7 L D
UK; raised 1920 by nursmn William Pope at Welford Park, Berks.
Cox's Orange Pippin X Worcester Pearmain. Introd 1935 as Winter King, renamed Winston 1944. RHS AM 1935, 1951; AGM 1993.
Aromatic, rich, quite like Cox, but sharper. More colourful by New Year and mellows to more delicate flavour.
Grown small extent for sale; popular garden variety; also valued Netherlands.
Frt *Col* dull purplish rd flush, brd stripes over grnish yell; becoming bright rd over yell; some russet; lenticels as russet dots. *Size* med/sml. *Shape* rnd-con to oblng-con. *Basin* med wdth, dpth; sltly puckered; ltl russset. *Eye* prt open to clsd; sepals, lng, nrw, qte downy. *Cavity* med/nrw, qte dp; ltl russet. *Stalk* shrt/med, thck. *Flesh* pale crm.
F 14. T². C gd, frt sml. P m-Oct. S Dec-Apr.
   WINSTON SPORT (Dingwall); uniform red flush. Arose c1952 Hort Centre, Loughall, N Ireland.

WINTER BANANA 5 L D
USA; arose 1876 on farm of David Flory, nr Adamsboro, Cass County, Indiana. Introd 1890. RHS AM 1912.
Yellow with slight blush. Sweet, scented in good year; juicy, melting flesh; in US develops 'aromatic flavour similar to bananas'.
Grown early 1900s in Washington, British Columbia and exported to UK. Currently being revived by American amateurs. Planted 1920s in English gardens, but needs warmer climate. Still grown small extent on continent; in Germany used for juice manufacture and valued garden fruit.
F 12. T². P e-Oct. S Nov-Jan/Mar.

WINTER CODLIN 1/5 M C
UK; recorded 1831; desc Hogg (1851), when believed to be old variety.
Large, oblong, resembling Catshead. Cooks to very lightly flavoured purée.
F* lrg, mauve/wht 15. T². P e-Sept. S Sept-Oct.

WINTER GEM 4 L D
UK; raised 1975 by H. Ermen, Faversham, Kent.
Grimes Golden X Cox's Orange Pippin. Introd c1993 by F. W Matthews Nursery, Tenbury Wells, Worcs.
Handsome, orange red over gold. Tasty and rich, good balance of sugar and acidity, crisp, juicy, pale cream flesh.
Popular garden apple; also introd to France.
F est 11. T²; prt res scab. C gd. P e-Oct. S Nov-Jan; commercially-Feb.

WINTER LEMON 5 L D
Ukraine; raised at Scientific Inst Hort, Kiev. Introd 1968.
Lemon yellow, markedly conical. Refreshing; crisp, juicy; high vit C content. Grown Ukraine.
F* 21. T²; v hrdy. C hvy. P m-Oct. S Dec-Feb/Mar.

WINTER MAJETIN 1/3 L C
UK; believed arose Norfolk. Recorded 1820; desc Lindley (1831).
Cooks to firm purée, good strong taste; some richness.
Hardy, never failing to crop according to Lindley; well known C19th in Norwich markets. Resistant to root woolly aphid; formerly used Australia as rootstock.
F 18. T². C frt sml unless thinned. P l-Oct. S Dec-Apr.

WINTER MARIGOLD (?) 4 L D
UK; acc may be variety of Hogg (1884).
Sweet, light flavour, slightly winey taste; crisp, juicy.
F* 10. T²; sprd. P l-Sept/e-Oct. S Oct-Nov/Dec.

WINTER PEACH (?) 5 L D
USA poss; received 1979 from collection of O. D. Knight, Norfolk; originally from Hanniford Nursery, Paignton, Devon in c1918. Acc may be variety known England 1853, desc by Hogg, Bunyard.

Large; turning pale yellow, slight blush. Brisk, firm flesh in Dec, but Bunyard found 'slight spicy flavour'.
F 13. T². P l-Oct. S Dec-Feb/Mar.

WINTER PEARMAIN (?) 7 L DC
UK; believed very old English variety, but origins confused. Syns many. Acc may be Winter Pearmain of Hogg (1884). In RHS Wisley as Winter Queening.
Quite rich, aromatic, deep cream flesh, but underlying astringency early in season. Cooked, slices keep shape, bright yellow with rich, sweet-sharp flavour; also makes stiff purée.
With keeping becomes sweeter, less interesting cooked.
Possibly variety grown in Kent until at least 1890s for London markets, where sold as Winter Pearmain and Winter Queening.
F 6. T². C gd. P e-Oct. S Oct-Jan/Mar.
PLATE 1 (blossom)

WINTER QUARRENDEN 6 L D
UK; introd by Pearson's Nursery of Nottingham; recorded 1896; desc 1920 by Bunyard.
Deep maroon flushed. Sweet but coarsely flavoured and textured.
F 8. T². P e-Oct. S Oct-Dec.

WINTER QUEENING *see* Winter Pearmain

WINTER STUBBARD (?) 5 L C
UK; received 1948 from Launceston, Cornwall; impossible to know if variety recorded 1883 Congress, or variety exhibited 1934 as Devon cider apple as descs too brief.
Large. Sweet, slightly vinous flavour; soft, juicy cream flesh. Cooked keeps shape; insipid.
F 16. T²; sprd. P e-Oct. S Oct-Dec.

WITHINGTON FILLBASKET 1/3 M C
UK; exhib 1883 Congress by RHS; desc Hogg (1884).
Large. Cooks to juicy lemon purée; very sharp, with sugar good intense flavour.
F 9. T³; sprd; tip. C prn bitter pit. P e-Sept. S Sept-Nov.

WOLF RIVER 6 M DC
USA; arose nr Wolf River, Fremont County, Wisconsin with W. A. Springer. Believed Alexander seedling. Recorded 1875; listed 1881 by American Pom Soc.
Very large, dark red flushed. Soft, juicy flesh. Dual purpose in States – 'one apple makes a pie'. Resembles and overtook Alexander as market fruit in western US; now only grown by amateurs.
F 10. T³; sprd; hrdy. C reported res scab, mildew. P m-Sept. S Sept-Oct.

WOODFORD 5 M C
UK; known c1900 in Essex.
Large. Light flavour; insipid cooked.
Known Colchester area.
F* dp pink, 14. T³. P e-Sept. S Sept-Oct.

WOOLBROOK PIPPIN 7 M D
UK; raised 1903 by J. H. Stevens & Son,
Woolbrook Nursery, Sidmouth, Devon.
Cox's Orange Pippin seedling. RHS AM
1929.
Lightly aromatic, like weak Cox, but
ripening earlier; rather coarse flesh.
F 10. T². P m-Sept. S Oct-Dec.
RHS AM 1930.

WOOLBROOK RUSSET 8 L C
UK; raised 1903 by J. H. Stevens & Son,
Woolbrook Nursery, Sidmouth, Devon.
Bramley's Seedling X King's Acre Pippin.
RHS AM 1930.
Large, netted, dotted with russet. Sharp,
mellowing to quite rich, sweet-sharp taste.
Cooked, quite good flavour.
F 12. T². P m-Oct. S Dec/Jan-Mar.

WORCESTER CROSS 6 M D
UK; raised 1920 by G. T. Spinks, LARS,
Bristol.
Cox's Orange Pippin X Wealthy. Introd
1932.
Sweet, fruity, aromatic flavour; can be very
attractive.
F 8. T². P e/m-Sept. S Sept-Oct.

WORCESTER PEARMAIN 6 E D
UK; arose with market gardener Mr Hale, of
Swan Pool, St Johns, nr Worcester.
Believed Devonshire Quarrenden seedling.
Richard Smith of St John's Nursery,
Worcester, bought grafts in 1873; showed it
to RHS; FCC 1875. Smith selling trees at
guinea each in 1876. RHS AGM 1993.
Bright red and sunbaked, densely sweet
with intense strawberry flavour; firm, juicy,
white flesh. Can be rather cloying;
reminding Morton Shand of 'boiled sweet
flavoured with synthetic pear juice'. Often
picked early and less colourful and tends to
be chewy with little definite flavour.
England's main early autumn commercial
variety since late 1800, but now only grown
small extent for market.
Remains widely popular garden apple. Its
bright fruits and distinctive blossom –
almond opening to silvery white – led to its
decorative use in shrubbery in 1890s.
Frt Col bright rd flush, some faint rd stripes,
grnish yell/pale yell background; lenticels
conspic as russet dots. Size med. Shape rnd-
con; sometimes sltly lop-sided; sltly ribbed.
Basin nrw, med/shall; sltly ribbed; often 5
beads. Eye sml, clsd; sepals sml, qte downy.
Cavity med wdth dpth; russet lined. Stalk
shrt, qte thck. Flesh wht.
F* 11. T²; prt tip. C hvy. P e/m-Sept.

S Sept-Oct.
PLATE 7

WORCESTER WOODSIL 6 M D
UK; raised prob late C19th; received 1992.
Attractive red flush; crisp, juicy cream flesh;
good fruity flavour.
F est 11. T². P l-Sept. S Oct-Nov.

WRIXPARENT 5 E CD
USA; discovered 1920 by Wrixham
McIlvaine in Magnolia, Delaware.
White Transparent seedling or sport. Introd
1940.
Resembles large White Transparent, but
ripening earlier.
Sharp, firm. Cooks to brisk purée, but not as
well-flavoured as White Transparent.
F 9; tetrap. T³. P use l-July-e-Aug.

WYKEN PIPPIN 5 L D
UK; either raised early 1700s from pip of
continental apple by Lord Craven at Wyken
nr Coventry or, according to Bunyard,
introd from Holland c1720. Well known
England by late 1700s.
Brisk, densely fruity taste of Golden Pippin
type; similar to Downton Pippin, yet also
certain aromatic quality.
Among top dozen garden varieties in 1883;
also grown for market in West Midlands. By
1920s popularity waned, but still considered
'fine old fruit worth preserving'.
Frt Col gold, sltly more brownish yell flush;
many fine russet dots, ltl netting of russet.
Size sml/med. Shape rnd to flt-rnd. Basin
brd, qte shal, puckered. Eye lrg, open;
sepals, shrt, sep at base, qte downy. Cavity
nrw, shal; russet lined. Stalk shrt, thck.
Flesh crm.
F 14. T²; sp; uprt. C gd. P m-Oct. S Nov-
Jan.

YAMBORKA 5 L C
Bulgaria; arose 1934 in Kjustendil, south
west Bulgaria; fruited 1945.
Pale yellow. Soft fleshed, but seems
undeveloped in England.
F 14. T²; some res scab, mildew. P m-Oct.
S Nov-Jan.

YELLOW BELLFLOWER 5 L DC
USA; arose nr Crosswicks, Burlington
County, New Jersey. Original tree very old
in 1817. Distributed over North America,
Europe. Syns numerous, include Belle
Fleur Jaune.
Long, conical yellow apple which hangs
like bell on tree. Rich taste; plenty of sugar,
acidity; firm almost yellow flesh. Cooks to
gold purée; sweet, quite rich.
Widely grown north east America in C19th,
taken to mid-west and California, where
became leading commercial variety.

Overtaken by Golden Delicious in USA,
but still favoured by amateurs. Grown in
Ukraine, also widely found France,
Germany, Switzerland; never popular
England.
F 9. T²; sprd. P e/m-Oct. S Nov-Mar.

YELLOW DELICIOUS (Stauffer Strain)
see Golden Delicious

YELLOW INGESTRIE 5 E D
UK; raised c1800 by T. A. Knight prob at
Elton Manor, nr Ludlow, Shrops.
Orange Pippin X Golden Pippin.
Named after Lord Talbot's estate, Ingestrie
Hall, Staffs. Sharp, densely fruity taste. Best
in early Sept, still firm and good in Oct.
In C19th widely grown in gardens,
especially in north; also recommended in
1890s as 'charming lawn tree apple on
account of its beautiful drooping habit'.
Grown in Kent for London markets; sold as
'Summers' which 'although small are so
well known their market value is higher
than larger more showy fruits'. Still
remembered in Kent. Highly decorative in
table displays, often keeping until
Christmas and ideal for wiring onto
evergreens to make Kissing Boughs and
sprays.
Frt Col grnish yell turning yell, gold cheek.
Shape oblng-con to oblng; slight ribs; flt
base. Size sml. Basin brd, shal. Eye hlf open.
Cavity brd, shal; ltl russet. Stalk shrt, thck.
Flesh yell.
F* 9. T¹; spd; hrdy. C gd. P e-Sept. S Sept-
Oct.
PLATE 4

YELLOW NEWTOWN PIPPIN *see*
Newtown Pippin

YELLOW PITCHER (?) 5 L D
Ireland; may be variety desc 1951 (Lamb).
Very sweet, juicy. Prob used as sweet cider
apple. Yellow Pitcher claimed roots from
cuttings - pitchers.
F 21. T². P l-Sept. S Oct-Nov/Dec.

YELLOWSPUR *see* Golden Delicious

YELLOW TRANSPARENT syn White
Transparent

YE OLD PEASGOOD 4 E D
UK; prob raised by George Carpenter, HG,
West Hall, Byfleet, Surrey. Received 1932.
Large; pink flush over pale yellow. Soft,
sharp, white flesh.
F 11. T². P eat e-Aug.

YORK-A-RED *see* York Imperial

YORK IMPERIAL 4 L D
USA; arose early 1800s on farm of Mr
Johnson, nr York, Pennsylvania. Propagated
by local nursmn as Johnson's Seedling, but

279

with little sale for trees these were discarded. Rescued and distributed by passing Pennsylvanian farmers. Downing called it 'Imperial of keepers' and renamed it. Represented in NFC by sports.
Sweet, quite honeyed, but rather coarse flesh.
Formerly grown in Pennsylvania, Virginia for export to England in 1920–30s. Still leading variety of Appalachian region, but now grown for processing as yellow flesh gives attractive colour to sauce, pie filling; also bonus of small core, less wastage.

> **PERRINE YORK** 4 L D; tetrap. Found by D. B. Perrine, Centralia, Illinois.
> **YORK-A-RED**; red flush. Discv 1931 by Paul Lingamfelter, Hedgesville, West Virginia; introd 1937 by Stark Bros Nurseries, Missouri.
> F 15. T^2. P i-Oct. S Jan-Mar/Apr.

YORKSHIRE AROMATIC 5 M C
UK; received 1945 from Tyninghame Gardens, East Lothian, Scotland.
Cooks to rich, juicy purée; quite sharp.
F 10. T^2. P m-Sept. S Oct-Dec.

YORKSHIRE GREENING 3 L C
UK; pres arose Yorks. Recorded 1803 by Forsyth, but name cited in 1769 list of nursmn William Perfect, Pontefract, Yorks; desc Ronalds, Hogg. Syns include Yorkshire Goosesauce.
Cooks to very strong, sharp, pale green purée; with sugar, fruity, good taste.
Remains popular northern garden apple.
F 9. T^1; sprd; hrdy. C gd. P m-Oct. S Dec-Mar/Apr.

YOUNG AMERICA 6 E
USA; arose New York; introd 1800s.
Ornamental flowering crab, bearing small, bright red small apples.
F 8. T^2. P l-Aug-e-Sept.

YOUNG'S PINELLO 4 L D
UK; raised by Miss E. L. Young, Letchworth, Herts. Received 1935.
Refreshing, fruity, crisp in Nov; can be sweeter, quite perfumed.
F 9. T^2. P m-Oct. S Nov-Dec/Jan.

ZABERGÄU RENETTE 8 L D
Germany/Belgium; obtained 1885 at Hausen on Zaber, Württemberg, distributed 1926 according to German pomologist, Petzold. French and Belgium pomologists, however, believe it to be Reinette Parmentier obtained by Parmentier, Belgium nursmn, in 1830 and widely distributed in northern Europe.
Bronzed russet, often large. Intense sweet-sharp taste, firm flesh. By Jan sweeter, quite

nutty, sometimes little of 'russet' flavour associated with Egremont Russet.
Grown Germany, where considered 'less sharp substitute for Belle de Boskoop'. As Reinette Parmentier known Belgium, France.
F* 10; trip. T^2; uprt. C gd. P m-Oct. S Dec-Mar.

ZELENKA KHARKOVSKAYA 4 L D
Ukraine prob; arose first half 1800s.
'Green Apple from Kharkov', former capital of Ukraine. Rich flavour, sugary acidity in Dec.
F 13. T^3. P l-Oct. S Dec-Mar/Apr.

ZHIGULEVSKOE 4 M D
Russia; raised at Pavlovsk Exp St, of Vavikov Inst, St Petersburg.
Borovinka (Duchess of Oldenburg) X Wagenar Prizovoe. Received 1976.
Named after place on Volga. Savoury, soft, juicy. Widely grown central Russia.
F 4. T^2. P l-Sept. S Sept-Nov.

ZIGEUNERIN 4 E DC
Latvia; poss arose Riga; introd to Netherlands by Vallen of Swalmen.
Dark red flushed over pale yellow; tall, ribbed, like Red Astrachan; its name means Gipsy. Refreshing, juicy soft flesh. Cooks to juicy purée, lightly flavoured.
Remains popular Netherlands, mainly for its appearance.
F 5. T^1. P use Aug.

ZLATAVA 4 L D
Czech Republic; raised 1974 at Exp Botany St, Strizovice.
Lord Lambourne X Blahova Oranzova.
Sweet, honeyed, crisp, juicy, but rather light on flavour.
Never developed as commercial apple in Czech, but grown small extent by amateurs.
F 12. T^2; mod res scab, mildew. C prn bitter pit. P e-Oct. S Oct-Dec/Feb.

ZOETE ERMGAARD 7 L D
Netherlands; known since 1864.
Pink flushed over cream. Tastes like sugared almonds; scented, very sweet, firm flesh, slightly nutty. Little juice or taste of fruit; often tannic undertones. Long popular in Holland for drying; after soaking, eaten mixed with mashed potatoes.
F 19. T^2. P m-Oct. S Dec-Jan.

ZOMER DELICIOUS 5 M D
Netherlands; raised at IVT, Wageningen.
Pale milky yellow, freckled in fine russet; reminiscent of Golden Delicious. Sweet, rather empty of flavour; soft juicy flesh; tough skin.
F 11. T^2. P e-Sept. S Sept-Oct.

ZUCCALMAGLIO see Von Zuccalmaglio's Renette

ZWEIGELTS ZINSAHLER syn Weigelts Zinszahler

The DEFRA Cider Collection contains many of the cider apple varieties which are widely grown in England and other lesser known varieties of merit. It was formed at Long Ashton Research Station (LARS) and moved to Brogdale in 1991.

A number of varieties have since been addded to this collection. The main cider variety collection for commercial growers is maintained by H. P. Bulmer & Co., Hereford.

Cider apple varieties are categorised on the basis of their juice, into bittersweets, bittersharps, sweets and sharps. Bittersweets and bittersharps are high in tannin.

Bittersweets have high sugar levels and bittersharps are high in acidity. Sweets are high in sugar and low in tannin and sometimes have also served as juicy, eating apples. Sharps are low in tannin and high in acid and similar to culinary apples.

Bittersweets: tannin levels above 0.2% and acid levels below 0.45%
Bittersharps: tannin levels above 0.2% and acid levels above 0.45%
Sweets: tannin levels below 0.2% and acid levels below 0.45%
Sharps: tannin levels below 0.2% and acid levels above 0.45%

Most ciders are made from a blend of a number of different varieties, but a few 'vintage' fruits will make a well-balanced cider on their own.

Abbreviatons as for main Directory.
F: flowering period. e: early, up to approx May 10. m: mid season, approx May 11–19. l: late, from approx May 20 onwards. Recent mild winters have advanced these dates.
T: tree vigour and habit; T^1 – weak; T^2 – medium vigour; T^3 – vigorous.
C: crop production; prec (precocious); bien (biennial).
H: harvest. e: early; m: mid; l: late.

ASHTON BITTER bittersweet
Raised 1947 by G. T. Spinks, LARS, Bristol. Dabinett X Stoke Red. Fruit conical, medium size; highly coloured with orange red flush and stripes. Widely planted 1990s in intensive cider orchards.
F m. T^2. C prec. H l-Sept.

ASHTON BROWN JERSEY bittersweet, vintage
Arose Somerset; one of local unnamed varieties tested at LARS in early 1900s.
Small; red flush and stripes over orange yellow. Produces high quality juice and full bodied, medium bittersweet cider. Found Herefords, Somerset in farm orchards planted in 1920–30s.
F lm. T^3; slow to bear; res scab. H e-Nov.

BACKWELL RED medium sharp, vintage
Widely grown early 1900s around Backwell village, north Somerset.
Small fruit; dark red flushed, striped. Produces acidic juice, little astringency and 'sharp, light, fruity, but rather thin cider'. Declined in favour due to irregular crops and difficulty in harvesting as mature fruit cannot be readily be shaken down.
F m. T^2. H l-Oct.

BELLE FILLE DE LA MANCHE mild bittersweet
Widely grown Normandy. Collected 1984 by LARS from nr Caen.
Fruit red flushed over yellow; medium size.
F e. T^2. H m-Sept.

BLACK DABINETT bittersweet, vintage
Arose Kingsbury Episcopi, Martock area of Somerset.
Resembles Dabinett, but darker colour; poss Dabinette seedling. Grown very small extent Somerset.
F m. T^2. H Nov.

BLACK VALLIS syn Vallis Apple' sharp
Originated in North Somerset where usually known as Redskins.
Striking, scarlet flushed. Quite large, flat-round fruits, which were sold as a sharp eating and cooking apple as well as used for cider. Produces 'pleasant but not outstanding cider'.
F l. T^3. H m/l-Oct.

BREAKWELL'S SEEDLING medium bittersharp
Arose 1890s on Perthyre Farm, nr Monmouth; propagated by Mr George Breakwell.
Dark red flushed small fruit; very dark foliage; seedling of Foxwhelp type. Mildy sharp juice, sometimes slightly astringent; produces 'rather thin, light cider of average quality'. Formerly grown Gloucs, Herefords; planted modern intensive orchards.
F e/m. T^2 hvy crops; v bien. H l-Sept/e-Oct.

BROAD-LEAVED HEREFORD bittersweet
Exceptionally large leaves give the name. Accs received 1992 from Taunton, Somerset; may be variety listed 1886 (Hogg & Bull), but insufficient details to be certain. Large, green fruit with netting of russet.
H l-Sept/e-Oct.

BROWN'S APPLE full sharp, vintage
Raised early 1900s by Mr Hill, cider maker and nursmn of Staverton nr Totnes, Devon; popularised 1930s by Hill.
Dark red fruit, often red stained flesh. Produces acidic juice with little astringency and scented, fruity cider, mildly bittersharp in character. Since 1920s widely planted; one of sharps of modern orchards.
F m. T^2; slow to crop, v bicn; rcs scab. H m-Oct/e-Nov.

BROWN SNOUT mild to medium bittersweet
Name derives from russet around eye; number of cider apples with this name have been known in Gloucs, Somerset and Devon. This variety prob arose c1850 on farm of Mr Dent, Yarkhill, Herefords; propagated and distributed by H. P. Bulmer & Co, Hereford.
Small, greenish yellow fruit with pronounced russet ring at apex. Produces sweet, slightly astringent juice and mild to medium bittersweet cider with 'soft tannin and average quality'. Formerly grown commercial orchards, but declined due to susceptibity to fireblight.
F l; slf fertile. T^2; slow to crop. H e/m-Nov.

BROWN THORN mild bittersweet
Syn of Argile Grise, which in C19th was claimed to be one of oldest and best varieties of Norman orchards. Introd 1884 by Woolhope Naturalists' Club; renamed.
Flecked with red, covered in network of russet. Produces sweet, slightly astringent juice and 'mildly bittersweet cider of high quality'. Formerly grown commercially, but declined due to suceptibility to fireblight.
F l. T^2; susc fireblight. H Nov.

BROXWOOD FOXWHELP medium bittersharp
Planted 1920s in Broxwood Museum Orchard, Herefords of H. P. Bulmer & Co. Prob sport of one of old Foxwhelps.
Small, round, red fruit. Produces full bodied juice, similar to Bulmer's Foxwhelp; valuable for blending.
F e. T^2. H Sept.

BULMER'S NORMAN medium bittersweet
Introd from Normandy in early 1900s as unnamed variety by H. P. Bulmer & Co,

Hereford; subsequently proved valuable and named.

Large, conical, greenish yellow fruit. Produces sweet, astringent, fast fermenting juice and mildly bittersweet cider with hard, bitter tannin. Formerly grown throughout cider counties, but now mainly used as 'stem builder' to form standard trees. F e/m; trip. T^3; hvy crops, bien. H m-Oct.

BURROWHILL EARLY full bittersweet
Brought to attention of LARS in early 1980s by Julian Temperley of Burrow Hill Cider, Kingsbury Episcopi, Martock, Somerset. True name unknown, but renamed after local landmark.
Bright red striped over yellow fruit; medium sized, conical. Produces 'full bodied early bittersweet cider with fruity aroma'.
F m. T^2. H Sept.

CAPTAIN BROAD syn John Broad, bittersweet
Formerly popular Cornish apple, presumably raised by John Broad; old trees found 1982 at Golant, nr Fowey, by James Evans.
Large, ribbed, green slightly flushed fruit. Propagated by sticking branches, 'pitchers', in ground.
H Sept.

CHISEL JERSEY full bittersweet
Arose Martock, Somerset in C19th. Name poss derives from 'Jay-see', term used to signify bitter apple or poss apple with a 'nose'; both features common to Jersey varieties.
Red flushed, conical fruit. Produces bittersweet, very astringent juice and 'full bittersweet cider, astringent with plenty of body and good quality'. Well known locally up to 1960s; planted intensive orchards 1970s, but declining in popularity.
F lm/l. T^3; prec, hvy crops. H Nov.

CIDER LADY'S FINGER mild sharp
Arose prob south west of England; variety desc 2001 (Copas); prob not variety of Hogg & Bull (1886).
Tall, large, cylindrical fruit.
F m. T^2. H Oct.

COAT JERSEY bittersweet
Arose Coat village, Martock, Somerset. Came to notice LARS 1950s.
Typical 'Jersey' cider apple. Widely planted in Taunton Cider Company orchard contracts in Somerset in 1970s.
F lm. T^3; prn scab. H e-Nov.

COLLINGTON BIG BITTERS bittersweet
Arose Herefords, prob in C19th; found growing in many West Midlands orchards, especially around Bromyard.
Medium sized fruit, green slight orange flush. Produces mildly bittersweet juice and 'moderate quality cider'. Formerly also used for cooking and especially valued for making mincemeat; hence its syn Mincemeat Apple. Cooked keeps its shape; sharp fruity flavour.
F l. T^2. H l-Oct.

COURT ROYAL pure sweet, vintage
Origin unknown, but prob arose east Devon.
Large, greenish yellow fruit with orange red flush and stripes. Produces pure sweet, fast fermenting juice and light cider. Also sweet, crisp, eating apple, sold to many industrial areas in early 1900s.
Formerly grown throughout Somerset and East Devon, where many old trees remain; also planted all over West Midlands. Now used mainly for cider.
F e; trip. T^3; gd crops; bien; prn scab. H e-Nov.

CRIMSON KING medium sharp, vintage
First propagated late C19th by Mr John Toucher of Bewley Down, Chardstock, Somerset, which gave syns John Toucher's, Bewley Down Pippin.
Vivid crimson. Produces acidic juice with no astringency and 'light, fruity cider of quite good quality'. Also culinary fruit. Found in farm orchards in south west Somerset, adjacent areas of Devon.
F e/m; trip. T^3. H m-Nov.

CRIMSON VICTORIA
Received 1992 from Taunton, Somerset; listed by Thornhayes Nursery, Devon, who found tree at Shute, Axminster, Devon.
H e-Sept.

CUMMY NORMAN mild bittersweet
Acc received 1992 from Taunton, Somerset; may be Cummy, syn Cummy Norman of Hogg & Bull (1886), which was believed to have arisen at Cummy, Radnorshire.
H l-Sept/e-Oct.

DABINETT full bittersweet, vintage
Found prob early 1900s, in a hedge in Middle Lambrook, Somerset, by Mr William Dabinett. Believed Chisel Jersey seedling.
Small, greenish yellow, flushed and striped in red; strong aroma when ripe. Produces sweet, astringent juice and bittersweet cider with 'soft, full bodied, astringency'. Grown all cider counties and widely planted intensive orchards.
F m; slf fertile. T^1; prec, gd crops. H Nov.

DOUX NORMANDIE very mild bittersweet
Arose Sarthe, Western Loire, where well known and esteemed in early 1900s.
Red flushed over yellow fruit. Produces sweet, perfumed juice.
Localised now in Normandy, Brittany, Western Loire, but no longer planted due to susceptibility to fireblight.
F m. T^3; prn fireblight. H Oct.

DOVE medium bittersweet
Arose Glastonbury area, Somerset; prob very old variety. Recorded 1899.
Small; green turning yellow fruit, red striped. Produces sweet, slightly astringent juice and medium bittersweet cider with soft, tannin quality. Very late flowering, which was advantage in low lying frost prone areas, but now rarely planted due to susceptibility to scab.
F l. T^2; v prn scab. H e-Nov.

DUFFLIN sweet
Received 1992 from Taunton, Somerset; impossible to say if old variety mentioned 1886 (Hogg & Bull), as no desc given.
Large, golden with sweet coarse flesh.
H l-Sept.

DUNKERTON'S LATE syn Dunkerton's Sweet, sweet
Found or raised 1940s by Mr Dunkerton, Baltonsborough, nr Glastonbury, Somerset, where original tree still stands.
Green turning yellow, slightly red flushed fruit; cylindrical shape. Produces 'sweet, low tannin juice' and 'light, fruity cider'. Quite widely planted.
F m. T^2. H e-Nov.

DYMOCK RED bittersharp, vintage
Prob arose late C17th; named after Dymock village in Gloucs, on Herefords borders.
Dark red flushed, streaked. Produces 'well balanced cider of high quality'; similar to Kingston Black. Formerly much grown around Ledbury, Herefords.
F e. T^2. H l-Sept.

ELLIS BITTER medium bittersweet
Prob arose C19th on farm of Mr Ellis, Newton St Cyres, Devon.
Large fruit, conical, bold flush and red stripes. Produces sweet, astringent juice and medium bittersweet cider with 'soft astringency and good quality'. After trials at LARS, became widely distributed throughout west country; also found modern orchards in all cider counties.
F lm. T^3; tip. H l-Sept/e-Oct.

FAIR MAID OF TAUNTON syn Moonshines, mild sharp
Prob originated in Taunton, Somerset; now found in orchards around Glastonbury. Desc 2001 (Copas); prob not variety of Hogg (1884).
Flat, round fruit with slight red flush over pale yellow; said to derive its syn because apples shine in the moonlight. Produces 'moderately sweet cider with fairly

agreeable aroma and flavour, but lacking in character'.

F e-May. **T**^{3}. **H** m/l-Oct.

FILLBARREL mild bittersweet
Arose prob late C19th at Woolston, Sutton Montis, Wincanton area, south-east Somerset.
Small, conical; red flushed, netted with russet. Planted small extent 1970s. Makes 'good medium cider, full bodied, marked astringency'.
F e. **T**^{2}; bien. **H** l-Oct.

FOUR SQUARE sharp
Received 1992 from Taunton, Somerset.
Large ribbed green, slightly flushed fruit; sharp, coarse flesh.
H e-Oct.

FOXWHELP bittersharp, vintage
Known C17th; prob arose Gloucs but soon spread to Herefords.
Brilliant red apple. Believed to have arisen near a fox's earth. Produced cider with characteristic 'musky flavour and strong aroma', which commanded high price in C17th and C18th; but more often used to give strength and flavour to blends. Several Foxwhelps now exist which are prob sports of old variety – Red, Improved, Bulmer's and Broxwood Foxwhelp are found in orchards.
PLATE 25

FREDERICK full sharp, vintage
Arose C19th in Forest of Dean, Mons.
Dark red flushed, small. Produces pure sharp juice with no astringency, and cider with a characteristic aroma and flavour of 'good to excellent quality'. Also made 'excellent apple jelly'.
Formerly extensively planted Mons, Gloucs, Herefords.
F m. **T**^{2}. **H** m-Oct.

FRÉQUIN TARDIVE DE LA SARTHE syn Tardive de la Sarthe

HANGDOWN mild bittersweet
Origin unknown, but poss arose Glastonbury area, Somerset; widely distributed throughout Somerset and North Devon where known also as Pocket Apple. Name derives from habit of tree when laden with fruit.
Small; green turning yellow fruit with slight flush. Produces 'average quality cider'. Formerly recommended for planting by LARS and found throughout cider areas, but no longer in favour because of small fruit.
F l. **T**^{2}; prn scab. **H** l-Oct.

HARRY MASTERS JERSEY medium to full bittersweet, vintage
Believed raised by Mr Harry Masters in

Yarlington village, south Somerset, prob late C19th.
Fruit is small to medium size, dark red flushed; also known as Port Wine in Glastonbury area. Produces sweet, medium tannin juice and full bittersweet cider, of 'very good quality with soft astringency'. Widely planted; grown modern orchards.
F m. **T**^{2}; tip; prec; gd crops, sltly bien. **H** l-Oct.

IMPROVED DOVE mild bittersweet
Prob one of many seedlings of well known Somerset variety Dove; prob arose Glastonbury area in early 1900s.
'Jersey' type; fruit small to medium size; red flushed and striped over greenish yellow. Larger and less susceptible to scab than Dove.
F l. **T**^{3}. **H** m-Oct/e-Nov.

IMPROVED LAMBROOK PIPPIN mild sharp
Arose Lambrook village, nr Martock, Somerset; prob seedling of Lambrook Pippin. Brought to notice LARS in 1960s.
Medium size fruit; pinky red flush and stripes over yellow. Planted modern orchards in Somerset, but not common.
F vl. **T**^{2}. **H** e-Oct.

IMPROVED REDSTREAK bitter sharp
Origins unknown, but known to LARS by 1940s. Not necessarily related to Redstreak; many red streaked apples were given its name.
F e. **T**^{3}; uprt. **H** e-Oct.

KINGSTON BLACK syn Black Taunton bittersharp, vintage
Prob arose village of Kingston, nr Taunton, Somerset in mid-late C19th.
Dark maroon flushed, small. Produces full bodied cider with distinctive flavour. Popularised early 1900s and widely planted in west country, but declined due to susceptible to canker, scab and poor crops.
F m. **T**^{2}. **H** e-Nov.
PLATE 25

LANGWORTHY sharp
Prob arose Devon. Also known as Wyatt's Seedling after Mr Wyatt of Kingweston, south Somerset, who often sent it to LARS for cider-making competitions; winning first prize for mild sharp in 1932. Subsequently planted Devon and Somerset, but no long popular because of small fruit.
Bright red, round fruit, making 'pleasant, brisk, light, sweet cider with good flavour'.
F e-May. **T**^{2}. **H** e-Nov.

LE BRET pure sweet
Found in garden of Mrs Le Bret, St Annes Park, Bristol and included in 1950s in LARS trials.

Widely planted in south west and Herefords in 1970s, but often mistakenly called Sweet Alford.
Bright red flushed over pale yellow; quite large fruit.
F e. **T**^{1}. **H** m/l Oct.

MAJOR full bittersweet, vintage
Poss arose Somerset; popular by 1880s. Formerly common variety of old farm orchards in south Somerset, Devon.
Prominently red striped over yellow; conical fruit. Produces sweet, astringent juice and 'excellent bittersweet cider with soft astringency'. Planted modern intensive orchards.
F lm. **T**^{3}. **H** l-Sept.

MAUNDY bittersweet
Collected 1974 by LARS from Whetton's Museum Orchard of H. P. Bulmer & Co, Herefords, before these were discontinued.
Small fruit, pea green, slight flush, stripes.
H l-Oct.

MÉDAILLE D'OR full bittersweet
Raised 1865 by M. Godard, nursmn of Bois-guillaume, nr Rouen, Normandy. Introd England 1884 by Woolhope Naturalists' Field Club.
Small, russeted, orange and yellow fruit. Produces sweet, heavily astringent juice and full bittersweet cider 'often high in alcohol, fruity and good quality'. Formerly widely planted Herefords and still relatively common. Remains well known northern France in Orne, Nord, Pas-de-Calais, Somme.
F vl. **T**^{2}. **H** m/l-Nov

MICHELIN medium bittersweet
Raised by nursmn M. Legrand of Yvetot, Normandy; fruited 1872. Named after M. Michelin, who did much to promote study of cider fruits. Introd Herefords 1884 by Woolhope Naturalists' Field Club.
Small; pale green turning yellow fruit with slight flush. Produces sweet, mildy astringent juice and cider of medium bittersweet character similar to Bulmer's Norman, but with softer tannin. Extensively planted in West Midlands since 1920s; now most widely grown variety in all cider areas. Remains well known Normandy.
F m. **T**^{2}; prec, gd crops; susc mildew, canker. **H** l-Oct/e-Nov.

MORGAN SWEET pure sweet
Prob arose Somerset in C18th.
Quite large; greenish yellow turning yellow, often prominently freckled with russett. Produces light, fruity cider ready to drink before Christmas. Also can be eaten fresh – sweet, juicy, but with definite cidery tang. Formerly widely grown Somerset, also Devon, Gloucs, where plantings of early 1900s were intended to supply Sept eating

apples to South Wales mining towns. Now only cider fruit.
F e-m; trip. T³. H l-Aug/l-Sept.

MUSCADET DE DIEPPE; bittersweet
Prob originated near Dieppe, Seine-Maritime, Normandy, where now localised. Collected 1984 by LARS.
Small fruit; orange red flush over yellow. Among recommended varieties in France.
F e. T². H Sept.

NÉHOU full bittersweet
French variety introd 1920s by H. P. Bulmer & Co, Hereford.
Small, conical, yellow fruit. Produces juice of pronounced bittersweet character, sweet, astringent and full bodied cider of 'excellent quality'. Planted modern intensive orchards.
F em. T²; prec; hvy crops; v bien; susc scab. H l-Sept/e-Oct.

NORTHWOOD pure sweet, vintage
Prob arose Crediton area, Devon, where common 1960s, but may date back to C18th.
Red flushed over greenish yellow fruit. Typical Devon sweet; produces sweet juice with little astringency and 'sweet, soft, fruity, good quality' cider.
F m. T². H e-Nov.

OMONT bittersweet
Raised by M. Omont at Bourghteroulde, Normandy. Collected 1984 by LARS.
Medium sized fruit; orange red flush over greenish yellow. Found in Normandy orchards.
F ve. T³. H Sept.

OSIER bittersweet
Collected 1974 from Whetton's Museum Orchard, Hereford.
Small to medium fruit; green turning yellow with slight flush.
H m-Oct.

PAIGNTON MARIGOLD medium bittersweet
Received from Thornhayes Nursery, Devon; claimed arose Paignton, Devon before 1834.
H e-Oct.

PENNARD BITTER bittersweet
Takes its name from West Pennard village, nr Shepton Mallet, Somerset; propagated late C19th by Harold Heal of Glastonbury.
Large; green turning yellow with flush and russet netting
F m. T³. H e-Oct.

PERTHYRE mild bittersweet
Arose Perthyre Farm, nr Monmouth; achieved wide attention 1920s.
Greenish yellow; conical. Produces sweet,

slightly astringent juice and mild bittersweet cider 'of variable quality, sometimes excellent'.
F m. T²; susc scab, canker. H l-Oct.

PORTER'S PERFECTION medium bittersharp
Arose C19th in orchard of Mr Charles Porter, East Lambrook, Somerset. In 1907 Prof B. T. P. Barker, first director LARS, tested fruit from original tree and was so impressed by quality of juice that it was propagated and became distributed throughout Somerset; commonly found now in Martock area.
Small fruit, flushed and striped in dark red. Produces sharp juice with little astringency and medium bittersharp cider of 'average to good quality'.
F em. T³; hvy crops; fused fruits. H m/l Nov.

RED FOXWHELP bittersharp
Foxwhelp type; prob sport of Foxwhelp.
F e. T². H m-Oct.

RED JERSEY bittersweet
Prob originated West Pennard, Shepton Mallet, Somerset, and recorded from this area at Bath & West Show 1895.
Typical 'Jersey' type. Small, conical, red flushed. Produces 'good quality cider, full bodied' with astringency varying from soft to strong bitter tannin. Trees still found in Somerset and small extent adjacent parts of Devon.
F l/vl. T²; bien; prn scab. H m-Oct.

REDSTREAK bittersharp
Believed raised early 1600s by Lord Scudamore of Holme Lacey, Herefords. Produced cider 'fit for Princes' and established Herefordshire's reputation as cider country. In decline by late 1700s; only one tree could be found by Woolhope Club in 1879, but many red streaked apples were given this name.
Original Redstreak could be propagated from cuttings.
(Not in Collection.)

REINE DES HÂTIVES mild bittersweet
Raised 1872 by M. Dieppois, nursmn at Yvetot, Normandy.
Introd England 1920s by Dr H. E. Durham and distributed by H. P. Bulmer & Co, Hereford.
Small, pale yellow fruit. Produces sweet, slightly astringent juice and 'soft, neutral but often rather thin cider'.
Grown modern intensive orchards. In France, formerly recommended for plantings in Eure and Seine Maritime (Normandy), also found Somme, but no longer popular.
F em. T²; prec; gd crops; bien; prn mildew, scab. H l-Sept/e-Oct.

REINE DES POMMES bittersweet
Arose northern France. Introd England 1903 by LARS.
Dark red flushed and striped fruit with russet netting. Produces sweet, astringent juice and full bittersweet cider – 'bitter and of good quality'. Grown Gloucs, Herefords, Mons and Worcs in 1920s. In France among recommended varieties until 1960s, but no longer popular. Now found Brittany, Western Loire, Picardy.
F e. T². H m-Nov.

ROUGETTE DOUCE mild bittersweet
Collected 1974 by LARS from Whetton's Museum Orchard, Herefords.
Crimson flush over yellow.
H m-Nov.

ROYAL SOMERSET (COPAS) medium sharp
Acc is variety now known Somerset, desc Copas (2001), but not Royal Somerset of Hogg (1884).
'Traditional dual purpose apple' of Somerset, used for cooking and cider. Produces 'first class medium sharp cider'.
F m; part tip. T³. H l-Oct.

SEVERN BANK (CIDER) sharp
Collected 1970s by G. Potter, orchard manager at H. P. Bulmer & Co. Hereford from an orchard in Much Marcle. Number of varieties known by this name in Severn Valley area; not same as Severn Bank of main Collection.
F m. T². H Oct.

SOMERSET REDSTREAK mild bitter-sweet
Prob arose Sutton Montis area, south east Somerset. After inclusion in LARS trial 1917, good orchard performance led to it being planted in many orchards; since 1970s planted extensively in all cider counties.
Fruit small to medium size, prominently red striped over greenish yellow. Produces sweet, astringent juice and medium bittersweet cider of 'average quality', usually blended with sharper varieties.
F m. T²; bien; prn apple sawfly. H e-Oct.

STABLE JERSEY bittersweet
'Jersey' type, received at LARS in 1987, but much older. Poss arose Shepton Mallet, Glastonbury area of Somerset, where old trees are found.
Marroon flushed over gold fruits. Produces 'full flavoured cider with hard tannin; useful for blending'. Limited modern plantings.
F m. T²; tip. H e-Nov.

STEMBRIDGE CLUSTER bittersharp
Arose with Mr Sam Duck of Stembridge, Kingsbury Episcopi, Somerset, where

several old trees remain. Introd 1957 by J. Stuckey of Stembridge, orchard manager of Taunton Cider Comp; then given present name, which derives also from fact that fruit is borne in clusters.
Fruit, small to medium size; yellow, slightly flushed. Planted 1970s, but declined in popularity because of susceptibility to disease.
F m/l. T^2; v bien; prn scab, canker, blossom wilt. H m/l-Oct

STEMBRIDGE JERSEY bittersweet
Arose prob Stembridge, Kingsbury Episcopi, Somerset. Introd by W. J. Stuckey of Stembridge; sent 1957 for trial to LARS by his son, J. Stuckey, orchard manager of Taunton Cider.
Small, crimson flushed, striped. Produces bittersweet juice. One of main varieties of intensive orchards.
F m. T^2; bien. H l-Oct.

STOKE RED bittersharp, vintage
Came to prominence 1920s when surveys found trees at Rodney Stoke, Somerset, but prob arose Wedmore area, where old trees were later found and known locally as Neverblight on account of its reputation for resistance to disease.
Small, 'sealing-wax' red fruit. Produces sharp juice with some astringency and very sharp and often scented cider. Formerly planted chiefly Somerset.
F l. T^2; slow to bear; hvy crops; bien; v res scab. H l-Nov.

SWEET COPPIN pure sweet, vintage
Arose Devon, prob in early C18th; typical low acid, low tannin Devon cider apple.
Small, pale yellow, slightly flushed fruit, with soft, sweet flesh. Produces sweet juice with no astringency and sweet to very mildly bittersweet cider of 'good quality'. Formerly common in farm orchards of Exeter area, also widespread in Somerset. Planted in modern orchards.
F m. T^2; bien; prn mildew. H l-Oct/m-Nov.

TALE SWEET sweet
Arose village of Tale, Devon; locally favoured seedling.
Medium sized fruit, prominently striped in scarlet over pale green; much scarf skin. Sweet with little acidity.
F m. T^3. H e-Nov.

TAN HARVEY bittersweet
Prob arose Tamar valley, Cornwall, where

trees found 1980 by James Evans. Small to med size fruit, orange pink flush over gold.
H e-Oct.

TARDIVE DE LA SARTHE syn Fréquin Tardive de la Sarthe bittersweet
Prob arose Sarthe, Western Loire; now found Morbihan (Brittany); Seine-Maritime, Orne (Normandy); Eure-et-Loire, Sarthe (Western & Val de Loire).
Small fruit, yellow with slight flush. Formerly recommended but no longer planted due to susceptibility to fireblight.
H l-Oct/Nov.

TARDIVE FORESTIERE
heavy bittersweet
Old French variety.
Small fruit, red flushed and striped over greenish yellow.
F ve. T^2; bien. H Nov.

TAYLOR'S syn Taylor's Sweet, sweet to mild bittersweet
Old variety, prob arose South Petherton area, Somerset, where propagated late C19th by Porter's Nursery.
Red flushed and striped over yellow fruit. Produces sweet juice and sweet or mildly bittersweet cider of 'fair quality'. Mainly used as pollinator for Tremlett's Bitter.
F e. T^2; tip. H e/m-Oct.

TOM PUTT sharp
Grown throughout west country in C19th as culinary as well as cider fruit. Also widely planted in west Midlands in 1920s. (*See also* Main Directory).
F em. T^2; some res scab. H l-Aug-e-Sept

TREMLETT'S BITTER full bitter sweet
Arose Exe Valley, Devon, prob in late C19th.
Deep red, conical fruit. Produces sweet, astringent, high tannin juice and full bittersweet cider with hard, bitter tannin. Formerly grown all cider counties; planted intensive orchards.
F e; res frost. T^2; prec; hvy crops; prn scab. H e/m-Oct.

VAGON ARCHER mild bittersweet
Collected 1974 by LARS from Whetton's Cider Museum Orchard, Herefords.
Pink flushed over yellow, small fruit.
F vl. T^3. H e-Nov.

VALLIS APPLE syn Black Vallis

VILBERIE syn Vilbery, full bittersweet
Arose around Dinan, Brittany; now localised in Cotes d'Amour, Ille-et-Vilaine. Introd late C19th to Herefords by Woolhope Naturalists' Field Club.
Orange red flush over green/yellow fruit. Produces full bittersweet juice and cider with 'good full bodied flavour'. Planted modern orchards, particularly in frost prone areas.
F vl; slf fertile. T^3; prec crops; bien; prn mildew. H e/m-Nov.

WHITE HEREFORD see White Norman

WHITE JERSEY medium to mild bittersweet, vintage
'Jersey' type; prob arose C19th nr Cadbury Castle on Somerset–Dorset border. Rec 1895.
Small, pale yellow. Produces sweet, mildly astringent juice and very mildly bittersweet cider of 'moderate quality'. Formerly widely grown in farm orchards in south east and north Somerset; planted intensive orchards.
F m. T^2; prec; hvy crops; prn mildew. H l-Sept/e-Oct.

WHITE NORMAN syn White Hereford, bittersweet
Prob arose Herefords. White Hereford of *Herefordshire Pomona*, but may be Blanc Mollet of Normandy. Commonly known as White Norman and widely spread throughout county in late C19th.
Small, pale yellow fruit. Produces 'high quality cider'. Found in many standard orchards, but no longer planted.
F m. T^2. H e-Oct.

YARLINGTON MILL medium bittersweet, vintage
Somerset 'Jersey' type. Arose at Yarlington, nr North Cadbury, where it grew out of wall by water wheel. Reputedly discovered by Mr Bartlett, prob early 1900s. Transplanted into gardens of Yarlington Mill; prob propagated by nursmn Harry Masters and soon achieved local fame for its high yields. Small, red flushed over pale yellow fruit. Produces sweet slightly astringent medium bittersweet cider of 'good aroma and flavour'. Widely distributed Somerset and West Midlands; planted intensive orchards.
F m. T^3; v bien. H l-Oct/m-Nov.

For help with this Directory we thank R. R. Williams and Liz Copas.

Appendix 1
Cooking with Apples

T HE APPLE'S CULINARY VERSATILITY has been one reason for its global popularity. The number of apple recipes is, therefore, almost infinite and it is only possible here to draw attention to some of the main 'themes' of traditional apple cookery and to the classic dishes. Entries in the Directory give the cooking properties of culinary and dual-purpose varieties and also note if the variety has been favoured for a particular dish.

APPLES AND MEAT

T HE MOST WIDELY EATEN COMBINATION of fruit and meat in Europe, Scandinavia and North America is that of apples and pork, although apples with duck and game are almost as familiar. It represents a happy coincidence of agricultural and botanical timing, that apples – brisk, juicy and refreshingly acidic – are ready to be eaten at the same time as the stubble-fattened geese go to slaughter at Michaelmas, and the pigs are killed in November. Apple sauce – for which there is a succession of perfect culinary varieties that only need to be peeled, chopped and simmered to make a smooth purée – is the traditional way for apples to accompany these meats in Britain. In Germany roast pork and goose may be served with baked apples; pork chops are slowly simmered with cider and apples in Asturias, northern Spain; and in Austria pork is served with a mixture of *sauerkraut* and apple seasoned with caraway.

Apples also find their way into, or are served with, many pork products. Sausages with fried apples and onions are on the menu of every German *Ratskeller*, and variations on this theme are found all over Europe. In Denmark, fried bacon is substituted for the sausages, and in Belgium, bacon and apples are baked together in the oven. Germany's *Himmel und Erde*, or 'Heaven and Earth', consists of fried bacon and a mixture of boiled potatoes and apples, while the Dutch version – *Hete Bliksem* or 'Hot as Lightning' – serves haricot beans alongside the bacon and apples.

Apples and liver is also a general association. *Liver Berliner*, which is a speciality of that city and also popular in Poland and Czechoslovakia, consists of slices of fried liver served with sautéed onions and apple rings. As an hors-d'oeuvre, lightly fried goose liver is served upon slices of apple and finished with a sauce flavoured with Calvados in parts of France, and Norman chefs may cook duck liver pâté in an envelope of apple slices. Turkey, chicken, duck, pheasant, partridge and guinea fowl are all served or cooked with apple in northern France, and any bird benefits from being roasted with an apple in the cavity, to help keep the flesh moist and supply extra flavour to the sauce. Chopped apple added to the usual stuffing of liver, breadcrumbs and parsley is also a great improvement, and pieces of apple wrapped in a slice of bacon and baked in the oven make a good alternative to the usual bacon and sausage rolls.

Apples are not usually served with other meats in Britain but in Austria it is customary to add grated apple to the horseradish sauce served with beef, and on special occasions the joint itself might be accompanied by apple sauce. The classic Polish dish, *Bigos*, which is a mixture of beef or lamb and *sauerkraut*, traditionally stored in casks and taken on hunting expeditions, often includes apple to give extra sharpness. In Middle Eastern cookery fruit can be combined with any kind of meat or pulse to make traditional sweet-sour dishes, such as the *tagines* of Morocco and *khoreshthas* of Iran. Northern Europe even makes apple soups in which puréed apples are added to a vegetable or chicken stock and served either iced or warm depending upon the season. Fruit soups, however, have never been popular in Britain, although nostalgic colonials, it appears, would substitute apples for mangoes in their muligatawny soup.

DOLMEH SIB – STUFFED APPLES

Usually served in the Middle East with rice, these make a good lunch dish served with bread. The following recipe is adapted from *In a Persian Kitchen* by Maidely Mazda. The dish calls for apples that will cook well, but not burst open. A brisk, savoury cooking apple provides a good contrast to the meat and the sweet-sour seasoning.

6 medium-sized cooking apples
For the filling: 2oz (50g) yellow split peas; 1lb (450g) ground beef; 1 onion; 1½ tsp cinnamon, black pepper, salt
For the sauce: 1¼pt (140ml) vinegar; ¼pt (140ml) water; 3 tbsp (75g) sugar

First hollow out apples by coring them to within ½in (1cm) of base. Scoop out flesh with teaspoon to leave shell about ½in (1cm) thick; take care not to break shell, and reserve flesh if wished.

Prepare the filling by simmering split peas in salted water for 30 mins or until they are cooked; drain, set aside.

Fry meat for about 10 mins until it is cooked. Assemble the filling by mixing together split peas, meat, chopped onion, spices, seasoning.

To prepare the sauce: put vinegar, water and sugar together in a saucepan, bring to the boil. Fill apples and set them in baking dish. Bake uncovered in a pre-heated oven 180°C (350°F). Gas mark 4 for about 30 mins. Halfway through the cooking baste the apples with the vinegar sauce. When the apples are cooked, serve with the sweet-sour sauce from the baking pan. If desired the apple flesh may be placed around the apples while they are baked and contribute to the sauce.

APPLES AND VEGETABLES

APPLE AND CABBAGE ARE THE TWO STAPLES of the northern European winter kitchen. They are frequently combined in dishes such as red cabbage and apples, and *sauerkraut* and apples. Braised cabbage and apples is the Swedish Christmas Eve dish, and in Spain a dish of cabbage, apple and potato serves to launch the festivities. Added near the end of cooking time, grated apple can also add zest to the modern stir-fry cabbage. Robust mixtures of raw vegetables and apples, together with cooked meats or fish, also occur in many cold dishes of Dutch, German and Scandinavian origin. Today's creations tend to be lighter and more closely associated with the 20th-century 'chefs' salads' of America – which were almost certainly inspired by immigrant memories of home. Apples bring crispness and colour, and a brisk juiciness to almost any combination of celery, celeriac, nuts, tomatoes, fennel, cold cooked potatoes, artichokes, green beans, rice and so on. The salads may be dressed with mayonnaise, vinaigrette or plain yogurt and can be served on their own, as an accompaniment to cold meats, or mixed together with chicken, ham or cheese. Any crisp, well-flavoured apple is suitable, but red-skinned varieties look the most attractive.

WALDORF SALAD

This famous combination of diced, unpeeled apples and sliced celery, dressed with mayonnaise, was invented by the maître d'hôtel of the Waldorf Astoria in New York; the walnuts were a later addition.

2 apples; 1 head of young celery; 4oz (110g) walnut halves; juice of half a lemon; salt and pepper; ¼pt (140ml) mayonnaise

Core the apples and cut into ½in (1cm) chunks. Wash the tender, inner stalks of the celery (tough fibrous stalks should be discarded) and cut into similar sized pieces. Mix apples, celery and walnuts with the lemon juice. Season and add the mayonnaise. Mix well and serve.

APPLE PUDDINGS

THE LIST OF APPLE PUDDINGS is almost infinite as the same basic dishes are adapted to local ingredients and preferences.

Apples baked inside a pastry envelope, for example, are eaten all over Europe, but the varieties used, as well as the kind of dough, flavourings and fillings, all vary. In Britain we know them as 'dumplings', in France they are called *Bourdons*, and in Austria *Apfel im Schlafrock* or 'Apples in Dressing Gowns'.

Most apple growing countries also have a particular speciality. The sweet, glazed tarts of France – *Tartes aux Pommes* – consist of a pastry base filled with apple purée, and topped with overlapping slices of apple. Usually the tarts are finished with an apricot glaze, but an apple glaze spiked with Calvados or cider brandy enhances the 'appley' flavour.

Austria has its *Strudel*, Italy and Germany their apple cakes, and Britain and America their pies. Making strüdel dough at home is a major undertaking, but the commercially available Filo pastry, although inferior in the eyes of any Austrian chef, is an acceptable alternative.

For a traditional British apple pie, or tart, the plainer the ingredients the better. The apples can be flavoured with a few pieces of lemon peel and the juice of half a lemon, or a couple of cloves, but no embellishments are needed if a good cooker is used. Apple tart is best eaten warm, fresh from the oven with cream, although it is still often the custom to serve it with a slice of cheese.

In Holland apple fritters are regarded as a festive dish and are served on the evening of St Nicholas night. Some recipes recommend that the apple rings are first steeped in cider brandy or brandy, or dusted with sugar and grated lemon or orange peel, before they are dipped in batter and deep fried, but with a ripe Bramley's Seedling or Golden Noble the flavour should be strong enough on its own.

APPLE PANCAKES AND GÂTEAUX À LA NORMANDE

Pancakes filled with an apple purée are especially good and one of the best of all apple dishes is a 'cake' made up of alternating layers of pancake, apple purée and ground almonds. The traditional recipe upon which this is based is given in *Mastering the Art of French Cookery* by Julia Child and Louisette Bertholle.

To make 10–12, 6in (15cm) pancakes:
5ozs (140g) white plain flour; ¼pt (140ml) milk; ¼pt (140ml) cold water; 2ozs (60g) melted butter; 3 egg yolks. Reserve egg whites.

Either liquidise all ingredients together, or gradually stir liquid ingredients into flour to make smooth batter.
Let mixture stand for several hours. Just before cooking pancakes, beat egg whites until stiff and carefully fold into batter. Lightly grease 6in (15ml) diameter pancake pan or frying pan and heat over moderate flame, until drop of mixture solidifies immediately. Pour in half cup of batter. Cook until browned and flip over to cook other side.
Set pancake aside and continue until all batter is used up.

Apple Filling:
2lb (900g) apples; sugar to taste; 1–2 tbsp (15–30ml) Calvados, or cider brandy

Make thick purée from peeled, chopped apples; add sugar to taste and then apple brandy. Cool.

Assembling the 'cake':
pancakes; cold apple purée; 2oz (60g) ground almonds; 2oz (60g) chopped almonds; 1oz (30g) melted butter; 2 tbsp (30ml) Calvados or brandy

Place a pancake in middle of oven-proof plate. Spread layer of apple purée on top and sprinkle ground almonds over.

Cover with another pancake and continue to build up alternating layers, finishing with a pancake.
Spread chopped almonds over top, sprinkle over some sugar and melted butter. Bake for 10–15 mins in an oven 180°C (350°F), Gas mark 4.
Just before serving pour over ladleful of warm Calvados or brandy, set alight and flambé cake. Serve with cream.

BROWN BETTY

A favourite American pudding of alternate layers of apples and breadcrumbs flavoured with cinnamon. There is a Danish variant of this dish known as 'Peasant Girl with a Veil', in which layers of apple purée and breadcrumbs are topped with cream and served cold.

1lb (450g) apples weighed after peeling and coring; 6oz (170g) fresh breadcrumbs; 2oz (60g) sugar; 1 tsp (5ml) ground cinnamon; 2oz (60g) butter cut into small pieces

Butter a deep oven-proof dish (about 2pt (1.2 litre) capacity) put in a generous layer of chopped apple. Sprinkle with sugar, a little butter and cinnamon. Cover with breadcrumbs, then more apples and continue until dish is full.
Finish with layer of breadcrumbs.
Cover, cook in moderate oven – 180°C (350°F), Gas mark 4 – for ¾ hour.
Uncover, allow to brown in oven for further 10 mins or so.
Serve warm with cream.

APPLE SNOW

A frothy, smooth cream, which makes a good summer sweet, although in Austria and parts of Eastern Europe it is known as 'Witches' Cream', and eaten on Walpurgis night, or Hallowe'en, when it is supposed to endow the eater with the power to see witches. Rum is used to give extra flavour, and in Normandy Calvados is added. Early codlins such as Early Victoria are ideal. Any apple that cooks well and is fairly juicy is suitable, but avoid very acidic apples as these will curdle the egg whites.

1lb (450g) apples; 2 large fresh egg whites; sugar to taste

Peel, core, slice apples; cook to purée with tablespoon of water or in rinsed saucepan.
Leave to cool, then add a quarter of the stiffly beaten egg white to purée and mix thoroughly. Lightly fold in the remainder. Spoon the mixture into tall glasses, chill and serve with thin cream.

APPLE BREAD, CAKE AND BISCUITS

I N NORMANDY AND AMERICA apple pulp or pieces of apple were – and still are – used in bread making, and in England, according to the eminent 19th-century horticulturist John Loudon, a bread formed of 'one third boiled apple pulp baked with two thirds flour having been properly fermented with yeast' was 'excellent, full of eyes and extremely palatable'.

Loudon's instructions, which appeared in his *Encyclopaedia of Gardening* of 1824 produce a savoury bread that goes well with cheese:

APPLE BREAD

The following proportions are sufficient for one loaf.

12oz (340g) wholemeal flour; 6oz (170g) apples, weighed after peeling and coring; ½oz (14g) fresh yeast; salt

Cook apples to purée.
Blend yeast to smooth liquid with about ¼pt (140ml) warm water.
Add this to the still warm, but not hot, apple purée. Stir apple and yeast mixture into flour and make into a moist dough. If too dry add a little more water.
Leave it to rise until doubled in volume: 1–2 hours.
Dust with flour and work into ball.
Put this into greased bread tin and leave to rise.
When the dough reaches the top of the tin put it in heated oven.
Bake 20 mins at 210°C (410°F), Gas mark 6 and 20 mins at 190°C (375°F), Gas mark 5.
Turn out of tin; let it cool on wire rack.

A PPLE COOKIES, in which apple sauce or chopped apples and chopped nuts are added to the biscuit dough are popular in North America, but in England and Europe apple cakes are more traditional. Dorset Apple Cake, which is about ¾in (2cm) thick, is usually eaten hot, split and buttered. Italian and German cakes are usually served with whipped cream.

DORSET APPLE CAKE

The following recipe is based on that given by Florence White in *Good Things in England*.

1lb (450g) apples; 8oz (225g) self-raising flour; 2oz (60g) butter; 2oz (60g) lard; 4oz (110g) sugar; pinch of salt; a little milk

Peel, core, chop apples.
Mix sugar and apples together. Set aside.
Rub fat into flour; add salt. Stir in apple and sugar mixture. Make into firm dough with milk. Turn into shallow, greased cake tin.

Bake at 180°C (350°F), Gas mark 4 for 45 mins and until nicely browned. Turn out and eat either warm or cold.

APPLE PRESERVES, SWEETS
AND PICKLES

I N 19TH-CENTURY AMERICA, Apple Butter was made from apples and cider, and 'after they have boiled down, molasses in the proportion of two quarts to every four pails of apples' was added. Some people would also include quinces: 'a peck of them gives a fine flavour to a large kettle'. The mixture was then set over the embers to gently simmer all night and in the morning it was 'a fine red color' and ready to be stored in 'firkins or stone jars'.

The apple jellies, suckets, pastes and drops which played a prominent role in the English fruit banquet were more refined preserves. Any small apple, such as Yellow Ingestrie and Ananas Reinette can be used to recreate suckets such as Hannah Glasse decribed in her *Compleat Confectioner* of 1770.

APPLE SUCKETS

12 small apples, plus 2 apples for syrup; sugar; ½lb (225g) for ½pt (300ml) of syrup; 1 lemon

Preparing the apples:
Peel apples very thinly and neatly; take out cores. Into centre of each apple put a long narrow strip of lemon peel. Set aside.

Preparing the syrup:
Boil up parings and whole unpeeled apples in about 1 pint (600ml) of water; simmer for 15 mins.
Strain through muslin into measuring jug.
Measure out half a pint (300ml), add sugar, bring to boil.

Poaching the apples:
Add apples about three at a time; poach gently until just tender. Set aside.

Assembling 'Suckets':
Reduce liquid until it gels on cold plate; add lemon juice.
Glaze poached apples with liquid jelly. Pour some more around apples; it will quickly solidify.

T HE PÂTÉ DE FRUIT of Normandy can be made with any type of apple that is not too juicy. The starting point is a thick, sweetened purée of apple pulp, which can then be cast into pots or moulds or poured over a plate to form a thin layer, which can later be cut into different shapes and rolled in granulated sugar. These may be eaten straight away, but if they are to be kept for any length of time they must be dried in a very low oven for several hours and stored in a screw-top container.

APPLE JELLY

All that is needed are apples or crab apples and sugar. Golden crab apples, such as Golden Hornet will give a pale jelly, but if you use apples with coloured skins then the jelly will be pink or even deep red. Veitch's Scarlet gives a claret-coloured jelly. The species, *Malus Niedzwetzkyana*, which has deep-maroon fruits and red-stained flesh produces a jelly the colour of port wine, and the variety Apple Sauce yields a salmon-pink jelly.

Apple or crab-apple jelly has a delicate flavour, and goes with genteel slices of white bread and butter. A spoonful added to plain yogurt is also good.

APPLE AND RAISIN PICKLE

Apple chutneys and pickles are good for using up the windfalls and will be ready to eat by Christmas.

This is a rich dark brown, sweet pickle.

3lb (1.35kg) cooking apples; 1lb (450g) raisins; 1½lb (680g) sugar; 3 onions; 1 lemon rind and juice; 1pt (600ml) cider vinegar; a few mustard seeds, pinch salt

Peel, chop apples into large bowl.
Add all other ingredients.
Mix well and transfer to large saucepan.
Bring to boil; simmer gently until apples are soft.
Pour into jars and seal.
Keep 6 weeks before using.

APPLE DRINKS

Apple juice can be freshly pressed at home, but it has to be frozen if you want to keep it. Domestic juicers are also available which both extract and pasteurise the juice enabling it to be easily stored in bottles, and apple juice will keep in this way for months, although the flavour is not as good. Apple juice is often sold mixed with other fruit juices but at home try fresh apple juice and gin, which is a whole new experience.

'Lammas Wool' or 'Lamb's Wool' was hot, sweet, spiced ale served with apple pulp foaming on the surface. Samuel Pepys sat, on 9 November, 1666, at 'cards til two in the morning drinking Lamb's Wool'. The bowl of ale or cider was warmed beside the fire, and apples hung up by string in front of the heat gradually roasted, and their pulp fell on to the surface of the ale.

The Wassail bowl, passed round on the eve of Twelfth Night, was usually served decorated with roast apples, and was sometimes even made of apple wood. Any small pretty apples can be used, and in America, Api or Lady apple continues to find a role as a garnish in hot Christmas punches.

WASSAIL CUP
3 quarts (3.5 litres) beer; 4 glasses of dry sherry; 11b (450g) sugar; lemon; nutmeg; fresh ginger; 6 small red apples; 6 squares toast

Grate the nutmeg and ginger on to the sugar and stir into 1 quart (1.2 litres) of warm beer. Add the sherry, 2 more quarts (2.3 litres) of beer, and 3 slices of lemon.
Simmer until thoroughly warm. Meanwhile bake the apples.
Serve with roasted apples sitting on top of piece of toast and floating on the surface of the drink.

Appendix 2
Growing Apples

 PPLES ARE THE EASIEST OF THE TREE FRUITS to grow, the most tolerant, the least demanding and the most rewarding of all for the amateur gardener. The choice of varieties to plant can seem almost limitless, but it is always a good idea to aim for a range of seasons. A good collection, of say ten varieties, would include one summer apple, another for ripening in September, two for October and the remainder late keepers to take the supply through to spring. A spectrum of flavours could be built into this selection and one or two culinary apples included. In addition, the choice could be limited to varieties local to the area, or to apples grown in the historic period appropriate to the house. Dozens of varieties are generally available from fruit nurseries and if not, many will propagate any particular variety you like, or you can graft or bud your own. Within two to five years an apple tree can be fruiting and will go on cropping for 20 or even 50 years, and longer if well looked after.

SITUATION

THE BEST SPOT TO GROW FRUIT is on a south-facing, gentle slope, open to the sun and sheltered from the prevailing wind, where the frost does not collect but filters down and away. As fruit trees need plenty of sunlight and warmth during the growing period to develop the fruits, they will not grow well, nor will the fruit ripen and colour properly in heavy shade.

Apple trees grow best in a slightly acid, deep, well-drained loam, but heavy clay can be lightened by digging in plenty of strawy manure, which will also increase the water-retaining capacity of light sandy soils. Soil pH – its degree of acidity or alkalinity – can easily be measured with a soil-testing kit. Very acid soils should be limed, and a dressing of sulphur will lower the alkalinity of chalky soils. On good deep soil of

about pH 6.5 there should be few problems with mineral and trace element deficiencies.

Trees should be planted in the dormant season, ideally in November-December (see Fig. 1). Container-grown trees can be planted at any time, although in dry spells they will need regular watering to get them established. If apples, pears or any 'pip' fruit has been grown in the ground before, it is advisable to use fresh soil or peat compost in the planting hole, or plant the new trees as far as possible from the sites of the old trees. This is because the ground around the roots of the old tree may have become impoverished and contaminated with soil-borne fungi which may retard the new tree's growth.

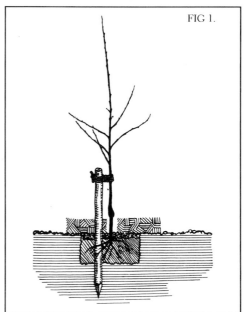

FIG 1.

Fig. 1 The finished planted maiden tree, staked, tied and mulched with straw. Stake should be driven into place, and fertiliser, manure or compost forked into the base of the hole before planting the tree.

In high rainfall areas, trees will be more prone to diseases such as scab and canker. Apple scab is most prevalent in a damp summer, but some modern varieties carry immunity to apple scab, and often local apples may do well in the particular climatic conditions of an area. Canker can be a problem also, particularly on poorly drained soils, and it is best to avoid planting canker-prone varieties such as Cox's Orange Pippin or McIntosh and many of their offspring. In hot, dry summers, trees may need to be irrigated, especially if they are on light soils and on dwarfing rootstocks. Mulching the trees with compost, straw or crushed bark before the soil starts to dry out in April or May will reduce moisture loss. Hot dry weather also encourages mildew but, again, it is best to avoid varieties known to be prone to mildew, such as Jonathan and many of its progeny.

ROOTSTOCKS

APPLES ARE USUALLY PROPAGATED by grafting or budding on to a suitable stock and before buying a tree it is essential to know which rootstock has been used. The more dwarfing the rootstock the smaller the tree, and the sooner it fruits – but the crop will, of course, be smaller. Dwarfing rootstocks tend to have brittle root systems and as a result need the support of a permanent stake. They will also need good, deep soil and to be grown in clean ground, as they cannot tolerate competition from grass or weeds. On poorer, thin soil a more vigorous rootstock is advisable. (See Table 1 and Fig. 2.)

TABLE 1
ROOTSTOCKS

M27	Extremely dwarfing, producing tree 4–6ft (1.2–1.8m) in height and spread, fruiting from 2nd–3rd year. Needs permanent 6ft (1.8m) stake. Best trained as pyramid or centre leader tree. Needs good soil, clean ground, does not tolerate competition from grass or weeds.	Plant 4–5ft (1.2–1.5m) apart; 6ft (1.8m) between rows. Fruit yield 10–15lbs (4.5–7kg).
M9	Dwarfing, widely used commercially and in gardens, producing tree 6–8ft (1.8–2.4m) in height, slightly wider spread. Cropping from 2nd–3rd year. Needs permanent stake, good soil, clean cultivation. Grow as centre leader, pyramid, bush tree or cordon.	Plant 8–10ft (2.4–3m) apart; 12ft (3.6m) between rows. Yield 30–50lbs (13.5–23kg). As *cordon* plant 2–3ft (60–90cm) apart; 6ft (1.8m) between rows. Yield 6–8lbs (3–3.5kg).
M26	Semi-dwarfing, producing 8–10ft (2.4–3m) tree in height, slightly wider spread, cropping within 3 years. Tolerates average soil conditions, can be grown in grass orchard. Grow as centre leader, pyramid, bush tree, cordon, espalier. Needs stake as bush tree for first 5 years; permanent stake as centre leader.	Plant 8–12ft (2.4–3.7m) apart; 8–12ft (2.4–3.7m) between rows. Yield 30–50lbs (13.5–23kg). As *cordon*, plant 2–3ft (60–90cm) apart; 6ft (1.8m) between rows. Fruit yield 6–8lbs (3–3.5kg). As *espalier* plant 15ft (4.5m) apart. Fruit yield from 3 tiers 25–40lbs (11–18kg).
MM106	Semi-dwarfing, producing tree 9–11ft (2.7–3.5m) in height, slightly wider spread, cropping within 3–4 years. Tolerates wide variety of soils, can be grown in grass orchard. Grow as centre leader, bush tree, cordon, espalier. Needs stake for first 5 years in exposed position.	Plant 8–12ft (2.4–3.7m) apart; 12ft (3.7m) between rows. Yield 50–100lbs (23–45kg). As *espalier*, plant 15ft (4.5m) apart. Yield from 3 tiers 25–40lbs (11–18kg).
MM111	Vigorous, producing 10–12ft (3–3.7m) tree in height, slightly wider spread, cropping within 4–8 years. Tolerant of wide range of soils, suitable for grassed orchard or lawn. Grow as bush, half-standard or standard tree.	Plant 15–20ft (4.5–6m) apart; 20ft (6m) between rows. Yield 100–400lbs (45–180kg).
M25	Very vigorous, producing tree at least 11–15ft (3.5–4.5m) high, cropping within 4–5 years. Tolerant of wide range of soils, suitable for grassed orchard, lawn or avenue. Suitable for full standard; often used to produce large ornamental crab apple tree.	Plant 20ft (6m) apart; 25ft (7.6m) between rows. Yield 200–over 400lbs (90–over 180kg).

Cropping and size of tree will depend also upon depth and quality of soil and variety.

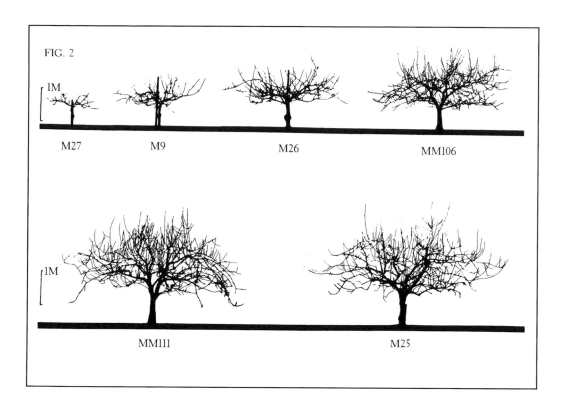

FIG. 2

1M

M27　　　M9　　　　M26　　　　MM106

1M

MM111　　　　　　　　M25

POLLINATION

IF YOU GROW HALF A DOZEN or more varieties, or there are apple trees in a neighbour's garden, there should be no problem achieving pollination and a good fruit set. The lack of a pollinator, or the presence of few pollinating insects due to cold weather or pollution can, however, often be the reason for a poor crop. Suitable pollinators are varieties which flower at the same time.

THINNING

ONCE THE CROP HAS SET, it may be necessary to thin the fruit in order to ensure that the apples are of good size and quality. Carrying a heavy crop can also impose a strain on the tree's resources and induce it to go into biennial bearing, which can then be difficult to correct. Thinning is done by pinching out the little fruitlets as soon as possible after the fruit has set, while they are just under ½in (1.25cm) diameter. Blemished and misshapen fruits should be taken out first. As a general rule eating apples are thinned to 4–6in (10–15cm) apart, and culinary varieties slightly further.

HARVESTING

EARLY VARIETIES ARE RIPE as soon as they are well coloured, and once picked need to be kept in a cool, dark place. This applies to all

Fig. 2 A selection of trees, each grafted with the same variety, to show the effect of the different rootstocks.

apples, and no apple will keep in good condition in a warm room. Mid-season apples ripening in September are also best picked in two or three stages, harvesting the most highly coloured first. Very late keepers need the benefit of autumn sunshine to build up their flavours, but they must not be left too long on the tree or the fruit will not last through the winter. An apple is ready to be picked if it comes away easily when the fruit is lifted and given a slight twist. A slight scattering of windfalls on the ground is also an indication that the crop needs harvesting, but experience of your particular trees in their situation is the only sure guide.

Fruit for storing needs careful handling as bruised apples will soon rot, and apples should be picked with the stalk intact. The ideal apple store is frostproof, with a cold, stable temperature. Garages and sheds are suitable, but need to be reasonably well insulated or a warm day in February or March can easily bring an end to the fruit. Centrally heated houses are out! Ripe and unripe fruit – mid-season and late varieties – should be kept apart, otherwise there is always the chance that the ethylene given off by the ripe fruit will cause premature ripening of the late apples.

The store also needs to be well ventilated to prevent any build-up of gases. Wrapping each apple in tissue or waxed paper will help prevent shrivelling and serves to isolate a fruit should it become rotten. Apples may also be stored in polythene bags with ventilation holes punched in them, although not all varieties respond to this treatment. Some people use an old refrigerator, which will keep small quantities in much the same way as a commercial cold store.

The simplest way to store fruit is probably to use old, clean, tomato trays, which can be stacked one on top of the other. Stocks should be sampled from time to time to catch each variety in its prime, and checked for rotten fruit which must be removed.

TREE FORMS

THERE ARE MANY WAYS TO TRAIN AND GROW apple trees, each suited to a particular situation. Bushes, standards, half-standards and centre leaders (often called spindle trees) are grown in the open as orchard trees, while cordons, espaliers and dwarf pyramids are restricted forms, which are often more suitable for small gardens. Cordons and espaliers can also be trained as decorative arches and apple trees can even be grown in pots. In all cases it is best to begin with a one-year-old 'maiden' that is 'well feathered', in other words has plenty of lateral branches. These laterals will form the framework of the new tree and bear the first fruit. Many varieties will naturally form feathers, but a number – notably Bramley's Seedling, Epicure, Grenadier, St Edmund's Pippin, Spartan and Warner's King – produce few laterals. These are sold as 'whips', and pruning back to the point at which you want the branches to start after planting will encourage them to grow out in the second year. Often the trees on sale in a garden centre will be two years old or more, but again, select specimens with plenty of lateral branches. Once the form of the tree is established, which

will take about three to four years, the shape is maintained by annual pruning, which will also regulate the quantity and quality of the crop, as well as helping to control pests and diseases.

Everyone has their own method of pruning. Rarely will two people entirely agree on the best approach! Pruning is best learnt by watching a demonstration.

THE REASONS FOR PRUNING ARE:
– to form the desired tree shape, fill the space available and obtain balanced growth
– to grow regular crops with fruit of good size and quality
– to obtain a measure of control over pests and diseases
– to repair and rejuvenate as necessary

NEGLECT OF PRUNING CAN RESULT IN:
– tangled over-crowded growth
– small fruit
– excessive cropping leading to biennial bearing
– increased pests and disease
– harvesting difficulties
– unstable tree

Pruning can be carried out in winter or summer, but the effect and purpose differs. Winter pruning is done to remove dead, diseased and badly placed branches, and also to maintain the tree shape. Hard winter pruning, carried out in the dormant season, will stimulate vigorous new growth in the next season, but there will be less fruit. Summer pruning reduces the number of leaves, and hence restricts growth. It lets in more light to ripen and colour the fruit and encourages the formation of fruit spurs. It is used mainly on closely trained forms – cordons, espaliers and pyramids. Summer pruning is normally carried out in early August, when new growth from the main stem is cut back to three leaves to form spur growth.

Figures 4–8 overleaf give instructions on training and growing the different tree forms.

Fig. 3 Different styles of growing and training apple trees.

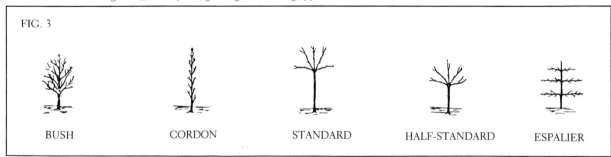

FIG. 3

BUSH   CORDON   STANDARD   HALF-STANDARD   ESPALIER

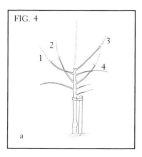

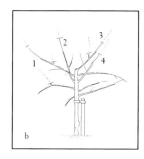

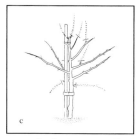

Fig. 4 Bush tree: a head of open branches above a short trunk. a. After planting in winter, cut leader back. Select laterals 1,2,3,4 to form framework and tip back, cutting to an outward facing bud. Lower laterals can be left to fruit and if necessary removed in later years. b. In year 2, leaders of main branches cut back in winter. Strong growing upright shoots are also cut out. c. By year 3, laterals of main branches will be filling up available space. Prune as necessary, cutting back to a fruit bud in winter, which will start to form a spur system on that branch. Tip bearers should only have a proportion of their laterals cut.

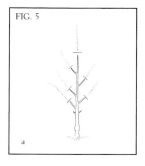

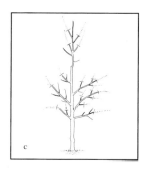

Fig. 5 Dwarf Pyramid tree: usually closely planted and summer pruned to maintain shape.
a. After planting in the winter, cut leader back to about 20in (56cm) from ground. Cut back laterals to about 5in (12.75cm). b. In second winter cut leader back to about 10in (25.5cm) of new growth. Remove or stub back strong growing verticals. Cut leaders of laterals to about 8in (20.5cm). c. In next and subsequent summers cut branch leaders to 5in (12.75cm), laterals from branches to 3in (7.5cm) and those from existing laterals to 1in (2.5cm). In following and subsequent winter years until the tree reaches 6ft (1.8m), cut leader back.

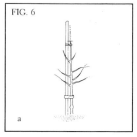

Fig. 6 Centre leader tree: most of the crop is carried on lower branches, while upper branches bear a modest amount of fruit and throw little shade.
a. After planting, tip centre leader and select 5 to 6 main branches, which will form the permanent lower framework. b. In following summer tie or weigh down these laterals if necessary to encourage them towards the horizontal. Early cropping will also bring the branches down. In winter tip centre leader, cut out any strongly growing vertical growth and remove weights or ties if branches have set at required angle. c. In second year after planting tip centre leader in winter. Remove any branches growing too low. Remove upright growth. In subsequent years remove vertical growth and cut back top branches so that none is growing higher than the centre leader. Each year, centre leader is cut back and replaced by a lower shoot; upper branches as soon as they have reached the thickness of a thumb are cut back and replaced by new growth. If tree is growing vigorously do not prune leader in winter, but if growing poorly cut hard back to increase growth in lower branches.

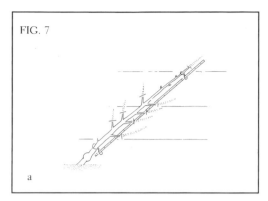

FIG. 7

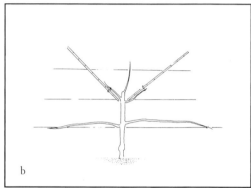

Fig. 7 Cordon: a single stem, bearing fruit spurs along its length, grown against a wall, fence or framework.
a. To form an oblique cordon, plant at an angle of 45° to the ground with graft uppermost. After planting in winter, cut back laterals to 3in (7.5cm) or 3 buds. Cut leader. b. In following early August summer prune. Cut new laterals arising from main stem to 3in (7.5cm) or 3 buds. Cut all other laterals back to 1in (2.5cm). In winter cut leader back by up to a third. Follow this regime for following years, until leader reaches maximum height. Continue annual summer pruning and when leader reaches top of the wire cut back in May.

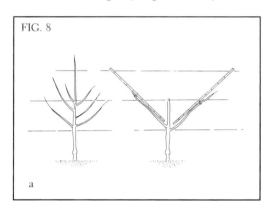

FIG. 8

Fig. 8 Espalier: a central stem from which pairs of fruiting arms arise at intervals, trained against a wall or framework.
a. Tree at planting. After planting in winter select branches for first tier and tie to canes at 45°.

b. In following early August lower first tier to the horizontal. Summer prune new laterals. In winter select branches for second tier and tie to canes. In following summer these will be lowered to the horizontal – Summer prune new growth on tiers.

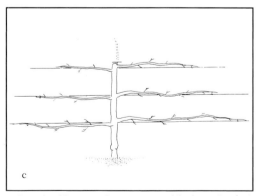

c. Completed 3 tier espalier in autumn of third year. Leader is cut back in May and each tier is summer pruned like a cordon.

HALF-STANDARD AND STANDARD TREE

Half-standards are usually grown on a 4ft (1.2m) stem and standards upon a 6ft (1.8m) stem. The basic framework of branches is usually already formed before they are sold and it is then a case of encouraging the structure to develop. If you commence with a one-year-old maiden, allow it to grow on for two seasons to reach the necessary height before cutting back the leader. The procedure is then the same as for a bush tree.

CORDON ARCHES AND PERGOLAS

Arches, pergolas and tunnels can be created by planting pairs of cordons on either side of an arched framework, or by training espaliers along and over the frame. Although it is more interesting to use a range of varieties to plant up a tunnel, management will be easier and the effect more spectacular if only one variety is used. Then the individual cordons will all be of the same vigour and come into blossom and fruit at the same time, but a pollinator needs to be included or planted nearby. If a selection is used, then it is important to choose varieties of approximately matching vigour, and the choice of rootstock will depend upon the height of the arches. For an arch of about 8ft (2.4m) high, M26 rootstock would be appropriate, but if the soil was poor or a taller structure required an MM106 rootstock would be advisable.

TREES IN POTS

Trees grown in pots to form attractive features for patios and small gardens, are usually trained as pyramids, and summer pruned to maintain the shape. For pot-grown trees, some people advise using the most dwarfing M27 rootstock, but those with experience recommend a more vigorous rootstock, such as M26 or MM106. The act of growing a tree in a pot is itself restricting, reducing the growth of the tree and using a stronger stock will mean that it is better able to withstand the occasional weekend without watering, whereas trees on very dwarfing rootstocks will need daily attention in dry weather.

PROPAGATION

APPLE TREES CAN BE PROPAGATED by chip budding in late July and by whip and tongue grafting in March (see Figs. 9 and 10 below). Either way, the first requirement is to have already established the appropriate rootstock.

GROWING ON OWN ROOTS

Some varieties, such as Fillingham Pippin, Burr Knot and Ben's Red, form burrs at the base of branches and these will readily root if branches are inserted into sandy soil in the open. With a little extra care it has been found that a number of other varieties can be rooted from cuttings.[1] The procedure is the same as for all hardwood

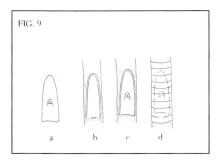

FIG. 9

a   b   c   d

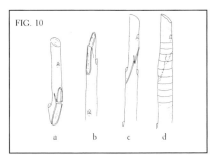

FIG. 10

a   b   c   d

Fig. 9 Chip-budding: bud sticks are cut from current season's growth in July– Leaves are cut off leaving about half an inch (1.25cm) of stalk. Select buds from centre of stick.
a. Thinly slice off bud and surrounding sliver of bark. b. Rootstock with sliver cut out the same size as the sliver of the chip bud. Cut made so as to leave a lip at base to hold chip bud. c. Chip bud fitted onto rootstock so that maximum contact between cambial regions is obtained. d. Chip bud securely bound into place with polythene tape. Remove tape about 4–6 weeks after budding. Bud will grow out next spring, when stock above should be cut back. It will produce a maiden tree by the following autumn.

Fig. 10 Whip and Tongue Graft: collect scion or graft wood from tree when dormant and if necessary store outside in ground with shoot buried for two thirds length. Cut scion about 4in (10cm) long with 3–4 buds.
a. Flat slanting cut made at base of scion. Downward pointing tongue made in top half of this. b. After cutting off root stock at about 12in (30cm) above ground, a slanting cut of similar length is made in upward direction. Upward pointing tongue is made in this. c. Cut surfaces placed together so that tongues interlock and maximum contact between cambial regions. When rootstock is of a larger diameter than the graft, contact will be possible on one side only. d. Graft bound in place with polythene tape or rafia and cut surfaces covered with wax. Remove tape about 8 weeks after grafting. The bud will grow out when the weather warms up and produce a maiden tree by the autumn.

cuttings, such as those of roses. Cuttings, not less than 18in (45cm) and up to 3ft (90cm) long, of straight one-year-old shoots with a heel should be taken. November and February are the best times to take cuttings. Dip the base of the cutting in hormone rooting powder and insert it in a heated propagated bed with a temperature between 20–30°C (68–86°F). The actual conditions and ease of rooting will depend upon the variety, and varies from season to season. Cox's Orange Pippin, Discovery, King of the Pippins, Golden Noble, James Grieve, for example, require bottom heat in a frame for up to a month. Once rooted they can be potted on in the usual way.

The resulting trees will, depending upon the variety, have a vigour approximately the same as that of tree grafted on MM106 rootstock. If the vigour of the tree is a problem, then it can be controlled by summer pruning or bark-ringing (see below), but the best way to check vigour is by regular cropping.

REJUVENATING OLD APPLE TREES

UNLESS THE TREE is exceptionally damaged and very old, it is rarely so debilitated that it cannot be saved. Careful pruning, together with annual barrow loads of compost or manure will usually revive its vigour and bring it back into producing reasonable crops.

Pruning should be undertaken in several stages over a period of about three to four years. Over-zealous pruning in the first year will encourage a forest of new growth – water shoots – which will grow vigorously up but produce no fruit. Very hard pruning on old trees can also lead to papery bark canker and the tree will die. In the first year all the damaged and diseased branches can be taken out, as well as any that are very badly placed – crossing over other branches, for example. Pruning cuts, if made neatly, will heal over naturally, but they are potential sources of infection and with large cuts it is a good idea to paint these over with a proprietary sealing compound. The following year work can begin towards creating an open centred form by cutting out further surplus growth and this can be continued in subsequent years with the aim of encouraging fresh growth and then bringing this into bearing.

If the tree is very vigorous and too large for its

TABLE 2

DISEASES

Disease	Symptoms	Conditions	Preventative Measures
Scab	Brownish-black scabs on fruit. Blisters on shoots become scab-like. Leaves show olive green blotches and fall.	Air-borne fungus. Prevalent in wet climates. Some varieties more prone than others.	Remove; burn scabby shoots. Rake up leaves, burn.
Powdery Mildew	White powdery covering to young shoot tips and leaves. Blossom falls.	Fungus overwinters in dormant buds. Prevalent in hot, dry conditions.	Cut and burn badly infected shoots and flower clusters. Mulch to conserve moisture.
Canker	Sunken cankers on shoots causing severe die-back if branches encircled. Red fruiting bodies seen on damaged branches.	Prevalent in damp climates, worst on badly drained soils. Some varieties more prone than others.	Remove and burn affected branches. Cut out diseased area from large branches, paint with canker paint.
Brown Rot	Rots apples on tree and in store.	Fungus enters through wounds.	Remove and burn fruits.
OTHER PROBLEMS			
Bitter Pit	Sunken pits on fruit surface, brown area under skin. Apples will not keep.	Disorder caused by deficiency of calcium in fruit. Induced by water shortage.	Reduced by calcium nitrate sprays. Mulch to conserve moisture. Water in dry spells. Prune in summer.

position, its vigour can be reduced by bark-ringing. This consists of taking out a semicircular piece of bark from the trunk in May, which thus reduces the flow of nutrients up the tree and inhibits its growth. The strip cut out will vary from ¼in (6mm) wide for a large tree to ⅛in (3mm) for a small tree and it should extend for about a third of the tree's circumference. A complete ring should not be made as this can kill the tree. Root pruning will also curb the vigour.

PESTS AND DISEASES

THE MAJOR PESTS AND DISEASES that are likely to be encountered in gardens are summarised in Tables 2 and 3. Most of these problems can be overcome with chemical sprays, but knowledge of the conditions that favour their development will help reduce their occurrence. Some varieties are more susceptible than others to disease and problems such as bitter pit, which prevents the fruit from keeping.

TABLE 3
PESTS

Cause	Symptom	Conditions	Cultural Control
Codling Moth caterpillars	Maggoty apples July onwards.	Eggs laid on fruit in June; caterpillars tunnel into fruit. Overwinter in loose bark and ties.	Destroy maggoty apples. Trap overwintering cocoons in corrugated cardboard tree bands; put on in July; destroy over winter. Use pheromone trap to capture moths in spring.
Apple Sawfly caterpillars	Tunnel, ribbon scars on apples. Fruit falls June.	Eggs laid May, caterpillars infest fruit early June, overwinter in soil.	Destroy affected fruit June. Rake soil to expose to frost.
Apple Aphids (several kinds)	Distorted leaves, fruits, caused by young aphids on buds, shoots, leaves.	Eggs overwinter on trees. Hatch March.	Destroy on sight with finger and thumb.
Woolly Aphid	White woolly substance. Galls on woody stems, branches.	Overwinter as young adults; visible in spring as white areas on branches.	Paint methylated spirits or prune out shoot.
Apple sucker	Sucks sap from blossom which looks as if frost damaged.	Overwinter as eggs on trees.	Pinch out affected blossom
Capsid bugs	Bumps, malformations on fruit. Tattered leaves.	Bugs feed on leaves, young fruit.	
Winter Moth caterpillar	Tattered leaves, flowers.	Caterpillars feed on opening buds, young leaves. Overwinter in soil, wingless females climb up trunks.	Grease bands on trunk, late Oct, to trap females on way to lay eggs.
Fruit Tree Red Spider Mite	Bronzed leaves, premature leaf fall.	Mites overwinter on tree. Hatch May onwards, build up on leaves early summer.	Controlled by natural predators. Mites become problem if predators killed by insecticides, especially tar-oil washes. Sow floral strips of corn cockle, marigold or cornflower to encourage predators.

Provided below are details of the main collections worldwide of apple varieties and information on amateur fruit organisations.

Fruit Collections are maintained at Research and Breeding Stations in all the apple growing counties. These stations occasionally organise open days and in some cases supply scion wood of varieties. Collections have also been established in a number of countries by amateur organisations and by individuals. Many of these organisations hold events, publish newsletters, supply scion wood, sell fruit and fruit products, give advice and instruction on fruit growing and can supply details of local nurserymen and fruit growers.

UK
COLLECTIONS
DEFRA National Fruit Collections (Department for the Environment Food and Rural Affairs, formerly Ministry of Agriculture Fisheries and Food, MAFF): Curator, Imperial College at Wye, Brogdale, Brogdale Road, Faversham, Kent ME13 8XZ (01795 532271); http://www.nfc.u-net.com

Collection of over 2000 apple varieties, including cider apples; also full range of temperate fruits.

Public access to Collections contact: Brogdale Horticultural Trust, Brogdale Road, Faversham, Kent ME13 8XZ (01795 535286/535462); http://www.brogdale.org.uk

Guided tours of Collections from Easter to late November. Fruit from Collections and fruit trees on sale; scion wood supplied; fruit demonstrations and events staged throughout year. Fruit identification available.

Royal Horticultural Society, Wisley Gardens, Surrey GU23 6QB (01483 224234); www.rhs.org.uk

Collection of over 500 apple varieties; also full range of temperate fruits. Gardens open throughout the year; fruit and fruit trees on sale; scion wood supplied. Fruit identification available. Exhibitions of fruit staged at RHS Show Halls, 80 Vincent Square, London SW1P 2PE (020 7649 1885).

Fruit varieties especially suited to West Country climate grown at RHS Rosemoor, Great Torrington, North Devon EX38 8PH (01805 624067).

H. P. Bulmer Ltd, The Cider Mills, Plough Lane, Hereford HR4 0LE (01432 352000). Cider and perry pear collection.

FRUIT GARDENS
Hatton Fruit Gardens, East Malling Trust for Horticultural Research, Horticulture Research International, East Malling, nr Maidstone, Kent ME19 6BJ (01732 843833).

Gardens created by Sir Ronald Hatton; contains the largest collection of trained forms of apples and pears in the UK. Occasionally open to the public.

ORGANISATIONS
Apple Day. A yearly event on 21 October (contact Common Ground for details).

Countrywide festival celebrating local apples and orchards. Events, tastings, exhibitions etc. organised by local groups throughout UK. Initiated 1990 by Common Ground, which has embraced the conservation of orchard landscapes and fruit varieties within its programme. Publishes information, including list of Apple Day events.

Common Ground, Gold Hill House, 21 High Street, Shaftsbury, Dorset SP7 8JE. (01747 850820).
www.commonground.org.uk

Bees & Trees Trust, Whitton Cottage, Leintwardine, Shropshire SY7 0LS.

Founded by Paul Hand to promote interest in relationship between bees, orchards and woodland. Exhibitions and lectures given throughout year.

Big Apple, Hereford. Contact: J. Denman, The Lodge, The Twerne, Putley, Ledbury, Herefordshire HR8 2RD (01531 670544).

Annually in the villages of Much Marcle Ridge – series of events, exhibitions, talks, tastings of apples and cider over a two-week period.

Friends of Brogdale, Brogdale Horticultural Trust, Brogdale Road, Faversham, Kent ME13 8XZ (01795 535286/535462). Membership £20 per yr at time of printing. Formed 1990 to help support Brogdale Horticultural Trust and foster the appreciation of all fruits. Quarterly newsletter *Fruit News*; Friends Days organised twice year at Brogdale plus Spring Blossom Walk; free entrance to Brogdale, concessions on Brogdale events etc.

Hereford Cider Museum, 21 Ryelands Street, Hereford HR4 0LW (01432 354207).

Exhibitions on cider production, also cider brandy. Starting point for the Herefordshire Cider Route, comprising eight producers who are open to the public. H. P. Bulmer's Cider Mills with organised tours is adjacent to the Museum.

Marcher Apple Network. Founded to conserve and reintroduce apples and pears of the Marcher Counties on the borders of England and Wales; especially those dating from before 1900.

Co-ordinator: S. Leitch, Wye View, Glasbury-on-Wye, Powys, Hereford HRS 5NU (01497 847354).

Norfolk Apples and Orchards Project. Founded to conserve and gather information on Norfolk fruits. Orchard established at Norfolk Museum of Rural

Life. Contact: Norfolk Apple and Orchard Project, Norfolk Rural Life Museum, Beech House, Gressenhall, Dereham, Norfolk NR20 4DR (01362 860563).

Northern Fruit Group. Founded to conserve and gather information on fruits of northern counties.
Northern Fruit Group, c/o Harlow Carr Botanical Gardens, Crag Lane, Harrogate, N. Yorks, HG3 1QB (01423 565418).

Royal Horticultural Society Fruit Group, Royal Horticultural Society, 80 Vincent Square, London SW1P 2PE (020 7834 4333).
Founded 1945 to encourage interest in fruit among amateurs. Regular meetings, visits, talks, newsletter. Membership of RHS (£29 per year at time of printing) plus £3 enrolment to Fruit Group.
Also: West Midlands Fruit Group centred on Pershore College, Worcs WR10 3JP (01386 556528); South West Fruit Group centred on RHS Gardens Rosemoor.

Coast & Countryside Service, South Hams, South Devon, Follaton House, Plymouth Road, Totnes, Devon TQ9 5NE (01803 861 140).
Service provided in conjunction with local authority to promote care and conservation for South Devon; this includes orchards, which form a distinctive part of the landscape. In 1998 Orchard Link was set up as an independent company to provide support for orchard growers and help in marketing their apples. The aim is to help make orchards more viable and in the long term conserve the local landscape.
Orchard Link, PO Box 109, Totnes, Devon TQ9 5XR (01803 861 183).

USA
COLLECTIONS
National Germplasm Repository for Apple, New York State Agricultural Experimental Station, Cornell University, Geneva, New York 14456.
http://www.ars-grin.gov/gen/apple.html

ORGANISATIONS
North American Fruit Explorers (NAFEX). Membership: NAFEX 1716 Apples Rd, Chapin, Ill., 62628. http://www.nafex.org
NAFEX is a network of individuals spread throughout the USA, Canada and elsewhere which aims to discover, cultivate and appreciate fruit. Many regional and specialist groups. Groups organise workshops, tastings, exchange of scions etc. Annual NAFEX meeting; quarterly newsletter, *Pomona*; lending library, source lists of fruit and fruit tree nurseries.

Western Cascade Fruit Society, 2625 13th Ave W, Unit 306 Seattle, WA 98119.

http://www.wcfs.org
Chapters throughout Western Washington. Quarterly newsletter, shows, sales of fruit trees, scion wood, lectures, demonstrations.

CANADA
National Plant Germplasm Repository, Agriculture Canada, Smithfield Experimental Farm, PO Box 340, Trenton, Ontario, K8V 5R5.

AUSTRALIA
Huon Horticultural Research Station, Grove, Tasmania, 7106.

ORGANISATIONS
The following provide information and scion wood:
Heritage Seed Curators Australia, PO Box 113, Lobethal 5241.
e-mail: seeds@ chariot.net.au
http//www.ozemail.com.au/~hsca
Seeds Savers Exchange, PO Box 975, Byron Bay 2481
Heritage Fruits Group of Permaculture Melbourne Association, PO Box 8018, North Camberwell.
e-mail: pcmelb@vicnet.net.au;
http://home.vicnet.net.au/~pcmelb/

NEW ZEALAND
Auckland DSIR (Department of Scientific and Industrial Research), Havelock North.

BELGIUM
Main and largest Collection at:
Station de Phytopathologie de L'Etat, 5030 Gembloux.
Ongoing trials of rediscovered or previously unknown varieties; those of merit released via network of nurserymen. Specialise in evaluating worthwhile trees in seedling orchards which still exist in Belgium.
Collections also at a number of Technical Horticultural Schools.

FRUIT GARDENS
Fruit Gardens of Gaasbeek Castle; complex of walled gardens with extensive collection of trained trees. Gardens lie within parkland of Castle Gaasbeek, near Brussels. Contact: Ministerie van de Vlaamse Gemeenschap, Afdeling Bos en Groen, Waaistraat 1 bus 6, 3000 Leuven.
Tel: 016/211226 Fax: 016/211230. Martine.Mellaerts@lin.Vlaanderen.be

ORGANISATIONS
De Nationale Boomgaardenstichting, (NBS) (National Orchard Foundation), Postbus 49, B-3500 Hasselt, Belgium.
Extensive collections maintained in standard orchards around country which members can visit. Quarterly newsletter – *Pomologia* (Boschelstraat 21, B-3724

Kortessem, Belgium); lectures, visits courses. Autumn show and sale in winter of large range of fruit varieties from 30 hectares of orchard, largely of standard trees; also sales of fruit trees. Every 5 years stages Europom, international exhibition with contributors from all over Europe.

FRANCE
COLLECTIONS
Main Collection at Station d'Amélioration des Espèces Fruitières et Ornamentales, Centre INRA, Angers, BP57 449071 Beaucouze. Part of Collection duplicated in other sites.
Conservatoire National Alpin de Gap Charance, Hautes Alpes, Domaine de Charence, 05000 Gap (04 92 53 56 82; fax: 04 92 51 94 58). Fireblight free area with accessions especially at risk.

C. R. G. Nord, Centre Régional de Ressources Génétiques, 59650 Villeneuves d'Ascq, nr Lille, Nord. Specialising in varieties of Northern France. Orchards located at Ferme du Héron in Regional park; guided tours by appointment. Every two years organises exhibition – Pomexpo – with contributors from several European countries.

GRPA Aquitaine (Group de Ressources Phytogénétiques d'Aquitaine) Ecomusée de la Grande Lande 40630 Sabres. (http://www.conservatoirevegetal.com)
Collections throughout Aquitaine, specialising in fruits of south west France; largest collection at Verger Conservatoire d'Aquitaine, Montesquieu (Domaine de Barolle 47130 Montesquieu). Contact: E. Leterme, 48 rue du Commandant Cléré-le Coteau, 40000 Mont de Marsan; (05 58 75 78 43; fax 05 58 75 07 45). Open to public, orchard walks; sales of fruit, scion wood and trees.

Association des Croquers de Pommes du Jarrez. Contact: G. Nicaise, Rue d'Auvergne, 42800 Rive de Giers (04 77 75 13 45).
Verger Conservatoire Régionale Ferme du Roc, Mairie de Puycelsi 81140 (05 63 33 19 41; fax: 05 63 33 19 41). Specialising in conservation of fruits of Mid-Pyrénées.

Jardin du Luxembourg, Paris. Small fruit garden of apples and pears. Curator, Conservatoir des Jardins du Luxembourg, 64, Boulevard Saint-Michel, 75006 Paris.

Société Pomologique du Berry, 36230 Neuvy St Sepulchre.

Maison de la Pomme et de la Poire, Musée Traditions et des Techniques Cidricoles, La Logeraie 50720 Barenton (33 59 56 22). Small fruit collection.

Exhibitions on cider and apple brandy production. Cider, pommeau, calvados and perry tastings and sales.

Many other Collections throughout France maintained by regional and amateur organisations.

ORGANISATIONS
Centre International de Recherche Pomologique et de Documentation Fruitière 'Christian Catoire', 30 rue des Acacias – 30100 Alès, Gard. (33 466 56 50 24; fax: 33 466 56 66 15). Documentation covering wide range of fruits; especially comprehensive for temperate fruits. Founded and maintained by Christian Catoire at La Mazière; recently transferred to Alès. http://www.bsi.fr/pomologie/english/pomology

L'Association nationale des Croquers de Pommes, BP 702, F-90020 Belfort Cedex.
Founded 1978, dedicated to conservation of fruit tree varieties. Organisation comprises 36 regional associations, many of which are very active, maintain conservation orchards, organise shows and produce publications on regional fruits. Croquers des Pommes produces newsletter and co-ordinates publications of the regions.

Regional Groups include:
Association des Fruits Oubliérs, 4, Avenue de la Résistance, 30270, St Jean du Gard. Founded to conserve fruit varieties of the Cévennes. Organises demonstrations, lectures and annual fruit festival in St Jean du Gard on last weekend in Nov; sales of scion wood, trees; maintains fruit collection.

Association Pomologique de Normandie, 12 Rue de Mai, 27200 Vernon.
Conservation orchard of Norman apple varieties

Société Pomologique du Berry, see above.
Croquers, Terroir du Jarrez, see above.

GERMANY
COLLECTIONS
Institut für Obstbau, D-O 8057 Dresden Pillnitz, Pillnitzer Platz 2.

Prüfstation Wurzen des Bundessertenamtes, D-04808 Wurzen, Torgauer Strasse 100.

H. Schwärtzel, Waldstrasse 4, D-0 1278 Mücheberg-Mark.

Friedrich Renner, Pomona Franconia, Hauptstrasse 56, D-91732 Merkendorf. Claims largest collection of tree fruits in Europe, especially of apples.

ORGANISATIONS
Pomologen – Verein e. V.; Chairman W. Müller, Brünlasberg 52, D-08280 Aue/Sachsen. Founded 1991. Collection of over 300 apple varieties. Aims to conserve old varieties of Germany and nearby countries; issues newsletter, year book.

ITALY
COLLECTIONS
Largest collections at: Instituto di Coltivazariono Arboree, Universita di Bologna; Instituto Sperimentale per la Frutticoltura, Trento, Forli.
Collections at other institutes and universities including: Tori, Verona, Padua, Cesna, Pisa, Viterbo, Ancona, Napoli, Sassari, Palermo.

NETHERLANDS
COLLECTIONS
Institut voor de Veredeling van Tuinbouwgewassen, Waginingen.

Proefstation voor Fruitteelt Brugstraat 51, NL-4475 Wilhelminadorf. Public open days.

ORGANISATIONS
Noordelijke Pomologische Vereniging, (Northern Pomological Society); Secretary C. Couvert, Sluisstraat 165, 9406 AX Assen (05920 55221).

SWEDEN
COLLECTIONS
Nordic Genebank; Box 41 S-23053, Alnarp. Formed 1979 for conservation of agricultural and garden plants. Collections in southern and central Sweden.
Collections also at University of Agricultural Sciences, Balsgård.

SWITZERLAND
COLLECTIONS
Swiss Federal Research Station of Fruit Growing, Viticulture and Horticulture, CH-8820 Wädenswil.

ORGANISATIONS
Fructus, founded 1985. Membership: Frau Sabine Vögeli, Glärnischstrass 31, CH8820 Wädenswil.

JAPAN
Morioka Branch, Fruit Tree Research Station, Morioka, 020-01, Japan.

CHINA
Zengzhou Fruit Tree Research Institute, Zengzhou, Henan, People's Republic of China.

Collections also maintained in Denmark, Czech Republic, Hungary, Romania, Russia.

CHAPTER 1: THE FRUIT OF PARADISE

1. Vavilov, N., 'Wild progenitors of the fruit trees of Turkistan and the Caucasus and the problem of the origin of fruit trees' in *Report of 9th International Horticultural Congress*, ed E. Bunyard (London, Royal Horticultural Society, 1930). We thank Dr V. V. Ponomarenko, of Vavilov Institute, St Petersburg, and Dr R. Watkins, formerly of Commonwealth Bureau of Plant Breeding and Genetics, Oxon, for information.
Djangaliev, A. 'The Wild Apple Tree of Kazakhstan'; trans I. N. Rutkovskaya; ed. R. Way, and E. Dickson. Unpublished manuscript, kindly made available by Prof. A. Djangaliev and through the agency of Dr. P. Forsline, Plant Genetics Resources Unit, Cornell University.
Luby, J., Forsline, P. J., Aldwinkle, H. S. Bus, V. and Geibel, M., 'Silk Road Apples – Collection, Evaluation and Utilisation of *Malus sieversii* from Central Asia'. *Hort Science* (2001) vol. 35; pp. 225–31.
Wagner, I. and Weeden, N. F., 'Isozymes in *Malus sylvestris*, *Malus domestica* and related *Malus* species', *ACTA 538, Eucarpia Symposium on Fruit Breeding and Genetics*, 2000, p. 51–55.
Robinson, J. P., and Juniper, B. E., *Plant Systematics and Evolution* (2001), vol. 226, pp. 35–58. Browning, F., *Apples; The Story of the Fruit of Temptation* (1998).
2. Postgate, J. N., in *Bulletin on Sumerian Agriculture*, vol. III, Cambridge (1987), pp. 115–144.
3. Gurney, O. R., *The Hittites* (London, Penguin Books, 1981), p. 83.
4. Romer, J., *Testament; The Bible and History* (London, Michael O'Mara Books Ltd, 1988), p. 25.
5. Darby, W., Ghaliounqui, P. and Grivetti, L., *The Gift of Osiris* (London, New York, San Francisco, Academic Press, 1977), vol. 2, pp. 697–8.
6. Wiseman, D. J., 'A new stela of Assur-Nasir-Pal II', *Iraq* (1952), **XIV**, pp. 24–32.
7. Marshall Lang, D., *Armenia, Cradle of Civilisation* (London, George Allen & Unwin, 1980), pp. 98–101.
8. Moynihan, E. B., *Paradise as a Garden; In Persia and Mughal India* (London, Scolar Press, 1982), p. 25.
9. Roden, C. A., *New Book of Middle Eastern Food* (London, Penguin Books, 1986).
10. Theophrastus, *Enquiry into Plants*, trans Sir A Hort (London, William Heinemann Ltd; Cambridge, Massachusetts, Harvard University Press, 1968), Book IV, section V.
11. Plutarch, *Moralia*, trans. (1969) 'Table Talk V'. Question 8.
12. Pliny, *Natural History*, trans. H. Rackham (London, William Heinemann Ltd; Cambridge, Massachusetts, Harvard University Press, 1960), Book XV, section XV.
13. Columella, *De Re Rustica*, trans. H. B. Ash (London, William Heinemann Ltd; Cambridge, Massachusetts, Harvard University Press, 1934), Book XII. Varro, *On Agriculture*, trans. W. D. Hooper (London, William Heinemann Ltd; Cambridge, Massachusetts, Harvard University Press, 1934), Book I, section LIX.
14. Curtin, J., ed., *Hero Tales of Ireland*, 1894; Rees, A. and B., *Celtic Heritage* (London, Thames & Hudson, 1978).
15. Roach, F., *Cultivated Fruits of Britain* (Oxford, Basil Blackwell, 1985), p. 79.
16. Finberg, H. P. R., ed., *The Agrarian History of England and Wales AD43–1042* (Cambridge, Cambridge University Press, 1972), p. 160.
17. Meyvaert, P., 'The Medieval Monastic Garden' in *Medieval Gardens, Dumbarton Oaks Colloquium on the History of Landscape Architecture IX*, ed. E. B. Macdougall (Dumbarton Oaks, 1986).
18. Harvey, J., *Medieval Gardens* (London, B. T. Batsford, 1981), p. 35.
19. See Wiliam, A. R., *Llyr Iorwerth* (Cardiff, 1960). The laws referring to 'imps' occur in one family of manuscripts and not in any compilation, which we can be sure was made before c1250.
20. See *Rectitudines Singularum Personarum and Gerefa* in Finberg, *op. cit.* pp.512–16.
21. Harvey, *op. cit.* p. 8, 75.
22. McLean, T., *Medieval English Gardens* (London, Collins, 1981), p 262, 23.
23. Luby, J., et al, *op. cit.*

CHAPTER 2: FOR PLEASURE, MEATE, AND MEDICINE

1. Harvey, J., *Medieval Gardens* (London, B. T. Batsford, 1981), p. 6, 8.
2. *ibid.* pp. 92–3.
3. McLean, T., *Medieval Gardens* (London, Collins, 1981), p. 92.
4. Leroy, A., *Dictionnaire de Pomologie* (Angers, Leroy, 1873), vol. III, pp. 21–5.
5. McLean, *op. cit.* p. 71.
6. Harvey, *op. cit.* p. 54.
7. John Russell's *Boke of Nurture* part I, p. 121, in F. Furnivall, ed., *Manners and Meals in Olde Time* (1868); see also Wilson, C. A., ed., 'Banquetting Stuffe', Food and Society, vol. 1 (Edinburgh, Edinburgh University Press, 1990).
8. Wilson, E., 'An unpublished alliterative poem on plant-names from Lincoln College, Oxford', *Notes and Queries*, new series, vol. 26, no. 6, 1979, pp. 504–8.
9. Masson, C., *Italian Gardens* (Woodbridge, Suffolk, Antique Collectors' Club, rev. ed. 1987), p. 587.
10. Lazarro, C., *The Italian Renaissance Garden* (New Haven and London, Yale University Press, 1991).
11. Coffin, D., *The Villa d'Este at Tivoli* (1960), p. 11.
12. Benporat, C., *Storia della Gastronomia Italiana* (Milan, Mursia, 1990), pp. 103–5.
13. Breviglieri, N., 'Il melo Decio' in

Rivista della Societa Toscana d'orticultura (1940) **XXV**, pp. 1–2. See also *Agrumi, Frutta e Uve Nella Firenze di Bartolomeo Bimbi Pittore Mediceo* (CNR, 1982), pp. 90–1; Leroy, *op. cit.* vol. III.

14. Woodbridge, K., *Princely Gardens* (London, Thames & Hudson, 1986), p. 40, 44.
15. *ibid.* p. 112, 71.
16. Ketcham Wheaton, B., *Savouring the Past; The French Kitchen and Table from 1300 to 1789* (London, Chatto & Windus, 1983), pp. 49–52.
17. *The Husbandman's Fruitful Orchard*, N. F. (1609) see Amherst, A., *Gardening in England* (London, Bernard Quaritch, 1895), pp. 98–9.
18. Harvey, J., *Early Nurserymen* (London and Chichester, Phillimore & Co. Ltd, 1974), p. 31.
19. Lambard, W., *A Perambulation of Kent* (1570, reprint 3rd ed., 1596, Bath, Adams & Dart, 1970). Weber, R., *Market Gardening* (Newton Abbot, 1972), p. 51.
20. Gothein, M. L., *A History of Garden Art*, trans Mrs Archer-Hind (London and Toronto, J. M. Dent & Sons, 1928), vol. 1, p. 443.
21. Prest, J., *The Garden of Eden* (New Haven and London, Yale University Press, 1981), chapt. IV.
22. Amherst, *op. cit.* pp. 317–28.
23. Parkinson, J., *A Garden of Pleasant Flowers; Paradisi in Sole Paradisus Terrestris* (1629, reprint Dover Publications Inc, New York, 1976), p. 537.
24. Rea, J., *Flora Ceres and Pomona* (1665), see 'Pomona'.
25. See Wilson, *op. cit.*
26. *The Compleat Cook and A Queen's Delight* (1665, reprint London, Prospect Books, 1984). See also *Mrs Eales's Receipts* (1733 ed., reprint London, Prospect Books, 1985); Glasse, H., *The Art of Cookery made Plain and Easy* (1747, reprint London, Prospect Books, 1983).
27. Le Gendre, *The Manner of Ordering Fruit Trees*, trans John Evelyn (1658).
28. Ketcham Wheaton, *op. cit.* pp. 133–4.
29. La Quintinye, J. de, *The Compleat Gard'ner*, trans. John Evelyn (London, 1693).
30. Schama, S., *The Embarrassment of Riches* (London, Collins, 1987), pp. 170–1.

CHAPTER 3: FOR GOD AND COUNTRY
1. Prest, J., *The Garden of Eden* (New Haven and London, Yale University Press, 1981), p. 81.
2. Sinclair Rohde, E., ed., *The Garden Book of Sir Thomas Hanmer* (1629, published Nonsuch Press, 1933), pp. XXIX–XXX, pp. 153–4.

3. Webster, C., *The Great Instauration; Science, Medicine and Reform 1626–1660* (London, Gerald Duckworth & Co., 1975), p. 71; pp. 477–83.
4. Worlidge, J., *Systema Agriculturae* (1675).
5. Thirsk, J., ed., *The Agrarian History of England and Wales*, vol. V, I, Regional Farming Systems, 1640–1750 (Cambridge, Cambridge University Press, 1984), pp. 167–9; 275–6.
6. *The Illustrated Journeys of Celia Fiennes c1682–c1712*, ed. C. Morris (London, Webb & Bower, Michael Joseph, 1988), p. 64.
7. *ibid.* pp. 236–7.
8. Switzer, S., *The Practical Fruit Gardener* (1724), p. 3; Langley, B., *Pomona* (1729).
9. Batey, M., Lambert, D., *The English Garden Tour* (London, John Murray Ltd, 1990), p. 106.
10. Cowell, J., *The Curious and Profitable Gardener* (1730), pp. 94–5.
11. Switzer, *op. cit.* p. 133.
12. Bradley, R., *New Improvements of Planting and Gardening* (W. Mears, London, 1718), part III, p. 51.
13. Lady Grisell Baillie's Household Book, 1692–1733, ed. R. Scott-Monterieff (Edinburgh, Edinburgh University Press, 1911), pp. 297–8.
14. *Thomas Milne's Land Use Map of London and Environs in 1800*, publications No. 118, 119 of London Topographical Society.
15. Adburgham, A., *Shopping in Style; London from the Restoration to Edwardian Elegance* (London, Thames & Hudson, 1979), p. 32.
16. La Quintinye, J. de, *The Compleat Gard'ner*, trans. John Evelyn (London, 1693).
17. *M. Misson's Memoirs and Observations in his Travels over England* (trans. 1719), see 'Table'.
18. Mennel, S., *All Manners of Food* (Oxford, Basil Blackwell, 1985), p. 87.
19. Glasse, H., *The Art of Cookery made Plain and Easy* (1747; reprint London, Prospect Books, 1983), pp. 112–13.
20. Fiennes, *op. cit.* p. 204.
21. Hedrick, U. P., *A History of Horticulture in America* (1950, reprint Oregon, Timber Press, 1988), p. 30.
22. Leighton, A., *American Gardens in the Eighteenth Century* (1976, reprint Amherst, University of Massachusetts Press, 1986), pp. 31–2.
23. Leighton, *ibid.* pp. 137–9.
24. Leighton, *ibid.* p. 218; Fischer, D. V., Upshall, W. H., *History of Fruit Growing and Handling in the United States of America and Canada* (Pennsylvania, American Pomological Society, 1976), p. 151.
25. Hedrick, *op. cit.* pp. 310–11.
26. American Heritage eds., *The American Heritage Cookbook; and Illustrated History of American Eating and Drinking* (American Heritage Publishing Co., Inc., New York, 1964), p. 160.
27. Hedrick, *op. cit.* pp. 60–1.
28. American Heritage, *op. cit.* p. 91.

CHAPTER 4: APPLES FOR THE FEW
1. See 'The Dessert Triumphant' in Morgan, J., and Richards, A., *A Paradise out of a Common Field; The Pleasures and Plenty of the Victorian Garden* (London, Century, 1990), pp. 104–27.
2. 'Early Apples', *Journal Horticulture and Cottage Gardener* (1876), **XXXI**, old series **LVI**, p. 184.
3. Fish, R., 'Trentham', *J. Hort. and Cottage Gardener* (1863), **V**, old series **XXX**, p. 354.
4. Robson, J., 'Archerfield', *J. Hort. and Cottage Gardener* (1865), **IX**, old series **XXXIV**, p. 426.
5. Wright, J., 'Barham Court', *J. Hort and Cottage Gardener* (1877), **XXXXI**, old series **LVII**, p. 50.
6. Bunyard, E. and L., *The Epicure's Companion* (London, J. M. Dent & Sons, 1937), pp. 156–7.
7. *J. Hort and Cottage Gardener* (1883), **VII**, 3rd series, pp. 316–321.
8. *J. Hort. and Cottage Gardener* (1890), vol. XXI, 3rd series, pp. 305, 314–17.
9. *Gardeners' Chronicle* (1896), XX 3rd series, p. 407.
10. Shand, Morton P., 'Older Kinds of Apples', *Journal of the Royal Horticultural Society* (1949), **74**, pp. 60–7, 88–97.
11. Penshurst, *The Garden* (1881), **19**, p. 383.
12. Lawson, W., *New Orchard and Garden* (1618), in Sieveking, A. F., *The Praise of Gardens* (London, Elliot Stock, 1885), pp. 61–3.
13. Pettigrew, H. A., 'Fruit Trees as Flowering Shrubs' in *The Fruit Garden*, Bunyard, G., Thomas O., (London, Country Life, 1904), p. 252. Wilks, W., *Journal of the Royal Horticultural Society* (1895), **XVIII**, p. 126.
14. Potter, J. M. S., *National Fruit Trials; A brief history: 1922–1972* (London, Ministry of Agriculture, Fisheries and Food, 1972). We thank the late Mr J. M. S. Potter for information.

15. We thank the late Sir Leslie Martin and the late Mrs J. Finzi for information. Also Peter Weaver and George Gilbert, formerly of Long Ashton Research Station, Miss B. Parfitt, formerly of National Fruit Trials, and Michael Wallis, of Scott's Nursery, Somerset.

CHAPTER 5: APPLES FOR THE MANY
1. See Marshall, W., Review and Abstract of the County Reports to the Board of

Agriculture, Vol. 3 Eastern Department, Vol. 4 Midland Department, Vol. 5 Southern and Peninsula Departments (1818, reprint 1969, Newton Abbot, David & Charles); Weber, R., *Market Gardening* (Newton Abbot, 1972), p. 73.

2. 'Historical Notes on Orchards', *J. Hort and Cottage Gardener* (1888), **LXXIX**, p. 531.

3. Rivers, T., *J. Hort and Cottage Gardener* (1877), **XXXIII**, old series **LVIII**, p. 343.

4. Hedrick, U. P., *A History of Horticulture in America to 1860* (1950, reprint Oregon, Timber Press, 1988), p. 309, 327, 315, pp. 386–5.

5. Hedrick, *ibid.* pp. 238–9.

6. Fischer, D. V., Upsall, W. H., *History of Fruit Growing and Handling in the United States of America and Canada* (Pennsylvania, American Pomological Society, 1976), p. 77.

7. See Rivers, T., and Spencer, J., *The Florist, Fruitists and Garden Miscellany* (1854); *The Cottage Gardener* (1854), **XII**, pp. 151–2 and XIII, p. 113. *J. Hort. and Cottage Gardener* (1877), **XXXIII**, old series **LVIII**, p. 343.

8. *Gardeners' Chronicle* (1858) p. 171, 216, 542.

9. 'Robert Hogg 1818–97', *J. Hort. and Cottage Gardener* (1897), **XXXIV**, 3rd series, pp. 232–5; *Gardeners' Chronicle* (1897), **XXI**, 3rd series, pp. 188–9; *Gardener's Magazine* series (1897), p. 167.

10. See Hogg, R., and Bull, H. C., *Herefordshire Pomona* (Hereford and London, 1877–85) and *Transactions of the Woolhope Naturalists' Field Club* (1877), pp. 239–44.

11. *J. Hort. and Cottage Gardener* (1883), **VII**, 3rd series, pp. 316–21; *Gardeners' Chronicle* (1883), **XX**, new series, pp. 442–3; Barron, A., *British Apples* (London, Royal Horticultural Society, 1884); Morgan, J., 'A historic centenary', *The Garden, Journal of the Royal Horticultural Society* (1983), **108**, pp. 383–88.

12. Hyams, E., and Jackson, A. A., eds., *The Orchard and Fruit Garden; A new Pomona of hardy and sub-tropical fruits* (London, Longmans Green & Co. Ltd, 1961), pp. 154–6.

13. Bear, W., *Flower and Fruit Farming in England*, reprinted from *Royal Agricultural Society of England* (1898, 1899).

14. Bunting, W. H., *Report of a Special Inquiry into Fruit Growing in Canada in 1911*, p. 33, 50.

15. Hyams, Jackson, *op. cit.* p. 183.

16. Fischer, Upsall, *op. cit.* pp. 208–9.

17. Kreech, D., Clifford, S., Kendall, J., King, A., Yurner, S., and Vines, G., *The Common Ground Book of Orchards* (London, Common Ground, 2000).

CHAPTER 6: THE CIDER STORY

1. Charley, V. L. S., *The Principles and Practice of Cider-Making* (1949), pp. 17–18; (trans. of *Le Pommier à Cidre* by G. Warcollier, 1928).

2. French, R. K., *The History and Virtues of Cyder* (London, Robert Hale Ltd, 1982), p. 12.

3. McLean, T., *Medieval English Gardens* (London, Collins, 1981), pp. 226–7.

4. Webster, C., *The Great Instauration; Science, Medicine and Reform 1626–1660* (London, Gerald Duckworth & Co., 1975); French, *op. cit.* p. 10.

5. Beale, J., in 'Pomona', *Silva*, J. Evelyn (1662, 5th ed. 1729, reprint 5th ed., London, Stobart & Son Ltd, 1979), p. 82.

6. Thirsk, J., *The Agrarian History of England and Wales*, V, I, Regional Farming Systems, 1640–1750 (Cambridge, Cambridge University Press, 1984), pp. 382–4.

7. Beale, J., and Taylor, S., in 'Pomona', *Silva, op. cit.* pp. 73–4, 98.

8. Johnson, H., *The Story of Wine* (London, Mitchell Beazley, 1989), pp. 192–7.

9. *The Poems of John Philips*, ed. M. G. Lloyd Thomas (1927).

10. Evelyn, J., 'Pomona', *op. cit.*

11. Legg, P., *Cidermaking in Somerset* (Somerset Rural Life Museum, 1984), p. 25.

12. Legg, *op. cit.* pp. 27–30.

13. French, *op. cit.* p. 51; quoted in Nott, J., *Cooks and Confectioners Dictionary* (1726; reprint London, Lawrence Rivington, 1980).

14. Hogg, R., Bull, H. G., *Apples and Pears as Vintage Fruits* (Hereford, Jakeman & Carver, 1886), pp. 4–5.

15. Marshall, W., *The Review and Abstracts of the County Reports of the Board of Agriculture*, vol. 2 Midland Departments (1818, reprint, Newton Abbot, David & Charles), p. 415.

16. Hedrick, U. P., *A History of Horticulture in America* (1950, reprint Oregon, Timber Press, 1988), p. 38, 215.

17. *Ibid.* p. 78, 157; Coxe, W., *A View on the Cultivation of Fruit Trees and the Management of Orchards and Cider* (1817, reprint Rockton, Ontario, Canada, Pomona Books, 1976), pp. 148–51.

18. Hogg and Bull, *op. cit.* pp. 87–9.

DIRECTORY OF APPLE VARIETIES

Curatorship of the Collections is part of the overall programme of scientific development agreed between DEFRA and Imperial College at Wye. *See also* Morgan J., Lamont E. J., Lean, A., 'Verification of National Apple Collection 2000, 2001, 2002-'. Deaccessions made since the first edition of this book and those accessions now held under a number are not included in the Directory.

Baldini, E., and Sansavini, S., *Monografia delle Principali Cultivar di Melo* (Bologna, Instituto di Coltivazioni Arboree dell'Universita di Bologna, 1967).

Barron, A., *British Apples* (London, Royal Horticultural Society, 1884).

Beach, S. A., *The Apples of New York*, vol. 1, 2 (Albany, J. B. Lyon Company, 1905).

Bivort, A. *Annales de Pomologie Belge et Étrangère* (Brussels, Commission Royale de Pomologie, 1853; reprint Naturalia Publications, Turriers 1998).

Boré, J. M., Fleckinger, J., *Pommiers à Cidre* (Paris, INRA, 1997).

Brookes, R. M., Olmo, H. P., *The Brooks and Olmo Register of Fruit and Nut Varieties*, 3rd ed. (ASHS Press, Alexandria, VA, 1997).

Brooks, R. M., and Olmo, H. P., *Register of New Fruit and Nut Varieties* (Berkeley, Los Angeles, University of California Press, 1972) and after in *Hort. Science, Proceedings of the American Society for Horticultural Science*.

Bultitude, J., Apples; *A Guide to the Identification of International Varieties* (London, Macmillan Press Ltd, 1983).

Bunyard, E. A., *A Handbook of Hardy Fruits; Apples and Pears* (London, John Murray, 1920).

Bunyard, G., and Thomas, O., *The Fruit Garden* (London, Country Life, 1904).

Calhoun Jr, C. L. *Old Southern Apples* (Blacksburg, Virginia, McDonald & Woodward Publishing Comp, 1995).

Catoire, C., Villeneuve, F., *A la Recherche des Fruits Oubliés* (Saint-Hippolyte-du-Fort, Espace-Ecrits, 1990).

Choisel, J-L, *Guide des Pommes du Terroir à la Table* (Paris, Éditions Hervas, 1991).

Consiglio Nazionale delle Ricerche (CNR) *Agrumi, Frutta e Uve Nella Firenze di Bartolomeo Bimbi Pittore Mediceo* (CNR, 1982).

Coxe, W., *A View of the Cultivation of Fruit Trees* (1817, reprint Rockton, Ontario, Canada, Pomona Books, 1976).

DEFRA, National Apple Collection, Records and Archives, Brogdale, Faversham, Kent.

Downing, A. J., *The Fruits and Fruit Trees of America* (1845, revised and enlarged by Charles Downing, New York, John Wiley & Son, 1870).

Evelyn, J., *Sylva* (1662, 5th ed. 1729, reprint 5th ed., London, Stobart & Son Ltd, 1979).

Ferenc, N-T., *Régi Erdélyi Almák* (Kolozsvár, Erdélyi Múzeum-Egyesület, 1998).

Fischer, D. V., Upsall, W. H., *History of Fruit Growing and Handling in the United States of America and Canada* (Pennsylvania, American Pomological Society, 1976).

Fischer, M., *Farbatlas Obstsorten* (Stuttgart, Eugen, Ulmer GmbH & Co, 1995).

Forsyth, W., A *Treatise on the Culture and Management of Fruit Trees* (London, Longman, Hurst, Rees, Orme & Brown, ed. 1810).

Guide Pratique de L'Amateur de Fruits, Simon-Louis Frères (Paris 1876, 1895)

Hogg, R., *The Apple and its Varieties* (London, Groombridge & Sons, 18 5 1).

Hogg, R., *The Fruit Manual; A Guide to the Fruits and Fruit Trees of Great Britain* (London, Journal of Horticulture Office, eds 1860, 1875, 1884).

Hogg, R., and Bull, H. G., *Herefordshire Pomona* (Hereford & London, 1877–85).

Hyams, E., and Jackson, A. A., *The Orchard and Fruit Garden; A new Pomona of hardy and sub-tropical fruits* (London, Longmans, Green & Co. Ltd, 1961).

Lamb, J. G. D., 'The Apple in Ireland; Its History and Varieties' in *Economic Proceedings of the Royal Dublin Society* (1951), vol. 4, No. 1, 1951.

Langley, B., *Pomona, or the Fruit Garden Illustrated* (1728).

Lape, F., *Apples & Man* (New York, Van Nostrand Reinhold Company, 1979)

La Quintinye, J. de, *Instructions pour les Jardins Potagers et Fruitiers*, trans. by J. Evelyn as *The Complete Gard'ner* (London, 1693).

Lateur, M. *Variétés Anciennes d'Arbres Fruitiers peu sensibles aux maladies, diffusées sous le sigle 'RGF'*, (Centre de Recherches Agronomiques de Gembloux, 1999).

Le Lectier Catalogue, reprinted in Leroy, see below.

Leroy, A., *Dictionnaire de Pomologie* (Anger, 1873), Vol. III & IV Pommes.

Les Croqueurs de Pommes, Bulletins, 1990, 1994, 1995, 1997, 1998 (Belfort Association des Croqueurs de Pommes).

Leterme, E., *Les Fruits Retrouvés* (Rodez, Éditions du Rouergue, 1995).

Lindley, G., *The Guide to the Fruit and Kitchen Garden*, ed. John Lindley (London, Longman, Hurst, Rees, Orme & Brown, 1831).

Mas, A., *Le Verger*, (Paris, 1865-1874).

Maund, B., *The Fruitist; Treatise on Orchard and Garden Fruits* (London, Groombridge & Sons), 1843–50.

Ministry of Agriculture, Fisheries and Food, National Agricultural Advisory Service, National Fruit Trials Reports, 1964–1979.

Ministry of Agriculture, Fisheries and Food (HMSO):
  Bulletin No. 15 *Fruit Growing in old Red Sandstone in West Midlands*, 1931.
  Bulletin No. 61 *West Cambridgeshire Fruit Growing Area*, 1933.
  Bulletin No. 80 *Fruit Growing in Lower Greensand Kent*, 1934.
  Bulletin *Fruit Growing in Evesham*.
  Bulletin No. 111 *Commercial Apple Production*, 1938.
  Bulletin No. 49 *Intensive Apple Culture*, 1958.
  Bulletin No. 207 *Apples*, 1977.

Molnar, E., *Pomologie Hongroise* (Budapest, 1900-09).

Nilsson, A., *Våra äpplesorter* (Stockholm, Allmänna Fuorlaget AB, 1987).

Parkinson, J., *A Garden of Pleasant Flowers*; *Paradisus in Sole Paradisus Terrestris* (1629; reprint New York, Dover Publications Inc., 1976).

Petzold, H., *Apfelsorten* (Leipzig, Neuman Verlag, 1979).

Pomologie de la France (France 1863–1873).

Rea, J., *Flora Ceres and Pomona* (1665).

Ronalds, H., *Pyrus Malus Brentfordiensis* (London, Longman, Rees, Orme, Brown & Green, 1831).

Royal Horticultural Society, *Apples and Pears: Varieties and Cutivation in 1934* (London, RHS, 1935).

Sanders, R., *The English Apple* (Oxford, Phaidon, 1988).

Sansavini, S., Bergamini, A., Camorani, F., Faesi, W., Mantinger, H., 3–*Melo* (Bologna, Societa, Orticola Italiana – Sezione Frutticoltura, 1986).

Simirenko, L. P., *Pomologiya* (Kiev, 1961).

Smith, M. W. G., *National Apple Register of the United Kingdom* (London, Ministry of Agriculture, Fisheries and Food, 1971).

Societe Pomologique de France, *Verger Français*, vol I, II (Lyon, Paris, B. Arnaud, 1947-8).

Spiers, V., *Burcombes, Queenies and Colloggetts* (St Dominic, Cornwall, West Brendon, 1996).

Stievenard, R., Lebrun, J-L., *Les Pommes du Nord* (Centre Régional des Ressources Génétiques d'Espace Naturel Regional, Villeneuve d'Ascq, 1996).

Switzer, S., *The Practical Fruit Garden* (London, Tho. Woodward, 1724).

Taylor, H. V., *The Apples of England* (London, Crosby Lockwood & Son Ltd, 1946, 1948).

Thompson, R., and Lindley, J., *The Pomological Magazine*, 1828–30.

Thompson, R., *Catalogue of Fruit in the Society's Garden at Chiswick* (London, London Horticultural Society 1826, 1827, 1832).

Van Cauwenberghe, E. *Pomologie*, vol I (Ecole d'Horticulture de l'Etat à Vilvorde, 1955).

Vercier, J., *La detérmination rapide des variétés de fruits: Poires, Pommes* (Paris 1948).

Whealy, K., Demuth, S., eds *Fruit, Berry and Nut Inventory*, 2nd ed (Iowa, 1993, Seed Saver Publications).

Worlidge, J., *Vinetum Brittanicum* (1676, 1678).

PERIODICALS

Cottage Gardener, 1848–60, continued as *Journal of Horticulture, Cottage Gardener*, 1861–1915.

Florist and Pomologist, 1862–84.

Fruit News, The Magazine of the Friends of Brogdale, 1990–

Fruit Varieties Journal; publication of American Pomological Society, 1959–90.

Gardeners' Chronicle, 1841–

Gardener's Magazine, 1865–1916.

Transactions of the Horticultural Society of London, 1807–48.

Journal of the Horticultural Society of London, 1846–55.

Proceedings of the Royal Horticultural Society, 1859–65.

Journal of the Royal Horticultural Society, 1866–

NURSERY CATALOGUES

A Brief Catalogue of Fruit Trees held by William Perfect, Nurseryman at Pontefract, Yorkshire, 1769, held at Wiltshire Record Office, County Hall, Trowbridge, Wilts.

Copies from period 1880–1939 of George Bunyard & Co. Ltd, Maidstone, Kent; Laxton Bros, Bedford, Bedfordshire; Rivers of Sawbridgeworth, Hertfordshire.

DIRECTORY OF CIDER APPLE VARIETIES

Boré, J. M., Fleckinger, J., *Pommiers à Cidre* (Paris, INRA, 1997).

Bulmer's Pomona (Hereford, H. P. Bulmer & Co., 1987).

Copas, L., *A Somerset Pomona: The Cider Apples of Somerset* (Wimborne, Dorset, The Dovercote Press; 2001).

Hogg, R., Bull, H. G., *The Apple and Pear as Vintage Fruits* (Hereford, Jakeman, Carver, 1886).

Spiers, V., Burcombes, Queenies and Colloggetts (St Dominic, Cornwall, West Brendon, 1996).

Warcollier, G., *Pommier à Cidre* (Paris, Librairie, J. B. Baillière et Fils, 1926).

Williams, R. R., *Cider and juice Apples* (Bristol, University of Bristol, 1988).

Williams, R. R., and Child, R. D., 'Cider Apples and their Characters' in *Annual Report of Long Ashton Research Station*, 1962–65.

APPENDIX 1: COOKING WITH APPLES
For further recipes see:

The Apple Source Book, Common Ground (London, Common Ground, 1991).

Ayrton, E., *The Cookery of England* (London, Penguin Books, 1977).

Cooking Apples, Ampleforth Abbey (York, Ampleforth Abbey, 1982).

Chamberlain, L., *The Food and Cooking of Eastern Europe* (London, Penguin Books, 1989).

Child, J., and Bertholle, L., *Mastering the Art of French Cookery* (London, Penguin

Books).

Gili, E., *Apple Recipes from A to Z* (Kaye & Ward Ltd, 1975).

Grigson, J., *Good Things* (London, Penguin Books, 1974).

Grigson, J., *Fruit Book* (London, Michael Joseph, 1982).

Grigson, J., *The Observer Guide to British Cookery* (London, Michael Joseph, 1984).

Haroutunian, A., *Middle Eastern Cookery* (London, Century, 1982).

Mazda, M., *In a Persian Kitchen* (Rutland, Vermont, Charles E. Tuttle Comp Inc., 1960).

Roden, C., *A New Book of Middle Eastern Food* (London, Penguin Books, 1986).

Ward, R., *A Harvest of Apples* (London, Penguin Books, 1988).

White, F., ed., *Good Things in England* (London, Jonathan Cape, 1968).

APPENDIX 2: GROWING APPLES

1. Ermen, H., *Propagation by Hardwood Cuttings, Ministry of Agriculture, Fisheries and Food, National Fruit Trials annual report*, 1978, pp. 40–2.

For more information on growing apples see *The Fruit Garden Displayed*, The Royal Horticultural Society, Harry Baker (London, Cassells Ltd, 1986) and *Pruning Hardy Fruits*, Jack Woodward (London, Cassells Ltd, Royal Horticultural Society, Wisley Handbook, 1990).

Adburgham, A., *Shopping in Style; London from the Restoration to Edwardian Elegance* (London, Thames and Hudson, 1979)

American Heritage eds., *The American Heritage Cookbook*; and *Illustrated History of American Eating and Drinking* (American Heritage Publishing Co., Inc., New York, 1964)

Amherst, A., *History of English Gardens* (1895)

Ampleforth Abbey, *Cooking Apples* (York, Ampleforth Abbey, 1982)

Ayrton, E., *The Cookery of England* (London, Penguin Books, 1977)

Baldini, E., and Sansavini, S., *Monografia delle Principali Cultivar di Melo* (Bologna, Istituto di Coltivazioni Arboree dell'Universita di Bologna, 1967)

Baker, H., Royal Horticultural Society, *The Fruit Garden Displayed* (London, Cassells Ltd, 1986)

Beach, S. A., *The Apples of New York*, vol 1, 2 (Albany, J. B. Lyon Company, 1905)

Barron, A., *British Apples* (London, Royal Horticultural Society, 1884)

Batey, M., and Lambert, D., *The English Garden Tour* (London, John Murray Ltd., 1990)

Bear, W., *Flower and Fruit Farming in England* reprinted from *Royal Agricultural Society of England* (1898, 1899)

Benporat, C., *Storia della Gastronomia Italiana* (Milan, Mursia, 1990)

Bivort, A., *Annales de Pomologie Belge et Étrangère* (Commission Royale de Pomologie, Brussels, 1853; reprint Naturalia Publications, Turriers, 1998)

Boré, J. M., and Fleckinger, J., *Pommiers à Cidre* (INRA, Paris, 1997)

Bradley, R., *New Improvements of Planting and Gardening* (London, W. Mears, 1718)

Brooks, R. M. and Olmo, H. P., *Register of*

New Fruit and Nut Varieties (Berkely, Los Angeles, London, University of California Press, 1972)

Brookes, R. M., and Olmo, H. P., *The Brooks and Olmo Register of Fruit and Nut Varieties* 3rd ed. (Alexandria, ASHS Press, VA, 1997)

Browning, F., *Apples; The Story of the Fruit of Temptation* (Harmonsworth, Middlesex, Penguin Books Ltd, 1998)

Bulmer's Pomona (Hereford, H. P. Bulmer & Co, 1987)

Bultitude, J., *Apples; A Guide to the Identification of International Varieties* (London, Macmillan Press Ltd, 1983)

Bunyard, E. A., *A Handbook of Hardy Fruits; Apple and Pears* (London, John Murray, 1920)

Bunyard, E. A., *The Anatomy of Dessert* (London, Dulau & Company, 1929)

Bunyard, E. A., *The Epicure's Companion* (London, J. M. Dent & Sons, 1937)

Bunyard, G. and Thomas, O., *The Fruit Garden* (London, *Country Life*, 1904)

Bunting, W. H. *Report of a Special Inquiry into Fruit Growing in Canada in 1911*

Calhoun Jr, C. L. *Old Southern Apples* (Blacksburg, Virginia, McDonald & Woodward Publishing Comp, 1995)

Catoire, C., and Villeneuve, F., *A la Recherche des Fruits Oubliés* (Saint-Hippolyte-du-Fort, Espace-Ecrits, 1990)

Chamberlain, L., *The Food and Cooking of Eastern Europe* (London, Penguin Books, 1989)

Charley, V. L. S., *The Principles and Practice of Cider-Making* (1949), (trans of *Le Pommier à Cidre* by G. Warcollier, 1928)

Child, J. and Bertholle, L., *Mastering the Art of French Cookery* (London, Penguin Books)

Choisel, J-L, *Guide des Pommes du Terroir à la Table* (Paris, Éditions Hervas, 1991)

CNR, *Agrumi, Frutta e Uve Nella Firenze di Bartolomeo Bimbi Pittore Mediceo* (CNR, 1982)

Coffin, D., *The Villa d'Este at Tivoli* (1960)

Columella, *De Re Rustica*, trans H. B. Ash (London, William Heinemann Ltd, Cambridge, Massachusetts, Harvard University Press, 1934)

Common Ground, *The Apple Source Book* (London, Common Ground, 1991)

Copas, L., *A Somerset Pomona: The Cider Apples of Somerset*, (Wimborne, Dorset, The Dovercote Press; 2001)

Cowell, J., *The Curious and Profitable Gardener* (1730)

Coxe, W., *A View of the Cultivation of Fruit Trees* (1817, reprint Rockton, Ontario, Canada, Pomona Books, 1976)

Curtin, J. ed., *Hero Tales of Ireland* (1894)

Downing, A. J., *The Fruits and Fruit Trees of America*, (1845, revised and enlarged by Charles Downing, New York, John Wiley & Son, 1870)

Darby, W., Ghalioungui, P. and Grivetti, L., *The Gift of Osiris* (London, New York, San Franciso, Academic Press, 1977)

Evelyn, J., *Sylva*, (1662, 5th ed 1729, reprint 5th ed, London, Stobart & Son Ltd, 1979)

Ferenc, N-T., *Régi Erdélyi Almák*, (Kolozsvár, Erdélyi Múzeum-Egyesület, 1998)

Finberg, H. P. R., ed., *The Agrarian History of England and Wales AD43–42* (Cambridge, Cambridge University Press, 1972)

Fischer, D, V., Upsall, W. H. *History of Fruit Growing and Handling in the United States of America and Canada* (Pennsylvania, American Pomological Society, 1976)

Fischer, M., *Farbatlas Obstsorten* (Stuttgart, Eugen, Ulmer GmbH & Co, 1995)

Forsyth, W. A., *Treatise on the Culture and Management of Fruit Trees* (London Longman, Hurst, Rees, Orme and

Brown, ed 1810)

French, R. K., *The History and Virtues of Cyder* (London, Robert Hale Ltd, 1982)

Furnivall, F., ed. *Manners and Meals in Olde Time* (1868)

Gili, E., *Apple Recipes from A to Z* (Kaye & Ward Ltd, 1975)

Glasse, Hannah, *The Art of Cookery made Plain and Easy* (1747, reprint London, Prospect Books, 1983)

Gothein, M. L., *A History of Garden Art* (trans 1928)

Grigson, J., *Good Things* (London, Penguin Books, 1974)

Grigson, J., *Fruit Book* (London, Michael Joseph, 1982)

Grigson, J., *The Observer Guide to British Cookery* (London, Michael Joseph, 1984)

Guide Pratique de L'Amateur de Fruits, Simon-Louis Frères (Paris 1876, 1895)

Gurney, O. R., *The Hittites* (London, Penguin Books, 1981)

Haroutunian, A., *Middle Eastern Cookery* (London, Century, 1982)

Harvey, J., *Medieval Gardens* (London, B.T. Batsford, 1981)

Harvey, J., *Early Nurserymen* (London and Chichester, Phillimore & Co. Ltd., 1974)

Hedrick U. P., *A History of Horticulture in America* (1950, reprint Oregon, Timber Press, 1988)

Hogg, R., *The Apple and its Varieties*, (London, Groombridge and Sons, 1851)

Hogg, R., *The Fruit Manual; A Guide to the Fruits and Fruit Trees of Great Britain* (London, Journal of Horticulture Office, eds 1860, 1875, 1884)

Hogg, R. and Bull, H. G., *Herefordshire Pomona* (Hereford and London, 1877–85)

Hogg R., Bull, G., *Apples and Pears as Vintage Fruits* (Hereford, Jakeman & Carver, 1886)

Hyams, E. and Jackson, A. A. *The Orchard and Fruit Garden; A New Pomona of Hardy and Sub-tropical Fruits* (London, Longmans Green & Co Ltd, 1961)

Ketcham Wheaton *Savouring the Past; The French Kitchen and Table from 1300 to 1789* (London, Chatto & Windus, 1983)

Ketch, D., with Clifford, S., Kendall, J., King, A., Turner, S., and Vines, G., *The Common Ground Book of Orchards; Conservation, Culture and Community* (London, Common Ground, 2000)

Johnson, H., *The Story of Wine* (London, Mitchell Beazley, 1989)

Lamb, J. G. D., *The Apple in Ireland; Its History and Varieties* (Dublin, Economic Proceedings of the Royal Dublin Society,

1951)

Lambard, W., *Perambulations of Kent* (1st ed 1570, 3rd ed 1596; reprint 3rd ed, Bath, Adams & Dart, 1970)

Langley, B., *Pomona, or the Fruit Garden Ilustrated* (1729)

Lape, F., *Apples & Man* (New York, Van Nostrand Reinhold Company, 1979)

La Quintinye, J., *Instructions pour les Jardins Potagers et Fruitiers*, trans by J. Evelyn as *The Complete Gard'ner* (1693)

Lateur, M., *Variétés Anciennes d'Arbres Fruitiers peu sensibles aux maladies, diffusées sous le sigle 'RGF'* (Centre de Recherches Agronomiques de Gembloux, 1999)

Lazzaro, C., *The Italian Renaissance Garden* (New Haven and London, Yale University Press, 1991)

Legg, P., *Cidermaking in Somerset* (Somerset Rural Life Museum, 1984)

Leighton, A., *American Gardens in the Eighteenth Century* (1976, reprint Amherst, University of Mass. Press, 1986)

Le Gendre, *The Manner of Ordering Fruit Trees*, trans John Evelyn (1658)

Leroy. A., *Dictionnaire de Pomologie*, vols III, IV (Anger, 1873)

Les Croqueurs de Pommes, Bulletins, 1990, 1994, 1995, 1997, 1998 (Belfort, Associaton des Croqueurs de Pommes)

Letarme, E., *Les Fruits Retrouvés* (Rodez, Éditions du Rouergue, 1995)

Lindley, C., *The Guide to the Fruit and Kitchen Garden*, ed. John Lindley (London, Longman, Hurst, Rees, Orme and Brown, 1831)

Lloyd Thomas, M. G. ed., *The Poems of John Philips* (1927)

London Topographical Society, *Thomas Milne's Land Use Map of London and Environs in 1800*, publications No 118, 119

Macdougall, E. B., ed., *Medieval Gardens, Dumbarton Oaks Colloquium on the History of Landscape Architecture* IX (Dumbarton Oaks, 1986)

Maund, B., *The Fruitist: Treatise on Orchard and Garden Fruits* 1843–50 (London, Groombridge & Sons)

Marshall Lang, D., *Armenia, Cradle of Civilisation* (London, George Allen & Unwin, 1980)

Marshall, W., *Review and Abstract of the County Reports of the Board of Agriculture* Vol 4 Midlands, Vol 5 Southern Counties (1798, reprint Newton Abbot, David & Charles, 1969)

Mas, A., *Le Verger* (Paris, 1865–1874)

Masson, G., *Italian Gardens* (Woodbridge, Suffolk, Antique Collectors' Club, revised edition 1987)

Mazda, M., *In a Persian Kitchen* (Rutland, Vermont, Charles E. Tuttle Comp. Inc., 1960)

McLean, T., *Medieval English Gardens*

(London, Collins, 1981)

Mennel, S., *All Manners of Food* (Oxford, Basil Blackwell, 1985)

Ministry of Agriculture Fisheries and Food, National Agricultural Advisory Service, National Fruit Trials Reports, 1964–79

Ministry of Agriculture Fisheries and Food, Bulletin No 15 *Fruit Growing in old Red Sandstone in West Midlands*, 1931; Bulletin no 61 *West Cambridgeshire Fruit Growing Area*, 1933; Bulletin No 80 *Fruit Growing in Lower Greensand Kent*, 1934; Bulletin *Fruit Growing in Evesham*; Bulletin no 111 *Commercial Apple Production*, 1938; Bulletin no 49 *Intensive Apple Culture*, 1958; Bulletin no. 207 *Apples*, 1977

M. Misson's Memoirs and Observations in his Travels over England (trans 1719)

Molnar, E., *Pomologie Hongroise* (Budapest, 1900–1909)

Morgan, J., and Richards, A., *A Paradise out of a Common Field; The Pleasures and Plenty of the Victorian Garden* (London, Century, 1990)

Morris, C., ed., *The Illustrated Journeys of Celia Fiennes c1682–c1712* (London, Webb & Bower, Michael Joseph, 1988)

Moynihan, E. B., *Paradise as a Garden; In Persia and Mughal India* (London, Scolar Press, 1982)

Mrs Eales's Receipts (1733 ed, reprint London, Prospect Books, 1985)

Nilsson, A., *Vara äpplesorter* (Stockholm, Allmänna Fuorlaget AB, 1987)

Nott, J., *Cooks and Confectioners Dictionary* (1726; reprinted London, Lawrence Rivington, 1980)

Parkinson, J., *A Garden of Pleasant Flowers; Paradisi in Sole Paradisus Terrestris* (1629, reprint New York, Dover Publications Inc., 1976)

Petzold, H., *Apfelsorten* (Leipzig, Neuman Verlag, 1979)

Pliny, *Natural History*, trans H. Rackham (London, William Heinemann Ltd, Cambridge, Massachusetts, Harvard University Press, 1960)

Plutarch, *Moralia*, trans (1969)

Pomologie de la France (France 1863-1873)

Potter, J. M. S., *National Fruit Trials, A brief history: 1922–1972* (London, Ministry of Agriculture Fisheries & Food, 1972)

Prest, J., *The Garden of Eden* (New Haven and London, Yale University Press, 1981)

Rea, J., *Flora Ceres and Pomona* (1665)

Rees, A. and B., *Celtic Heritage* (London, Thames and Hudson, 1978)

Roach, F., *Cultivated Fruits of Britain* (Oxford, Basil Blackwell, 1985)

Roden, C. A., *New Book of Middle Eastern Food* (London, Penguin Books, 1986)

Romer, J., *Testament; The Bible and History*

(London, Michael O'Mara Books Ltd, 1988)

Ronalds, H., *Pyrus Malus Brentfordiensis: or A Concise Description of Selected Apples*, (London, Longman Rees, Orne, Brown & Green, 1831)

Royal Horticultural Society, *Apples and Pears: Varieties and Cultivation in 1934* (London, RHS, 1935)

Sanders, R., *The English Apple* (Phaidon, Oxford, 1988)

Sansavini, S., Bergamini, A., Camorani, F., Faesi, W., and Mantinger, H., *3 – Melo* (Bologna, Societa, Orticola Itakiana – Sezione Frutticoltura, 1986)

Schama, S., *The Embarrassment of Riches* (London, Collins, 1987)

Scott-Montcrieff, R., ed., *Lady Griselle Baillie's Household Book, 1692–1733*, (Edinburgh, Edinburgh University Press, 1911)

Sieveking, A. F., *The Praise of Gardens* (London, Elliot Stock, 1885)

Simirenko, L. P., *Pomologiya* (Kiev, 1961)

Sinclair Rohde, E., ed., *The Garden Book of Sir Thomas Hanmer* (1629, published Nonsuch Press, 1933)

Smith, M. W. G., *National Apple Register of the United Kingdom* (London, Ministry of Agriculture, Fisheries and Food, 1971)

Societe Pomologique de France *Verger Français*, vol I, II (Lyon, Paris, B. Arnaud, 1948)

Spiers, V., *Burcombes, Queenies and Colloggetts* (St Dominic, Cornwall, West Brendon, 1996)

Stievenard, R., and Lebrun, J-L., *Les Pommes du Nord* (Centre Régional des Ressources Génétiques d'Espace Naturel Regional, Villeneuve d'Ascq, 1996)

Switzer, S., *The Practical Fruit Gardener* (London, Tho. Woodward, 1724)

Taylor, H.V., *The Apples of England* (London, Crosby Lockwood & Son Ltd, 1946)

The Compleat Cook and A Queen's Delight (1665, reprint London, Prospect Books, 1984)

Theophrastus, *Enquiry into Plants*, trans Sir A. Hort (London, William Heinemann Ltd, Cambridge, Massachusetts, Harvard University Press, 1968)

Thirsk, J., ed., *Agrarian History of England and Wales*, V, vol I, Regional Farming Systems, 1640–1750 (Cambridge, Cambridge University Press, 1984)

Thompson, R., and Lindley, J., *The Pomological Magazine* 1828-30

Thompson, R., *Catalogue of Fruits in the Society's Garden at Chiswick* (London, London Horticultural Society 1826, 1827, 1832)

Van Cauwenberghe, E. *Pomologie*, vol I (Ecole d'Horticulture de l'Etat à Vilvorde, 1955)

Varro, *On Agriculture*, trans W. D Hooper (London, William Heinemann Ltd, Cambridge, Massachusetts, Harvard University Press, 1934)

Vercier, J., *La détermination rapide des variétés de fruits: Poires, Pommes* (Paris 1948)

Warcollier, G., *Pommier à Cidre* (Paris, Librairie J. B. Baillière et Fils, 1926)

Ward, R. A., *Harvest of Apples* (London, Penguin Books, 1988)

Weber, R., *Market Gardening* (Newton Abbot, David & Charles, 1972)

Webster. C., *The Great Instauration Science, Medicine and Reform 1626–60* (London, Gerald Duckworth & Co., 1975)

Whealy, K., and Demuth, S., eds *Fruit, Berry and Nut Inventory*, 2nd ed (Iowa, Seed Saver Publications, 1993)

White, F., *Good Things in England* (London, Jonathan Capc, 1968)

Wiliam, A. R., *Llyr Iorwerth* (Cardiff, 1960)

Williams, R. R., *Cider and Juice Apples* (Bristol, University of Bristol, 1988)

Wilson, C. A., ed., 'Banquetting Stuffe', *Food and Society*, vol 1 (Edinburgh, Edinburgh University Press, 1990)

Woodbridge, K., *Princely Gardens* (London, Thames and Hudson, 1986)

Woodward, J., *Pruning Hardy Fruits* (London, Cassells Ltd, Royal Horticultural Society, Wisley Handbook, 1990)

Worlidge, J., *Systema Agriculturae* (1675)

Worlidge. J., *Vinetum Brittanicum* (1676, 1678)

Page references relate to the main text and appendices, and not to the Directory on pages 163-285, where apple and cider apple varieties are listed alphabetically. Italic page references refer to illustrations.

ACKNOWLEDGEMENTS

The authors would like to thank the Department for the Environment, Food and Rural Affairs (DEFRA, formerly the Ministry of Agriculture, Fisheries and Food) for allowing them to use the National Apple Collection and its archives. We would also like to express our thanks to the Earl of Selborne for kindly writing the foreword for this edition and for supplying us with additional information for the Directory.

We would especially like to thank Drs Alison Lean and Emma-Jane Lamont of Imperial College at Wye, DEFRA National Fruit Collections for their help in the preparation of the revised edition of the Directory.

Especial thanks are also given to the following for their help with both editions of the book:
Hugh Ermen, Howard Stringer, the late Bob Sanders, Brian Self (former Chairman of Royal Horticultural Society Fruit & Vegetable Committee); Dr David Pennell (former Director Brogdale Horticultural Trust), Harry Baker (former RHS Fruit Officer, Wisley); the late Fred Jansen; the late Prof L. Tukey (Penn University, Pennsylvania, USA); John and Joan Tann of Crapes Fruit Farm, Essex; James Armstrong Evans and his Collection of Cornish apples.

We would also like to most gratefully acknowledge the assistance of, in particular, Christian Catoire of Centre International de Recherche Pomologique & de Documentation Fruitière, Alès, France; also Jean Pierre Billen and Ludo Royen of Nationale Boomgaaren Stichting and Jacobus Bosschaerts in Belgium; Dr Jan Blazek, Research & Breeding Institute of Pomology, Horice, Czech Republic; Dr Nicolae Braniste, Research Institute for Pomology, Mararineni, Romania; Dr V. Djouvinov, Fruit Growing Institute, Plovdiv, Bulgaria; Dr P. Forsline, Plant Genetics Resources Unit, Cornell University, USA. Dr J. D. G. Lamb, formerly Kinsealy Research Centre Dublin, Ireland; Csiszár Lászlo, Research & Experimental Centre for Fruit Growing, Ujfehértó, Hungary; Jean Nicolas, Association Pomologie de Normandie, France; Prof A. Nilsson, formerly of Nordic Gene Bank, Alnarp, Sweden; Dr. V. V. Ponomerenko, Vavilov Institute, St Petersburg, Russia; Prof S. Sansavini, University of Bologna, Italy and Chairman of Fruit Section, International Society for Horticultural Science; Prof R. D. Way, New York State Agricultural and Experimental Station, Cornell University, Geneva, USA; Dr R. Watkins formerly Commonwealth Bureau of Plant Breeding and Genetics, Wallingford, Oxon.

For help with the Cider Directory and cider chapter, we would like to thank R. R. Williams and Liz Copas, formerly of Long Ashton Research Station, Bristol; the late N. J. Howe, formerly of Weston's Cider, Hereford; Julian Temperley, Burrow Hill Cider, Somerset; and in Normandy, cider and Calvados producers E. Dupont and the 'Père Jules' family.

We would also like to thank:
Mary Pennell, Gill Ivison, Russell Williams and all staff, past and present, of the Brogdale Horticultural Trust; Peter Dodd, formerly of Wye College; Penny Hale, Pippa Palmer of Imperial College at Wye. Peter Assiter, David Burd, Michele Chiarini, Morgan Clarke, Parry Clarke, Dr Stephen Constantine, Nigel Dawson, Dunja Dhillon, English Apples & Pears Ltd, Chris Going, Dr Alan Hall, Paul Hand, Maria Hällqvist, Therle Hughes, Tom La Dell, Catherine Olver, Fred Roach, Michael Shotton, Dr Alan Turner, Kim Wilson-Gough, the late Peter Gow and Edward Wilson. Mary Lucas and staff of Imperial College at Wye Library; Dr Brent Elliot of the RHS Lindley Library and the ladies of Wye Village library.

Finally our thanks to our families and in particular to Michael whose appetite for good apples was insatiable.

Book credits

Extract from *Metamorphoses*, Ovid, trans. M. M Innes, 1955, reproduced by permission of Penguin Books Ltd.

Picture Credits

Every effort has been made to contact the copyright holders of the illustrations reproduced, but if any has been inadvertently overlooked the necessary corrections will be made in any future editions of this book.

The colour plates in the book were specially commissioned, and are watercolours by Elisabeth Dowle. Also by the same artist were the line drawings on pp 115 (redrawn), 174, 178, 179, 295, 296, 297.

All pictures not credited below are from the author's collection. For permission to reproduce pictures the authors and publisher gratefully acknowledge the following:
p 16 Akademische Druck-u, Verlgsanstalkt, Graz, Austria; p 73 The Bettmann Archive, New York, USA; p 162 Bodleian Library, Oxford; p 51 Peter Brears; pp 24 (Lauros-Giraudon), 36, 79, Bridgeman Art Library, London; p 39 British Museum, London; p 157 J. M. Dent & Sons; pp 43, 44 ET Archives, London; p 75, 148 Mary Evans Picture Library, London; p 28 Fotomas Index, London; pp 141, 143, 150 Hereford Cider Museum Trust, Hereford; pp 71, 110 Hulton Deutsch Collection Ltd, London; p 115 redrawn from original held at Lindley Library, Royal Horticultural Society, London; p 18, Mansell Collection, London; p 29 illustration by Kenneth Hauff in People of Britain 6 (1962), *Fruit and Hop Growing in Kent* by Mary Waugh, by permission of Oxford University Press; p 159 Royal Bath & West Society; pp 146, 163 Somerset County Museums Services; p 118 Twelve Trees Publishing Company, Tasmania, Australia; pp 19, 20, 52, 55, 96, 119, 140 Imperial College at Wye, University of London.